Wasmann, Erich

Modern biology and the theory of evolution

Wasmann, Erich

Modern biology and the theory of evolution

Inktank publishing, 2018

www.inktank-publishing.com

ISBN/EAN: 9783747761076

MODERN BIOLOGY

AND THE

THEORY OF EVOLUTION

BY

ERICH WASMANN, S.J.

TRANSLATED FROM THE THIRD GERMAN EDITION

BY

A. M. BUCHANAN, M.A.

SECOND IMPRESSION

B. HERDER BOOK CO.
17 SOUTH BROADWAY,
ST. LOUIS, MO.
1923

106

PREFACE TO THE SECOND EDITION

At the present day it is incumbent upon every educated man to familiarise himself to some extent with the progress made and the results attained by modern science, and especially by biology. Only in this way will he be in a position to form any opinion regarding the intellectual contest that rages round certain important philosophical problems arising out of biology, namely, the comparative psychology of man and beasts and the theory of evolution. I have already dealt with the former of these two problems in two special works, intended for general reading, viz.: 'Instinkt und Intelligenz im Tierreich' ('Instinct and Intelligence in the Animal Kingdom') (third edition, Freiburg im Breisgau, 1905), and 'Vergleichende Studien über das Seelenleben der Ameisen und der höheren Tiere' ('Comparative Studies regarding the intelligence of ants and the higher animals') (second edition, Freiburg im Breisgau, 1900). My aim in the present work is to comply with wishes expressed in various quarters, and to render my articles on biology and evolution accessible to readers in general.

These sketches appeared originally as a series of articles in the magazine entitled *Stimmen aus Maria-Laach*, 1901-3. Even in their present considerably expanded form they are still sketches, with no pretensions to completeness,[1] as they are intended chiefly for readers who have no special knowledge of the departments of science with which I have dealt. I hope,

[1] The chapter on the relation between cellular division and the problems of fertilisation and heredity has been rewritten. For much information on the subject of botany I am deeply indebted to my colleague, Father J. Rompel, S.J., Professor at the Stella Matutina Gymnasium at Feldkirch. I have received very valuable suggestions from other specialists in various branches of science, and I take this opportunity of expressing my gratitude to them.

however, that these dissertations will be of some use also to students attending lectures on biology and the theory of evolution; they will find many facts presented to them from a fresh point of view, and this is particularly true of the last four sections on the modern theory of evolution. The chapter headed 'Theory of Permanence or Theory of Descent' is based almost exclusively upon the results contained in my previous 150 special articles on inquilines or guests among ants and termites, and may be of interest to my colleagues who have made a special study of zoology.

I trust that this work will be received in as friendly a spirit as were the two previously mentioned psychological works. In all three alike I have spoken as a Christian engaged in scientific research, and I am firmly convinced that natural truth can never really contradict supernatural revelation, because both proceed from one and the same source, viz. the everlasting wisdom of God. Therefore the study of modern biology and of the theory of descent, if carried on without prejudice, can tend only to the glory of God.

THE AUTHOR.

LUXEMBURG,
Feast of St. Ignatius, 1904.

PREFACE TO THE THIRD EDITION

This new edition contains many corrections and additions, which our increased knowledge of this branch of science has enabled me to make. The chapter on the physiology of evolution and the section on the history of slavery amongst ants are entirely new. The former throws some light on the problem of determination, and the latter illustrates the application of the theory of descent to the development of instinct.

In its present form the book possesses more unity than it did before. The two chief parts, those, namely, on cytology, or the study of cells, and on the theory of evolution, are now connected harmoniously with one another. The branch of science with which I had to deal is, however, vast in itself, and is being enriched almost daily by the publication of fresh works, so that it is quite impossible to give an exhaustive account of it in a limited space. Similar considerations led even E. B. Wilson to have the new editions of his classical work 'The Cell' (1900 and 1902) reprinted without alteration, and so I may, perhaps, be forgiven for having made only the most absolutely necessary corrections and additions.

I wish to emphasise the fact that it is not my intention that this work should serve as a complete textbook of the theory of descent. The chapters on this subject are intended only, on the one hand, to help the reader to form a clear conception of the meaning of the theory of evolution, the philosophical and scientific principles underlying it, and its limits and causes; and, on the other hand, to lay before him fresh evidence, derived from my own special department of biology, which tends to prove that the theory of evolution is really better supported than that of permanence. This theory of evolution, which I regard as a well-founded hypothesis, must be polyphyletic and not monophyletic, if it is to correspond with known facts.

With regard to the application of the theory of descent to man, I abide by my previous opinion, and maintain that the mental evolution of man from brutes is impossible, and that his bodily descent from brute ancestors presents, from the scientific standpoint, difficulties that have hitherto not been solved.

In the chapter on the Division of Cells new diagrams have been substituted for those which appeared in the earlier editions, and in other places also fresh diagrams have been added (fourteen in all), which are almost all original. Three extra plates have been added, viz. Nos. II, VI, VII.

Since the appearance of the second edition it has been translated into Italian by Fra Agostino Dott. Gemelli, O.M.[1]

The worthy translator has inserted a long introduction in which he states his own opinions on the theory of evolution,[2] and throughout his translation he has inserted many remarks of his own.[3]

The Italian edition, therefore, for which Gemelli alone is responsible, is in many respects a totally new work, and I trust that it will meet with as friendly a reception in Italy as that accorded to the German edition on this side of the Alps.

I am deeply grateful to all my colleagues who, by supplying information or suggesting additions, have helped me in bringing out this new German edition ; and I am especially indebted to Father Robert de Sinéty for some valuable remarks on the most recent discoveries regarding the problem of reduction in Chapter VI. Father H. Muckermann, S.J., was kind enough

[1] *La biologia moderna e la teoria dell' evoluzione*, Florence, 1906.

[2] Gemelli does not call his theory the theory of evolution, but prefers to speak of polyphyletic evolution (Polifilogenesi). As I also have expressed myself in favour of polyphyletic evolution, there is no actual discrepancy in our opinions, although I have retained the name 'theory of evolution.' The chief difference between us and the Monists on the subject of evolution is not so much whether it is polyphyletic or monophyletic, but it affects rather the fundamental principles underlying it, for we accept the Christian cosmogony, which is in direct opposition to that of Monism.

[3] These remarks are in many cases added to my statements, in such a way as to make it difficult to decide who is answerable for them. This remark, however, does not apply to Chapter X.

to lend me the excellent photographs which are reproduced on Plates VI and VII in this edition.[1]

THE AUTHOR.

LUXEMBURG,
Feast of St. Ignatius, 1906.

[1] These and many other original photographs have been prepared by Dr. Wm. Gray at the U. S. Army Medical Museum in Washington for his new English textbook on physiology, that will shortly be published. (Cf. the list of plates in this edition, p. xxxii.) Any other reproduction of Plates VI and VII is forbidden.

A FEW WORDS TO MY CRITICS

THESE sketches on biology and the theory of evolution appeared in book form barely two years ago, and I could hardly expect that an edition of 2000 copies would be so soon exhausted. My friends had in fact told me bluntly that the book was too dry to find many readers, and that it made too great demands upon the power of thought possessed by our educated classes.

It is true that the book has not sold so quickly as Haeckel's 'Riddle of the Universe,' but it is not a popular scientific polemic aiming at the overthrow of Christianity, and therefore peculiarly welcome to those lower classes which are especially interested in this overthrow. It is rather an attempt at conciliation, based upon an objectively scientific foundation, and it aims at harmonising the ideas of modern biology with the Christian cosmogony, and thus it was not likely to prove acceptable except to men of culture and intelligence. Nevertheless the comparatively quick sale of the book, and the numerous discussions to which it has given rise, show that it has awakened considerable interest among educated men in Germany.[1]

The kind of interest thus awakened varies according to the personal views of those in whom it exists. They may be divided into three classes, viz. (1) supporters of Christianity, (2) scientific specialists, and (3) opponents of Christianity. The classification is not quite accurate, because there are many scientific men, and especially many zoologists, among the readers of the first class, and among those of the third class zoologists form a considerable majority. Under the second category I include those only who confine themselves

[1] Germany is here used to include Austria and all countries where German is spoken.

to considering the biological contents of my book, without allowing their philosophical pre-suppositions to transpire. Apart from some few expressions of opinion on points of minor importance, the book has been very favourably received by the supporters of Christianity in Germany, both Catholic and Protestant. Some have even described it as a 'rescue from bondage,' because it has shown the right tactics to adopt in the struggle between Christianity and the monistic doctrine of evolution. I will not allude further to the various reviews of it that have appeared in the German Catholic papers. In the *Reformation* of February 26, 1905, there is an article entitled 'Ein Jesuitenpater als Anhänger des Darwinismus?' ('A Jesuit as a supporter of Darwinism?') by E. Dennert, a Protestant reviewer, well known as an opponent of Darwinism, who expresses his complete agreement with my views on the subject of evolution. Of the reviews by Catholic writers in other countries, I will mention only three of the most important. The first appeared in a North American periodical, *The Review*, of November 24, 1904, and the reviewer's opinions coincided on all points with my own. The second, which is very thorough, appeared in the number for April and May 1905 of the Spanish *Razón y Fe*, and although the writer at the close of his article says that he prefers for the present to abide by the theory of permanence, still his verdict as to the author's position with regard to the theory of evolution is favourable. The third review, 'L'Haeckelisme et les idées du Père Wasmann sur l'évolution,' may be found in the Belgian *Revue des Questions scientifiques* for January 1906. The French critic, himself an eminent biologist, in the course of a very careful article, shows that it is not possible to oppose the monistic doctrine of evolution with success, unless we acknowledge the claims of the scientific theory of evolution; on this point he agrees fully with the author's opinions.

Reviews written by critics belonging to what I have called the second class deal with the book from the scientific aspect. On the whole they are appreciative and favourable, although some few objections have been raised. I will mention only the articles contributed by Professor Dr. C. Emery to the *Biologisches Zentralblatt* (February 15, 1905); by Dr. R. Hanstein to the *Naturwissenschaftliche Rundschau* (February

2, 1905); by J. Weise to the *Deutsche Entomologische Zeitschrift* (1905, part I); by Dr. K. Holdhaus to the *Verhandlungen der Zoologisch-botanischen Gesellschaft von Wien* (1905, parts 5 and 6); and by Professor H. J. Kolbe to the *Naturwissenschaftliche Wochenschrift* (July 2, 1905).[1]

The critics of the third class are those who seek to maintain their own monistic theory in opposition to the author, and to prove his position as a Christian untenable. It was easy to foresee that there would be many reviews written from this standpoint, as unfortunately most of the zoologists of the present day have monistic tendencies; and the fact that my book called forth such vigorous opposition may be regarded as far more satisfactory evidence of its success than the most appreciative comments proceeding from the Catholic party. Why have the monists thought it necessary to pay so much attention to my work? The only psychological explanation of their action is that they see in it a certain amount of danger to the supremacy of their anti-Christian views. For this reason they do their best to draw as sharp a distinction as possible between the author as scientist and as theologian. They cannot help recognising the merits of the book, and the only objections they can raise refer to minor points, or are based on misunderstandings and misrepresentations, but naturally they refuse to acknowledge that the author has succeeded in reconciling biology in its recent developments with the principles of Christianity, for such an acknowledgement would at once deprive modern unbelief of one of its chief weapons in the conflict with Christianity.

Of these hostile criticisms I can only refer here to the most important, those, namely, of K. Escherich, H. von Buttel-Reepen, Ernst Haeckel, August Forel, J. P. Lotsy

[1] On pp. 426 and 427, where Kolbe has attempted to give a summary of the 'results' of my opinions, there are some misstatements, that are probably due to some extent to Escherich's review, to which reference will be made later. Kolbe's fourth point, that 'polyphyletic origin of closely allied forms is more likely than monophyletic,' is exactly the opposite of my assertions. The remark on the sixth point regarding 'the great number of primitive types' is, to say the least, inaccurate. The statement on the ninth point that the assumption of a 'creation' of primary types is 'a dualism irreconcilable with the principles of natural science' is devoid of all proof. The reviewer, however, seems to have had in his mind some notion of 'creation out of nothing,' because in discussing the tenth point he says emphatically that 'nevertheless' in another place I have assumed 'that the primary types must originally have been formed out of matter.'

and F. von Wagner. They are not all written in the same spirit, as the following examination of them will show.

'Kirchliche Abstammungslehre'—the Church's teaching on descent—is the title of a long article by Dr. K. Escherich, lecturer on zoology, in the supplement to the *Allgemeine Zeitung* of February 10 and 11, 1905. He speaks very appreciatively of my position with regard to the theory of evolution, and especially of the ninth chapter, in which I have dealt with the inquilines or guests among ants and termites from this point of view. But, on the other hand, he believes that 'theological reasons' have led me to assume a polyphyletic evolution, which distinguishes as many 'natural species' as there are lines of evolution, independent of one another, and he thinks that I have done this in order the better to reconcile the doctrine of evolution with that of creation. My opinions regarding the origin of life and the creation of man seem to him inadmissible, for they contradict the most important postulates of the monistic doctrine of evolution. Escherich sums up the results, which he thinks he can deduce from my opinions, and arranges them under nine chief headings, whence he draws the conclusion 'that any reconciliation of the doctrine of descent with ecclesiastical dogmas is impossible.'

My reply to Escherich's review appeared in the supplement to the *Allgemeine Zeitung* of March 9, 1905. In it I showed that the reviewer's imaginary opposition between an ecclesiastical and a non-ecclesiastical doctrine of descent indicated a biased misrepresentation of facts. He ought to have proved that the doctrine of evolution as a scientific hypothesis and theory was incompatible with the Christian cosmogony, but instead of doing so, he had recourse to the postulates of a monistic philosophy, which are neither based on science nor philosophically correct. I drew attention also to a number of actual misunderstandings with regard to the 'natural species' and the 'inner laws of evolution,' &c. These, I believe, were accidental, but of the nine points which Escherich ascribes to me as summing up my opinions, three at least were wrongly so ascribed, and these were the very three which might have been challenged from the scientific standpoint.

In the 'Closing Word' appended to my reply by Escherich,

he acknowledged several of the misunderstandings as such, but he adhered to his assertion that my doctrine of descent ought to be described as 'illogical' in contrast to the 'logical' theory. Unhappily he forgot to add that the logical character of the monistic view, which he maintains, has no scientific basis, but rests upon the unproved postulates of a false philosophy. He concluded by recommending my book to all readers who had had a scientific education, but warned the general public against reading it! I am grateful to him for this recommendation, as I wrote expressly for educated people.

In the *Archiv für Rassen- und Gesellschaftsbiologie* (March–April, 1905) there appeared a very careful criticism of my book, contributed by Dr. H. von Buttel-Reepen, who is a specialist on the subject of social insects. The review is, on the whole, written in a friendly spirit, but it forces into prominence the question of cosmogony. 'Where does science end, and the Jesuit begin?' This is the subject for discussion. The 'science' which the book contains is praised by von Buttel, but he prefers to have nothing to do with 'that web of inconsistency, which, solely in order to save a number of dogmas, draws its illogical and untenable threads over Wasmann's scientific work, obscuring the results of research.' By this 'web of inconsistency' he means my views on the theory of creation, on spontaneous generation, and on the descent of man. That in these points I have not been 'consistent' in the reviewer's monistic sense, may soothe my conscience, not only as a theologian, but also as a scientific man and a philosopher.

By means of his lectures at the Berlin Singakademie (April 1905), Professor Ernst Haeckel, the well-known prophet of Darwinism, undoubtedly did very much to increase the circulation of my 'Biology and the Theory of Evolution.' Special importance may be attached to his criticism, as he states expressly, both in the preface and in the supplement to the printed edition of his lectures on the theory of evolution, that he was induced to deliver them chiefly through the publication of my book. What was the result of this official criticism, which Haeckel as the champion of German monism felt bound to pronounce? On the one hand he welcomes my work as a satisfactory proof that the Catholic Church has ceased to oppose the doctrine of evolution, and on the other hand he calls it a masterpiece of Jesuitical distortion and sophistry. He bestows upon it the highest praise that could proceed from his lips, when he says that the ninth chapter (The Theory of Permanence or the Theory of Descent) might be incorporated as a valuable addition in one of Darwin's works, but at the same time he regards it as one of the achievements of 'the marvellous system of falsification invented by the Jesuits.' I cannot but be grateful to Haeckel for the contradictory eloquence with which he has denounced my book as a dangerous 'snare' for all who are not yet perfectly convinced monists, for I believe that his very denunciation has led no small number of victims into that snare, and has induced them to read the book which he has solemnly placed on the index for Monism.

It would be superfluous for me on this occasion to discuss Haeckel's statements in detail. In an 'Open Letter to Professor Haeckel,' which appeared on May 2, 1905 in the *Germania* and in the *Kölnische Zeitung,* I answered his assertions clearly and decisively.

'Wissenschaft oder Köhlerglaube?' ('Science or charcoal-burner's Faith?') is the title of an article antagonistic to me, that appeared in the *Biologisches Zentralblatt* for 1905, Nos. 14 and 15. It was written by the well-known authority on ants, Professor August Forel. He does not discuss ants in this article, in which in fact he pays a high tribute to my scientific knowledge, but he challenges my 'charcoal-burner's faith,' by which he means my energetic defence of Christianity against the attacks of Monism. Two years previously I had contributed to the same paper (Nos. 16 and 17, 1903) a calm and courteous criticism of Forel's monistic theory of identity,[1] and this was his reply to it, expressed however in by no means the same appropriate terms, but in language that showed irritability, occasionally bordering on fanaticism. In the introduction to his article he states plainly why his reply was so long delayed, and why it displays so much hostility; he says: 'In the meantime Wasmann has worked out and favoured us with a doctrine of descent *sui generis*. . . . Now

[1] See my *Instinkt und Intelligenz im Tierreich,* Freiburg im Breisgau, 1905, 3rd edit., chap. xii.

that Wasmann is beginning to be the apostle of a new doctrine,[1] I regard it as my duty to answer him.'

Forel was therefore annoyed by my attempt to show that the theory of evolution was not irreconcilable with Christianity, and instead of impartially disproving my opinions, he showed a partisan spirit in trying to distort them, and allowed his imagination free scope in ridiculing the 'natural species,' whose primitive forms I assumed to have been created by God. His charges against 'charcoal-burner's faith,' or rather against the Christian standpoint, are based upon a confusion of ideas, such as one would hardly expect in a critic who has been trained in philosophy. Finally, to crown his arguments, he ingeniously makes fun of the letters S.J. (Societatis Jesu) after my name; he says S stands for scientist and J for Jesuit, and advises me to put an end to the unhappy union of the two letters. He goes even further and enlarges upon this distinction in the following words: 'Wasmann S. is a scientific man, whom I respect for his acumen and conscientious work; Wasmann J. is a scholastic Jesuit. But Wasmann S. is a slave under the control of Wasmann J., and can be free and independent only when he deals with matters on which he does not come into conflict with Wasmann J. As soon as any dispute arises, Wasmann S. ceases to think as a man of science and Wasmann J. begins with his syllogisms and scholasticism and all the war of words.'

Such an attack did not really require any answer at all, as it revealed its character plainly enough. Nevertheless, I wrote a short article in reply, entitled 'Wissenschaftliche Beweisführung oder Intoleranz?' ('Scientific Proof or Intolerance?') which appeared in No. 18 of the *Biologisches Zentralblatt* for 1905. I had no difficulty in showing that it would have been better for Forel to have said nothing than to have come forward with such weapons as the champion of Monism.

In their attacks upon my book, both Haeckel and Forel have had many followers in popular scientific circles of the same tendency. There is nothing surprising in this fact, and it does not call for any further comment.

[1] These words allude to my lectures on evolution delivered in Germany and Switzerland.

It is more significant that Forel's joke about Wasmann S. and Wasmann J. has been imitated even in highly learned university lectures.[1]

Lotsy praises the author of 'Biology and the Theory of Evolution' very highly, and says: 'Wasmann is a Jesuit, but at the same time he is one of the best zoologists of the present day, and we must feel the deepest admiration for his investigations into the life of ants. This very eminent man writes on p. 271: "Of two hypotheses in natural science or natural philosophy, put forward as offering an explanation of one and the same series of facts, it behoves us always to choose the one which succeeds in explaining most by natural causes, and on this principle we can hardly hesitate to choose the theory of descent in preference to that of permanence." But as soon as we have to consider man. . . .' Lotsy goes on to refer to p. 283 of my book, where I have limited the scope of zoology with regard to man to his body, declaring it and its attendant sciences incompetent to deal with him on his spiritual side. On this subject Lotsy remarks: 'These words remind me of Lamarck's saying, "Telles seraient les réflexions que l'on pourrait faire, si l'homme n'était distingué des animaux que par les caractères de son organisation, et si son origine n'était pas différente de la leur." Are we to accuse Wasmann of prevarication? Certainly not. I fully agree with what Forel said a few days ago in the *Biologisches Zentralblatt.* Forel sees in Wasmann two distinct personalities, the scientist and the theologian, whom I shall designate by A. and B.' Then follows verbatim Forel's distinction that I have already quoted, the only difference being that for Wasmann S. and Wasmann J., Lotsy writes A. and B.

Lotsy might easily have perceived the weakness of this argument of Forel's, if he had really considered the passage quoted from Lamarck, who agrees with me in declaring zoology alone incompetent to deal with the question of the origin of man. If Lotsy were consistent, he would have to see two personalities, viz. a scientific man and a 'scholastic Jesuit,' in Jean-Baptiste Pierre Antoine de Monet, Chevalier de Lamarck!

[1] J. P. Lotsy, *Vorlesungen über Deszendenztheorien, mit besonderer Berücksichtigung der botanischen Seite der Frage* ('Lectures on theories of descent, with especial reference to the botanical side of the question'), at the Imperial University of Leiden, Part I, Jena, 1906, pp. 328, 329.

Special reference is due to a very detailed criticism of my book that appeared in the *Zoologisches Zentralblatt*, a scientific periodical (1905, No. 22). The review was written by F. von Wagner of Giessen, professor-extraordinary of zoology, yet it is not of a purely scientific character, but shows a partisan spirit, although the author's anti-Christian bias is not so bluntly expressed as is the case in Haeckel's and Forel's articles. It is, however, perceptible throughout the review, which is consequently quite unlike the impartial criticisms that we usually find in the *Zoologisches Zentralblatt*.

In the introduction to the nine pages in which he deals with my book, von Wagner remarks that not a few of his fellow-zoologists have been induced to believe that Wasmann's attitude towards the theory of evolution indicates a 'change of front on the part of the Catholic Church with regard to modern biology.' The reviewer does his best to deliver his colleagues from this 'illusion,' and I am grateful to him for doing so, as, like Haeckel and Forel, von Wagner does not mean by 'modern biology' merely its scientific results, but also the monistic postulates which the opponents of Christianity have insisted upon attaching to these results. I gladly agree with the reviewer, and confess that my views do not coincide with the postulates of a false philosophy, by no means free from hypotheses. This is, however, all that he has really succeeded in proving.

Von Wagner himself acknowledges that within my own field of research I 'apply the principles of evolution in a scientific spirit' (p. 691), and he describes my account of modern cytology, or the study of cells, from the scientific standpoint as 'very successful' (p. 693). He is, moreover, particularly 'grateful' for those parts of the book which contain 'an excellent summary of the important results of Wasmann's investigations from the standpoint of the principle of descent.' The historical account, too, of the development of biology 'describes it accurately in its general outlines.'

We must now consider the reviewer's objections, which can be summed up in one sentence (p. 692): 'The book in question has *one* author, but *two* editors, a scientific man engaged in research work and a theologian. Consequently, the whole is a joint production; the theologian takes the lead, and the scientific man may assert himself only so far as the former gives permission.' The conclusion derived by von Wagner from this statement is that the book is written with a bias from beginning to end.

The answer to this is obvious; we need only apply the just quoted words of the reviewer to his own review. 'The review in question has *one* author, but *two* editors, a scientific man engaged in research work and a monistic philosopher. Consequently, the whole is a joint production; the monistic philosopher takes the lead, and the scientific man may assert himself only so far as the former gives permission.' The conclusion that we derive from this statement is that the review is written with a bias from beginning to end.

Let us now examine my book more closely and see how far the 'bias' imputed to it by the reviewers really exists, and how far they are mistaken.

Even in my account of the historical development of biology von Wagner discovers a bias, for he says that I have singled out for praise none but Christian representatives of this science. I do not understand why, if this were the case, I spoke, as he says, with remarkably scant appreciation of Cuvier's achievements in comparative anatomy, and mentioned Bichat's work in more eulogistic terms,[1] whereas if my opinion were really biased, I should have extolled Cuvier rather than Bichat, as being an eminent Christian as well as a scientific man. This fact shows that von Wagner's desire to discover a particular bias in my work is the outcome of his own imagination.

The bias of the book, as von Wagner has discovered (p. 694), is revealed especially 'in what it does not contain.' The author is accused of having purposely withheld from his readers the more general biological evidence in favour of the theory of evolution. I feel inclined to ask whether the reviewer has really read the eighth and ninth chapters of his edition. I am supposed not to have referred to Darwin, Lamarck and Geoffroy St. Hilaire, whereas they are all mentioned on p. 169. He seems not to have noticed the more general relations of the

[1] In speaking thus I relied upon M. Duval's statements in his *Précis d'histologie*, a book with which von Wagner seems not to be acquainted.

theory of evolution to the Copernican theory of the universe, to modern geology and palæontology (pp. 179–85), and the long dissertation following them on the limits and causes of the hypothetical phyletic evolution, but he notices my statements regarding 'natural species' and their connexion with the theory of creation, for these statements give him another opportunity of joining Escherich, Haeckel and Forel in imputing to me a theological bias. On pp. 219, 220, I referred expressly to the mass of indirect évidence supporting the theory of evolution to be derived 'from comparative morphology, comparative history of evolution, comparative biology and especially from palæontology,' but I said that I had no intention on this occasion of writing a textbook of the theory of descent. No one could discover in this any intentional concealment of evidence, who did not wilfully misinterpret my words by imputing to them a bias that is not there. Such a critic is plainly incapable of forming a just and objective opinion.

Let us for a moment regard the matter from the point of view of an extreme supporter of the theory of permanence. He would have quite as much justification for discovering a bias in favour of the theory of evolution from those very statements and omissions, in which a fanatical advocate of the theory discovers a bias hostile to it. He might, for instance, try to account for the fact that I have not discussed in detail the ordinary evidence in favour of the theory of evolution, by declaring that this evidence has lost most of its weight through Fleischmann's criticism, and therefore I have been obliged to establish the scientific justification of the evolution hypothesis upon the new and independent basis of my own research. Moreover, when I have expressed my preference for 'natural species' rather than 'systematic species,' he might discover an intention to set aside the theory of permanence and replace it by that of evolution, under the pretext that the latter is more easily reconciled with the Christian doctrine of creation, &c. I maintain, therefore, that, where it is possible to see in the same statements of any author two totally opposite tendencies, it is plain that both imputations are alike objectively without foundation. I need say no more regarding von Wagner's method of treating my book, as, whilst imputing a biased tendency to me, he shows the same himself.

I must acknowledge that with regard to the doctrine of creation, the hypothesis of spontaneous generation and the application of the theory of descent, I had a bias, and one that is directly opposed to that of my reviewer. I had the intention of proving that a reasonable theory of evolution necessitates our assuming the existence of a personal Creator, and I wished further to show that 'spontaneous generation' was scientifically untenable, and, therefore, could not be a postulate of science. Finally, I desired to prove that to regard man from the purely zoological point of view is a one-sided and mistaken proceeding. I was, however, forced to adopt this threefold bias by the monists, who were exerting themselves with a much greater bias to establish false philosophical postulates in the name of biology, and to force them as 'monistic dogmas' upon all interested in science. I considered it my duty as a Christian and as a scientific man to protest vigorously against these attempts at a fresh subjugation of the human intellect.

It is, moreover, psychologically very interesting to observe how a reviewer, himself an ardent advocate of Monism, seeks to discover throughout my book Christian tendencies, in order to destroy as far as possible its scientific objectiveness. A criticism undertaken on these lines cannot be truly free from prejudice, and the absolutely biased character of von Wagner's review appears most plainly in his closing words (p. 699): 'There is always the same discord, when science is only on a man's lips and not in his heart.' Because I do not accept the unscientific postulates of Monism, all love of science is to be denied me! Is not that plainly monistic intolerance? According to my opinion, science has its abode neither on the lips nor in the heart, but in the intellect or, as von Wagner would say, the brain, which he regards without doubt as the real organ of thought in a human being.

And now I take leave of my critics,[1] and commend the present edition to their kind attention. In it, as far as lay in

[1] A short reply to von Wagner's review has already appeared in *Beispiele rezenter Artenbildung bei Ameisengästen und Termitengästen* (written in hon ur of J. Rosenthal, Leipzig, 1906, pp. 45-58; *Biologisches Zentralblatt*, 1906, Nos. 17 and 18, pp. 565-580), 55 (577) et seq.

my power, I have taken into account all the really well-founded objections to statements in the previous editions, whether these objections were raised by friends or by opponents. It is in vain, however, to call upon me to conform to the tyrannical requirements of Monism, and such a demand will remain unsatisfied in the future, as it has done in the past.

CONTENTS

(A more detailed outline of contents is prefixed to each chapter)

CHAPTER IV

Cellular Life

CHAPTER V

The Laws of Cell-division

CHAPTER VI

Cell-division in its relation to Fertilisation and Heredity (Plates I and II)

CHAPTER VII

The Cell and Spontaneous Generation

CHAPTER VIII

The Problem of Life

CHAPTER IX

Thoughts on Evolution

CHAPTER X

Theory of Permanence or Theory of Descent?
(Plates III—V)

CHAPTER XI

THE THEORY OF DESCENT IN ITS APPLICATION TO MAN
(Plates VI and VII)

CHAPTER XII

Conclusion

APPENDIX

(Plate VIII)

LIST OF ILLUSTRATIONS

LIST OF PLATES

At the End of the Book

MODERN BIOLOGY

AND

THE THEORY OF EVOLUTION

CHAPTER I

THE MEANING AND FIRST DEVELOPMENT OF BIOLOGY

'Knowledge is inexhaustible in its source, unlimited by time or space in its force, immeasurable in its extent, endless in its task, unattainable in its aim.'—K. E. v. Baer.

At the close of any considerable epoch it is of peculiar interest to look back upon the historical development of nations and states during that period; to compare their position a century ago with that which they now occupy; to observe the rise and fall of their political power, and the fluctuations in their political and intellectual importance amidst the pressure of contemporary events, and to trace the causes of these fluctuations. In the same way it is most interesting at this juncture to look back at the development of a science. The history of science is a branch of universal history, not indeed accompanied by the thunder of cannon, like the great battles of the world, but, in spite of its silent working, it sometimes has more influence than war upon the destiny of nations and of humanity as a whole.

No one, I think, would deny that during the past century the development of chemistry and physics, and of the technical arts depending upon them, has been of the utmost importance in advancing the growth of civilised nations, and so has played no small part in the history of the world. Modern physics have enabled men to avail themselves of the forces of fire and water, and the discovery of steam power has altered the face of the earth, for now it is covered with a network of railway lines, upon which trains rush to and fro, whilst the sea too is constantly traversed by sea monsters built of steel and driven by steam, which bring the farthest ends of the world into communication, and convey to still uncivilised nations the achievements of modern progress. By means of physics, too, has the human intellect succeeded in subjugating the mysterious waves of ether, both visible and invisible, and now through the electric light we have new suns ; electric telegraphs and submarine cables have triumphed over the old limitations of time and space, while Röntgen-rays penetrate even the human body, and fix the outline of its skeleton on photographic plates. The development of physics and chemistry has enabled men to construct innumerable motors and machines, and to devise chemical compounds used in various branches of industry, resulting, on the one hand, in a complete revolution in the economical conditions of the people, and, on the other hand, supplying our armies with terrible guns and deadly explosives, in the invention and perfection of which each nation strives to outstrip its neighbours, in order to annihilate them more speedily, should an opportunity occur.

It is obvious that astronomy and biology owe very much to their kindred science—physics, and especially to optics and mechanics, without which the extraordinary progress made in recent times would have been impossible. Optics and mechanics have supplied the astronomer and the biologist with their instruments, and, in conjunction with chemistry, have given them technical methods, bringing the infinitely distant near to the investigator's eye, enlarging the infinitely small, and even rendering the invisible visible on the astronomer's photographic plate and in the coloured sections of the microscopist, revealing to the one the marvels of the heavens, and to the other the secrets of the most diminutive living beings.

It is not, however, my intention now to dwell upon the development of the physical sciences and their influence in changing the various circumstances of human life; I purpose to deal only with the development of biology, which cannot boast of such wide-reaching triumphs. Nevertheless, the history of biology in the nineteenth century forms part of the history of the human intellect, and is an instructive piece of what may be called internal history, of greater importance to mankind than a merely superficial examination might lead us to suppose.

1. Meaning and Subdivisions of Biology

We must begin by clearly understanding what we mean by *biology*. What is biology? As the name tells us, it is the science of life and of living creatures. This is biology in the widest sense of the word, and it coincides with its oldest historical signification, as it occurs in scholastic philosophy. Biology, or the study of living creatures, is closely connected with cosmology, or the study of the bodies composing the universe, for, strictly speaking, the study of living creatures includes the whole study of plants, animals and men, but this is so vast a territory that we generally apply the name biology to one comparatively small subdivision of it, and speak of the biology of plants and animals in contradistinction to their morphology, physiology, and morphogeny. *Morphology* deals with the forms and component parts (organs, tissues, and cells) of organisms. The history of individual development, or *Morphogeny*, deals with the growth of the organic forms from the egg to maturity. *Physiology* discusses the functions of the various parts of the organism, and establishes their relations to the process of life and also the chemical and physical laws regulating their activity. Finally, *Biology* is concerned with the external activities affecting the organisms as individuals, and consequently governing their relation to all other organic beings as well as to the inorganic world. In this respect biology differs from *Psychology*, the proper subjects of which are the processes of sensitive and intellectual life—essentially internal activities, although these frequently

B 2

come within the scope of biology in virtue of their outward manifestations.

In the narrower sense of the word, therefore, biology may be defined as the science dealing with the mode and relations of life in animals and plants. Human biology forms a distinct branch of knowledge, forming a part of anthropology, and is no longer regarded as belonging to biology in the more restricted sense of the word, now generally accepted by scientific writers.

With regard to the meaning of the word 'biology' and the most convenient definitions to be assigned to it, there are many different opinions, only a few of which can be mentioned here briefly. Almost all scientific men agree in retaining the old name 'biology' (in the wider sense) to denote the whole mass of knowledge regarding life and living creatures.[1] But there is great diversity of opinion as to the designation of the special branch of that science, which we have called biology in the narrower sense. German zoologists used to call it simply biology, until Ernst Haeckel suggested the name *Œcology*. Œcology means 'study of dwelling' or 'science of keeping house,' it approaches the more restricted meaning of biology, but does not cover it. This new name has found favour not only with many zoologists, but also with botanists. Fr. Delpino,[2] F. Ludwig,[3] and J. Wiesner[4] speak of the phenomena of plant life as the biology of plants, whereas other botanists, such as R. v. Wettstein,[5] prefer the name œcology of plants.

Fr. Dahl was the first German zoologist to suggest the adoption of *Ethology*, or science of the habits of life, a word first introduced by French scientific writers to replace biology in the narrower sense.[6]

This new name would certainly be more applicable to animal biology than Haeckel's œcology, but it is not applicable at all to plants, as we can speak of 'habits of life' only with reference to creatures that possess instinct and psychological life. If we are to have a new name, it ought to be applicable both to plants and to animals with regard to their phenomena of life.

An eminent botanist, J. Reinke,[7] is of opinion that we can dispense with the word 'biology' in the narrower sense, and, in order to avoid confusion when it is used in its wider sense, he suggests the simple expression 'Mode of life among animals and

[1] Cf. for instance, O. Hertwig's *Entwicklung der Biologie im* 19 *Jahrhundert*, Jena, 1900.

[2] *Pensieri sulla Biologia vegetale, &c.*, Nuovo Cimento, XXV, Pisa, 1867.

[3] *Lehrbuch der Biologie der Pflanzen*, Stuttgart, 1895.

[4] *Biologie der Pflanzen*, 1902, I.

[5] *Leitfaden der Botanik für die oberen Klassen der Mittelschulen*, 1901, 1.

[6] Cf. Wasmann, 'Biologie oder Ethologie?' (*Biolog. Zentralblatt*, XXI, 1901, No. 12, pp. 391–400).

[7] 'Was heisst Biologie?' (*Natur und Schule*, I, 1902, part 8, p. 449, &c.).

plants' as a substitute for the word in its more restricted signification. This designation is clear and convenient enough, but I scarcely think that it fulfils the requirements of science, for we need some internationally intelligible word for 'mode of life' or 'Lebensweise,' formed from Greek roots on the analogy of 'Morphology,' 'Physiology,' &c.

To supply this deficiency the word *bionomy* or *bionomics* has been introduced in England[1] and North America,[2] and this is perhaps the best word yet suggested to designate the mode of life of animals and plants, for it denotes the laws governing life' (βίος-νόμος), and so means exactly what we defined as biology in the narrower sense, and at the same time it avoids the ambiguity of the word biology. I should have no objection to accept this new name *Bionomics*, to designate the mode of life among animals and plants; but as it is not yet current in Germany, I may be permitted to retain the old name.

The experimental study of the laws of heredity and variation has recently been called *Biometry*.[3] In 1901 a new periodical appeared in Cambridge (England) entitled *Biometrica: A Journal for the Statistical Study of Biological Problems*. Biometry is, therefore, synonymous with Statistical Biology.

The following simile may serve to illustrate more clearly the original meaning of the word *biology*, and the various modifications which it has undergone owing to the progress made by science in the nineteenth century.

Biology, in its widest signification, embraces all that we know about living creatures, and we may compare it with a lofty tree having three main boughs, but many branches, and its stem, boughs, and branches are the biological sciences. The tree is crowned by twigs shooting from the main trunk, and this crown represents the science dealing with man, or anthropology, and the topmost of its twigs, rising up into the domain of the intellectual sciences, is the psychology of man and nations. Below it is human biology in the narrower sense, then human physiology, human morphology and the history of human development, all having many subordinate twigs,

[1] Cf., e.g., G. K. Marshall and E. B. Poulton, 'Five Years' Observations and Experiments on the Bionomics of South African Insects' (*Transactions of the Entomological Society*, London, 1902, part 3).

[2] Cf. Ch. S. Minot, 'The Problem of Consciousness in its Biological Aspects' (*Proceedings of the American Association for the Advancement of Science*, XXXI, p. 272).

[3] Cf. Chr. Schröder, 'Eine Sammlung von Referaten über neuere biometrische Arbeiten' (*Allgemeine Zeitschrift für Entomologie*, IX, 1904, Nos. 11 and 12, p. 228, &c.).

bearing, for the most part, the same names as the corresponding ramifications of the zoological stem. Some few branches belonging to the crown have names of their own, to which zoology supplies analogies only; such are ethnology and archæology, psychopathology, and medicine.

Below the crown a great bough springs from the main trunk of the biological sciences: this is zoology. Its chief offshoots are animal psychology and animal biology (animal bionomics) and the physiology, morphology, and morphogeny of animals. In the course of the nineteenth century a great number of little twigs grew out of each of these branches, of which only a few can be mentioned here. Out of animal biology or bionomics sprang trophology, or the science dealing with the food of animals; œcology, or the science dealing with their habitations; animal geography, dealing with their distribution; and, further, their parasites have been studied, and the tendency of certain animals to live with other animals or near to some particular plants (symbiosis). This has given rise to investigations of a biological nature into the way of life of ants and termites, and one of the most fertile offshoots of modern biology is the study of the inquilines among ants and termites. We cannot do more than name nervous physiology which, with its offshoots, cerebral physiology, physiology of the external organs of sense and of the nerve tracks, threatens to take the place of animal psychology, now said to be out of date.[1]

Modern morphology has even more ramifications, branching out in one direction into systematics, or the science of systematic classification, and in the other into morphology proper, which latter is subdivided into exterior and interior morphology, the interior comprising topographical anatomy, histology or study of the tissues, and cytology or study of the cells—all three well-developed offshoots of morphology. Moreover, all these branches of morphology have their counterparts on the physiological side, in the physiology of the organs, tissues, and cells.

Morphogeny, or the history of the development of animals,

[1] On this subject cf. my article 'Nervenphysiologie und Tierpsychologie' (*Biolog. Zentralblatt*, XXI, 1901, No. 1. pp 23–32) and also *Instinkt und Intelligenz im Tierreich*, 1905, chap. ii

has two great branches, viz. ontogeny, or the history of individual growth, and phylogeny, or the history of the race development. Ontogeny is divided into embryology and post-embryonic development, which includes the phenomena of metamorphosis, metagenesis, &c. Finally we must allude to animal pathology as a branch of zoology. Reference has already been made to animal geography as a branch of animal bionomics.

Nearer the root of the tree springs the lowest bough of biology, viz. botany. Nothing is found on it corresponding to the most dignified offshoot of the zoological bough—animal psychology, because plants have no consciousness, and even the most sensitive of them show only a faint resemblance to conscious life.[1]

There are, however, on the botanical bough a good many off-shoots corresponding to the other parts of zoology; we have the biology (bionomics) of plants, which includes plant-geography, and we have also plant-physiology and morphology, plant-anatomy and cytology, and finally phytopathology.[2] The botanical branch is further distinguished by possessing one suspiciously luxuriant and poisonous looking offshoot, which boldly rises up to the branch of the crown that we have called 'medicine,' and this is bacteriology. Fortunately it has a less poisonous side in the phenomena of fermentation and assimilation of nitrogen, which are in many respects beneficial to man.

To our astonishment we see that our tree bears one or two apparently dead branches of considerable size; they spring from the same point of the main trunk as the zoological and botanical boughs respectively, and they are called *palæozoology* and *palæophytology*. They are, however, by no means really dead, although they deal with the extinct ancestors of the animal and vegetable kingdoms of the present day.

In the main trunk supporting the crown and the branches

[1] Many modern botanists regard this analogy as constituting real identity (homology), but they are certainly mistaken. Cf. for instance, Haberlandt, *Die Sinnesorgane im Pflanzenreich zur Perzeption mechanischer Reize*, Leipzig, 1900. For a criticism on these views, see J. Reinke, *Philosophie der Botanik*, 1905, 66, &c., 83, &c.

[2] The distinction between anatomy and histology is less marked in the case of plants, as their tissues do not differentiate themselves so sharply into organs as do those of animals.

of the tree of biological knowledge with all their offshoots and twigs rises a stream of sap, representing the comparative and generalising elements belonging to all the biological sciences; these connect all the parts of the tree with one another and enable us to view them intelligently as a whole, and at the same time they enlighten us as to its growth. Comparative psychology effects a close connexion between the zoological branch and the crown of the tree; comparative biology and physiology, comparative morphology, anatomy and histology, comparative cytology and comparative morphogeny send streams of life through all the branches and twigs of the great tree, and show that they are all living parts of one vast whole.

Chemistry and physics, too, and especially mechanics of organic structures, are represented in the roots of the tree, as biochemistry and biophysics, and they connect it with the surrounding domain of the inorganic sciences. But the quintessence of all the sap flowing in the tree of biological knowledge is the scientific conception of life, and the trunk of the tree, which supports and nourishes all these branches and twigs, is the science of life.

2. The Earliest Development of Biology

We have just seen how the tree of biological sciences grew rapidly in the nineteenth century, and produced an indescribable abundance of offshoots, leaves, blossoms and fruit on branches previously bare. Let us now consider the origin of this tree and how it fared whilst still an insignificant seedling.

It was not planted first in the year 1800, nor did it suddenly develop on New Year's Day, 1801, into a trunk sturdy enough to support all the branches and twigs which the new century was destined to add to it. It is far older than this, and we can trace its history for several thousand years. The seed, whence this tree has grown, was planted when God breathed into the first man the breath of life, as we read in the beautiful figurative language of Holy Scripture. The breath of God's spirit, dwelling in man, its all-embracing power of understanding and its never satisfied thirst for knowledge, form the hidden motive power, the inner living force of this tree. Man has always been possessed by a thirst for knowledge, both among civilised nations and among the wild children of nature. The Eskimo of the present day adorns the walrus ivory implements used in shooting his arrows with dogs' heads and outlines of reindeer, birds and human beings, showing that the shapes of the living creatures around him have deeply impressed themselves upon his mind; and, in the same way, the cave-dwellers of Central Europe scratched rough sketches of fish, horses and other animals on reindeer bones. Even if the famous representation of a long-haired mammoth with a long mane, which was found on a piece of a mammoth's tooth, proves not to be genuine, and the much finer engraving, on a reindeer antler from the cavern at Kessler, of a reindeer grazing, is in all probability a modern forgery, still, as J. Ranke says,[1] it is difficult to say exactly when the germ of biological research latent in the mind of man first assumed a scientific form, and appeared as a young plant above the ground. We know, however, one famous gardener, who tended the little tree most skilfully, and that is Aristotle the Stagirite.

Aristotle had predecessors, no doubt; the animal system devised by the followers of Hippocrates of Cos had already prepared the way for him,[2] yet he certainly deserves to be called the Father of Biological Science. His classical works 'Historia animalium,' 'De partibus animalium,' and 'De generatione animalium' are the foundations of our scientific systematic classification and biology, of morphology, anatomy, and morphogeny.[3] In his writings he actually mentions 500 kinds of animals.[4] As he does not allude to many other varieties that are very common and occurred in ancient Greece in his day, we must assume that he did not think it necessary to speak of all the animals with which he was familiar. He divides animals into two chief classes, *ἔναιμα* or with blood (more correctly red-blooded), and *ἄναιμα* or bloodless, and

[1] *Der Mensch*, II, Leipzig and Vienna, 1894, 459, &c.

[2] Cf. R. Burckhardt, 'Das koische Tiersystem, eine Vorstufe der Zoologischen Systematik der Aristoteles' (reprinted from the *Verhandl. der naturf. Gesellschaft in Basel*, XV, 1902, part 3, pp. 377–414).

[3] R. Burckhardt, 'Das erste Buch der aristotelischen Tiergeschichte' (*Zoologische Annalen*, I, Würzburg, 1904, part 1). Also 'Zur Geschichte der biologischen Systematik' (*Verhandlungen der Naturf. Gesellschaft in Basel*, XVI, 1903, 388–440).

[4] We cannot here discuss their division into different classes. Günther remarks that the number of varieties of fish known to Aristotle seems to have been 115 (*Handbuch der Ichthyologie*, 1886, p. 3).

this division practically answers to the modern classification into vertebrates and invertebrates. The eight γένη μέγιστα, or chief classes of the Aristotelian system, agree roughly with our chief classes in the animal kingdom. The conception of the εἶδος or species, introduced by Aristotle, underlies our modern conception of it. But the great philosopher was not only a pioneer in systematic classification, he was equally eminent as a morphologist, an anatomist, a biologist, and an embryologist. He compared animals with regard to their form and structure, and studied their mode of life and the history of their development.

How great a biologist Aristotle was is proved by the fact that some of his discoveries were rediscovered in the nineteenth century, and were regarded as brand-new triumphs of modern science. Aristotle knew that many sharks do not only produce their young alive, but that in their case the young before their birth are nourished by a process closely resembling that of mammals (development of a placenta). This fact was rediscovered by Johannes Müller, a famous anatomist and zoologist (1801–58). Moreover, Aristotle was aware of the difference between male and female cephalopods, and had observed that young cuttlefish possess a vitelline sac near the mouth. The accuracy of these old observations has been completely proved by modern research. Bretzl has thrown an astonishing light upon the extent and importance of the botanical knowledge possessed by Greeks of Aristotle's time.[1]

When we consider the well-merited prestige enjoyed by Aristotle as founder of biology, when we remember the enormous wealth of knowledge, interspersed though it be with many errors, contained in his works, we cease to wonder that for two thousand years everyone, who studied biology at all, studied Aristotle almost exclusively, quoted Aristotle, made extracts from Aristotle, and wrote commentaries on Aristotle. The work of the Younger Pliny in this department is insignificant in comparison with that of his great predecessor, and even in some respects shows a falling off. Pliny, however, has been the chief source of information for most of the students of nature both of antiquity and of the Middle Ages, who derived

[1] *Die botanischen Forschungen des Alexanderzuges*, Leipzig, 1903. Cf. the review in the *Botanisches Zentralblatt*, XCIII, 1903, p. 97, &c.

from him their biological knowledge, and adopted as genuine all the stories found in Pliny's 'History of Animals,' without in any way testing their truth. A standard work of this description is the famous 'Physiologus' or 'Bestiarium,' in which all the legends connected with zoology are collected, with edifying morals appended to them.

It would be unfair not to acknowledge that, among the great scholastic philosophers of the thirteenth century, there were a number of men who did their best to carry on independent scientific research. Besides St. Thomas Aquinas, the Dominican Order produced in that century three great men, conspicuous not so much for their scholasticism, as for their proficiency in another department of knowledge.

These were Thomas of Chantimpré, Vincent of Beauvais, and Albertus Magnus or Albert the Great (1193–1280),[1] of whose treatise upon animals Victor Carus says, in his 'Geschichte der Zoologie,' p. 226, that, in comparison with the works of the two previously mentioned writers, it is far more thorough and composed with greater self-confidence.

Thomas of Chantimpré was a pupil of Albertus Magnus,[2] and that Vincent of Beauvais used his books is proved by his numerous quotations from them. Although, like all his predecessors, Albert the Great based his work on Aristotle,

[1] Cf. F. A. Pouchet, *Histoire des Sciences naturelles au moyen-âge, ou Albert le Grand et son époque considérés comme point de départ de l'école expérimentale*, Paris, 1853. Cf. also Fr. Ehrle, S.J., 'Der selige Albert der Grosse,' in *Stimmen aus Maria-Laach*, XIX, 1880; G. v. Hertling, *Albertus Magnus, Beiträge zu seiner Würdigung*, written in honour of the 600th anniversary of his death, Cologne, 1880; E. Michael, S.J., *Geschichte des deutschen Volkes vom 13 Jahrhundert bis zum Ausgang des Mittelalters*, III, 1903, pp. 445–460; Arthur Schneider, *Die Psychologie Albert des Grossen: Nach den Quellen dargestellt*, I, 1903, Vorwort VIII.

[2] He describes himself as an *auditor eius per multum tempus*. (Thomas Cantipratanus, *Bonum universale*, Duaci, 1627, l. 2, c. 57, § 50, p. 576. Cf. E. Michael, S.J., 'Albert der Grosse,' in the *Zeitschrift für Katholische Theologie*, 1901, part 1, p. 43.) Borman is therefore probably mistaken in thinking that Thomas of Chantimpré's work was one of Albert the Great's chief sources of information in the compilation of his book on animals. V. Carus falls into the same mistake in his *Geschichte der Zoologie*, p. 227. Cf. also Alex. Kaufmann, *Thomas von Chantimpré*, Cologne, 1899. Thomas was a canon regular in the Augustinian monastery at Chantimpré before he entered the Dominican Order in 1232. His book, entitled *Liber de rerum natura*, was subsequently translated into German by Konrad Megenberg, who belonged to the cathedral chapter at Ratisbon. Its German title is *Buch der Natur* (Book of Nature), and it records the results of much independent research. The same author's work on bees (*Bonum universale de apibus*) is a pious picture of manners rather than a treatise on natural history.

he took more pains than any of them to make independent observations of his own. His treatise on animals consists of twenty-six books, of which nineteen correspond to the writings of Aristotle, whilst seven are of independent origin.[1]

Book XX, the first of those containing his own results, deals with the nature of animals' bodies in general, and Book XXI with the degrees of perfection attained by them (*de gradibus perfectorum et imperfectorum animalium*), a quite modern idea in classification, on the lines of comparative morphology of animals. The remaining five books deal with animals singly, arranged alphabetically within the larger groups. These seven books show conclusively that the author was not content to write a commentary on Aristotle, but aimed at rendering his work more complete by adding the results of his own investigations.

Albert the Great's seven books 'De vegetabilibus et plantis,' which contain his views on botany, have been carefully studied and justly appreciated by E. Meyer, in his 'Geschichte der Botanik,' IV, Königsberg, 1857, but the more important work on zoology has hitherto met with far too slight recognition among scientific men. An attempt to display its merits, made by Karl Jessen in 1867, was frustrated, owing to the defective state of most editions of Albert the Great's works.[2]

E. von Martens subsequently published some observations on several of the mammals mentioned by him, and Victor Carus has devoted a few pages to Albert the Great in his 'Geschichte der Zoologie,' but without discussing his work in detail.[3] Although Carus is by no means a partisan of the Church, he feels bound to confess, on p. 224, that 'Albert, to whom the cognomen "Great" may justly be conceded, is undoubtedly the chief writer of the thirteenth century on the subject of natural science.' If Carus had adhered to the principle which he himself laid down, and had foreborne to judge Albert the Great as a zoologist by the standard of a modern writer on

[1] In the complete edition of Albert the Great's works, published in Paris by Vivès, the treatise on animals is contained in vol. xi (*De animalibus pars prior*) and vol. xii (*De animalibus pars altera*).

[2] 'Alberti magni historia animalium' (*Archiv für Naturgeschichte*, XXXIII, vol. i, 1867, pp. 95-105).

[3] Munich, 1872, pp. 224-237.

science, he would probably have spoken in more favourable terms of his achievements in zoology.

Although Albert the Great could not completely disentangle himself as a zoologist from the prejudices and fancies of his predecessors, his merit lies, not merely in his having gone back from Pliny to Aristotle, but also in his having led the way to independent research, which does not rely blindly upon authority, but looks for itself.[1]

R. Hertwig is perfectly correct in stating in the most recent edition (seventh) of his 'Lehrbuch der Zoologie' (1905, p. 7) that Albert the Great even began to collect his own zoological observations. In many passages of his work on animals he refers to his own investigations, and, when he describes anything, he frequently adds a remark to the effect that he has himself seen the thing in question, and even possesses it in his collection. He devotes several chapters to the habits of falcons, which he seems to have studied with particular interest. In one place he tells us that he took a short sea voyage for zoological purposes, and on the shore of an island he collected ten or eleven kinds of 'bloodless sea-beasts.' After recording the various tales told about the propagation of fish, he adds: 'I believe that none of all this is true, for I have myself made diligent investigations, and have questioned the oldest fishermen engaged in salt and fresh water fishing,' and he proceeds to give the results of his observations and inquiries. He declares that by personal observation he has disproved the popular theory that the left legs of a badger

[1] Men such as Albert the Great are enough to refute the discovery made by certain followers of Darwin, that Christianity has 'stifled the spirit of scientific research' and has 'caused a kind of hostility to the idea of busying the mind with natural objects.' It is unfortunate that such prejudiced statements have found their way into even our modern text-books of zoology. See, for instance, R. Hertwig, *Lehrbuch der Zoologie*, 1900, p. 7. The following words, which I quote from Hertwig, cannot be applicable to Albert the Great: 'The question how many teeth a horse has was discussed in many controversial treatises, in which the authors used all the heavy artillery at their disposal, but it did not occur to one of the learned men to look inside a horse's mouth and see for himself.' It is to the credit of the author of the above-mentioned excellent text-book of zoology, that the words just quoted have been omitted in the two last editions of his book (1903 and 1905). It is satisfactory to observe that the achievements of mediæval scholars in the domain of natural science are gradually receiving fairer treatment, and are being judged by a more unprejudiced standard. Cf. also J. Norrenberg, 'Der naturwissenschaftliche Unterricht in den Klosterschulen' (Scientific Instruction in Monastic Schools), in *Natur und Schule*, III, 1904, part 4, pp. 161-169

are shorter than the right legs, and he relegates the stories of geese growing on trees, and other zoological marvels, into their proper sphere as fictions of the imagination.[1] It is true that his statements are interspersed with a good many mistakes. He is right in saying that flies have two wings, but wrong in giving them eight legs—and his famous pupil, Thomas Aquinas, is falsely accused of having reckoned ants among the *reptilia quadrupedia*, and thus of having fallen into an opposite error.[2] It is hardly necessary to point out how impossible it was for him to correct the old legends with reference to exotic animals, and so he says that the porcupine shoots its quills at its enemies, that the wild unicorn grows tame when caressed by a maiden, &c. We ought to bear in mind that to a German student of nature in the thirteenth century no other source of information about foreign animals was accessible than the old fabulous stories. What pains Albert the Great took to obtain trustworthy information about animals that he had never seen, is proved by his admirable account of the methods then in use in the whalefishery.

Careful studies in another quarter have recently shown that Albert the Great followed an independent method of investigation. Dr. R. Hertwig, Professor of Zoology at the University of Munich, suggested to Dr. H. Stadler to make a critical examination of Albert's zoology and botany. The full result of this examination has just been published in the *Forschungen zur Geschichte Baierns*, XIV, 1906, first and second parts, pp. 95–114, but Stadler communicated a good deal of it previously, at a lecture delivered on March 20, 1905, to the 'Verein für Naturkunde' in Munich. The title of the lecture was: 'Albert the Great as an independent student'; I subjoin some extracts from it:—

This very prolific writer was a scholastic, but he occupies a position on a level with Aristotle rather than subordinate to him,

[1] The story of the geese growing on trees probably originated in the fact that the barnacle goose (*Lepas anatifera*) often attaches itself to floating tree trunks.

[2] In the *Summa Theologiae*, I, q. 72, ad 2. In Vivès' edition (1871) the passage reads as follows: 'Per reptilia vero (intelleguntur) animalia, quae vel non habent pedes . . . vel habent breves, quibus parum elevantur ut lacertae et tortucae.' There is a note on the word *tertucae*: 'Sic codices, sed nescio qua incuria in Parmensi et in omnibus editionibus *formicae*.' *Tortuca* is *tartaruga*, *tortue*, *tortoise*, and is rightly reckoned among the reptiles, only a constantly repeated misprint has turned tortoises into ants!

and did not simply reproduce Aristotle's statements, but, as far as he could, explained, completed and expanded them. He displayed great shrewdness and keen intelligence in carrying on his favourite observations on the animals and plants of Germany, whence he derived the evidence for his scientific statements that he based upon Aristotle. His writings therefore contain all the information on natural history possessed by the people of Germany in his day; he describes the life of animals as observed by intelligent huntsmen and farmers, fishermen and bird-catchers; everywhere the biological element and his own personality are prominent, and for this reason his writings form a sharp contrast to the dry book-learning of the periods preceding and following his lifetime. It is true that in dealing with botany he follows the lines of the pseudo-Aristotelian work 'De plantis,' really written by Nicholas Damascenus, but under the form of excursus he gives a far better account of the subject, based upon his own observations. He describes very correctly the vascular bundles of the plantain leaf and the medullary rays of the vine, and divides plants into two classes, cortical and tunical, a division approximately corresponding to that of monocotyledonous and dicotyledonous. He distinguishes parenchyma and bast-fibres in the large stinging nettle, hemp and flax; he knows the difference between the inner and outer bark, and the importance of each to the life of a plant. He has observed the square stem of the deadnettle, and the diversity in growth between plants in isolation and when cramped for space. He describes very clearly the difference between a thorn and a sting; he attempts a classification of leaves according to the shape, notices that plants with woody stems have bud-scales, and herbaceous plants have naked buds, and he recognises, as a peculiarity of the grape vine, the fact that fruit and tendrils are opposite to the foliage leaves.

In speaking of blossoms he draws attention to their various forms of insertion, and mentions stamens, pistil and pollen, although he confuses the pollen with wax. He comments upon the deciduous calyx of the poppy, tries in a very primitive fashion to classify the forms of the corolla, insists upon the importance of the seed in preserving the species, and gives a very fair classification of fruits. The position and the significance of the ovules and of the tissues connected with nutrition did not escape his notice. The sixth book, 'De vegetabilibus,' contains many admirable descriptions of single plants, especially of the mistletoe, the hazel, the alder, the ash, the date-palm, the poppy, borage and rose, and in the case of the last-mentioned he gives an excellent account of the æstivation of the calyx and of the alternation of the parts of the flower, and suggests the true explanation of their significance.

We may speak in similar terms of his work on zoology, for which, however, we are unfortunately obliged to use the very unsatisfactory edition published by Auguste Borgnet in Paris, 1891, so that much in it appears open to question. Of animals known in Germany, Albert begins by describing the German marmot and the earless marmot, the two kinds of marten, the garden dormouse and the common dormouse, and he is the first writer who alludes to the chamois, the badger, the rat, the ermine and the polecat.[1]

He gives charming accounts of the mole, the marmot and the squirrel; he knows the *Lepus variabilis* of the North and the polar bear; he describes a whaling expedition and remarks that in his day the elk, the bison, and the aurochs were to be found only in the extreme east of Germany. His description of the cat displays great sympathy with animals and very sharp powers of observation.

In dealing with birds, he discusses the various falcons in the greatest detail, but he is well acquainted with the other birds of prey. He speaks of the peculiar structure and purpose of the woodpecker's claws, and considers the distribution of the hooded crow and the habits of migratory birds.

Blackcock, grouse, and heathcock were familiar to him, and he knew many kinds of singing birds (four varieties of finches, two of sparrows and three of swallows), also the nutcracker and kingfisher; he describes the nest of the magpie and the habits of the cuckoo with great accuracy. The lecturer proposed to speak of Albert the Great's knowledge of fishes on another occasion; he stated that Albert had dissected insects and had perhaps recognised the digestive system and heart. He gives a correct account of the development of cockchafers and wasps, and also of caterpillars and their spinning process, and of the habits of the ant-lion. Of other creatures, the best description given as the result of his own observation is perhaps that of the jelly-fish.

Among the learned Franciscans of the thirteenth century, Roger Bacon, the *doctor mirabilis*, deserves special mention,[2] as he is in many respects the equal of the great Dominican, Albertus Magnus. His chief services to science are in the domain of physics, chemistry and medicine, rather than in that of the descriptive natural sciences. Considering the age in which he lived, he had wonderfully advanced opinions regarding physiology. Much attention has been paid to Bacon by Émile Charles,[3] who declares that the results stated in his

[1] In the printed text of the lecture there is a query after the word *rat*, but having had some correspondence with Stadler, I infer from a letter dated December 4, 1905, that the query ought to be omitted, as Albert the Great was really the first to describe the rat.

[2] See Dr. H. Felder, O. Cap. *Geschichte der wissenschaftlichen Studien im Franziskanerorden bis um die Mitte des 13 Jahrhunderts*, Freiburg i. B., 1904, pp. 379-402.

[3] *Roger Bacon, sa vie, ses ouvrages, ses doctrines d'après des textes inédites*, Paris, 1861.

work 'De vegetabilibus' surpass those of Albert the Great. We receive an impression of something quite modern, in fact almost anti-vitalistic, when the mediæval Franciscan speaks thus of the relation in which chemistry (which he calls *alchimia speculativa*) stands to the other natural sciences:[1]

> Because students are not acquainted with this science, they also know nothing of its bearing upon natural history, for instance, the origin of living creatures, plants, animals and men. . . . For the constitution of the bodies of men, animals and plants depends upon an intermingling of elements and fluids, and proceeds in accordance with laws similar to those governing inanimate bodies. Consequently whoever is ignorant of chemistry, cannot possibly understand the other natural sciences, nor theoretical and practical medicine. . . .

3. The Development of Systematic Zoology and Biology

As soon as the age of discoveries began in modern times, much more interest was taken in the study of nature, and the tree of biological knowledge put forth one branch after another, all of which were full of vigorous life. In our historical sketch we must follow this process of division, and we will begin by considering the growth of systematic classification, leaving for the present the development of some other branches.[2]

It was natural that external differences in form should be the first things to attract the attention of a student, in the case both of plants and of animals; later on he tried to learn something about the mysteries of their constituents, of their configuration, and of the vital phenomena of living organisms. It was natural, therefore, for systematic zoology and that *scientia amabilis*, systematic botany, to develop earlier than the other branches of biology. We cannot do more than mention the chief pioneers in systematics. Edward Wotton, an Englishman, wrote in 1552 a book called 'De differentiis

[1] *Opus tertium*, c. 12, ed. Brewer, 39: Et quia haec scientia ignoratur a vulgo studentium, necesse est ut ignorent omnia quae sequuntur de rebus naturalibus; scilicet de generatione animatorum, et vegetabilium et animalium et hominum: quia ignoratis prioribus necesse est ignorari quae posteriora sunt. Generatio enim hominum et brutorum et vegetabilium est ex elementis et humoribus et communicat cum generatione rerum inanimatarum. Unde propter ignorantiam istius scientiae non potest sciri naturalis philosophia vulgata nec speculativa medicina nec per consequens practica. . . .

[2] Cf. R. Burckhardt, 'Zur Geschichte der biologischen Systematik,' Bâle, 1903 (*Verhandlungen der Naturf. Gesellschaft in Basel*, XVI).

C

animalium,' in which he returned to Aristotle's system, which he developed by adding to it the group of zoophytes. Another Englishman, John Ray (1628–1705),[1] defined the Aristotelian idea of species more clearly. His works, 'Methodus plantarum nova' (1682) and 'Historia plantarum' (1686–1704), are very important in systematic botany, whilst his synopses of various classes of animals, especially of quadrupeds and snakes (1693), mark an epoch in systematic zoology. In this way Ray, the son of an English blacksmith, facilitated the work done by the great Swedish knight Karl v. Linné (Linnæus), who was born in 1707, being the son of a Protestant pastor in Råshult. A year after the birth of Linnæus died his chief forerunner in botanical research, the eminent Frenchman, Joseph Pitton de Tournéfort (1656–1708), who in his 'Éléments de botanique ou méthode pour connaître les plantes' laid the foundation of our present classification of plants.

The work of Linnæus (1707–78) marks a fresh stage in the growth of the tree of biological knowledge, and caused it to become a vigorous trunk with many branches. Under his influence it grew strong enough to support the wealth of offshoots which were destined to spring from it during the nineteenth century. He made many journeys to Central Europe in order to study the chief collections of his day, and with unflagging industry he acquired the material for his great work, the 'Systema naturae,' which stands alone of its kind and is of the utmost importance in the history of biology. The first edition appeared in 1735, the fifteenth (which was the last revised by Linnæus himself) in 1766–8. The most complete and best known is the seventeenth edition of the 'Animal Kingdom' brought out by Gmelin, 1788–92.

The chief value of the 'Systema naturae' lies not so much in the fact that Linnæus has in it formed systematic groups of all previously described varieties of animals and plants, adding many fresh ones to those already known, but rather in his having introduced in his binary nomenclature a fixed scientific terminology, so that exact statements of laconic brevity thenceforth took the place of long-winded descriptions. This work of Linnæus had as important a bearing upon the development of descriptive natural science, as the introduction of a

[1] Ray died on January 17, 1705, not, as is generally stated, in 1704.

written language has upon the development of a nation. Until a language possesses a grammar and a vocabulary, it is only a scientific embryo ; its elements lack sharpness and clearness ; it has, so to say, no framework to which they can be attached in orderly fashion.

There is no need for a long explanation of the binary nomenclature. It is enough to say briefly that to every species of animal and plant a scientific double name is assigned, consisting of a generic and a specific name, both latinised in form, and as these names are constant, universally current and unchanging, they are free from arbitrary fluctuations in use, such as are of common occurrence in the case of popular names. To the generic name, which is a noun, the *differentia specifica* is added by connecting with it the specific name, which is an adjective. *Canis familiaris*, *Carabus auratus*, and *Carabus nitens* may be taken as typical examples. Whoever gives a name of this kind adds a concise description of the animal to serve as a means of identifying its species, and a writer using the name appends to it in abbreviated form that of the author who first gave it and described the animal in question, so that, when in future any one reads *Carabus auratus*, L. (Linnæus), he knows exactly once for all what form it is intended to designate. In this way a name such as *Carabus auratus*, L., becomes a generally recognised scientific appellation, leaving nothing to be desired in the way of clearness and simplicity. Through the use of the binary nomenclature, the whole zoological and botanical system has been reduced to a classified catalogue, well arranged and visible at a glance, and in devising it Linnæus conferred an inestimable boon upon biology. The inspiration thus in so simple a manner to arrange logically the vast multiplicity of forms in the animal and vegetable kingdoms is like Columbus' egg—before Linnæus appeared, no one knew how it could be made to stand at all, but after Linnæus had once for all set it upright, no one had anything to do but to follow his example.

On account of his 'Systema naturae' Linnæus is to be reckoned as the founder of modern systematic science. His system of nomenclature is still the standard one, and will probably continue to be so. The laws of zoological nomenclature, as elaborated at the close of the nineteenth century by a committee, specially appointed for the purpose at recent zoological congresses,[1] and universally adopted in scientific circles, are only a logical carrying out and detailed specialisation of the principles laid down by Linnæus. At the annual meeting of the German Zoological Society in 1891, it was decided to appoint a committee to lay down rules securing uniformity in zoological nomenclature.[2] In order to have a firm basis on which to decide disputed points of priority, the German Zoological Society caused a reprint of the tenth edition of Linnæus' 'Systema naturae' to be issued thus marking the year 1758, in which the tenth edition first appeared, as the date when systematic zoology originated, and fixing as the standard generic names those used at that time by Linnæus.

The International Botanical Association is now dealing with the question of botanical nomenclature at the International Botanical Congresses, of which the first was held in Paris in 1900, and the second at Vienna in 1905.

Linnæus' 'Systema naturae' is a monumental work, such as could be accomplished only at one period, at least by a single individual. By means of the further development of systematic zoology and botany, effected by a closer study of European fauna and flora, as well as by the exploration of foreign countries, which has supplied a boundless and ever-increasing wealth of material, systematic science has now attained such gigantic proportions, that no single human intellect, not even the genius of an Aristotle, would be capable of grasping and assimilating it in all its details. In the year 1901 the total number of species of animals known to science amounted to at least 500,000, of which more than half are insects. In giving the number of species of beetle at 100,000 we are probably rather understating it. In the vegetable kingdom it is estimated that there are about 200,000 species scientifically described, divided into 11,000 genera—there are 50,000 species of cryptogams alone.

[1] *Règles de la Nomenclature des êtres organisés, adoptées par les Congrès Internationaux de Zoologie, Paris,* 1889 *et Moskou,* 1892 (Paris, 1895); *Report on rules of Zoological Nomenclature, to be submitted to the fourth International Congress at Cambridge by the International Commission for Zoological Nomenclature* (Leipzig, 1898); *Règles de la Nomenclature Zoologique adoptées par le cinquième Congrès International de Zoologie* (Berlin, 1901).

[2] *Verhandlungen der Deutschen Zoolog. Gesellschaft,* 1891, p. 47; 1892, p. 13; 1893, p. 89, &c.

In order to collect the enormous mass of information on systematic zoology which is now scattered in numberless articles in numberless scientific periodicals and books, the German Zoological Society determined, at their first general assembly in 1891, to issue a great systematic work entitled 'Species animalium recentium' or 'Das Tierreich' ('The Animal Kingdom'), which should contain systematically arranged descriptions of all the existent kinds of animals as far as they are at present known. This great plan, which in Linnæus' time was not beyond the power of one man, can now only be carried out by a scientific society having at its disposal many workers and abundant means; and even so it is doubtful whether the new 'Animal Kingdom' will be completed by the year 2000. I have made a careful calculation with regard to entomological literature, the results of which will perhaps be of interest here.[1]

Every number of the work is to be arranged according to the same detailed plan, therefore, from the nineteen numbers that had appeared in 1894, we can form some idea of the probable extent of the whole.[2] Assuming that the same method is followed in subsequent numbers as in those that have already appeared, for the Order of Coleoptera alone, according to a moderate estimate, 111 volumes of 500 pages each will be required, for the whole class of insects at least 300 volumes of 500 pages, and for the whole animal kingdom at least 500 volumes of 500 pages. These 500 volumes would contain approximately 15,625 signatures, so that if the work is to be completed in 100 years, 156 must be issued yearly. But, as a matter of fact, since 1897 on an average less than fifty signatures have appeared each year.

It is not my wish to take a pessimistic view of the matter, but to give the reader some idea of the advance made in biological knowledge. Let us hope, therefore, that the whole enormous task will be completed within a reasonable period, before the 'Twilight of the Gods' foretold by Wala sets in, for

[1] Cf. my discussion of the first numbers of the 'Tierreich' in *Natur und Offenbarung*, XLIII (1897), 508; XLIV (1898), 635.

[2] Cf. the annual reports submitted to the meetings of the German Zoological Society by Professor F. E. Schulze, the general editor. The publication of the work has now been undertaken by the Berlin Academy of Science. By the summer of 1905 twenty-three numbers had appeared.

this would probably be a twilight of zoologists also; let us hope that the zoology of the future will derive much pleasure and satisfaction from this creation of the German Zoological Society; in any case, the calculation I have made will serve to give my readers some approximate conception of the enormous strides made by systematic zoology in the course of the nineteenth century.

Modern botanists, too, have undertaken the publication of vast systematic works, continuing the enormous task of systematisation on Linnæus' principles. One of these works is 'Die natürlichen Pflanzenfamilien nebst ihren Gattungen und wichtigeren Arten,' von A. Engler und K. Prantl ('The natural families of plants together with their genera and more important species,' by A. Engler and K. Prantl). The Phanerogams were completed before the end of the nineteenth century, in a space of about twenty years, and are contained in eleven stately volumes, but the Cryptogams are not finished yet.

Another huge work on botany, the counterpart of the 'Species animalium recentium,' is being brought out by A. Engler for the Royal Academy of Science in Berlin, under the title 'Regni vegetabilis conspectus.' It has been appearing at intervals since 1900, and numerous collaborators in all parts of the world are engaged on it. We may trust that there are fewer hindrances in the way of its completion than in that of the 'Tierreich,' in the case of which the enormous class of insects presents great difficulties, though it is to be hoped that these will eventually be overcome.

There is one respect, however, in which the systematic advance of modern zoology and botany is not on the lines of Linnæus' 'Systema naturae.' Linnæus was unable to avoid using external differences as the distinctive marks of his systematic groups, and in this way he was led to unite in an artificial system forms that bore no natural relationship to one another. In describing and classifying plants and animals modern systematic science can avail itself of the assistance of other biological sciences, especially of anatomy and of morphogeny, or the history of individual development, and thus it attains to a more or less successful *natural* classification of organic forms. In spite of this difference, however, it is true that modern systematic science is based upon

Linnæus and his 'Systema naturae,' for without this achievement of his powerful intellect we should at the present time have had no natural systems of plants and animals.

The fact that the German Zoological Society regarded it as necessary to issue a fresh edition of Linnæus' 'Systema naturae,' and to undertake the publication of a great work on systematic zoology on the same lines, is testimony enough to the importance of systematics or the science of classification in the development of biological knowledge. It shows at the same time how deeply indebted the representatives of modern science are to Linnæus, and it is to be regretted that in some of the more recent books on zoology Linnæus is mentioned as the founder of the 'unintelligent zoology of species,' and this in more or less plain language.[1]

To a certain class of Haeckelists, systematic science seems like an inconvenient old man, who threatens to check them in their bold intellectual tricks and fantastic speculations, precisely because the actual multitude of forms in the animal world does not coincide with their ideas, and because they are too impatient to be willing to master the subject-matter of

[1] R. Hertwig is however justified in stating in his *Lehrbuch der Zoologie*, 7th edit., 1905, p. 9, that post-Linnæan zoologists, and especially entomologists, have made it their sole aim to describe the greatest possible number of new species, making quantity rather than quality the measure of their achievements. Unfortunately, even at the present day this class of pseudo-systematic biologists is not quite extinct, and there are still some who flood the scientific periodicals with superficial or even 'provisional' descriptions, and thereby put obstacles in the way of studying some groups of animals, for other, more thorough workers, who can make nothing of these superficial descriptions, are hindered by being obliged by the law of priority to take them all into account. An almost incredible story is told of a 'scientific worker' who was employed about fifty years ago at a great museum, and was paid £1 for each new genus and 1*s*. for each new species that he established. In order to work more quickly, he had two bags beside him, one filled with Greek and the other with Latin names. If he wanted a name for a new genus, he put his hand into the Greek bag and pulled out a name haphazard, and bestowed it upon his genus. If, on the other hand, he wanted a name for a new species, he had recourse to the Latin bag, and labelled it with the first adjective that he caught up. It can easily be imagined how applicable the new names thus assigned were to the genera and species, and the descriptions which he appended as 'original' to these names were equally suitable. Such work as this was really 'unintelligent zoology of species,' but it would be unfair to regard zoology of species as responsible for such lack of intelligence. There are excrescences in every branch of knowledge, and they do not occur more frequently in the systematic zoology of the Linnæan school than in the modern doctrine of evolution. Ernst Haeckel's famous book, *The Riddle of the Universe*, affords a striking instance of unintelligent blunders on the part of the Darwinian supporters of this doctrine. See my criticism of the same in *Stimmen aus Maria-Laach*, LX, 1901, p. 428, &c.

systematics before beginning their speculations. They completely forget that but for this stern old father they would have no existence at all.

Mere systematics is certainly by no means the ideal of biological knowledge ; it is not an end in itself, but is only an indispensable aid to biological research. It bears the same relation to the other biological sciences as the dry heart-wood of a tree bears to its tissues permeated by life-giving sap ; it forms the skeleton or scaffolding for other sciences. But just as in the human body the eye has no right to reproach the bones of the foot for not responding to the vibrations of ether, so modern morphology and morphogeny ought not to look down upon systematics for not perceiving many things that these branches of science can discover. In science, as in the living organism, the principle of the subdivision of labour holds good, and the greater the perfection attained by any science, and the more numerous its departments, the more indispensable is it to distinguish clearly the subject-matter with which each single subdivision deals, if any solid progress is to be made.

Let us apply this consideration, the truth of which no modern scientific man will question, to Linnæus' position with regard to biology. Scientific classification or systematics was his speciality, and it was a boon to science that Linnæus with his vast intellect devoted himself to it rather than to anatomy and physiology, for the formation of a strong systematic science was the first and most necessary starting point for all the other branches of biological science, if they were to thrive at all. Without it zoology and botany would have remained a hopeless chaos of forms, through which no one could have found his way.

In order to produce a great systematic work like Linnæus' 'Systema naturae,' even at that time a man was required who should devote his whole ability to this end, for otherwise it would have been unattainable. When his pygmy successors, who have inherited the achievements of his genius, reproach the great Linnæus with being merely a one-sided systematist, they show themselves to be both short-sighted and ungrateful.

CHAPTER II

THE DEVELOPMENT ON MODERN MORPHOLOGY AND ITS BRANCHES INVOLVING MICROSCOPICAL RESEARCH

1. The Development of Anatomy before the Nineteenth Century

We have already shown how Aristotle may justly be regarded as the founder of modern systematics,[1] and he may with equal right be called the first morphologist in the modern sense, because he carried on a comparative study of the varieties of form among animals. Aristotle laid the foundation of the science of morphology in his work 'De partibus animalium,' and Galen (131–201 A.D.) continued what Aristotle had begun, for his famous work on human anatomy is based chiefly upon post-mortem investigations on the higher animals, and so should be called *animal* rather than *human* anatomy. The real originator of human anatomy was Vesalius (1514–64), who dissected human bodies, and thus was able to correct many errors arising out of Galen's studies of animals.

[1] Cf. also on this subject Professor R. Burckhardt, 'Zur Geschichte der biologischen Systematik' (reprinted from the *Verhandlungen der Naturf. Gesellschaft in Basel*, XVI, 1903, pp. 388–440).

Marco Aurelio Severino (1580–1656), a Calabrian, was the author of the first book on general anatomy. It was published in Nüremberg in 1645, and bears the title: 'Zootomia Democritaea, id est anatome generalis totius animalium opificii libris quinque distincta.' Severino treats the 'lower animals' in a very curt fashion; they fare better at the hands of writers towards the close of the seventeenth century. Marcello Malpighi, a Bolognese physician (1628–94), wrote a 'Dissertatio epistolica de bombyce' (1669) on the anatomy of the silkworm, and this work opened the way to the anatomical study of insects, for the discovery of the Malpighian tubes, of the heart, nervous system, tracheae, &c., for the first time revealed insects as organic masterpieces, whose wonderful construction is scarcely inferior in perfection to that of the higher animals, and is more worthy of admiration, because of its diminutive size.

Johann Swammerdam (1637–85), who lived at Amsterdam, in his 'Bijbel der natuure' (*Biblia naturae*), published 1737–8, describes with astonishing accuracy the internal structure of bees, ephemera, snails, &c.; and whoever is acquainted with the excellent anatomical discussion of the larva of the goat-moth, published in 1760 by Pieter Lyonet of Maastricht, cannot fail to recognise its merits even at the present time, when we can avail ourselves of greatly improved instruments and technical methods in dealing with the same subject.

The great scientists mentioned above inaugurated a new era in anatomical knowledge, yet morphology was still not a systematically organised science, but only a collection of interesting monographs. It was raised to the rank of a special science at the beginning of the nineteenth century, by Bichat, a Frenchman, who introduced the idea of systems of organs and systems of tissues. Bichat's 'Traité des membres en général' (1800) and his 'Anatomie générale' (1801) created comparative anatomy, for he divided the constituent parts of the bodies of animals into organs and tissues, and into systems of organs and tissues, thus fixing a firm basis for the comparison of the constituent parts of various animals. It is true that this idea of Bichat's was not altogether new; Aristotle, Galen, and Albert the Great distinguished heterogeneous and homogeneous parts among the constituents of the bodies of animals. The heterogeneous parts are the individual organs, the homogeneous are the tissues, which may be found in various organs, and of which the organs are composed.

A famous Italian anatomist, Gabriele Antonio Fallopius (1523–62), as early as the sixteenth century wrote 'Tractatus quinque de partibus similibus,' in which he distinguished and described a considerable number of tissues. In 1767 Bordeu, a Frenchman, devoted an entire work to one kind of tissue, viz. the mucous connective tissue; his book bears the title 'Recherches sur le tissu muqueux ou organe cellulaire.' Still it was Bichat who first arranged the homogeneous tissues as a scientific whole, distinguishing them from organs and systems of organs. A system of organs is a complex of organs working together to discharge the same vital function and so forming one physiological whole. A system of tissues is a complex of tissues consisting of the same morphological elements, and so forming one logical whole, from the point of view of comparative morphology. Two examples will explain this distinction. The digestive system in man is a system of organs, for it is made up of several organs which unite to produce one and the same physiological result, though they are formed of various kinds of tissue; for, in addition to epithelial tissue, both connective and muscular tissues enter into their structure. But the glandular system in man is a system of tissues, for it consists of essentially similar tissues, viz. modifications of the epithelium, which serve very various physiological purposes; such are the gland of the intestine, the renal gland, the salivary gland, the sweat gland, &c. In other cases the distinction between a system of organs and a system of tissues is not so strongly marked as in those to which I have just referred. For instance, when we speak of the nervous system of man, we are alluding to both a system of organs and a system of tissues. Nevertheless, in theory the two systems are totally distinct even here.[1]

A far greater man, and one who had much more influence on the development of comparative morphology, was Georges Cuvier (1769–1832). He was born at Mömpelgard and educated

[1] Textbooks on zoology treat chiefly of systems of organs, and those on histology chiefly of systems of tissues, therefore a writer on zoology is apt to ignore the histological point of view, and *vice versa*, which is disastrous to perspicuity.

at the Karlsakademie in Stuttgart. Whilst he was professor of comparative anatomy at the Jardin des Plantes in Paris, he published numerous important works. In 1812 he established a new classification of the animal kingdom, which is known as Cuvier's Theory of Types, and is based upon the anatomical comparison of the various groups of animals. According to it animals are divided with reference to their structure into four main classes, which Cuvier called *embranchements,* but Blainville subsequently substituted the name *types.* These are vertebrata, mollusca, articulata, and radiata. Cuvier's Theory of Types was expanded and elaborated by Karl Ernst von Baer (1792–1876), an Esthonian, the founder of comparative embryology, whose theory of germinal layers reduced the embryology of animals to a scientific system.

Cuvier's Theory of Types was not by any means his sole contribution towards the development of modern zoology. His comprehensive work 'Le règne animal' (1816),[1] in the compilation of which he was assisted by many collaborators, is the most important achievement in the domain of systematics since the time of Linnæus. His 'Histoire des sciences naturelles,' published after his death in Paris (1841–5), as R. Burckhardt aptly remarks,[2] presents the history of zoology and the natural sciences in one vast frame, and is a monumental work of wide scope. Cuvier devoted much attention also to fossil animals, and between 1795 and 1812 he brought out several works on the subject, laying down definite morphological principles to be followed in comparing fossils with still existing animals of the zoological system, and he thus became one of the chief founders of modern palæontology. His chief service to comparative biology was that he established the law of correlation, i.e. he was the first to formulate the regular connexion of the organs of any animal with one another, and with its habits and environment. Although Cuvier did not regard as essential the variations of form within his four great types, he was an adherent of the theory of permanence, and in 1798 for the first time he gave a clear concise statement of the meaning of the 'systematic species,' a definition that still holds good. His views on the permanence of species brought him into

[1] The fourth edition in eleven volumes appeared 1836–49.
[2] 'Zur Geschichte der biologischen Systematik,' 390.

conflict with his contemporaries, Jean Lamarck and Etienne and Isidore Geoffroy St. Hilaire, who upheld the transmutation theory. The scientific struggle carried on by the members of the French Academy ended for a time in the victory of Cuvier's opinion, but we shall have to recur in the ninth chapter to the further history of the theory of evolution.

2. The Early History of Cytology

Hitherto, in speaking of the development of anatomy, we have referred chiefly to macroscopic anatomy, which is not dependent upon the microscope ; it is, however, to this instrument that most of the progress made by modern morphology is due.[1]

It was invented some hundreds of years ago, but not until the nineteenth century did the real age of microscopical research begin. As early as the year 1100 the Arab, Alhazen ben Alhazen, described the magnifying power of a convex lens. The English Franciscan, Roger Bacon, who lived 1214–1294, and whom we have already mentioned (p. 16), seems to have constructed complicated optical instruments. He is said to have ground a piece of glass so that people saw wonderful things in it, and ascribed its action to the power of the devil. If this glass deserves to be called a microscope, the honour of inventing this instrument would have to be ascribed to Roger Bacon, but various nations claim to have given birth to the inventor of it. The Italians say that either Galileo or Malpighi invented it, but most people consider two Dutchmen, Hans and Zacharias Janssen (1590), to be more justly entitled to the credit of the invention. The name 'microscope' was first applied to the new instrument by Giovanni Faber in Rome in 1625, and many improvements in it were made about 1646 by the astronomer Francesco Fontana in Naples. Malpighi and Swammerdam certainly used the microscope in their scientific work, and the Dutchman Anton Leeuwenhoek of Delft (1632–1723), the 'Father of the Microscope' as Schlater calls him, used it in examining the ova and stings of bees, and many other things connected with the anatomy of insects.

[1] Cf. Dr. J. Peiser, 'Die Mikroskopie einst und jetzt,' in *Natur und Schule*, IV, 1905, parts 10, 11.

By its aid he discovered infusoria, and drew the attention of scientific men to a new world of diminutive creatures, our knowledge of which was greatly increased by Christian Gottfried Ehrenberg in the middle of the nineteenth century. By means of the microscope Leeuwenhoek was enabled to discover the red-blood corpuscles and the transverse striation of the muscular apparatus, and Hamm to perceive spermatozoa, the key to those mysterious problems of heredity which the greatest biologists of the present day are so eager to solve.

Thus we see that microscopical anatomy made steady progress, and advanced towards the marvellous triumphs of modern histology and cytology. It was, however, a long time before scientific men generally made use of the microscope ; it is a surprising fact that even in 1800 it was altogether neglected by Bichat, to whom we have already referred as the founder of comparative anatomy. Consequently he could give no account of cells, the smallest constituents of animal tissues, although they had long before been recognised by other scientific men who used the microscope.

Who discovered cells and the structure of organic tissues out of cells ? In plants it is much easier to find the cells, as they possess, as a rule, a more independent existence in plants than in animals. It is therefore only natural that cells were discovered first in botany. An Englishman, Robert Hooke, gave cells their name because of their resemblance to the cells of the honeycomb. In his 'Micrographia,' which appeared in 1667, he gave the first illustration of a plant cell, or rather cell-wall. The figure represents a bit of cork, along which lengthwise run rows of black specks or cells. Hooke's purpose in speaking of cells was not so much to add to the scientific knowledge of botany, as to display the power of his microscope, and so it is usual to ascribe the discovery of cells to two other scholars, the Italian Malpighi (1674), whom we have already mentioned, and the Englishman Nehemiah Grew (1682). Their works on this subject appeared at almost the same time, a few years after Hooke's 'Micrographia.' Ninety years elapsed before another great scientist continued their work. In 1759 Kaspar Friedrich Wolff published his remarkable book 'Theoria generationis,' in which he propounded new ideas on morphogeny, and threw much light on the morphology of organisms. His descriptions and illustrations show plainly that he had studied the cells in both animal and vegetable tissues; he calls those in the former 'globules' or 'spheres' and those in the latter 'utriculi' or 'cells.' With regard to botany, clear evidence that the vascular system of plants consists of cells was adduced by Treviranus in his work 'Vom inwendigen Bau der Gewächse' ('The internal structure of vegetables'), 1808. The honour of having been the first to discover and mention the nucleus of the living cell is generally ascribed to an Italian-Tyrolese, Abbé Felice Fontana, 1781. However, H. Bolsius, S.J.,[1] has recently proved that the discovery was made by Leeuwenhoek, the Dutch scientist already mentioned, in 1686, about a century earlier.

The English botanist, Robert Brown, was the first to discover (1833) the regular significance of the nucleus in its relation to the cell, and for this reason many people regard him as the real discoverer of the nucleus.[2]

It was not until Joseph von Fraunhofer in 1807 constructed the first achromatic lenses, and thus greatly increased the capabilities of the microscope, that modern cytology was able to develop. It is a remarkable fact that just at this time (1809) Mirbel, a Frenchman, began again to apply the name 'cell' to the smallest elements in living organisms; Malpighi's word *utriculus* had long taken its place, but now, at the dawn of modern cytology, the old name was revived, which Hooke had given to these organic elements 150 years before. The word 'cell' is still in use, in spite of various attempts to substitute some more modern name, such as *protoblast* (Kölliker) and *plastid* (Haeckel). The study of the organic tissues composed of cells was first designated *Histology* by Karl Mayer in Bonn in 1819. Germany is therefore the real home of both histology and cytology, and, as even the French scientists acknowledge, both have grown and developed chiefly in Germany.[3]

[1] Antoni von Leeuwenhoek et Félix Fontana, 'Essai historique sur le révélateur du noyau cellulaire,' Rome, 1903 (*Memorie della Pontificia Accademia Romana dei Nuovi Lincei*, XXI).

[2] Cf. O. Hertwig's *Allgemeine Biologie* (1906), pp. 5 and 27. Hertwig's account of the history of the cell theory is very valuable, pp. 4, &c.

[3] Cf. M. Duval, *Précis d'Histologie*, Paris, 1900, p. 12.

Everyone who has ever opened a modern book on zoology or botany must know the names of Schleiden and Schwann.

Matthias Jakob Schleiden, born 1804 in Hamburg, became the founder of modern botanical cytology when, in 1838, he published his 'Beiträge zur Phytogenesis' in Müller's 'Archiv.'[1] The zoologist, Theodor Schwann, born 1810 in Neuss, applied the same principles to animal tissues in 1839, when he published his 'Mikroskopische Untersuchungen über die Übereinstimmung in der Struktur und dem Wachstum der Tiere und Pflanzen,'[2] and he added so much to Schleiden's work that we generally speak of Schwann-Schleiden's theory of cells, or cytology.[3]

In the case of every object of sense perception, human knowledge invariably proceeds from the exterior to the interior, from the shell to the kernel, and this is true of our knowledge of cells. The dry walls of dead plant cells were what Hooke called cells 250 years ago. Malpighi also studied particularly the plant-cell, which is, as a rule, much larger and has thicker and more conspicuous walls than the animal cell, and hence it became the custom to regard the cellular membrane as the essential part of the cell. Malpighi and Wolff represented the cell as being practically an empty tube or bag—and this was equivalent to mistaking a snail shell for a snail. Schleiden and Schwann had a deeper insight into the truth, for they had better aids to research at their disposal; they discovered that each tube or bag is filled with a fluid, and they noticed the nucleus, though this had been discovered long before. Their opinion was that the cell is a little vessel filled with fluid in which a nucleus is suspended. Subsequent examination of young cells has shown that they have no real walls, and the membrane appears to be an accidental part of the cell, and thus the scientific idea of the cell advanced to the third stage, at which it still practically remains. Franz Leydig in

[1] Cf. Jos. Rompel, S.J., 'Der Botaniker Matthias Jakob Schleiden' (1804–81), in *Natur und Offenbarung*, I (1904), parts 4–7; see especially pp. 393–410.

[2] 'Microscopical researches into the accordance in the structure and growth of animals and plants.'

[3] The botanists Treviranus and Meyen ought to be mentioned as having prepared the way for Schleiden. Their works were published in 1808 and 1830 respectively.

1857[1] and Max Schultze in 1861[2] defined a cell as a mass of living protoplasm containing one or more nuclei.

The fluid contents of the cell were called *protoplasm* by Hugo von Mohl in 1846, and the name has been universally adopted, for it conveys an idea fundamental in biological research.[3] Dujardin in 1835 had named the same substance *sarkode*, but no one now uses this word.

Von Mohl drew the attention of scientists to the movements of protoplasm within the cells of plants, but they had been noticed long before by Bonaventura Corti (1774) and C. L. Treviranus (1807), and described as 'rotatory movements of the cellular fluid.'

At this point the question naturally arises: What are the chemical constituents of protoplasm? In the first part of his 'Studien über das Protoplasma' (1881), J. Reinke describes it as 'a mixture of numerous organic compounds.' Von Hanstein, however, in 1879 defined protoplasm as an albuminous compound or a mixture of albuminous compounds, and he proposed to call it *protoplastin*. In his 'Lehrbuch der Zoologie,' R. Hertwig says in a resigned way that we must acknowledge our inability to determine the chemical characteristics of protoplasm. 'It is not known whether protoplasm is a definite chemical body, which from its constitution is capable of infinite variation, or whether it is a varying mixture of different chemical substances. So, also, we are by no means certain whether or not these substances (as one is inclined to believe) belong to those other enigmatical substances, the proteids. We can only say that the constitution of protoplasm must, with

[1] The year 1859 or 1861 is generally given as the date when cytology entered upon its third stage, therefore I will quote here a passage from Leydig's *Lehrbuch der Histologie des Menschen und der Tiere*, published at Frankfurt a. M. in 1857. He writes as follows (p. 9): 'To the morphological conception of a cell belongs a more or less soft substance, originally almost globular in form, containing a central body called the nucleus.' This, therefore, according to Leydig's opinion in 1857 was the essence of the cell—he had already discarded the membrane as non-essential—for he continues: 'The substance of the cell *frequently* hardens so as to form a more or less independent outer layer or membrane, and *when this takes place* the cell is technically said to consist of membrane, substance, and nucleus.'

[2] 'Über Muskelkörperchen und das, was man eine Zelle zu nennen habe' (*Archiv für Anatomie und Physiologie*, 1861).

[3] Cf. O. Hertwig, *Allgemeine Biologie*, p. 7, &c., for the history of the protoplasm theory; p. 12, &c., for investigations regarding the meaning and nature of protoplasm.

D

a certain degree of homogeneity, have a very extraordinary diversity.'[1]

We may be satisfied to endorse J. Reinke's[2] remark that our conception of protoplasm has always been morphological, i.e. all we know about it is that it forms the primary substance common to every living cell. A detailed account of all the information hitherto acquired on the subject of the chemical composition of protoplasm, as well as on that of the organisation of the cell and nucleus, and their reciprocal chemical relations, will be found in E. B. Wilson's 'The Cell in Development and Inheritance,' New York, 1902, chapter vii; also in O. Hertwig's 'Allgemeine Biologie,' Jena, 1906, chapter ii, pp. 12, &c. On pp. 18 et seq. Hertwig has shown very clearly that the discovery of the substance and process of life is a vital problem, and not merely an affair of chemistry and physics. This subject will be discussed more fully in Chapters VII and VIII.

Our knowledge of tissues and cells has been vastly increased by means of microscopical research since the middle of the nineteenth century. The names of the scientific men distinguished in this branch of research would make a long list; we can mention only the most eminent—Henle, Gerlach, Reichert, Remak, Leydig and Kölliker—some of the more recent zoologists will be noticed later on. Botanists have been no less zealous than zoologists in studying cells under the microscope. We may refer to W. Hofmeister, A. Zimmermann, de Bary and Sachs, as well as to the more recent students—Pfeffer, Wiesner, and Strasburger.

3. Methods of Staining and Cutting Sections for Use under the Microscope

Microscopical research has been greatly facilitated by the discovery of the modern methods of chemical colouring.

As soon as definite colouring matters were applied to animal and vegetable tissues, their structure became more plainly visible, and the structure of the cell itself was revealed, for the nucleus was found to absorb readily certain colouring

[1] English translation, 1903, p. 61.
[2] *Einleitung in die theoretische Biologie*, Berlin, 1901, p. 221.

matters which do not affect the protoplasm of the cell. The nucleus was then seen to contain some darker coloured granules or filaments or nucleoli, which suggested the idea that the nucleus was not a simple but a composite body. In the same way there appeared in the protoplasm darker coloured granules or a network of filaments against a lighter background, and the observation of these led to the discovery of the cell framework. When the colouring process was applied to cells and nuclei in course of division, pictures of wonderful beauty were revealed, from which the laws of the division of the nucleus and of fertilisation were learnt.

Gerlach in 1858 first used carmine as a stain for microscopical purposes, and since his time the number and variety of colouring methods have increased almost indefinitely. Gerlach used carminate of ammonia, others have employed alum-carmine, borax-carmine or carmalum, picro-carmine, &c.

The carmine stains were, however, discarded in favour of haematoxylin, an excellent stain prepared from logwood (*Haematoxylon campechianum*), which is applied in various solutions and combinations, and is still much used in microscopical work. The double stains obtained by using haematoxylin in conjunction with eosin or Congo red or saffranin have lasted admirably, and have produced beautiful and instructive plates, so that haematoxylin has not yet been displaced by its numerous rivals prepared from coal-tar, and known as aniline dyes. The colouring methods just mentioned, and especially the use of haematoxylin and its combinations, are of universal application, and can be employed for almost all histological purposes, but there are also certain special methods of staining particular tissues, especially those of the nerves. Golgi, Ramón y Cajal, and Ranvier used solutions of nitrate of silver, chromate of silver, and formic acid with chloride of gold, in their attempts to overthrow the long-established theory of a central nervous system, and thus extended our knowledge of ganglion cells and their processes.

When Waldeyer formulated his theory of neurones in 1891, and when soon after the theory of fibrils was put forward in opposition to it,[1] the chief arguments adduced in this scientific

[1] At the seventy-second meeting of German naturalists and physicians at Aix-la-Chapelle in 1900, a lively discussion of the two theories took place.

contest were supplied by observations on the nervous system, rendered possible by the use of stains,—methods which Apáthy, Bethe, Nissl, Held, Bielschowsky and others have carried to the utmost perfection. The anatomical and physiological study of nerves owes much to Ehrlich, Retzius and others, who have succeeded in staining the nervous system of a living animal with methyl blue, so that it has become possible to trace the action of the finest fibres and terminations of the nerves.

Quite recently Carnoy and other cytologists at Louvain have used methyl green, and have shown it to be of great service in the development of biology, for it gives a vivid colour to the nucleus of a cell still living, thus rendering visible the most minute details of its structure.

As special stains, used in studying the stages of division of the nucleus in the process of mitosis, we may mention particularly Heidenhain's use of iron alum with haematoxylin and Plattner's metallic nuclear black.

All these colouring methods would avail but little, however, if scientists had not at their disposal a means of cutting organic tissues, as well as entire animals and plants, after artificially hardening them, into layers so thin that light can penetrate them and make their wonderful construction visible under the microscope. The art of cutting sections is as indispensable as the art of staining, and it is by means of both in conjunction that microscopic anatomy has been enabled to make its extraordinary progress in recent times. It owes the one to chemistry, and the other to modern mechanics, which created the microtome and placed it at the service of biology.

The microtome is a mechanical apparatus which passes an extremely sharp knife in a definite direction over an object embedded in paraffin or celloidin or some similar embedding substance, and at the same time a movable plate provided with a scale automatically regulates the thickness of each section.

As at each turn of the plate, about a given angle, the knife is lowered, for instance, $\frac{1}{100}$mm., or (in other microtomes) the object is raised $\frac{1}{100}$mm., a skilful worker is able to obtain an

M. Verworn supported the theory of neurones in his lectures, 'Das Neuron in Anatomie und Physiologie' (reprinted at Leipzig, 1901). See also Fr. Nissl, *Die Neuronentheorie und ihre Anhänger*, Jena, 1903; M. Wolff, 'Neue Beiträge zur Kenntnis des Neurons' (*Biolog. Zentralblatt*, 1905, Nos. 20–22); Wasmann-Gemelli, *La Biologia Moderna*, Florence, 1906, p. 44 note.

unbroken series of sections, each $\frac{1}{100}$mm. in thickness. In the same way he can obtain sections of $\frac{1}{200}$mm., $\frac{1}{300}$mm., $\frac{1}{500}$mm., if he requires them. The microtomes most generally used at the present day are those made by R. Jung in Heidelberg. Microtomes on another system were devised by Professor Hatschek and made by Jensen in Prague; in these the knife does not move up and down along an inclined surface, as it does in Jung's apparatus, but it moves backwards and forwards over a horizontal surface. With the latter I have succeeded better than with the former, and have even prepared very thin and regular sections cut through the hard chitin integument of beetles and other insects. There are also lever microtomes, English microtomes with a pointed spindle, and Minot's new American microtomes intended to cut sections of larger objects. The construction of these ingenious instruments has in the last few years become a special branch of mechanics, and interesting accounts of their great perfection may be found in the illustrated price-lists issued by R. Jung and Walb in Heidelberg, Reichert in Vienna, and others.

4. The Microscopic Study of the Anatomy and Development of a Diminutive Fly

(*Termitoxenia.*) (Plate V)

I should like to illustrate the great advance made in biological research through the adoption of modern methods of staining and cutting sections, and my illustration, derived from my own work, will take my readers out of the gloom of theories into the cheerful atmosphere of practical results.

I am at this moment studying some extremely small insects only 1–2 mm. in length, belonging to the order of Diptera. They have a relatively enormous white abdomen, and in the course of the last few years have been found in the nests of termites in South Africa, the Soudan and India, by G. D. Haviland, Dr. Hans Brauns, J. B. Heim, J. Assmuth, S.J., and Y. Trägårdh.[1]

[1] In subsequent chapters I shall have occasion to refer repeatedly to this remarkable fly, belonging to the family of *Termitoxeniidae*. An account of it is given in Chapter X, 'Theory of Permanence or Theory of Descent,' and illustrations will be found on Plate V.

Diptera of the normal type have two wings, but in their stead this little creature (which I have described under the generic name *Termitoxenia*)[1] has peculiar appendages to the thorax (Plate V, figs. 1, 2, 4, 5) which are morphologically homologous with wings, but have actually so developed as to serve quite other purposes than that of flight, for which their narrow, club-shaped or hooked form and their horny structure render them altogether unsuitable. They are, however, well adapted to perform a number of new functions, closely connected with the insect's habit of living among the termites. The appendages to the thorax of the *Termitoxenia* serve as organs of transport, by which these little inquilines are picked up and carried about by their hosts; they serve to maintain the fly's equilibrium and enable it to balance itself when it walks, as otherwise the enormous size of its body would render walking very difficult; they are sense organs, supplying the creature with a great many percepts by way of touch; they are organs of exudation, through which it emits a volatile element in its blood as a pleasing stimulant to the greed of its hosts; finally they resemble supplementary spiracles, that to some extent are like the tracheal gills of the insect's earliest aquatic ancestors.

These little termitophile Diptera are indeed a store-house of anomalies, whether we consider them from the point of view of morphologists, anatomists, evolutionists, or biologists. They are exceptions to the laws of entomology. They are not merely Diptera without wings, but they are flies without the larval and pupal stages, and are actually insects having neither male nor female!

In order to shorten the lengthy and complete process of metamorphosis undergone by other Diptera, the *Termitoxenia* lays comparatively enormous eggs, from which is hatched not a larva, as is the case with other flies, but a perfect insect,

[1] '*Termitoxenia*, ein neues flügelloses, physogastres Dipterengenus aus Termitennestern,' Part I (*Zeitschrift für wissenschaftliche Zoologie*, LXVII, 1900, pp. 599–618 with plate XXXIII); Part II (*ibid.* LXXX, 1901, pp. 289–98); 'Zur näheren Kenntnis der termitophilen Dipterongattung *Termitoxenia*' (*Verhandl. des V. internationalen Zoologenkongresses zu Berlin*, August 1901, pp. 852–72 with one plate); 'Die Thorakalanhänge der *Termitoxeniidae*, ihr Bau, ihre imaginale Entwicklung und phylogenetische Bedeutung' (*Verhandl. der deutschen Zoolog. Gesellschaft*, 1903, pp. 113–120, with plates II and III).

the imago form, still in a stenogastric or thin-bodied condition. To compensate for the absence of metamorphosis, the *Termitoxenia*, as imago, undergoes a postembryonic development, for its organs of generation, especially the single-tubed ovaries, its fat-body, consisting of large cells joined together end to end, its abdominal muscular system, and even the outer skin of the abdomen, receive their final form only in the course of a long process of growth. Each of these insects is moreover a complete hermaphrodite, there are no distinct males and females at all. The youngest imagines have some quite undeveloped ovaries, such as occur in the larvae of other Diptera, but even in the youngest specimens the male generative glands and the bundles of spermatozoa connected with them are well marked, although they subsequently become atrophied, when the spermatozoa have ripened, whilst the ovaries develop. We have, therefore, here an instance of what is called protandric hermaphroditism, which regularly allows first the male and then the female generative glands to develop in the same individual, so that the *Termitoxenia* is something quite unique in insect biology.

It is most interesting to trace the development of the ovaries. (See Plate V, fig. 6.) Each one consists of a single egg-tube—a phenomenon long sought in vain among insects by the upholders of the theory of evolution, until Grassi discovered it occurring in the very rudimentary ground-flea (*podura*), belonging to the genus *Campodea.*

This single egg-tube on each side of the *Termitoxenia*'s body is, in the case of the youngest specimens, merely one single long terminal chamber, filled with apparently undifferentiated little nuclei.[1]

In course of time the egg-tube contracts in between the eggs, and forms a long series of ovarian chambers, those at the lower end of the ovary being the largest. In each of these chambers the elements of the ovary differentiate themselves into nutritive cells and true egg-cells, so that each chamber eventually contains several large cells, one of which develops

[1] I use the word 'apparently' advisedly, for in one of his recent works ('Untersuchungen über die Histologie des Insektenovariums,' in the *Zoologische Jahrbücher*, Section for Anatomy, 1903, part 1), Gross has proved that the epithelial cells and those that eventually become germ-cells differ from one another even in the terminal chamber.

more rapidly than the rest and becomes the egg. The other cells in the same chamber serve as its food, or, in scientific language, a fusion takes place of the egg-cell with the nutritive cells, the substance of the latter being gradually absorbed into that of the former, and transformed into tiny yolk-capsules collected round the germinal vesicle of the young egg. Thus the egg is nourished and it continues to grow until it occupies about a quarter of the entire abdomen of the full-grown insect. (Plate V, fig. 6 ov.) By this time it has taken up enough yolk-material to serve for the whole embryonic development until it reaches the stage of imago, when it must make its own way in the world. It is fertilised, and, passing along the ovarian duct, it is laid among the eggs of the termites.

The history of the development of a fly belonging to the sub-genus *Termitomyia* is somewhat different, but still more extraordinary. In this case the egg, whilst still within the parent's body, becomes an embryo, which develops until it reaches the form of a stenogastric imago. Therefore this sub-genus lays no eggs at all, but brings forth its young alive. These viviparous insects are a worthy contrast to the oviparous mammals, such as the ornithorhynchus and the Australian ant-eating Echidna.

There is a regular correlation between all the points on which the remarkable anatomy and development of the *Termitoxenia* differ from those usual among insects. The fact that each ovary has only one egg-tube facilitates the formation of eggs few in number, but large and rich in yolk. The large size and richness in yolk of the eggs render the omission of the larval and pupal stages possible, and so the whole process of development is conveniently shortened and simplified, and the imago is produced out of the egg or rather out of the embryo.

Moreover, in the case of the *Termitoxenia*, the complicated process of assigning sex to the individual is simplified in a form that is perfectly ideal for insects, as each individual fulfils both functions. And all these wonderful peculiarities in the morphology, development, and biology of the *Termitoxenia*, its physogastria and its ametabolia, its growth as an imago and its hermaphroditism, the shape of its appendages to the thorax and the formation of the parts of its mouth—

for it has a long proboscis for sucking the tender, juicy young of the termites—all these are closely connected with and dependent upon the affection of these Diptera for the termites!

And how, it may be asked, do we know all this? Have observations been made in India and Africa regarding the habits of these diminutive creatures, and has their development been studied for years in artificial nests of termites? By no means. The discoverers of the six known varieties of *Termitoxenia* merely established the fact that they always are found in the nests of certain kinds of termites and among their eggs and larvae. The inquilines and their hosts were sent to me in alcohol or formol. But the further question arises, how can it be possible, in that case, to make such definite and apparently rash statements as to the habits of these creatures? They are so small, that even a powerful magnifying glass scarcely enables us to distinguish the details of their exterior configuration; even under the microscope it is difficult to make out the halteres or balancers, which are placed behind the thoracic appendages, and prove that the latter morphologically correspond to the wings of Diptera and do not point to a coalescence of wings and halteres.

What scientific evidence is there, then, in support of the account just given of the anatomy, development, and biology of *Termitoxenia*?

The account is based on the results obtained by modern methods of using stains and cutting sections. The series of sections of *Termitoxenia* supply us with material for studying its anatomy, development, and biology.

So far I have obtained by means of the microtome complete series of sections of sixty specimens of five species of *Termitoxeniidae* of various ages, and I have also cut sections of a number of eggs of various species; as a stain I have generally used a double preparation of haematoxylin (Delafield's method) and eosin.[1]

The total number of sections thus prepared amounts to 10,000. Each specimen submitted to microscopical examination furnishes a series of from 80 to 200 sections of $\frac{1}{100}$ mm. in thickness; the number varies according as the sections are

[1] Or a double stain obtained by using haemalum (Meyer's method) and orange eosin, &c.

longitudinal or transverse. **Each** series of sections therefore forms a book of from 80 to 200 pages, on which are recorded in unbroken sequence the whole exterior and interior morphology of the specimen, and this record is legible under the microscope. If the sections of various kinds of *Termitoxenia* at different ages, and also of their respective eggs, are compared with one another, the morphological volumes come to form a library containing an account of the *Termitoxenia*'s development. As, however, almost every point in the anatomy

I	II	III	IV	V	VI	VII	VIII
1	7	13	19	25	31	37	43
2	8	14	20	26	32	38	44
3	9	15	21	27	33	39	45
4	10	16	22	28	34	40	46
5	11	17	23	29	35	41	47
6	12	18	24	30	36	42	48

↓ I.

I	II	III	IV	V	VI	VII	VIII
49							
							96

↓ II.

Fig. 1.—Scheme of a series of sections of *Termitoxenia Heimi* Wasm.

and development of these tiny creatures is of significance in their habits, this library supplies also trustworthy information for their whole biology.

The accompanying illustration (fig. 1) represents a series of sagittal sections of *Termitoxenia Assmuthi*. It consists of the longitudinal sections of specimen No. 13 of this variety, arranged upon two slides (i and ii). The Roman numerals on each slide refer to the sequence of the rows of sections, the Arabic numerals to the sequence of the sections in each row. Thus the series begins with No. 1 on the first slide and ends with No. 96 on the second. No. 49, the first on the second slide, is a section cut from the middle of the creature's body—a photograph of it will be found on Plate V, fig. 6, at the end of the book.

I need hardly say that a great expenditure of time and trouble is needed, not merely to make such series of sections, but far more to study them with success. The instances of morphological and biological conformity to law, which a scientist can discover, seem to be written in a mysterious cipher, the key to which is found only by careful study. No one, therefore, will be astonished to hear that I have spent years on my study of the *Termitoxenia*, especially as I had not only to describe my microscopical results in words, but to reproduce them by means of drawings or photographs upon a series of carefully executed plates.[1]

The marvellous beauty of the various sections is no less noticeable than their scientific value in biological research. The material for several series of sections of *Termitoxenia Heimi* and *Assmuthi* was supplied me by J. B. Heim, S.J., Missionary in India, and J. Assmuth, Professor at St. Francis Xavier's College in Bombay. The creatures reached me in very good preservation, having been killed and hardened in a mixture of alcohol and formalin. The sections, stained with haematoxylin and eosin, or some similar double stain, are so beautiful that they cannot fail to arouse admiration in any one who sees them, even in the mind of one who regards all insects alike as 'vermin.' Eosin stains the protoplasm of the tissues various shades of light red, whilst the nuclei, which chiefly serve to differentiate the various kinds of tissue, are coloured light or dark blue by means of haematoxylin or haemalum; the whole picture displays a delicacy of design and a beauty of colouring such as no artist's skill could reproduce in perfection. The most complex and most highly coloured pictures are formed by sections showing the various stages of development in which the mysterious biological processes of cell-division, cell-multiplication, and cell-growth—those elementary functions of life—are most active.

Modern microphotography will, perhaps, succeed in fixing

[1] A fuller account of my work will appear in the *Zeitschrift für wissenschaftliche Zoologie*. A *résumé* of the results obtained hitherto was given in an address delivered at the fifth International Zoological Congress in Berlin, August 1901.

microscopical sections with all their gorgeous colouring directly upon photographic plates. If this is ever done, it will be of the utmost scientific importance, as the precise shades of colour in the nuclei and other parts of the tissues often give a trustworthy clue, of great assistance in histological and cytological research.

A learned professor of theology, on seeing some series of sections of the *Termitoxenia*, remarked very aptly that microscopical research, by means of modern methods of staining and cutting sections, had become a second creation, *creatio secunda*, revealing to us for the first time all the marvels which God at its first creation had concealed within the body of this diminutive fly.

In order to give my readers a wider idea of the application of microscopical study to our investigations into animal biology, the following remarks may be added. Let us suppose that some one asks: '*Why* do ants and termites show such energy and pleasure in licking their "true inquilines"? Upon what does the satisfaction depend which they derive from so doing?'

Before this question can be answered, a reply must be given to another, viz.: 'What tissues underlie the external exudatory organs, which lead to the process of licking the inquilines?' With a view to answering this latter question I have, in the course of the last ten years, prepared about 20,000 sections of various kinds of inquilines among ants and termites (they are chiefly beetles), and studied their tissues under the microscope. In this way I have arrived at the following conclusion:—the exudation of true inquilines, with which they repay their hosts for their hospitality, is partly a direct and partly an indirect product of adipose tissue; when it is indirect, it is partly a glandular secretion and partly an element in the blood plasm of the inquiline.[1]

We are therefore now in a position to divide the genuine inquilines among ants and termites into various classes according to their exudatory tissues, and thus have made a perceptible step towards solving the mystery of true guest-relationship.

[1] Articles on this subject appeared in the *Biologisches Zentralblatt*, 1903, Nos. 2, 5, 6, 7, 8, under the heading: 'Zur näheren Kenntnis des echten Gastverhältnisses (Symphilie) bei den Ameisen- und Termitengästen.'

5. Recent Advance in Microscopical Research

After this little digression let us return to the historical development of modern histology and cytology.

Improvements in the microscope itself, the chief implement in our research work, have kept pace with the adoption of better methods of staining and cutting sections.

As a result of very careful physical studies, Abbe of Jena devised an apochromatic objective, calculated exactly with reference to its refractive and dispersive power. This was worked out by Schott & Co., in Jena, and then further perfected by Karl Zeiss, the able optician in Jena. The apochromatic objective has been imitated with various degrees of success by other German and foreign firms. Its introduction, and that of the corresponding compensating ocular or eye-piece, mark an important stage in the development of the microscope. Speaking from my own personal experience, I can safely assert that the pictures produced by this system of lenses are infinitely clearer than those produced by the achromatic objectives and Huygenian oculars previously in use. It is now possible to see every detail in the structure of tissues even when magnified 1500–2000 times.

This advance in optical appliances has enabled modern cytologists to study the most delicate construction of a resting cell, as well as the processes of division and fertilisation, and to discover the laws governing these most important phenomena of life.

Histology and cytology made great progress during the latter half of the nineteenth century in other countries as well as in Germany, where they had their birth, and where they grew to the rank of independent sciences, in consequence of the research work done by Schleiden, Schwann, Remak, Leydig, and Max Schultze.

I can mention the names of only a few of the more recent workers in this department of science; in Germany, besides Leydig and Max Schultze, we have Strasburger, Weismann, Flemming, Bütschli, Henking, Heidenhain, Boveri, A. Brauer, Reinke, the two Hertwigs, Haecker, Erlanger, O. vom Rath, Schaudinn, Rhumbler, &c.; in Bohemia, Rabl; in Hungary,

Apáthy, who has made nerve-cells his special study; in Switzerland, Fol; in France, Ranvier, Balbiani, Giard, Maupas, Kunstler, Guignard, Armand Gautier, and Yves Delage; in Belgium, van Bambeke, E. van Beneden, and the great cytologists of the Catholic University of Louvain, viz. Abbé Carnoy, the author of 'Biologie cellulaire,' and his pupils, of whom G. Gilson, A. van Gehuchten, and Abbé Janssens are well known through their important publications; in Spain, Ramón y Cajal; in Italy, Giardina; in Great Britain and Ireland, A. Sedgwick, Moore, McGregor and Dixon; in Sweden, Retzius and Murbeck; in Russia, Kowalevsky, Tichomirow, Nawaschin and Sabaschnikoff; in North America, Ch. Sedgwick Minot, Chittenden, E. B. Wilson, Th. H. Montgomery and Osborn; lastly, in Japan, Chiyomatsu Ishikawa, director of the zoological institute of the Imperial University of Tokio.

We may therefore well say that all civilised nations of the present time have contributed to the development of modern histology and cytology.[1]

In order that my readers may not regard the Jesuits as 'mediæval obscurantists' trying to stem the advance of science, I may be allowed to add that a Dutch Jesuit, H. Bolsius,[2] has done much to increase our knowledge of the microscopical anatomy of Hirudines or leaches, and has shown himself an authority of the highest rank on this subject. A modern morphological and biological

---

[1] This is of course true, not only with regard to the morphology of the cell, with which we are now chiefly concerned, but also with regard to its vital phenomena, especially the processes of cell division and fertilisation, to which we shall have to refer later. I should like to draw particular attention to Carnoy's *Biologie cellulaire*, 1884, which unhappily was never completed; also to Oskar Hertwig's *Allgemeine Anatomie und Physiologie der Zelle*, 1893; and Max Verworn's *Allgemeine Physiologie*, the third edition of which appeared in 1901, and deals mainly with cellular physiology. I regret that Verworn's work is not altogether free from phrases suggestive of Haeckel's influence and wanting in scientific dignity. For instance, on p. 214, in speaking of parthenogenesis among the lower animals, he refers to 'the ancient legend of the Immaculate Conception.' The author seems to be as far as Haeckel from a comprehension of the dogma of the Immaculate Conception.

[2] 'Nouvelles recherches sur la structure des organes segmentaires des Hirudinées,' 1890; 'Les organes ciliés des Hirudinées,' 1891; 'Le sphincter de la Nephridie des Gnathobdellides,' 1894; 'La glande impaire de l'Haementaria officinalis,' 1896; 'Recherches sur l'organe cilié de l'Haementaria officinalis,' 1900 (this article appeared in *La Cellule*). I might also mention a number of other articles which the same author contributed to the *Annales de la Société scientifique de Bruxelles*, to the *Memorie della Pontificia Accademia dei Nuovi Lincei*, to the *Zoologischer Anzeiger* (Leipzig), and the *Anatomischer Anzeiger* (Jena), &c.

work, universally regarded as a masterpiece, has been written by J. Pantel, a French Jesuit, on the larva of *Thrixion halidayanum*;[1] and no less excellent are an anatomical and histological study of the anal glands of beetles by a Belgian Jesuit, Fr. Dierckx,[2] and a biological and anatomical study of walking stick insects by a French Jesuit, R. de Sinéty.[3]

These publications, as well as most of the works of Carnoy, Gilson, van Gehuchten and Bolsius, appeared in *La Cellule*, a periodical published by the Cytological Institute of the Catholic University of Louvain, a society founded by Abbé Carnoy. This periodical is highly esteemed by German scientists, and forms a complete refutation of the old fiction that Catholics, and especially those of Romance nations, must needs be bad men of science. In the sixth chapter I shall have to refer to some articles on the chromosomes in the eggs of Selachii and Teleostei by J. Maréchal, a Belgian Jesuit, and among Italian scientists, a Franciscan, Dr. Fra Agostino Gemelli, has written some excellent works on anatomy and histology during the last few years.

---

[1] 'Le *Thrixion halidayanum*, Rond.: Essai monographique sur les caractères extérieurs, la biologie et l'anatomie d'une larve parasite du groupe des Tachinaires,' 1898 (*La Cellule*, XV).

[2] 'Étude comparée des glandes pygidiennes chez les Carabides et les Dytiscides,' 1899 (*La Cellule*, XVI); 'Les glandes pygidiennes des Coléoptères,' 2nd mémoire, 1900 (*ibid.* XVIII).

[3] *Recherches sur la biologie et l'anatomie des Phasmes*, Lierre, 1901. This work contains splendid illustrations; in the eighth chapter the author discusses the karyokinetic processes in the spermatogenesis of Orthoptera, a subject of peculiar interest as throwing light on the accessory chromosomes.

CHAPTER III

MODERN DEVELOPMENT OF CYTOLOGY

1. The Cell, a Mass of Protoplasm with one or more Nuclei

On p. 33 we have seen that Franz Leydig in 1857 and Max Schultze in 1861 defined the cell as a small mass of protoplasm containing one or more nuclei. This has remained to the present day the fundamental idea of the cell, as we may see on referring to the definitions of it given by Richard Hertwig in the seventh edition of his 'Lehrbuch der Zoologie,'[1] Matthias Duval in the second edition of his handbook of histology,[2] and Oskar Hertwig in his 'Allgemeine Biologie.'[3] With regard to this definition there is almost unanimous agreement on the part of the chief cytologists of various nations, and this is a very significant fact, especially as modern cytology is a much debated subject. If it is possible in any branch of knowledge to speak of a *sententia communis doctorum,* we may regard

[1] Jena, 1905, p. 50: 'The cell is a little mass of protoplasm containing one or more nuclei.'

[2] *Précis d'Histologie,* Paris, 1900, p. 26: 'La cellule est essentiellement une petite masse de protoplasma avec un noyau.'

[3] 1906, p. 27: 'The nucleus is just as essential to the existence of a cell as is the protoplasm.' Cf. also the more detailed account given by O. Hertwig in the third chapter of the same work.

the definition of a cell as such in a very conspicuous degree.

I must acknowledge, however, that this unanimity exists among zoologists and histologists more than among botanists.[1]

In many of the smallest forms of plant life, especially in many bacteria, the presence of a true, clearly differentiated nucleus has not yet been established.[2] I use the words 'true, clearly differentiated nucleus' advisedly, for cytologists are more and more adopting the opinion that even in those micro-organisms previously regarded as devoid of nucleus the nuclear substance is present, though divided into smaller particles, which R. Hertwig has designated *chromidia*.[3] This opinion gains support from the discovery of a true nucleus existing at a definite stage in the formation of the spores of the *Bacillus Bütschlii*.[4]

We shall have to return later on (Chapter VII) to the most recent investigations made by biologists on the subject of the absence of nucleus in these extremely small forms of life. For the present it is enough to say that the idea of a living cell involves that of a nucleus, either as a whole or in parts, but the chromatophores that exist in most plant cells besides the cytoplasm and the nucleus are certainly not essential to the existence of the cell, for they are absent in Bacteria and fungi, and in all animal cells.[5]

Let us now proceed to study the structure of a cell more in detail.

In shape and size cells vary greatly. The normal shape of a free cell, not united with others of the same kind to form a tissue, is spherical, but even the unicellular plants and animals are seldom quite round, and cells united to form tissues still less often approach a spherical shape; they are rounded, or oval, or cylindrical, or cubical, or pentagonal, or hexagonal;

[1] Cf. *Lehrbuch der Botanik für Hochschulen* by Strasburger, Noll, Schenk and Karsten, 6th edit., Jena, 1904, pp. 46–7, 270, 274, where it is stated that the presence of a nucleus in the lowest plants (Cyanophyceae and Bacteria) is still uncertain. (English translation, 3rd edit. 1908, pp. 53 and 332.)

[2] Cf. J. Reinke, *Einleitung in die theoretische Biologie*, 1901, pp. 256, &c.

[3] R. Hertwig, 'Die Protozoen und die Zellentheorie' (*Archiv für Protistenkunde*, I, 1902, pp. 1–40).

[4] Fr. Schaudinn, 'Beiträge zur Kenntnis der Bakterien und verwandter Organismen,' I. *Bacillus Bütschlii*, n. sp. (*Archiv für Protistenkunde*, I, pp. 306, &c.).

[5] Cf. Strasburger, &c., pp. 46, 47 (Eng. trans. p. 53).

D

sometimes they are of almost the same thickness in all three dimensions, at other times they are flattened out like those of the pavement epithelium (fig. 2d), or extraordinarily long, like

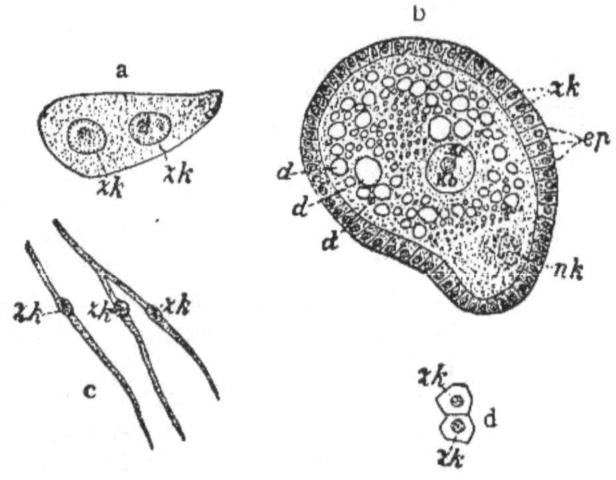

FIG. 2.

Magnified 230 times [Zeiss D, Ocul. 2].

All the figures have been prepared with the camera lucida from series of sections.

KEY TO FIG. 2.

a = Giant cell containing two nuclei from the abdominal fat-body of a physogastric specimen of *Termitoxenia Heimi* Wasm.
zk, *zk* = nuclei.
b = young egg of *Termitoxenia Heimi* Wasm. The egg-cell is still enclosed within the follicular epithelium of the ovary. (From a sagittal section of a physogastric specimen of *Termitoxenia Heimi.*)
ep = epithelial cells of the one-layered follicle.
zk = nuclei of the epithelial cells.
kb = germinal vesicle of the egg.
kf = nucleolus of the germinal vesicle.
dd = vitelline spherules.
nk = remains of the nucleus of a nutritive cell, the material of which has served to form the yolk.
c = three unicellular muscular fibres from the cutaneous muscular apparatus of the abdomen of a stenogastric specimen of *Termitoxenia (Termitomyia) mirabilis* Wasm.
zk = nucleus.
d = two epithelial cells from the hypodermis of the abdomen of a stenogastric specimen of *Termitoxenia Heimi.*
zk = nucleus.

the spindle-shaped cells of the smooth muscular fibres, and the still more slender cells that form the transversely striated muscular fibres (fig. 2c).

As a rule, the cells that make up tissues have no prolongations, but in making this statement I am not challenging Heitzmann's discovery (1873) of protoplasmic cell-bridges.[1] Many cells, however, possess long offshoots, which give them a ramified appearance ; this is particularly the case with nerve-cells, and is closely connected with their telegraphic functions.

The shape of the nucleus varies less than that of the cell,[2] it is mostly round or oval, although other shapes not infrequently occur. Very remarkable are the branching nuclei of the Malpighian tubes in certain caterpillars, and the nuclei resembling a string of beads in some unicellular Stentors.

In speaking of the size of a cell, we must have a standard by which to measure it. In this respect little cells resemble so-called tall men ; we cannot measure either by any usual method, an old-fashioned foot-rule and a modern metre measure are equally out of place. Cells have to be measured under the microscope, and the following method is the simplest. The number of times that the object is magnified is carefully noted, and a sketch of the cell is made on paper by means of a camera lucida. This sketch is then measured with a very exact millimetre measure, and the number thus obtained is divided by that of the magnifying power. For instance, if a cell, magnified 230 times, measures 23 mm., its real magnitude is 0·1 mm. This would be a giant cell if it belonged to animal tissue. Such giant cells as this (cf. fig. 2a) compose the abdominal fat-body of the *Termitoxenia*, a variety of Diptera living among termites, as we have already seen (pp. 37, &c.). Most cells in animal tissues are dwarfs in comparison, and dwarfs among dwarfs are the average blood corpuscles, especially of insects, and the spermatozoa of most animals. Therefore, as a constant unit for microscopical measurement of cells, the thousandth part of a millimetre has been adopted, which is known as a micromillimetre or micro, and is designated by the letter μ. The giant cells of the *Termitoxenia*'s fat-body have a diameter of 100μ. Cells of 10μ (e.g. figs. 2d

[1] A further account of these protoplasmic cell-bridges will be found in Wilson, *The Cell*, pp. 56, 60, where there is a careful discussion of the evidence for their existence among very various kinds of plant and animal cells. See also O. Hertwig's *Allgemeine Biologie*, pp. 400, &c.

[2] For the shape, size, and number of nuclei, see O. Hertwig, *Allgemeine Biologie*, pp. 28, &c.

E 2

and 2b, &c.) are of medium size, so the former may well be called gigantic.

But there are some animal cells far larger than these, viz. the egg-cells. These are the largest in the animal kingdom.[1]

The ripe egg-cell of a diminutive insect such as the *Termitoxenia*, barely 2 mm. in length, measures almost 1 mm., i.e. it is half as long as the creature's whole body. The eggs of this fly are reckoned, therefore, among the relatively largest in the entire animal kingdom; the absolutely largest occur among birds; it is in fact possible to use a yard measure to ascertain the size of the eggs of the ostrich or moa. A bird's egg before fecundation consists of one huge cell, but to the egg-cell belong in this case not only the germinal vesicle, which represents the nucleus of the protoplasmic part or formative yolk of the egg-cell, but also a quantity of nutritive yolk or deuteroplasm,[2] which is really the yolk of the bird's egg. The white of the egg and the shell appear only after tecundation, and are outer coverings, and not parts at all of fhe egg-cell. Animal egg-cells owe their conspicuous size to the presence of deuteroplasm or nutritive yolk, which is found in the eggs of all creatures that are oviparous and not viviparous. In the case of the former a considerable quantity of nutritious matter must be stored up in the egg itself, in order that the embryo may develop. My readers must not, however, fancy that, when they see a new-laid hen's egg, they have only one huge egg-cell before them; for, quite apart from the above-mentioned exterior coverings, which grow before the egg is laid, the egg itself is already fertilised, its germinal vesicle has become a germinal disc, i.e. a still very diminutive embryo chick, consisting of numerous segmentation cells, and the huge egg serves as its lodging and store-room during its further development.

In order to illustrate the various shapes and sizes of the cell by examples, I have reproduced some cells of *Termitoxenia* on p. 50. To the explanations already given I may add that,

[1] Very large cells constitute the plasmodia of the Mycetozoa, which are also reckoned among the lower orders of plants and called Myxomycetes, whilst by others again they are classed with the Protozoa. Cf. R. Hertwig, *Lehrbuch der Zoologie*, 7th edit., 1905, pp. 49 and 168 (Eng. trans. pp. 60, 61, 198).

[2] E. van Beneden called the nutritive yolk 'deutoplasm,' to contrast it with protoplasm; 'deuteroplasm' is a more correct form of the word.

with a view to economising space, I chose for Fig. 2b not a ripe and fully developed egg-cell, but a young cell, still surrounded by a thick follicle of epithelial tissue, and having at its lower end the remains of an incompletely consumed nutritive cell. As the latter is already incorporated with the substance of the egg, the young cell (without the epithelium) measures 135μ in length and 95μ in breadth. A ripe egg-cell of the same kind of *Termitoxenia* would, if drawn on the same scale (magnified 230 diameters), occupy a space of 2 dm., and cover a whole page of this book.

Some plant cells are also very large; for instance, there are bast-cells 2 dm. in length and of considerable breadth. Among the lower plants too, such as the *Caulerpa* (one of the Algae), there are cells several decimetres in length; in fact, according to J. Reinke and other botanists, the whole plant with its root, stem, and leaves consists of one cell with many nuclei.[1]

The dwarfs among plant cells are many of the Bacteria, which have a longitudinal diameter of not quite 1μ ($\frac{1}{1000}$mm.). The petal of a violet consists of about 50,000 cells which are comparatively large.

By far the greater number of cells have but one nucleus, and if they are found to contain more than one, it is generally because the process of cell-multiplication by division is just beginning. There are, however, some cells that always contain several nuclei; such are, for instance, those in the marrow of vertebrates, and partly also those known as *syncytia* in the adipose tissue of insects and other Arthropods.[2]

In his classical and suggestive work on cell-division among the Arthropods,[3] Carnoy expresses the opinion that these are all multinuclear giant cells, not masses of cells formed by the fusion of others. This view cannot be adopted without reservation, as there are undoubtedly cases in which syncytia arise from a gradual breaking down of the cell-walls. This takes place, for instance, in *Termitomyia*, a sub-genus of *Termitoxenia*. In the sub-genus *Termitoxenia* (in the narrower sense)

[1] See Reinke, *Einleitung in die theoretische Biologie*, p. 213, and his *Monographie der Gattung* Caulerpa. See also Frank, *Synopsis der Pflanzenkunde*, III, Hanover, 1886, § 890; van Tieghem, *Traité de Botanique* (1891), pp. 9, 10.

[2] On the subject of *syncytia* or cell-fusions see also O. Hertwig, *Allgemeine Biologie* (1906), pp. 378–381.

[3] 'La Cytodiérèse chez les Arthropodes' (*La Cellule*, I, 1885, n. 2, p. 235, &c.).

these adipose cells are very large, but they are distinct one from the other, though in full-grown physogastric specimens, in which no further cell-division occurs, there are frequently two nuclei (cf. fig. 2a) instead of one. According to Weismann[1] multinuclear cells occur also in the festooned columns of cells found in the larvae of flies. I have myself found cells with two or more nuclei in the halteres of *Termitoxenia*, and Bolles Lee discovered them before me in those of common Diptera.[2] In many of the lower orders of plants, such as the Thallophyta, cells containing several or even many nuclei are of frequent occurrence, and among the Siphonaceae, a subdivision of the Algae, there are plants (*Caulerpa*, *Vaucheria*, &c.), which consist of one huge multinuclear cell, as has been already stated.

Just as in the tissues of living organisms there may be, and actually are, cells which contain several nuclei, but still do not divide into more cells, so, in the lowest forms of animal life, the Protozoa, there are unicellular organisms containing two or more nuclei, but not forced on that account to split up into several individuals.

The reader must, however, carefully distinguish the multinuclear cells just mentioned, from others which contain beside or in the true nucleus one or more little round bodies known as *nucleoli*. The founders of cytology, Schleiden and Schwann, noticed these bodies and regarded them as having some essential importance in the structure of the cell. This opinion has proved to be erroneous, and most nucleoli seem to be merely differentiations of the ordinary substance of the nucleus. For this reason I have purposely refrained from referring to them until now, when we are concerned with the more detailed morphology of the cell.

2. The Structure of the Cell examined more closely

In an account of the origin of modern cytology, Gustav Schlater writes as follows:[3] 'The cell is a little mass of protoplasm, endowed with all the properties of life. This was the

[1] *Die Entwicklung der Dipteren*, Leipzig, 1864, p. 132 and Plate 8, fig. 10.

[2] 'Les balanciers des Diptères' (*Recueil Zoolog. Suisse*, II (1885), 389 et pl. XII, fig. 18).

[3] G. Schlater, 'Der gegenwärtige Stand der Zellenlehre' (*Biolog. Zentralblatt*, XIX, 1899, Nos. 20–24, p. 667).

definition given by Max Schultze, and at the time our idea of a cell seemed to have reached its full development. Thenceforth, we had only to submit cells to examination from many points of view, and the representatives of every branch of biology did in fact turn their attention to the cell. The word "Protoplasm" was ever on their lips, and the number of works devoted to the examination of the structure and life of this elementary unit in living substance is so great that it would be quite impossible for anyone to read them all. This examination has proved very fertile in results; every step has supplied fresh evidence supporting the general biological importance of the cell-theory; every book written has proved that we must start from the cell in order to extend our knowledge of nature. The reputation of the cell increased; it revealed itself as more and more complex in its formation. Within it, in this little mass or drop of living substance, modern research has discovered a complicated structure, and more and more details of this structure, and each day adds to the interest taken by men of science in the whole complicated vital processes that go on in the small compass of the cell.'

The interesting question arises here: Are we to consider the cell simple or complex? Is it the ultimate biological unit in the structure of organisms, or is it itself a diminutive organism made up of subordinate units? This is a weighty question, having an important bearing on the problem of life, and students are apt to overlook its twofold character. In order to emphasise it, let us divide the question into two, and ask: (1) Is the cell morphologically simple? (2) Is it the ultimate biological unit of organic life, or is it an aggregation of lower elementary units? It is possible to deny the simplicity of the cell and at the same time to affirm its unity, for, according to the unchanging laws of thought which are still binding upon the *Homo sapiens* of the twentieth century, simplicity and unity are two quite different ideas. Modern research will never attain to assured philosophical results regarding the nature of life, if it confuses unity and simplicity. Let us try to give to both questions an answer based upon facts.[1]

[1] Cf. O. Hertwig, *Allgemeine Biologie*, 1906, chapters ii and iii; Wilson, *The Cell*, 1902; Yves Delage, *La structure du protoplasma et les théories sur l'hérédité*, Paris, 1895.

Is the cell simple? No, it is not simple, but extremely complex in many cases, a true microcosm. It consists of a number of parts that differ morphologically, chemically,[1] and physiologically, and yet on their harmonious connexion depends the biological unity of the vital process of the cell. Although all parts of the cell participate more or less in its vital activities, still the nucleus is of chief importance in the principal processes.[2]

Such are briefly the results of the most recent investigations of cytology, and we have now to consider them more in detail.[3]

The two chief morphological constituents of the cell are the cell-body and the nucleus, and this has been universally acknowledged ever since Leeuwenhoek discovered the nucleus (see p. 31). At the present time everyone regards them as essential to the cell, whilst the membranous covering of the cell and the nucleoli within the nucleus are not essential.[4] In 1882 Strasburger suggested the name *cytoplasm* to designate the protoplasm of the cell-body, and his suggestion has generally been adopted.[5]

It was originally regarded as absolutely homogeneous, but after Dujardin's study of it (1835) little granules were noticed in it, and further examination revealed a structure variously described as filar, reticular, or alveolar. There are many modern theories regarding the structure of cytoplasm. All students, with the exception of those mentioned first, agree in recognising in the protoplasm of the cell-body two distinct substances, one being transparent and forming the foundation of

[1] The chemical constituents of protoplasm and the morphological variety of the parts of the cell are not discussed here in detail, because very little is as yet known with certainty about them. (Cf. Chapter II, p. 33.) How complicated the chemical composition of the nucleus is may be seen on reference to Dr. Hans Malfatti's work, 'Zur Chemie des Zellkerns' (*Berichte des naturwissenschaftlich-medizinischen Vereins*, Innsbruck, XX, 1891-2).

[2] This fact is acknowledged even by those who, like J. Reinke, regard it as not essential to differentiate the nucleus as a distinct morphological formation. (See Reinke's *Einleitung in die theoretische Biologie*, 1901, p. 256.)

[3] An excellent account of the morphology of cells and of the various theories regarding the structure of the cell-body and the nucleus will be found in Wilson's *The Cell*, pp. 19-62.

[4] The subject of the centrosomes will be reserved for discussion in Chapter V. See O. Hertwig, *Allgemeine Biologie*, pp. 45-49.

[5] O. Hertwig prefers to retain the older meaning of the word protoplasm, in which it was originally used by von Mohl, Max Schultze and Leydig, to designate the substance of the cell-body as distinct from the nucleus. Strasburger's cytoplasm is thus identical with the protoplasm of these earlier writers.

the cell (*hyaloplasm*, as Leydig calls it), and the other granular, consisting of microsomes, which form the framework of the filar, reticular, or alveolar structure (*spongioplasm*, as Leydig calls it). The former is also very suitably called *cytoplasm*, and the latter *cytomitom*, but a great number of names have been given to both,[1] names calculated to astound any ancient Hellene who heard the modern derivatives coined from the wealth of old Greek words.

Those who believe cytoplasm to be homogeneous do not recognise the presence in the living cell of two morphologically distinct substances, but they regard the granules and threads and meshes of the so-called cell-framework as merely artificial products, resulting from the chemical reactions and the use of stains for microscopical purposes.

There are, however, good reasons why this theory does not find many supporters at the present day,[2] for recent microscopical research has revealed in the living cell a structure, which is not produced by the processes of fixing and staining, but is only rendered visible by means of them. This is especially true of the filar structure of spongioplasm, which is practically identical with the reticular structure or framework. It was discovered first by Karl Frommann in 1875, but Flemming recognised it as filar,[3] and his observations have been confirmed by those of many other scientists, such as Klein, Leydig, E. van Beneden, Carnoy, Heidenhain, Zimmermann, &c., and are now regarded as of unquestioned accuracy. It is of secondary importance to decide whether, as Flemming thinks, the protoplasmic threads are of greater significance, or, in agreement with Klein, Carnoy, &c., we should lay stress particularly on the network formed by these threads.

Bütschli's alveolar theory represents another view of the structure of the cell. According to it the protoplasm of the

[1] See Bütschli, 'Über die Struktur des Protoplasmas,' 19 (*Verhandl. der deutschen Zoolog. Gesellsch.*, 1891, pp. 14–29).

[2] A. Fischer, whose theory regarding the polymorphic character of protoplasm will be discussed later on, must not be reckoned among those who uphold the homogeneity of protoplasm.

[3] See W. Flemming, 'Über den gegenwärtigen Stand unserer Kenntnisse und Anschauungen von den Zellstrukturen,' a paper read at the opening of the thirteenth meeting of the Anatomical Society at Tübingen on May 22, 1899 (*Naturwissenschaftliche Rundschau*, XIV, 1899, Nos. 35 and 36).

cell has a structure resembling honeycomb or foam, due to the mechanical mixture of the various fluid constituents of protoplasm. That suspended in the fluid hyaloplasm there are often vacuoles, filled with another kind of fluid, is a fact not questioned even by the opponents of this theory, but they deny that the minute structure of the protoplasm depends merely upon the presence of these vacuoles; for, whereas spongioplasm, treated according to Bütschli's methods, appeared to reveal an alveolar structure, closer examination has shown that a reticular structure really underlies it. The chief evidence brought forward by Bütschli in support of his alveolar theory is derived from artificial mixtures of various fluids, which bear a superficial resemblance to cell-structures, but cannot of themselves prove anything about the real structure of the cell.

I have no wish, however, to condemn Bütschli's alveolar theory, for we ought, in speaking of it, to distinguish between his view of the honeycomb structure of the cell, and his explanation of that structure by assuming a mechanical mixture of various fluids. The latter hypothesis is extremely doubtful, and has been thoroughly discussed by Oskar Hertwig in his 'Allgemeine Biologie' (p. 23). On the other hand, Bütschli's theory of the alveolar structure of many cells has been strengthened by recent research. In very thin microscopical sections very highly magnified, what appears as a network seems in fact often to be only a section of a framework consisting not of meshes but of closed chambers; and, if this is true, in these particular cells the protoplasm has really not a reticular but an alveolar structure. In my series of sections of the large gland-cells in the wing-covers of a termitophile beetle (*Chaetopisthes Heimi*) I have occasionally perceived a distinctly alveolar structure of the spongioplasm.[1] It seems, therefore, that the alveolar theory may stand beside the reticular theory, although latterly it has been attacked by those who are inclined to regard the alveoli seen under the microscope as an artificial product, or as a pathological vacuolisation of the protoplasm.[2]

[1] Cf. 'Zur näheren Kenntnis des echten Gastverhältnisses bei den Ameisen- und Termitengästen' (*Biolog. Zentralblatt*, XXIII, 1903, Nos. 2–8, p. 269).

[2] Cf. A. Degen, 'Untersuchungen über die kontraktile Vakuole und Wabenstruktur des Protoplasmas' (*Botanische Zeitung*, 1905, Part I, pp. 163–225).

Less satisfactory than Bütschli's alveolar theory is Altmann's granular theory,[1] which is based upon the granular structure of protoplasm. If Altmann merely asserted that numerous granules, now generally termed microsomes, are embedded in the transparent hyaloplasm of the cell, there would be no objection to his theory, for it would rest on actual observations. But he goes on to deny the fibrillar or reticular structure of the spongioplasm, and thinks that it may be explained as a close series of granules. Flemming, on the other hand, rightly points out that the microsomes are often arranged like beads on the reticular framework, but do not actually form that framework. Moreover, a large proportion of Altmann's famous granules have been proved not to be microsomes at all, but merely artificial products accidentally resulting from chemical reaction ; in fact, they are metaplasmic bodies and consist of protoplasm and foreign substances embedded in it, and were mistaken by Altmann for his granules, and the scientific value of his theory is greatly diminished in consequence. Its chief defect, however, is that it regards the granules contained in protoplasm as alone forming its essential active basis, and that it boldly accepts them as elementary organisms out of which the cell, as a secondary formation, is composed. This view is devoid of all real foundation in facts, and has been rejected by most scientists. We shall have to refer to it again later, in discussing the unity of the cell.

There is great diversity of opinion as to the relative importance of the two morphologically distinct constituents of the cell-body, viz. hyaloplasm (cytoplasm) and spongioplasm (cytomitom). Heitzmann, van Beneden, Reinke, Carnoy, Ballowitz and others agree in thinking the latter, which forms the framework of the cell, its really living, moving and contractile element, whereas others, and especially Leyden, ascribe these qualities to the former, and regard the hyaloplasm as the living substance. As Flemming saw, these two opinions ought probably to be united, for, as no living cell contains hyaloplasm exclusively or spongioplasm exclusively, both must be considered essential constituents of protoplasm, although most scientists agree with Flemming in assigning

[1] Cf. Richard Altmann, *Die Elementarorganismen und ihre Beziehungen zu den Zellen*, 1894.

greater importance to spongioplasm than to hyaloplasm. It is obvious that for the present we must be content to accept hypotheses of various degrees of probability, and these various theories regarding the more minute structure of the cell are all more or less of a hypothetical character.

Quite recently, in 1895–6, another theory as to the structure of the cell has been brought forward by Friedrich Reinke and elaborated by Wilhelm Waldeyer, and Gustav Schlater calls it the newest achievement of modern research into the morphology of the cell.[1] This theory attempts to reconcile the various views as to the structure of protoplasm. According to it, in the homogeneous ground-substance of the cell (i.e. in the *cytoplasm*, as other writers call it) there is embedded a reticular framework (*cytomitom*); the formation of the latter varies, but in the main it is alveolar and in its walls lie very small granules (*microsomes*), which in certain cases are aggregated, so as to form filaments and network. The chief framework of the cell owes its alveolar structure to the larger vacuoles and granules which it contains. Reinke-Waldeyer's theory thus harmonises the views of other scientists, and we may regard it as summing up all that was known of the structure of the cell in the year 1900; there is, however, one drawback to it theoretically, for it lays too little stress upon an essential element, viz. the meshwork or alveolar structure of the cell-framework, with the rows of microsomes arranged along it, and it lays comparatively too much stress upon an unessential element, viz. the vacuoles and larger granules which the cell contains.

3. The Minute Structure of the Nucleus

Hitherto we have discussed only the details of the cell-body, now we must consider the structure of the nucleus. Here again we find two chief substances, which, however, differ morphologically, physiologically, and chemically far more from one another than do the spongioplasm and the hyaloplasm of the cell-body. It is often possible to discover in the nucleus not only two, but three or four protein substances differing under chemical and microscopical examination. The nucleus is

[1] *Biolog. Zentralblatt*, XIX, 1899, No. 20, p. 676.

therefore, as O. Hertwig rightly remarks, a very complex[1] formation, so far as its constituents are concerned. According to their behaviour when stains are applied to them to facilitate their microscopical examination, the two chief substances in the nucleus have been called *chromatin* and *achromatin*; according to their chemical properties they are called *nuclein* and *linin* respectively. Chromatin or nuclein takes a brilliant colour when treated with carmine, haematoxylin, &c., whereas achromatin or linin is either not stained at all or takes a colour only under special circumstances. Achromatin resembles in structure the protoplasm of the cell-body, for it contains a fluid known as *karyoplasm*, and a fibrillar or reticular or alveolar framework known as *karyomitom*. These are analogous to the cytoplasm and cytomitom of the cell-body. Large nuclei are bounded on the outside by a peculiar nuclear membrane.

Chromatin has been mentioned as one of the chief substances in the nucleus; the parts that are readily stained are formed of it, and it is composed of nuclein.[2]

Closely connected with it, though differing chemically both from chromatin and from achromatin or linin, is another substance, less readily stained, known as *plastin* or *paranuclein*. Nuclein and plastin together form the chromatin nucleoli, the chromatin nuclear framework, or the chromatin skein-like nuclear filaments; these are only different names for the different forms assumed by the nuclein-plastin elements in the nucleus.

With regard to the relation in which they stand to the achromatic nuclear framework, many theories have been propounded by Flemming, Carnoy and others, but we cannot discuss them in detail now. For the present let it suffice to say that two distinct kinds of nucleoli have been discovered, the one kind very readily stained, the other less so, but both consisting of combinations in different proportions of nuclein and paranuclein, wkilst on the other hand the *true nucleoli* or *plasmosomes* are not susceptible to any stain, consist only of paranuclein (pyrenin), and form more or less transparent vacuoles.

[1] *Allgemeine Biologie*, p. 29. For further details as to the constituents of the nucleus, see pp. 29–44.
[2] Cf. J. Reinke, *Philosophie der Botanik*, 1903, pp. 69 and 72.

It may be asked why different parts of the cell behave in such different fashions, when the same stain is applied to them, and so render it possible for us to penetrate into the mysteries of its structure. Two theories have been put forward to account for this behaviour. According to one, which is known as the chemical theory of stains, it is assumed that the degree of readiness with which the various parts of the cell take a stain depends upon the amount of chemical affinity existing between the various albuminous compounds and the stain applied. According to the other and newer theory, certain parts of the cell are susceptible to stain, only because of the changing physical qualities of the thing stained, and, as a result, its powers of absorption vary. Alfred Fischer is the chief supporter of this physical theory.[1] It seems probable that both theories are more or less true, and that the staining capacity of the various morphological elements of the cell may be ascribed partly to chemical and partly to physical causes.

In close connexion with his examination of the effects of fixing and staining upon the substance of a living cell, A. Fischer has propounded a new theory, which he designates that of the polymorphism or pleomorphism of protoplasm.[2] He believes protoplasm to be in general viscous, containing structures of various shapes, granular or reticular, some of which remain permanently, whilst others are of a transitory nature. All these varieties in the cell-framework are due to definite albuminous compounds fluctuating between a fluid and a solid condition. Moreover, Fischer is of opinion that protoplasm is often homogeneous on the surface, but in the interior occur granules, filaments, reticular framework, and occasionally also Bütschli's alveolar structures. Fischer is not a supporter of the absolute homogeneity of protoplasm, for in the face of ascertained facts this can no longer be defended, but he admits that the various cellular structures observed by modern scientists are, at least to a great extent, not artificial products, i.e. the results of staining and fixing, but occur also in the living cell. He does not, however, believe that

[1] *Fixierung, Färbung und Bau des Protoplasmas*, Jena, 1899.

[2] We find similar ideas in Yves Delage's *La structure du protoplasma et les théories sur l'hérédité*, pp. 30 and 31.

these structures point to any chemical difference in the parts of the cell, but are the outcome of the physical conditions affecting the protoplasm at any given moment. Fischer obviously does not intend to deny the complex chemical composition of living substance, but he doubts whether there is any necessary connexion between the chemical constitution of the parts of the cell and their staining capacity—such a connexion as would justify our assuming that a chemical difference exists between parts that show a different staining capacity.

Although Fischer's theory of the polymorphism of protoplasm has a good deal that is hypothetical about it, there is far more actual foundation for it than for Altmann's granular theory; in fact, the latter bears the character of a phylogenetic speculation rather than that of a scientific theory. The theory of the polymorphism of protoplasm has one great advantage, viz. that it reconciles the conflicting opinions regarding the morphological structure of the cell with one another, and supplies one uniform explanation of the actual variety of phenomena.

4. Survey of the Historical Development of the Morphology of the Cell

What, then, is the morphology of the cell in the light of modern research? This question can be answered best, if we glance back at the views regarding the structure of the cell that have been current at various stages of cytological research. They may be represented by the diagram on p. 64 (figs. 3–6).[1]

Fig. 3 is a cell as Malpighi (1678) and Wolff (1759) conceived it; it consists simply of the enclosing membrane, and so is nothing but an empty sac.

Fig. 4 is a cell such as Schleiden and Schwann described (1838–9). The membrane is still an essential part, but it is now partly filled with fluid, in which is suspended another essential part, viz. the nucleus, with one nucleolus.

Fig. 5 is the cell according to Leydig (1857) and Max Schultze (1861). The viscous fluid fills the whole sac, and

[1] Cf. M. Duval, *Précis d'Histologie*, 1900, pp. 25, 31. Also G. Schlater, 'Der gegenwärtige Stand der Zellenlehre' (*Biolog. Zentralblatt*, XIX, 1899, p. 756).

surrounds the nucleus and its nucleolus, but the membrane has disappeared as not essential to the existence of the cell. Subsequently the finer structure of the cell was more closely examined, and the mass of apparently homogeneous protoplasm was seen to be a compound formation, consisting of framework and fluid, whilst the nucleus, too, was found to contain, besides the nucleolus, an achromatic framework embedded in nuclear fluid, and also a chromatin framework that assumes various forms. We may connect the names of

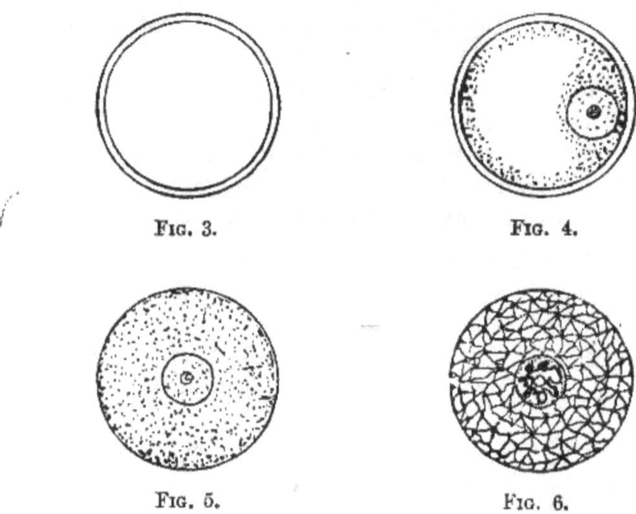

FIG. 3. FIG. 4.

FIG. 5. FIG. 6.

Schlater, Reinke, and Waldeyer with this stage of cellular morphology (1894–5).

Fig. 6 represents it according to Carnoy,[1] who regards the cellular framework as reticular, and the chromatin nuclear framework as consisting of a coil of nuclein-plastin thread.[2] This conception of the cell harmonises best with my own cytological examination of the huge pericardial[3] cells of the *Termitoxenia (Termitomyia) mirabilis.*

[1] Carnoy's valuable work in the development of cytology has been already mentioned. See p. 46.

[2] Cf. also E. B. Wilson, *The Cell*, p. 35. Fig. 13A is an admirable representation of a permanent spireme nucleus, showing chromatin in a single thread (*Balbiani*).

[3] This is the name given to some peculiar cells, allied to the adipose cells, and connected with the 'heart' of the insect, i.e. with its vas dorsale.

Within the chromatin thread of the nuclear framework it is possible in many cases to perceive a still finer morphological differentiation. In the American salamander *Batrachoseps* the threads are plainly divided and each *pronucleus* contains, according to Gustav Eisen, twelve chief parts or *chromosomes*.[1] Each chromosome as a rule is subdivided into six *chromomeres*, in each of which on an average six of the most diminutive bodies or *chromioles* can be traced. There are therefore about 400 distinguishable parts in the chromatin thread of the nucleus!

There are also other animal and vegetable cells, which, before division, show only a coil of chromatin thread, or a chromatin framework, but, in the course of indirect or mitotic division, this develops into definite groups of chromatin knots or chromosomes; whilst within the achromatic framework, that was previously scarcely visible, there now appear as organs of cell-division tiny round *centrosomes*, in the midst of which rises an achromatic spindle. All these phenomena will be discussed more fully in Chapters V and VI, for they do not properly belong to the morphology of the resting cell, or cell not in process of division.

The cell is therefore far from being a simple formation; it is, on the contrary, composed of parts differing widely from one another, and having different functions in its life. We have now to consider the chief kinds of activity in the cell, and the parts taken in this activity by the morphologically different elements of it, and then we shall be in a position to discuss the question whether the cell is the ultimate unit in organic life, or whether it is equivalent to an aggregate of still more simple and elementary units. A result of this discussion will be to show us what ought to be our attitude, as students of natural science, towards the famous theory of the spontaneous generation of organic beings.

[1] *Pronucleus* is the name given to the nucleus of both the egg- and sperm-cells immediately after their union in the process of fertilisation. See Chapter VI.

CHAPTER IV

CELLULAR LIFE

1. The Living Organism as a Cell or an Aggregation of Cells

Cells are the bricks composing the whole building of the organic world. Therefore to them also is the Creator's command addressed: 'Increase and multiply,' for without growth and multiplication of cells no organic life is conceivable. All living creatures consist of one or more cells; if they are unicellular, increase is possible only if from one cell several cells are formed; if they are multicellular, growth and increase are possible only by way of growth and increase of the cells composing their organs and tissues.

In the previous chapter we discussed the structure of the resting cell, as revealed to us by modern microscopical research; we have now to turn our attention to the cell as active and alive. In the case of unicellular animals and plants, the diminutive mass of protoplasm with its one nucleus is the one organ that has to discharge all the functions of life; it is, to compare small with great, a Jack-of-all trades in the economy of life. Nutrition and multiplication, as well as independent movement and sensation (as far as these latter manifest themselves in unicellular creatures), all depend upon one and the same atom of living substance. It is true that here, in spite of the diminutive size of the creature under consideration, we have something analogous to what is called 'organisation' in higher animals, for, as we shall show later on, the morphologically different parts of the cell have various functions. Still, strictly speaking, the parts of the cell ought not to be called *organs*, although, perhaps, we may follow some recent writers and call them *organellae*, at least when speaking of the multicellular animals known as metazoa. In their case, whenever we use the word organ, we mean some part consisting of definite tissues and serving as an instrument in the vital activity of an individual. As the tissues are made up of cells, which are therefore the ultimate constituents of the organs, it would be logically wrong to apply the same word 'organs' to the smallest parts of the cells themselves. It has lately become too much the custom to disregard the connecting membrane which unites cells together to form tissues, and tissues to form organs. The result of this has been that, in both the higher animals and plants, the cell has come to be regarded as having an independent existence, as being an individual of a lower order. This view is however, altogether mistaken, and it is no less wrong to apply the name 'organs' to the minute constituents of the cell, which differ morphologically and physiologically. If they are organs at all, they are so only in a loose, metaphorical sense.

It is only in the case of unicellular organisms that this theoretical opinion corresponds with facts, for in them the constituent parts of the cell really discharge the vital functions of the individual, and so are equivalent to the organs of multicellular organisms. For this reason the unicellular organisms form the lowest rung of the ladder of organic perfection. The higher we ascend, the more are the various parts differentiated to perform distinct functions, and the greater is the perfection of the organisation. A vertebrate animal, or even a tiny insect, is a well-ordered and regulated state, whose inhabitants and officials are thousands and tens of thousands of cells.[1]

[1] The reader must notice that this expression is figurative. In reality, as has been already pointed out, the cells of a multicellular organism are not individuals, because they are not physiological units complete in themselves, as are unicellular organisms. On this subject see Chapter VII, § 1 : 'The cell as the ultimate unit in organic life.' Cf. also O. Hertwig, *Allgemeine Biologie*, 1906, chapters 14–17.

All are democrats, for none is of higher origin than the others; the nerve-cell of the brain, which exercises control, like the ruler of the state, is a cell in exactly the same way as the glandular cell of the stomach, or the epithelial cell of the skin. But in spite of their genuinely democratic disposition, the cells are by no means anarchists; there prevails among them a most perfect harmony, based upon a regular division of labour between the various organs, tissues, and cells.[1]

Just as in every well-ordered state different duties are assigned to different officials, so to various organs are assigned the functions of nutrition, digestion, circulation of the blood, respiration, propagation, movement and all the work done by the nerves and senses. But these organs, which resemble the heads of departments in the state, are themselves made up of different kinds of subordinate tissues, and each tissue consists of a more or less varied combination of cells, differing in the case of the different tissues. All these millions of cells compose what we call an organism, and in spite of their vast number and endless variety they all have the same origin, for they all proceed from an egg-cell fertilised by means of a spermatozoon ; such at least is the ordinary process of development of any higher organism.[2]

The continuation of the process of cleavage, begun in the first cleavage or segmentation nucleus, leads eventually to a differentiation of the living creature into various cells, tissues and organs, until it attains its full development, and then the work of propagation renews the cycle of life. But even the egg-cells and the spermatozoa, although they carry on the task of propagation, differ in no respect from other cells, as far as their origin is concerned ; in the course of embryonic development they are differentiated from common cells, into which the fertilised egg split up at the formation of the periphery of the embryo.[3]

[1] On the subject of the division of labour in an aggregation of cells, see O. Hertwig, chapter 17, pp. 417, &c.

[2] I say 'ordinary,' because of the phenomena of parthenogenesis among insects, &c., where the egg-cell develops without fertilisation. (See Chapter VI, § 6.)

[3] See Chapter VI, § 3, for the most recent results of investigations regarding the distinction between somatic and germ cells, which is either very early or even original.

All the cells, therefore, in the organism enjoy absolute 'equality before the law,' but it is an equality, not of death but of active life, inasmuch as from cells, at first similar, the mysterious laws of organic development produce the living being in all its wonderful, complete, and complex structure.

Such is in outline the cellular life of the multicellular organism, which we cannot now discuss in greater detail. What has been said will suffice to show that the cell must be called the lowest unit of organic life in multicellular animals and plants. Let us now study more closely the vital processes affecting cells as such, whether they are united to form tissues of a higher order, or lead an independent existence as unicellular beings. This study will give us a deeper insight into the real nature of the cell, this marvel of creation.

Life is, in its physiological aspect,[1] an uninterrupted process of movement, every phase of which tends to the preservation of the individual and of the species. The interior movements, which form the really essential processes of vegetative life, tend to the assimilation of fresh material, and so to the growth of the individual. These processes of assimilation, depending as they do upon nutrition and respiration, are necessarily closely connected with analogous phenomena of dissimilation,[2] for the building up of what is new requires a tearing down of what is old, and the reception of fresh nutritive matter and its transformation into living substance necessitate a removal of what is worn out. Growth is based upon assimilation and leads naturally to numerical increase. As soon as a cell has reached a definite maximum size, it divides and forms new cells; if these remain united in one aggregate of tissues, the division of the cell promotes the growth of the individual; if, however, the new cells separate from the parent organism, so as to form new independent individuals, then the division of the cell is a process of propagation, and furthers the preservation of the species. To these interior processes of movement in the living substance correspond other exterior

[1] For further details regarding the physiology of the vital processes, the nutrition and transmutation of energy of cells, and the processes of assimilation and dissimilation, see Bunge, *Physiologische Chemie*, and J. Reinke, *Einleitung in die theoretische Biologie*, chapters 26–29.

[2] The word *dissimilation* was introduced by Hering as an euphonious abbreviation of *des-assimilation*, which, being a clumsy word, is now but little used.

movements, due to the susceptibility of protoplasm to definite external stimuli; these latter movements tend to procure the material necessary to support the interior vital processes, whether it be by the assimilation of food to promote individual growth, or by the union of individuals to promote the preservation of the species; finally, the exterior movements protect the organism from its enemies. Thus all the exterior movements are subservient to the interior, even when, as voluntary, they belong to conscious existence, and therefore are on a higher level than the vegetative processes, for the whole conscious life of an animal aims at the preservation of the individual and of the species; it stands to living matter in the position of a slave; its sole aim is material, and it has no power to rise above the material, as the intellectual life of man enables him to do.

2. Activity of Living Protoplasm

The foregoing general observations will enable us to understand the phenomena that we are now about to consider.

Oskar Hertwig in his 'Allgemeine Biologie,' pp. 108, &c., recognises several distinct kinds of movement in protoplasm, and we may safely follow him on this point. Real protoplasmic movement either belongs to a complete protoplasmic body, such as an amœba, or it takes place in the interior of a cellular membrane. This latter form of movement occurs chiefly in plants, and is divided into rotatory and circulatory movements. The rotatory movement was discovered by Bonaventura Corti as early as 1774. We must distinguish these genuine movements of protoplasm from those due to exterior appendages on the cells, such as cilia and flagella, with which we shall deal in the next section of this chapter. We must refer also to the movements of pulsating vacuoles in unicellular animals, and to the manifold passive alterations in shape and position undergone by the cells of an organism in consequence of the vital process going on within it as a whole. At present, however, we are concerned only with a few instances of true protoplasmic movement.

The protoplasm of a living cell is in a state of constant activity, and moves on definite lines inside the cell, its course being apparently determined by the framework of spongioplasm. At the end of the eighteenth and at the beginning of the nineteenth century Corti and Treviranus noticed (see p. 33) that the chlorophyll granules, which give plants their green colour, are frequently in vigorous movement within the cells; later on, in 1848, von Mohl discovered this granular movement not to be active, but passive, and due to the power of contraction possessed by protoplasm. In many of the lower animals protoplasm appears capable of active movement, but we must be careful to distinguish two forms of activity—the active movement of the protoplasm framework, that manifests itself especially in external changes of shape, and a more passive flow of the granules in the cell-sap, which is a result of the contraction and expansion of the protoplasmic framework. It is obvious that these processes of movement cannot always and everywhere be traced with the same clearness in living cells. They can be seen very well in various little unicellular creatures possessing no enclosing membrane, such as the *Amœba proteus*,[1] and still better in other animals belonging to the same class of Rhizopods, but having a thin shell, through the openings of which the so-called pseudopodia protrude, as, for instance, in the case of the *Gromia oviformis*.[2]

The body of the Amœba is subject to constant changes of shape, whence the creature has received its name. It can protrude protoplasmic continuations of its substance in all directions and again withdraw them. The pseudopodia are outstretched to catch food and to effect a change of place; they are withdrawn when any danger threatens. If the pseudopodia of an Amœba are fed with very small grains of carmine, these grains are at once surrounded by the protoplasm of the pseudopodia and absorbed by it, and then they share in the interior flow of the protoplasm and render it visible under the microscope. In Amœbae there is no sharp distinction between interior and exterior movements, for both are nothing but the same flow of the same protoplasm. When the pseudopodia discover anything edible they close round it, and it at once becomes the centre of a vortex of

[1] The changes of shape undergone by this little Amœba were described as early as 1755 by Roesel von Rosenhof.

[2] Within the pseudopodia of true Amœbae no movements can be discerned, although they occur in the other Rhizopods.

protoplasm, for the creature's whole body contracts round its prey. The same protoplasm, which sought and captured its food, now proceeds to assimilate it, and digests as much of it as is digestible, and then rejects the rest by uncoiling the enclosing ring of protoplasm.

More vigorous movements than those of the Amœba can be observed, as already stated, in the pseudopodia of many other Rhizopods, especially the Foraminifera and Radiolaria, which possess a solid skeleton of chalk or silica, and through its openings protrude the long pseudopodia in quest of food or to effect change of place.

Amœboid movements as well as the granular flow of protoplasm may be produced, checked, and altered by mechanical, chemical and thermal stimuli, and this constitutes the chief proof of the irritability of living protoplasm.

Analogous to the action of the Amœbae and their relations in the water is that of some cells in the organism of multicellular animals, especially of the white blood-corpuscles or *leucocytes*. They too possess amœboid prolongations, enabling them to move and traverse all the tissues of the body. In order to pass through a narrow crevice, they put out a pseudopodium first, and gradually the whole body of the cell follows it. Cohnheim, who discovered the power of the leucocytes to wander through the tissues of the body, bestowed upon it the very suitable name of *Diapedesis*. These wandering cells have an almost insatiable appetite; they are like tramps, always hungry and thirsty, and they attack other cells, as well as any extraneous substances that have penetrated into the body, and encounter them on their way. The leucocytes surround these on all sides and devour them, hence their other name of *Phagocytes*. Their voracity gives them a high degree of importance in the life of the organism. The white blood-corpuscles discover the red blood-corpuscles that are old and incapable of taking up oxygen, and seize them and carry them off, and thus, by consuming the useless members of the community of cells, the leucocytes are able to impart the nourishment so obtained to other active formative elements of the body. They are the police, appointed to keep order in the cell-republic that we call an organism. They go to and fro through all the tissues and purify them from hostile bacilli and other wrongdoers. Whenever they light upon anything harmful, they simply close round it and devour it; or, if it is altogether inedible, e.g. a speck of coal dust, they arrest it and drive it over the frontier. The leucocytes are therefore real sanitary inspectors in the organisms of man and the higher animals. Many authors ascribe to their agency the assimilation of the nutritive matter absorbed in the intestinal glands, as well as the diffusion of nourishing lymph throughout the whole body,[1] and from this point of view the wandering leucocytes appear as nurses, supplying food to the other cells and tissues. On the other hand, however, under certain morbid conditions, leucocytes increase with such overpowering rapidity as to become dangerous. They then attack cells that ought to be left in peace, and so excite a kind of revolution resulting in inflammation and suppuration of the tissues, and tending to the eventual destruction of the whole organism. In spite, therefore, of their physiological merits, leucocytes have acquired a bad reputation in cellular pathology. Moreover, the most recent investigations carried on by Ehrlich, Metchnikoff and others have deprived leucocytes of many of the police functions generally ascribed to them. According to the most modern views, the struggle between health and disease is fought out chiefly by toxins and antitoxins, the former being chemical substances injurious to the organism, and given off by harmful bacteria, &c., whilst the latter are the chemical antidotes, produced by the organism itself as a protection against toxins. Modern processes of inoculation aim at causing immunity from certain diseases by producing specific antitoxins.

A harmless counterpart to the pathological action of leucocytes in the bodies of men and the higher animals occurs in the phagocytes of those insects which undergo a complete metamorphosis. To these cells is assigned the pleasing task of devouring the old tissues of the larval body during the pupal stage, in order to impart the stored-up nutritive matter to other cells concerned in the formation of the new tissues of the imago.

A flow of protoplasm occurs also in cells where it has deposited an exterior membrane and cannot therefore protrude

[1] Cf. M. Duval, *Précis d'Histologie* (1900), p. 42.

pseudopodia, but in this case the movements are limited to the interior of the cell. This movement of protoplasm in plant cells has long been known to botanists and often described, for instance, in the leaf cells of the *Elodea canadensis* and in the stamens of the *Tradescantia*, &c.

3. Exterior and Interior Products of the Cell[1]

Just as the activity of the protoplasm inside a cell enables it to form a solid membrane as its envelope, so it can produce movable processes on the surface of the cell, such as cilia and flagella, which facilitate the locomotion of the cell. In this way ciliated and flagelliform cells arise. The latter have either one or a few long, thick processes, whilst the former have rows of delicate hair-like threads. Among the Infusoria there is a class of unicellular creatures called Flagellata, from their having these flagelliform processes, and another class of Protozoa is known as Ciliata, because their cell-walls are provided with cilia, which enable them to move about in the water. Cilia are important in the ingestion of food, for these creatures, though unicellular and of diminutive size, have voracious appetites. The ring of cilia surrounding the oral aperture of an infusorian by its rhythmical motion produces a vortex in the water, at the centre of which is the mouth of the little animal. If a tiny diatom or another of the Algae is caught in this vortex, it has no chance of escape; it is sucked down and vanishes in this Scylla, and only its indigestible remains are eventually thrown up.

Flagelliform and ciliated cells occur also in multicellular animals. Spermatozoa are simple flagelliform cells, of which the nucleus forms the head, and a long thread of protoplasm the body and tail. Ciliated cells occur chiefly in the respiratory and digestive apparatus, and in this case the cilia do not assist in the movement of the cell to which they are attached, but in that of the substance passing over them. The cilia of the trachea serve to expel small foreign bodies that have entered the respiratory orifices, and those of the œsophagus help to carry down the nutritive fluids taken in through the mouth, and to keep them in steady movement towards the digestive

[1] See O. Hertwig, *Allgemeine Biologie*, 1906, pp. 79, &c., pp. 100, &c.

organs. In many of the higher and lower animals ciliated cells occur in the real digestive canal. I have seen very beautiful ones, magnified 1500 times, in the transverse sections of the mesenteron of the *Termitoxenia* (*Termitomyia*) *Braunsi*.

The outward or exoplasmic products of the cell are the external results of the internal activity of the protoplasm. They may take the form of a cellular membrane, whether it is homogeneous with the protoplasm (as is the case with most animal cellular membranes), or whether it is a chemical product of protoplasm, as is the case with the cellulose cell-walls of plants,[1] or the shells of many of the lower animals (e.g. the Foraminifera) or the coverings of plants (e.g. the Diatomaceae) which have been hardened by taking up silicic acid or carbonate of lime. Further exoplasmic products of the cell are the elastic intercellular bridges uniting cells with one another, and the cilia and flagella which protrude from the cellular membrane.

The internal or endoplasmic products of the cell are contained in its interior. They are of most frequent occurrence in the vegetable kingdom. In the chemical laboratory of the living plant cell grains of starch are being prepared which supply the world with sugar, either directly, or indirectly through the activity of the plant. Starch is the form in which the plant stores up the carbo-hydrates that produce sugar. The protoplasm of plants was believed to form chlorophyll under the influence of light, thus giving its colour to the foliage;[2] but recently many scientists have inclined to the opinion that chlorophyll is not a cellular product, and that its presence, not only in many lower animals, such as the *Hydra viridis*, but also in plants, is due to a symbiosis of special chlorophyll cells with other vegetable or animal cells.[3]

[1] The young membrane of a plant cell consists always of cellulose, but in many instances the cell-walls harden later on into cork or wood.

[2] The granules which convey the colouring matter originate in the plant cell even without the influence of light, although the green colour, which can be extracted from them, only develops as a rule when light is admitted. Young fir trees are green, however, and full of chlorophyll, even when grown in the dark, and several cryptogams become green in spite of complete exclusion of light.

[3] Cf. C. Mereschkowsky, 'Über Natur und Ursprung der Chromatophoren im Pflanzenreiche' (*Biolog. Zentralblatt*, XXV (1905), No. 18, pp. 593–604). He believes the Cyanophyceae to be independent chromatophores, and tries to account for the origin of the vegetable kingdom, and its difference from the animal kingdom, by assuming that they have penetrated into animal cells. In fact a lion, sleeping under a palm tree, would change places with it,

Animal and vegetable fat is a product of the interior activity of the cell, and is stored up in its empty spaces. In the animal kingdom this biochemical branch of industry is of great importance, and a special class of fat-forming cells, called adipose cells, often make up large quantities of tissue. In their vacuoles little drops of fat collect and grow, until finally the whole cell resembles a ball of fat surrounded by a membrane. The neighbouring cells that are not of this class can feed upon this stored-up fat by way of endosmosis. The protoplasmic product that we call fat is of great importance in the nutrition of the animal organism. It used to be regarded as the material for supplying heat in the process of combustion connected with respiration. In insects fat is closely connected with the formation of blood, for which reason, in speaking of them, we often call the adipose tissue simply the blood-forming tissue. I found many instances of this connexion between fat and blood in the course of my microscopical study of the inquilines among ants and termites, and especially in the physogastric guests of the termites, which rejoice in an extraordinary abundance of fat. In the larvae of the termitophile beetle of Ceylon, known as *Orthogonius Schaumi*, the outer edge of the huge adipose tissue may be seen just at the spot where it touches the hypodermal masses of blood, and it is frequently in a state of disintegration, and being absorbed almost imperceptibly by the diminutive corpuscles of the insect's blood. I observed similar phenomena in other genuine inquilines among the termites, which become physogastric through their abundance of adipose tissue; the same transition from adipose to blood tissue appeared on a series of sections of a termitophile insect, *Xenogaster inflata* of Brazil. The ants and termites seem to appreciate the advantages of their guests' adipose tissue, and hold to the dictum *Omne pingue bonum*; for all their true inquilines, belonging to the class of beetles, possess a great deal of fat, and it is this tissue which directly or indirectly emits the volatile exudation that attracts them so greatly and induces them to lick their guests.[1]

provided the cells in his body were filled with chromatophores (p. 604). This is certainly a very bold theory.

[1] Cf. on this subject 'Zur näheren Kenntnis des echten Gastverhältnisses bei den Ameisengästen und Termitengästen' (*Biolog. Zentralblatt*, XXIII, 1903, Nos. 2, 5, 6, 7 and 8, p. 68).

There are a number of other products of the interior of the cell which might be mentioned; some of them occur in animal cells and some in vegetable, and take the form of essential oils, colouring matters, nectar, caoutchouc and india-rubber, resin, tannic acid, poisons of various kinds, digestive ferments, &c., thus serving the most manifold and interesting biological purposes.

In vertebrate animals the haemoglobin of the red blood-corpuscles is one of the products of the interior of the cell. This haemoglobin, to which blood owes its colour, carries the life-giving oxygen which we breathe in; the molecules of oxygen are brought through the lungs into the blood, and accompany the red blood-corpuscles over the whole extent of the arterial circulation, making their way through the finest capillary vessels to the single cells of the tissues, where they give out their oxygen and so oxydise the existing organic connexions. The free carbonic acid, which is the chief combustion product of the vital process, has now to be expelled from the body by the same means; so the red blood-corpuscles are accompanied by carbonic acid molecules on their way back from the capillary vessels, through the whole extent of the venous circulation, until they reach the lungs, where the carbonic acid is breathed out into the air, and at the next inspiration fresh oxygen is taken up, to join the red blood-corpuscles on their next journey through the body. The arterial and the venous blood differ in colour because the haemoglobin of the red blood-corpuscles forms a soluble chemical combination with the oxygen, producing bright red oxyhaemoglobin, whilst the same blood-corpuscles, after giving off their oxygen to the cells of the body, resume their previous dark bluish-red tint.

4. The Predominance of the Nucleus in the Vital Activities of the Cell

We have now considered some characteristic instances of the processes of cell-nutrition, cell-growth, and cell-motion. Before passing on to a new and important class of phenomena of cellular life, viz. the process of multiplication by cell-division, we must examine more closely the part played by the nucleus in the manifestations of cell life already described.[1] We have to answer this question: Are the nutrition and growth of the cell and the formation of its interior and exterior protoplasmic products to be ascribed to the cell-body, or does the nucleus participate in them as an essential element?

R. Hertwig says, in his 'Lehrbuch der Zoologie,' 7th ed. p. 55 (Eng. trans. p. 67), that 'for a long time the functional significance of the nucleus in the cell was shrouded in complete darkness, so that it began to be regarded, in comparison with the protoplasm, as a thing of little importance.' In fact, a merely superficial consideration of the phenomena already described might easily lead us to doubt any participation in them on the part of the nucleus. If, for instance, a little Amœba grasps its still smaller prey with its pseudopodia and devours it, we can observe a series of movements about and in the viscous protoplasm of the creature's body, but we can perceive no change in its nucleus. If, on the other hand, a plant cell is trying to thicken a definite portion of its enclosing membrane by depositing layers of cellulose, the nucleus may be seen to quit its former position in the centre of the cell, and to approach that part of the periphery where the depositing action of the protoplasm is at its height, and, when the task is accomplished, the nucleus comes back to the middle of the cell. In the same way the nuclei of certain unicellular plant-hairs approach the offshoot as long as it is in process of formation, but when its growth is complete they return to their original place. The eggs of the threadworm (*Rhabdonema nigrovenosum*) have been observed during the process of cleavage, and the nuclei of the newly formed cells moved towards the surface of the cell, where the fresh membrane was forming, and after remaining there for some time, on the completion of its formation, they withdrew into the centre of the cells.[2]

[1] Cf. on this subject especially O. Hertwig, *Allgemeine Biologie* (1906), chap. 10, pp. 249, &c.

[2] Cf. L. Rhumbler, 'Über ein eigentümliches periodisches Aufsteigen des Kerns an die Zelloberfläche innerhalb der Blastomeren gewisser Nematoden' (*Anatomischer Anzeiger*. XIX, 1901, pp. 60-88). See also the address delivered by the same scientist at the seventy-sixth assembly of German naturalists at Breslau, on September 23, 1904, and printed under the title 'Zellenmechanik und Zellenleben' in the *Naturwissenschaftliche Rundschau*, 1904, Nos. 42 and 43. See especially pp. 546 and 548.

Numerous similar phenomena, pointing to a participation of the nucleus in the processes of nutrition and formation, were described in 1887 by Haberlandt, an eminent botanist,[1] and in 1889 by Korschelt, a zoologist.[2] These two scientists deduced the following conclusions from their observations :—

1. The fact that the nucleus occupies a definite position only, as a rule, in a young cell in course of development suggests that its functions are connected primarily with the processes of cell-development.
2. From its position we may assume that the nucleus is especially concerned, during the growth of the cell, with the thickening and spreading of the cellular membrane ; but it is quite possible that in a fully grown cell the nucleus has other functions to discharge.
3. The nucleus is concerned not only with the cell's power of secretion, but also with its nutrition. We can infer this both from its position and also from the fact that it sends out numerous branches, thus increasing its surface on the side nearest to the place where secretion or nutrition is going on.[3]

We must refer here also to the correlation between the size of the protoplasmic body and that of its nucleus, which R. Hertwig calls the *Kernplasmarelation*.[4] It can be explained by the interior reciprocal action of the cell-body and cell-nucleus. What actual observation pronounced probable has been confirmed by experiments. Gruber, Nussbaum, B. Hofer, Verworn, Balbiani, Lillie, Klebs and others had recourse to

[1] 'Über die Beziehungen zwischen Funktion und Lage des Zellkerns bei den Pflanzen,' Jena, 1887.

[2] 'Beiträge zur Morphologie und Physiologie des Zellkerns' (*Zoolog. Jahrbücher*, Section for Anatomy, IV, 1889).

[3] This accounts for the occurrence of nuclei with corners or even branches in the gland-cells of certain insects when in a state of active secretion. I noticed such nuclei on my series of sections of the ant-inquiline *Paussus cucullatus*, which has a strongly marked layer of gland-cells in its antennae. Similar nuclei occur in the large frontal glands which open through an exudatory pore of the forehead. Cf. 'Zur Kenntnis des echten Gastverhältnisses bei den Ameisengästen und Termitengästen' (*Biolog. Zentralblatt*, 1903, pp. 240, 241, 244, 245).

[4] Cf. R. Hertwig, 'Über Korrelation von Zell- und Kerngrösse für die geschlechtliche Differenzierung und die Teilung der Zelle' (*Biolog. Zentralblatt*, 1903, Nos. 1 and 2). See also O. Hertwig, *Allgemeine Biologie*, p. 257

merotomy, and cut unicellular creatures into several parts,[1] and the results of these investigations are extremely instructive.[2]

If an Amœba be cut into several pieces, the part that is fortunate enough to contain the nucleus continues its previous way of life; it moves about and feeds, and so it replaces what it lost in living substance and recovers its normal size. The other parts, however, which contain no nucleus, soon cease to move, and in course of time the network of protoplasm that forms their body begins to disintegrate, until nothing is left of them. A non-nucleated fragment of an Amœba is as incapable of feeding as it is of moving. It can no longer contract so as to enclose any particle of nourishment and absorb it into its own body. If a portion of an Amœba had already begun such a nutritive movement before its separation from the main body, its action is soon arrested and the inactivity of death sets in. In the case of unicellular Rhizopods, which deposit a chalky shell, this process of secretion, being analogous to the formation of membrane, becomes impossible as soon as the nucleus is removed, but the nucleated fragments are able to secrete a shell wherever a wound has been inflicted.

With regard to plants, too, Klebs has shown[3] that only the nucleated portions of a plant cell are able to form a new cellulose membrane, and so to close an opening cut in the cell-body.

Balbiani has succeeded in establishing,[4] by means of merotomical experiments on Infusoria, the precise part taken by the chromatin of the nucleus in the nutrition and growth of unicellular creatures. In a previous chapter (pp. 60, &c.) we discussed the morphological importance of chromatin or nuclein in the finer structure of the nucleus; its physiological importance is now to be revealed.

In many Infusoria the chromatin is arranged in numerous

[1] Merotomy must not be confused with merogony, which is a name given to attempts to fertilise or develop ova that have been cut up or otherwise artificially mutilated. We shall refer to this subject again in Chapter VI, § 8.

[2] Cf. Wilson, *The Cell*, pp. 342, &c. Also O. Hertwig, pp. 254, &c.

[3] *Untersuchungen aus dem botanischen Institut zu Tübingen*, 1888, II, p. 552.

[4] 'Recherches expérimentales sur la mérotomie des Infusoires ciliés' (*Revue Zoologique Suisse*, V, 1889); 'Nouvelles recherches expérimentales sur la mérotomie des Infusoires ciliés' (*Annales d. Micrographie*, IV, 1892 and V, 1893).

somewhat coarse granules in the interior of the nucleus. Balbiani succeeded in cutting a ciliated Infusorian (Stentor) into three pieces in such a way that the nucleus was also cut, each segment containing a part of it (fig. 7).

The upper division containing the mouth received four granules of chromatin, the middle portion received one, and the lowest three. All three parts of the Stentor continued to live, and in twenty-four hours each had become a fresh individual. The one formed from the middle piece of the

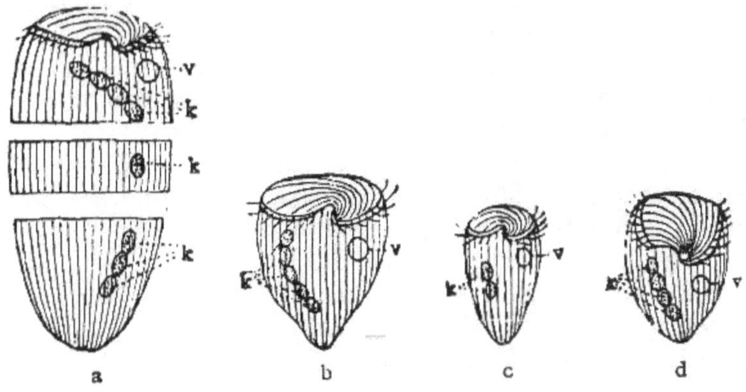

FIG. 7.—Stentor.

On the left (a) is the specimen cut into three parts; on the right (b, c, d) the new specimens formed by regeneration.

k = nucleus; v = vacuole.

original specimen was, however, considerably smaller than the other two, because its nucleus had possessed only one chromatin granule.

In 1896 Lillie succeeded in dividing a Stentor into as many pieces as he wished, by simply shaking the glass vessel containing it.[1] In this way he was able to show that fragments consisting of only $\frac{1}{27}$ of the creature's volume were capable of regeneration, provided they contained a particle of the nucleus; all non-nucleated portions perished.

In other merotomical experiments made by Balbiani, the Infusorian was only partially severed, so that the two parts remained connected by the protoplasm of the cell-body. If the

[1] 'On the smallest parts of Stentor capable of regeneration' (*Journal of Morphology*, XII, Part 1).

6

nucleus was not cut, the wound healed quickly and the creature recovered its previous appearance; it never happened that two individuals were formed in consequence of a division of this kind. If, however, the nucleus also was severed, each part of the Infusorian grew into a new animal, and, as they were connected by a piece of the protoplasm, the result of this division was the production of a monstrous double creature that reminds one of the famous Siamese twins. In course of time, however, the two individuals began to approach one another, their nuclei came together and coalesced, and the monstrosity became one normal specimen.

Other experiments, carried on by Verworn in 1891,[1] and Balbiani in 1892 and 1893, have led to a modification of views based on the experiments just described, inasmuch as they have thrown additional light on the participation of the protoplasm in the life of the cell, and so put us on our guard against overrating the importance of the nucleus. Verworn chose as the subject of his experiments a spherical Protozoon, *Thalassicola,* which measures half a centimetre across, a gigantic size for a unicellular creature. He succeeded in isolating the nucleus from the protoplasm of this huge cell-body, and demonstrated unequivocally that the nucleus cannot live alone without a particle of protoplasm; it died and did not form a new cell-body. On the other hand the non-nucleated cell-bodies continued alive for a considerable time and went on feeding, but they were unable to multiply by means of division, and so they too eventually died. In his more recent experiments Balbiani compared very exactly the varying behaviour of nucleated and non-nucleated portions of Infusoria. He came to the conclusion that nucleus and cytoplasm are each the complement of the other in discharging the most important functions of life, although the nucleus plays the chief part. Cytoplasm alone was able for some time to produce the movements of the body and of its ciliated envelope, the ingestion of food and the contraction of the pulsating vacuoles of the body. The nucleus was, however, indispensable to secretion, regeneration, and the processes of division, without which the cell-plasm must inevitably die.

[1] 'Die physiologische Bedeutung des Zellkerns' (*Pflüger's Archiv für die gesamte Physiologie,* LI).

Not only zoologists, but also botanists, have recently been making careful experiments with a view to determining the part taken by the nucleus and the cell-body respectively in the vital processes of the cell. The results show that in plants too the value of the cell-body must not be underestimated, although the nucleus actually controls the vital activity of the cell.[1]

I have already (p. 80) quoted Klebs' assertion that fragments of vegetable protoplasm containing no nucleus are incapable of forming a cellulose membrane. This statement has been challenged by Palla and others, who think that they have traced the formation of a new cell-wall in non-nucleated fragments, although other botanists regard this as very doubtful.[2]

Klebs himself mentions the fact that non-nucleated fragments of Algae remained alive for weeks, but eventually died. I may therefore on this point agree with J. Reinke, the botanist, when he says: [3] 'The nucleus is unquestionably the most important organ in the cell-body.'

The total results of these merotomical experiments may be summed up shortly as follows:—Nucleus and cytoplasm are both essential to the life of a cell. A cell-body without a nucleus has no more practical value than a nucleus without a body of protoplasm. In a normal cell the nucleus is to a certain extent the central point, the organising principle of the living matter, or, as Wilson aptly expresses it, 'the controlling centre of cell-activity.' [4] Nevertheless, after the nucleus has been removed, the cytoplasm alone is in many cases able for a time to continue the vital processes already begun, but it is incapable of producing any notable new formations, and is absolutely unable to divide and to perpetuate the species. The nucleus is, as will be shown more clearly in other chapters, the real bearer of heredity, and within the nucleus in its turn the chromatin is chiefly concerned with heredity.

The division of an Infusorian into a definite number of nucleated pieces results in the formation of the same number

[1] Further information on this subject will be found in Chapters V and VI, where I shall deal with cell-division and fertilisation.

[2] Cf. Pfeffer, *Pflanzenphysiologie*, I (1897), pp. 45, &c.

[3] *Einleitung in die theoretische Biologie*, 1901, p. 256.

[4] *The Cell*, p. 30.

G 2

of fresh animals, therefore we are justified in calling the nucleus the principle of individuation of living matter ; and here again, within the nucleus, it is to the chromatin that this property must especially be ascribed, for just as many new individuals are formed as there are fragments of nucleus containing chromosomes. If an Infusorian is partially severed, a double animal is formed only if the nucleus be cut in half.

That the protoplasm of the cell-body is not, however, without importance in the formation of a living unit seems to be proved by Balbiani's experiment with the double Stentor. The nuclei of the two creatures gradually approached one another, and one normal animal resulted from their coalescence. If there had been no living bond to unite them, they would not have grown together again into one animal.

Later on I shall have to discuss the important part played by the nucleus and its chromatin in the processes of cell-division and fertilisation. In this place I may, however, quote a passage bearing on our subject from R. Hertwig's 'Lehrbuch der Zoologie,' 1905, p. 55 (English translation, p. 67). He is insisting upon the significance of the nucleus, and says: 'The evidence that the nucleus plays the most prominent rôle in fertilisation has altered this conception (of its secondary importance). Then arose the view that the nucleus determines the character of the cell ; that the potentiality of the protoplasm is influenced by the nucleus. If from the egg a definite kind of animal develop, if a cell in the animal's body assume a definite histological character, we are, at the present time, inclined to ascribe this to the nucleus. From this, then, it follows further that *the nucleus is also the bearer of heredity* ; for the transmission of the parental characteristics to the children (a fact shown to us by our daily experience) can only be accomplished through the sexual cells of the parents, the egg- and sperm-cells. Again, since the character of the sexual cells is determined by the nucleus, the transmission in its ultimate analysis is carried on by the nucleus.' [1]

[1] For the biological and physiological importance of the nucleus, see also Wilson, *The Cell*, pp. 358, 359

CHAPTER V

THE LAWS OF CELL-DIVISION

In a previous section (p. 66) we spoke of the cells as the bricks composing the building of the organic world. But they are at the same time the architects, always rebuilding the organic world in an unbroken series of generations. They are *living* constituents, growing and multiplying in virtue of the laws of development imposed upon them, and they unite to form tissues, organs, and living creatures of various kinds. The fundamental process upon which the architecture of the cell depends in all multicellular organisms is that of cell-division. What the delicate scalpel of the scientist effects violently, when he vivisects unicellular organisms (see p. 80), is done automatically under certain circumstances, in accordance with the interior laws of organic growth ; and one cell, by dividing, forms two or more.

Let us now study this natural cell-division and the interesting processes that attend it.

1. Various Kinds of Division of the Cell and Nucleus

Whenever the development of an individual requires an increase in the number of cells, whether to make new tissues, or to enlarge those already existing, or to form new creatures and carry on the process of propagating the species,—in every case the cells concerned have to divide. In cells containing one nucleus, the first step is the division of the nucleus. Then the protoplasm of the cell-body either divides too, or remains undivided;[1] in the latter case a uninuclear cell becomes multinuclear; in the former, which is much more common, one cell becomes several. If the cellular membrane is divided and fresh cell-walls are formed, we have *exogenous* cell-division; but if the daughter-cells remain within the membranous covering of the mother-cell, we have what is called *endogenous* cell-division.[2] When exogenous cell-division takes place, the new cells either remain side by side, so that a cellular tissue is formed, or they leave their homes and migrate. Again, when a cell divides, it may form two or more cells of equal size, and this is simple cell-division; or the new cells cut off from the mother-cell may be much smaller than it is; this kind of division is called *gemmation*—it occurs in the growth and multiplication of many of the lower animals, for instance, in the *Podophrya*, the *Hydra*, &c., and in some plants, such as the yeast fungus. Whatever be the form of cell-division, its chief feature is invariably the division of the nucleus, and we must therefore devote attention particularly to it. We here touch upon a subject with regard to which modern microscopical research has been most successful; in fact, it would be difficult to name any other subject in dealing with which microscopical research has produced more brilliant results, so great have been the delicacy and intelligence with which the investigations have been conducted, and so bold and shrewd the conclusions deduced from their results, although these conclusions are to a large extent still hypothetical. Modern cytology has succeeded in some degree in solving the mysteries of heredity, by means of microscopical research. If we are careful to distinguish the actual results from the conclusions deduced from them, we shall be able

[1] The process of division which affects only the nucleus and does not result in a cell-division is sometimes called 'free nuclear division.' (Cf. Strasburger, *Lehrbuch der Botanik*, 1895, pp. 55, &c. Eng. trans. 1893, pp. 89, 90.) This free nuclear division must not be confused with 'free formation of the nucleus,' to which I shall refer later.

[2] On the subject of endogenous increase of nuclei, resulting in the presence of several nuclei in one cell, see O. Hertwig, *Allgemeine Biologie*, 1906, pp. 213, &c.

subsequently to form a true opinion of the modern theories of heredity.

Nuclear division is either direct or indirect. In the former, the division of the nucleus takes place without causing any essential change in its structure; but in the latter it is accompanied by a complicated mechanism, involving great changes in the structure of the nucleus, and partially also in the protoplasm of the cell. These changes are chiefly in the position and arrangement of the chromatin constituents of the nucleus, viz. the nuclear thread and its chromosomes; but there are also no less regular formations of fibres and asters out of the achromatic nuclear substance.

On account of the characteristic movements of the chromatin in the nucleus, the indirect nuclear division is sometimes called *karyokinesis* (nuclear movement), while the transformation

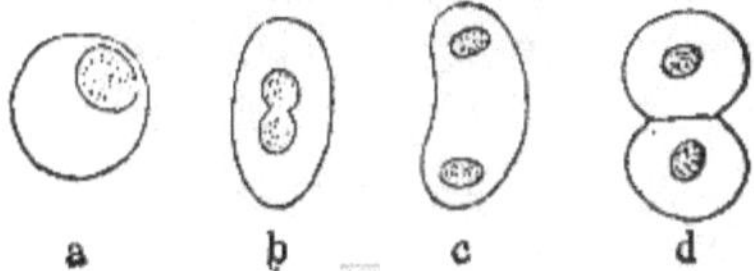

FIG. 8.—Direct division of the nucleus in red blood-corpuscles.

and breaking up of the chromatin thread and the simultaneous appearance of achromatic spindle fibrils have given rise to the name *mitosis* (μίτος = thread) or mitotic division, whereas the direct division is called amitotic. Let us begin by considering the latter, as it is the simpler form, and will help us to understand the more complex process of indirect division.

Direct division of the nucleus was observed by Remak in red blood-corpuscles as early as 1841. Young corpuscles contain one nucleus, the division of which leads to their multiplication. The process is very simple, as the accompanying figure will show.

The nucleus in the cell is at first spherical, then it elongates, gradually contracting in the middle. At the same time the cell itself assumes an oval shape, having previously been round. The nucleus next splits in half, and the two halves retire from one another; then the protoplasm of the cell-body contracts in the middle, the indentation deepening until finally two spherical blood-cells are formed, each with a round nucleus in its centre. Therefore, in the course of direct cell-division, the nucleus by simply contracting breaks into two, and then the protoplasm of the cell-body and the cellular membrane divide likewise. This form of division of the nucleus and cell occurs frequently among Protozoa, especially among those possessing a nucleus that is rich in chromatin.

There is some uncertainty as to the discoverer of indirect division. Wilson (' The Cell,' p. 64) ascribes the discovery of mitosis to Anton Schneider, a zoologist, in 1873. Sachs thinks J. Tschistiakoff,[1] a botanist, has a better claim to the honour, as his work, published in 1874, gave the first impulse to modern research on this subject. Others again mention E. Strasburger, the botanist, as the discoverer of this complicated form of cell-division. There is no doubt that the German anatomist, Walter Flemming, was the first to formulate and expound the process of mitosis in his ' Beiträge zur Kenntnis der Zelle und ihre Lebenserscheinungen ' (1878–82).[2] Abbé Carnoy, a Belgian, has thrown much light upon the subject in his ' Biologie cellulaire ' (1884), and by means of his admirable study of cell-division in Arthropods.[3]

It would be superfluous to mention more names, for the study of mitosis has now become a favourite branch of cytological research, and we know that, in the case of very different kinds of tissue, indirect division of the nucleus occurs far more generally than direct. The two great forms of division of nucleus and cell are, however, connected by various intermediate forms.

A very thorough discussion of all the phenomena observed in mitosis may be found in Wilson's ' The Cell,' pp. 65–121, a book that I have frequently had occasion to mention. My own account of the process must be limited to the barest outlines.

2. Various Stages of Indirect Division of the Nucleus

We have seen that in direct division of the nucleus, or amitosis, the division of the chromatin elements of the nucleus

[1] Sachs, *Vorlesungen über Pflanzenphysiologie*, 1887, p. 115, note 4. Tschistiakoff's work to which Sachs refers is his ' Matériaux pour servir à l'histoire de la cellule végétale ' (*Nuovo Giornale Botan. Ital.* VI). See particularly Plate VII, figs. 11–13.

[2] *Archiv für mikroskopische Anatomie*, XVI–XIX.

[3] ' La Cytodiérèse chez les Arthropodes ' (*La Cellule*, I, 1885, No. 2).

in the mother-cell, so as to form the nuclei of the two daughter-cells, is effected by means of a rough partition of the mother-nucleus, which first contracts in the centre and then splits in half. In indirect cell-division, or mitosis, there is a complicated series of phenomena, all aiming at dividing the chromatin of the mother-nucleus in a most exact and regular fashion between the two daughter-nuclei. This may be called the fundamental idea underlying the whole process of karyokinesis or mitosis, and all the other incidents are subordinate to it.

It is, however, as E. B. Wilson rightly remarks, difficult to give a connected general account of mitosis, because the details vary in many respects in different cases, and especially because great uncertainty still hangs over the nature and functions of the so-called centrosome. In German textbooks of zoology we generally find the process of karyokinesis exemplified by the nuclear divisions of the epithelial cells of the spotted salamander (*Salamandra maculosa*), and my own experience shows that these supply us with an excellent means of tracing the process of karyokinesis conveniently. It is only necessary to cut off a piece of the epidermis from the tail of a salamander or triton larva, to treat it in the usual way with carmine or haematoxylin, so as to prepare it for the microscope, and then it is possible to see a series of karyokinetic figures in the cells of the epithelium. In order to be able to distinguish the single chromosomes, we generally have recourse to some special staining methods, and Heidenhain's stain with iron-haematoxylin can still be recommended. In discussing the subject, however, I shall refrain from alluding to differences in single instances and in staining methods, and shall follow Wilson's admirable account of karyokinesis in 'The Cell,' pp. 65–72.

We may distinguish four groups of phenomena as four successive stages in karyokinesis. There are :—(1) the *Prophase* or preparatory changes ; (2) the *Mesophase* or *Metaphase*, in which the chromatin substance of the nucleus is actually divided ; (3) the *Anaphase*, in which the divided nuclear elements are rearranged so as to form the daughter-nuclei ; (4) the *Telophase*, in which the cell finally divides and the daughter-nuclei return to the state of rest.

These four stages are, of course, not sharply marked off from one another, but one gradually passes into another.

In all four we see a double series of changes going on simultaneously in the cell. The first involves the chromatin figures of the nucleus, formed by the change in position and the halving of the chromatin substance of the nucleus; the second series involves the achromatic nuclear figures, resulting from changes in the achromatic nuclear framework, and to some extent also from changes in the achromatic cell-framework. The first series of changes effects the actual division of the nucleus; the second series is subsidiary, and consists of a radiating arrangement of the protoplasm, rendering possible the movements that occur in the first series.

Let us now examine some diagrams (figs. 9–16) which will give us a better idea of the marvellous mechanism of karyokinesis.

1. *Prophase.*—The first step towards indirect division of the nucleus is a change in the chromatin substance. When the cell was resting, this appeared as a coil of thread or as a reticular or alveolar framework, but now it thickens into a skein. Fig. 9 represents a cell at rest, with its reticular chromatin framework of the nucleus. The dark spot *n* within the network is a nucleolus (see pp. 54 and 61), but its presence is not essential; *c* is the centrosome already in process of division—it is a spherical body, only slightly susceptible to stains, which is also called the polar body, from its position. Boveri terms it the organ of cell-division, and he is probably right in so doing, as we shall see later.[1]

In Fig. 10 the prophase of karyokinesis has begun, and the chromatin thread of the nucleus has thickened and contracted, so as to form one unbroken skein. The nucleolus *n* is still visible, the centrosome has divided, so that there are now two, which are moving apart and beginning to send out delicate rays of protoplasm to form the attraction-sphere *a*. This is sometimes called the chromatin skein or spireme stage of cell-division, from the arrangement of the chromatin substance of the nucleus. As it often forms a kind of rosette, it has also been described as the chromatin monaster (single star) stage.

[1] This polar body must not be confused with the directing or polar globule of the egg-cell. See Chapter VI, § 2.

Lastly, as the achromatic centrosome figure (*a* in fig. 10) resembles a double star, it is sometimes called the achromatic amphiaster stage. The farther apart the two centrosomes move in order to take up their position at the opposite poles

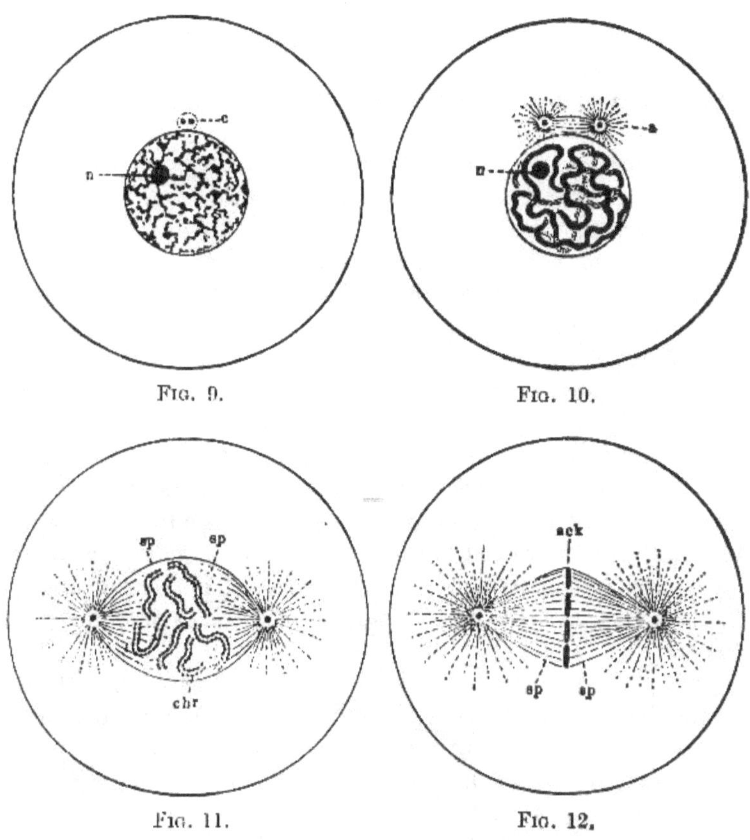

Fig. 9. Fig. 10.

Fig. 11. Fig. 12.

Fig. 9.—Cell with resting nucleus.

Figs. 10–12.—Prophases of mitosis (Wilson).

c = centrosome; *n* = nucleolus; *a* = amphiaster; *sp* = spindle; *chr* = chromosomes; *aek* = equatorial plate.

of the nucleus, the more applicable becomes the name *amphiaster* to this achromatic figure.

Fig. 11 represents the second stage of prophase. The double star or amphiaster now forms an achromatic spindle, and the chromatin figure shows remarkable changes. The chromatin spireme thread has broken up into a number of regular segments, which form the chromosomes. They originally composed the chromatin network of the nucleus, and at each cell-division they appear in the same shape and number.[1]

The chromosomes of the same nucleus are generally all of the same size and shape, but occasionally they form a series of pairs, and in some very rare cases superfluous or accessory chromosomes appear. They have, as a rule, the shape of a fairly regular U or V, sometimes however they are rod-like or even spherical. In certain cases the lengthwise division of the chromosomes, which takes place in the metaphase, is suggested previously, as each splits lengthwise into two parallel parts, which remain connected by delicate transverse fibres. (See the chromosomes in fig. 11.)

As we shall see in the next chapter, the chromosomes are of very great importance in the propagation of the race and in the transmission of hereditary characteristics, and therefore we must devote a little more attention to them. In all plants and animals propagated by the union of two sexes, the number of chromosomes in every cell is invariably *even*, one half being derived from each of the parents. Further, with very few exceptions, every species of plant and animal has always the same fixed number of chromosomes in every cell.[2]

Only the germ-cells are an important class of exceptions, as we shall see in the next chapter, for they contain only half as many chromosomes as the other cells of the body.

The number of chromosomes in each cell varies very greatly in different species of animals and plants. It ranges from 2 to 168. Sometimes there is a considerable difference in the number of chromosomes of closely related species, whilst on the other hand those of unconnected species are often identical in number. Any one who is interested in the subject may find the chromosome numbers of sixty-two species of

[1] Boveri has based his theory of the individuality of chromosomes upon this fact. See Chapter VI, § 9.

[2] The threadworm, *Ascaris megalocephala*, has two varieties, one of which contains four, and the other two, chromosomes in the cells of its body. For other instances see Korschelt and Heider, 'Lehrbuch der vergleichenden Entwicklungsgeschichte der wirbellosen Tiere' (*Allgem. Teil*, part 2, p. 612).

plants and animals tabulated on p. 206 of Wilson's 'The Cell.'[1]

I quote from it a few numbers by way of example; they are those of the chromosomes in the somatic cells of each species; in the ripe germ-cells, as has been said before, only half the number of chromosomes occurs.

In many worms there are 2 or 4 chromosomes; in others 8; in some *Medusae*, grasshoppers and Phanerogams, 12; in one *Hydrophilus*, a snail, the ox and man, 16; in the sea-urchin and a sea-worm (*Sagitta*), 18; in an ant (*Lasius*), 20; in the lily, the salmon, the frog and the mouse, 24; in the torpedo, 36; in a worm (*Ascaris lumbricoides*), 48; and in a little fresh-water crab (*Artemia*), 168.

Let us now turn to fig. 11, and follow the movements of the chromosomes during karyokinesis. We see that the chromatin within the nucleus now appears as an independent formation. The nuclear membrane enclosing the nucleus has meantime disappeared, and so has the nucleolus (*n* in figs. 9 and 10).[2]

The two centrosomes, which in fig. 10 are still above the nucleus, have now taken up their position at its two poles. The protoplasmic rays proceeding from them have grown longer, and now meet in the centre of the nucleus forming the nuclear spindle (*sp*). This is also called the direction spindle, because it serves to direct the chromosomes in their movement both before and after the actual division. The chromosomes now lie apparently free in the middle of the cell, but in reality they are connected with the fibres of the achromatic spindle, which are, as a rule, formed out of what was previously the achromatic nuclear framework, but in some cases out of the cell framework, or out of both together.[3]

This stage (fig. 11) is called, from the chromatin nuclear figure, the stage of chromatin loops, or, from the achromatic figure, the stage of the direction spindle.

[1] Cf. also O. Hertwig's *Allgemeine Biologie*, 1906, p. 203, where the same table is given with some additions.

[2] On the behaviour of nucleoli in different cases, see Wilson, *The Cell*, pp. 67, 68.

[3] There was for a long time great divergency of opinion regarding the origin of the protoplasmic spindle-fibres. Modern research seems to show that we ought to distinguish three kinds of spindle: (*a*) those that are formed of the nucleus alone; (*b*) those that are formed of the cell cytoplasm; and (*c*) those that are of mixed origin. Cf. O. Hertwig, *Allgemeine Biologie*, 1906, pp. 193–195.

Fig. 12 depicts the third part of the prophase, which leads on to the metaphase. The chromosomes are moving along the spindle-fibres towards the centre, and finally group themselves in the form of a ring in a plane passing through the equator of the spindle, which is known as the equatorial plate.[1]

From the chromatin nuclear figure, this stage is called that of the equatorial plate, or rather crown (*aek* in fig. 12), because the chromosomes remain distinct from one another, and only group themselves in the shape of a ring. The achromatic nuclear figure, the spindle (*sp*), is best seen in this stage.

2. *Metaphase.*—The middle stage, or metaphase, now begins, and is the culminating point of the whole karyokinesis, because in it the actual division of the nucleus takes place (fig. 13). In 1880 W. Flemming discovered that this division consists of the splitting of the chromosomes lengthwise into two exactly similar halves. If each chromosome had originally the shape of a V, it now becomes a W ; if it was a simple rod, it is now a double one. This division of the chromatin nuclear substance takes place with such extraordinary exactitude, that it is impossible to avoid regarding it as of great importance to the processes affecting heredity. As W. Roux showed in 1883, the entire chromatin of the nucleus in the mother-cell is divided according to the strictest rules of distributive justice, so that the nuclei of the daughter-cells receive precisely equivalent portions, and each portion is arranged in exactly the same number of chromosomes as there were in the mother-cell. It is a matter of indifference whether the lengthwise splitting of the chromosomes in the metaphase was anticipated by a longitudinal division of each single chromosome (fig. 11), or whether the whole process takes place at once. The nucleolus *n* may remain visible during the metaphase (as in fig. 13) or it may disappear. Its behaviour is of minor importance.

This central stage of indirect cell-division, which we have just described, is known as the stage of doubling the equatorial crown.

3. *Anaphase.*—In this stage the daughter-nuclei of the

[1] For the sake of simplicity, the chromosomes on the diagram are represented as rod-like rather than curved, although the latter is the more usual form. Each loop points to the centre of the equatorial plate.

new cells are built up. After splitting lengthwise in the metaphase (fig. 13), the two halves of each chromosome begin to draw apart. Those on the right group themselves about the right pole of the spindle, and those on the left about the left pole, the spindle-fibres serving as guides. Fig. 14

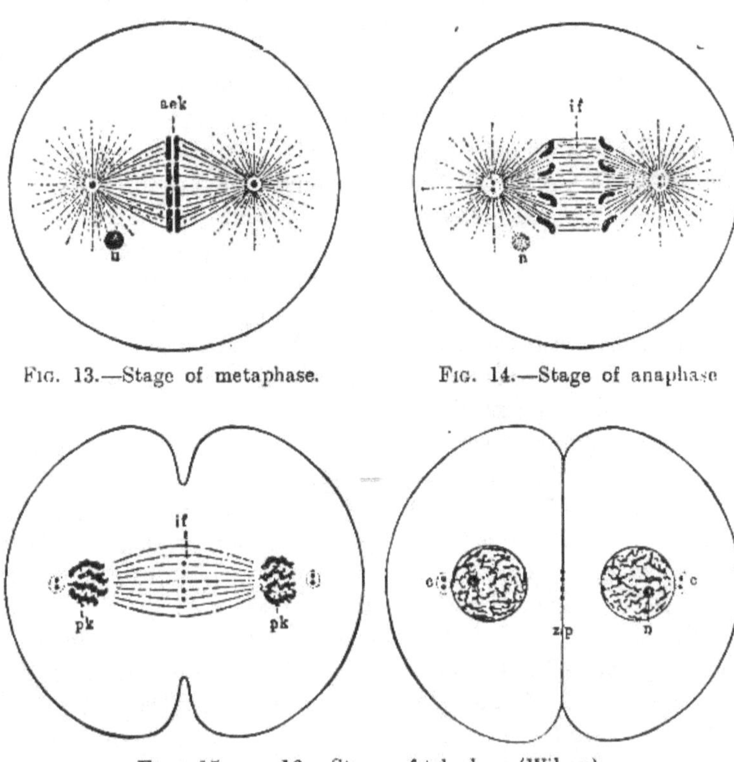

Fig. 13.—Stage of metaphase.

Fig. 14.—Stage of anaphase

Figs. 15 and 16.—Stages of telophase (Wilson).

c = centrosome; n = nucleolus; if = interzonal fibres;
ep = equatorial plate; pk = polar caps; zp = cell-plate.

represents this stage of the anaphase. It is known as that of dicentric orientation of the daughter-chromosomes.

4. *Telophase.*—The process of karyokinesis now advances rapidly through its final stages or telophase. Fig. 15 represents the transition from the anaphase to the telophase. The chromosomes of the daughter-nuclei have now reached the two opposite poles of the spindle, have grouped themselves together and sent out delicate fibres, which bind them together

and will eventually enable them to unite and form the chromatin framework of the daughter-nuclei. In some cases the chromosomes do not directly coalesce to form the new nuclear framework, but it is produced by the fusion of vesicles to which the chromosomes have given rise (vacuolisation).[1] From the chromatin nuclear figure, which forms a dark coloured ring round the two poles of the cell in course of division (fig. 15), this stage has been called that of the two polar caps or crowns. If these crowns assume a stellate shape, it is called the stage of the *chromatin diaster* or double star. When, as in the epithelial cells of Amphibia, the egg-cells of *Ascaris* and many plant cells, the chromatin framework of the new daughter-cells is not produced by vacuolisation of the chromosomes, but by their thickening and growing together, the chromatin diaster stage is followed immediately by that of the *chromatin dispireme.* We can form some idea of this, if we imagine the ends of the chromosomes within the future daughter-cells in fig. 15 to be united. This would produce two skeins similar to that which we noticed in the prophase (fig. 10) as the beginning of the division of the chromosomes.

The fibres of the spindle, which appear in fig. 15 uniting the two chromatin asters, have now another name. They are called interzonal or connecting fibres (*if*). In almost all plant cells, and occasionally in animal cells, they are thickened in the middle, and these thickened portions subsequently make up the cell-plate (*zp*) or mid-body of the dividing cells.

At the end of the telophase we reach the last stage of indirect division of the nucleus (fig. 16). The two chromatin skeins of the daughter-nuclei have surrounded themselves with a membrane, within which the new framework has been formed. We can again perceive the nucleolus (*n*) in the nucleus. Each daughter-nucleus has brought with it a centrosome into the new cell, where it will divide, and the two fresh centrosomes will move from the poles to the two sides of the equator of the original karyokinetic figure and take up their position there. This is, however, not always the case. Sometimes they vanish altogether, and reappear only when the process of division is to begin again. The fate

[1] For further information regarding the growth of the nucleus, see Wilson, *The Cell*, p. 71.

of the interzonal fibres (*if*), which remind us of the spindle of the former achromatic karyokinetic figure, varies greatly. In plant cells they remain, and by thickening they help to build up the new cell-walls formed by the secretion of cellulose.[1] Fig. 16 gives us an instance of this. The perpendicular line in the middle represents the cell-plate (*zp*) or mid-body of the cell in course of division. In animal cells, on the contrary, the interzonal fibres generally disappear early and no trace of them remains, as they are not in this case needed to form a cell-plate. Fig. 15 shows the mother-cell with deep indentations above and below; these increase until it finally splits in half, and the two daughter-cells are formed, and thus the process of indirect division of the nucleus and cell is completed.

3. General Survey of the Process of Karyokinesis

Let us review once more the phenomena of karyokinesis. The first two stages of the prophase, those, namely, of the chromatin spireme and the chromatin monaster, correspond exactly to the last two stages of the telophase, those of the chromatin diaster and the chromatin dispireme. The stages lying between these two extremes belong to the doubling of the equatorial plate or crown. This culminating point is connected on the one hand with the prophase, by the breaking up of the chromatin monaster into V-shaped segments, and by their grouping to form a simple equatorial plate; it is connected on the other hand with the anaphase, by the dicentric orientation of the daughter-segments in the double equatorial plate, and with the telophase by their withdrawal to the poles and formation of the two polar caps or crowns. Indirect karyokinesis is therefore a process that is at once marvellously complex in its conformity to law, and wonderfully simple in design. Its object is to divide the chromatin of the nucleus in the mother-cell into two absolutely equal parts, in such a way that the nucleus of each of the two daughter-cells shall receive the half of every chromosome in the mother-cell, and that the number of chromosomes in each daughter-nucleus shall be the same as that of the chromosomes in the mother-nucleus.

[1] Cf. Strasburger, *Lehrbuch der Botanik*, 1895, p. 52.

The account just given of indirect karyokinesis and the diagrams illustrating it must be regarded as in some degree theoretical, for many modifications occur in various kinds of animals and plants.[1]

Reinke says very truly in his 'Einleitung in die theoretische Biologie,' p. 260 : 'To variations in the structure of the nucleus in different organisms correspond variations in the course of mitosis, as will be seen by comparing them. But we find everywhere four fundamental phenomena, viz. the formation of the chromatin and achromatic figures out of the resting nucleus ; the splitting of the chromosomes ; the movement of the divided chromosomes to the poles of the mitotic figure ; and the rearrangement of the parts so as to reproduce the configuration of the resting nucleus. The persistence of the number of chromosomes from generation to generation in nuclei of the same species may be added as a fifth point.'

The polar bodies called centrosomes were discovered by Flemming in 1875,[2] and I have designated them and the spindle radiating from them a biomechanical contrivance for securing a regular division of the chromatin. This view is confirmed by the account of karyokinesis given by the best authors. We may therefore follow Boveri, Weismann, and others in calling the centrosomes the especial organs of cell-division.[3]

R. Bergh is inclined to ascribe even greater importance in the process of cell-division to the achromatic than to the chromatin nuclear figure.[4] E. van Beneden, Flemming, Guignard and others are also, perhaps, disposed to overrate the importance of the centrosomes.[5]

[1] This is true of the normal processes concerned in karyokinesis, but there are other modifications which are matters of pathology, and which we cannot discuss here. See O. Hertwig, *Allgemeine Biologie*, pp. 214, &c.

[2] On the subject of centrosomes see O. Hertwig, *Allgemeine Biologie*, pp. 45–49, 195, &c., and E. B. Wilson, *The Cell*, pp. 50, &c., 74, &c., 101, &c., 208, &c., 354, &c.

[3] In the next chapter we shall have to examine Boveri's opinion regarding the importance of the centrosomes as fertilising elements. Cf. also Boveri, *Zellenstudien*, Part 4. 'Über die Natur der Centrosomen' (*Jenaische Zeitschrift für Naturwissenschaft*, 1901).

[4] 'Kritik einer modernen Hypothese von der Übertragung erblicher Eigenschaften' (*Zoologischer Anzeiger*, XV, 1892, No. 383).

[5] See also V. Haecker, 'Über den heutigen Stand der Centrosomenfrage' (*Verhandl. der Deutschen Zoologischen Gesellschaft*, 1894, pp. 11–32). This work is a standard one, but only for the state of knowledge on the subject when it was written.

Fol's famous 'Quadrille of Centres,' which the two halves of the male and female centrosomes were supposed to dance round the segmentation nucleus of the fertilised egg-cell, has proved to be erroneous. Strasburger and his followers[1] think that centrosomes are wanting in the higher kinds of plants, and in the division of Protozoa they are either altogether absent or of rare occurrence. They are present in the segmentations of the nucleus which lead to the formation of spindle-poles before fertilisation in the sun-animalculae (*Actinosphaerium*).[2]

If centrosomes were absolutely essential to the action of heredity, they would inevitably be present whenever cells divide, or at least whenever those cells divide which are connected with the preservation of the species, and this is not the case.

The whole question of the function of centrosomes is still involved in much obscurity, and Strasburger sums up the difficulties admirably in the following words : [3] 'At the present moment and at the present state of our investigations, I must content myself with the thought that individualised centrosomes disappear in the more highly organised plants. Why otherwise should we fail to trace them in any of the Pteridophyta and Phanerogams, whilst we succeed in the Bryophyta, (Mosses)? I am quite willing to agree with Flemming, who thinks it possible that in the future centrosomes will be found also in the higher plants. . . . No one as yet has been able to form a conclusive opinion regarding the origin, structure, function, persistence or disappearance of the centrosomes whilst the cell is at rest, nor is much known as to their distribution, although the reasons brought forward by Flemming for believing them to occur everywhere seem very weighty, when considered separately. Carnoy, however, takes a decidedly opposite view.'

We must refer our readers to Wilson and O. Hertwig for further information on the subject of centrosomes. These two writers have collected a quantity of material involving

[1] *Histologische Studien aus dem Bonner Botanischen Institut*, Berlin, 1897.
[2] O. Hertwig, *Allgemeine Biologie*, 1906, p. 189.
[3] 'Über Reduktionsteilung, Spindelbildung, Centrosomen und Cilienbildner im Pflanzenreich' (*Histolog. Beiträge*, 1900, Part 6, pp. 170, 171).

much research. Strasburger concludes with a reference to a theory based on recent research, according to which the centrosome is a mass of kinoplasm, not only serving the purpose of cell-division, but also concerned in the movement of the flagella and cilia of many cells and especially of the spermatozoa. O. Hertwig has adopted this view in his 'Allgemeine Biologie,' 1906, p. 122, &c.[1]

As Strasburger says in the above quotation, we still know very little as to the origin of the centrosomes. Some regard them as composed of the protoplasm of the cell; others, with more probability, think that they are a product of the nucleus. A new theory is that the centrosomes are not permanent constituents of the cell,[2] but are merely microsomes, representing a part of the achromatic framework of the cell or nucleus, which have a temporary importance during the processes involved in karyokinesis, inasmuch as such a microsome, by taking up its position at the pole of the nucleus in course of division, becomes the focus of the protoplasmic rays from which the spindle proceeds. If this theory is true, the centrosomes, and the attraction sphere which they form, are perhaps not the causes of nuclear division, but a result of the beginning of the process. Mitrophanow tried to prove this theory as early as 1894, in his 'Contribution à la division cellulaire indirecte chez les Sélaciens' (*Journal international d'anatomie et de physiologie*, XI).

Wasilieff thinks that the centrosome is only a temporary product of the joint action of nucleus and protoplasm;[3] and this theory is supported by experiments (to which reference will be made in the next chapter) by Morgan, Loeb and Wilson, who succeeded in artificially producing centrosomes in the unfertilised eggs of sea-urchins by means of salt solutions.

The astral rays of the nuclear spindle may all be formed of

[1] See also Ikeno, 'Blepharoplasten im Pflanzenreich' (*Biolog. Zentralblatt*, XXIV, 1904, No. 6, pp. 211–221). Recent investigations made by Russo and di Mauro in 1905, and by Gemelli in 1906, seem however to show that the flagella and cilia are not connected with the centrosomes, but with special basal bodies formed by a thickening of the cell-wall.

[2] Cf. the views expressed by Brandes and Flemming in the *Verhandlungen der Deutschen Zoolog. Gesellschaft*, 1897, pp. 157–162.

[3] 'Über künstliche Parthenogenesis des Seeigeleis' (*Biolog. Zentralblatt*, XXII, 1902, No. 24, pp. 758, &c.).

the achromatic nuclear framework, or of the spongioplasm of the cell-body, or they may have a mixed origin.[1]

We really know nothing of the cause producing this radiation, nor do we know what makes the V-shaped loops of chromatin split in half lengthwise.[2]

The only certain facts are that karyokinesis depends upon the partition of the chromosomes, and that the protoplasmic rays of the nuclear spindle determine the direction in which the chromosomes move. We are also convinced that great importance in the processes of evolution must be assigned to the persistence in the number of chromosomes contained in the somatic cells of individuals belonging to one and the same species, which number is most accurately preserved during karyokinesis by the longitudinal division of the chromatin loops. If we compare this normal form of mitosis with the method of dividing the chromatin in the germ-cells (cf. the next chapter) we shall lay still greater stress upon the importance of this point. We must, however, remember that the science of the present day is quite unable to tell us anything about the inner causes that produce the wonderfully complicated phenomena observed in indirect karyokinesis.

'We must acknowledge that we are not in a position to form any plausible theory at all as to the kind of reciprocal

[1] Cf. Henking, 'Über plasmatische Strahlungen' (*Verhandl. der Deutschen Zoolog. Gesellschaft*, 1891, pp. 29-30); also Yves Delage, *La structure du protoplasma*, 1895, p. 75; O. Hertwig, *Allgemeine Biologie*, pp. 192, etc.

[2] Cf. also H. E. Ziegler, 'Untersuchungen über die Zellteilung' (*Verhandl. der Deutschen Zoolog. Gesellschaft*, 1895, pp. 62-83.) A great number of theories have been advanced to account for the nuclear figures in karyokinesis, but none of them can claim a high degree of probability. This remark applies to Ziegler's own comparison of these figures with the lines of force in a magnetic field. Yves Delage (pp. 310-314) gives a good summary and criticism of the various theories regarding the causes of cell-division and of the formation of karyokinetic figures. He says with much truth of the comparatively best of these theories—that, viz., advanced by Henking—that it would be just as reasonable to see in the lion, the scales, and the fish of the zodiac a real lion, real scales and real fish, as to act like the propounders of these theories, and pretend that their mechanical representations of cell-structures and karyokinetic figures are real cell-structures and real figures. Another attempt, no more satisfactory than its predecessors, at explaining the mechanism of cell-division has been made quite recently by V. Schläpfer in his article 'Eine physikalische Erklärung der achromatischen Spindelfigur und der Wanderung der Chromatinschleifen bei der indirekten Zellteilung' (*Archiv für Entwicklungsmechanik*, XIX, 1905, pp. 107-128). It is an undoubted fact that many physical and chemical influences are at work in the process of karyokinesis, but we possess as yet very little real knowledge of their power to direct and further the biological aim of the division of cell and nucleus.

action existing between the cell-body and the nucleus. We have no foundations of facts upon which to construct a theory.'[1]

Whoever cares to see a summary and criticism of the various hypotheses regarding the mechanism of mitosis propounded by E. van Beneden, Heidenhain, R. Hertwig, Fol, &c., may refer to Wilson, 'The Cell,' pp. 100–111. His *résumé* of the whole discussion is as follows: 'A review of the foregoing facts and theories shows how far we still are from any real understanding of the process involved either in the origin or in the mode of action of the mitotic figure' (p. 111).[2]

The secret physiological causes that motive cell-division are unknown to the scientist, whose microscope reveals to him only their morphological action. They are a problem of cellular physiology, a problem containing in itself the whole mystery of life. We have now to trace this mystery in the phenomena of fertilisation and heredity, and we shall be able to approach its solution in Chapter VIII, where we shall deal with the processes of organic development.

[1] Korschelt and Heider, *Lehrbuch der vergleichenden Entwicklungsgeschichte* (Allgem. Teil, Part I, pp. 153, 154).

[2] See also Wilson's chapter on 'Some problems of cell-organisation.'

CHAPTER VI

CELL-DIVISION IN ITS RELATION TO FERTILISATION AND HEREDITY

(See Plates I and II)

9. REVIEW OF THE SUBJECT OF FERTILISATION AND CONCLUSIONS.

The essence of normal fertilisation is the union of the egg- and sperm-cells (p. 156). Normal fertilisation compared with abnormal and with parthenogenesis (p. 157). Is the essential part of the new organism contained in the egg-cell alone or in the sperm-cell alone, or in both? (p. 158). Why must the nuclei of two germ-cells unite to effect fertilisation? Twofold purpose of fertilisation (p. 160). First, to stimulate the production of a new individual. Various theories regarding rejuvenescence of the organic substance through the process of fertilisation (p. 161). Second purpose of fertilisation, to transmit to the offspring the combined properties of both parents (p. 163). Final significance of the process of reduction (p. 164). Final significance of the distribution of chromatin at the union of the germ-nuclei (p. 165). The nuclear chromosomes the chief material bearers of heredity. Boveri's theory of the 'Individuality' of chromosomes (p. 167). Its connexion with Mendel's Law (p. 170). Object of the combination of qualities effected by the chromosomes in the process of fertilisation (p. 173). Criticism of Weismann's views regarding amphimixis (p. 174). The chromosomes probably are the bearers of the interior laws of development governing organic life (p. 177).

INTRODUCTORY REMARKS. AIDS TO THIS INVESTIGATION

EVER since the time of Aristotle the minds of men have busied themselves with the problem of fertilisation, and with the way in which the characteristics of the parents are handed down from generation to generation of their descendants. In the last few centuries the ovulists and the animalculists have argued with one another as to whether the ovum or the sperm-cell was alone, or at least chiefly, responsible for the phenomena of fertilisation and heredity; the matter was discussed with much energy and varying success, and was finally left undecided, for neither party possessed the actual knowledge necessary to enable them to arrive at a decision—it was reserved for modern microscopical research, with its extremely delicate and ingenious methods of investigation, to supply a more or less adequate basis for the solution of these problems. Let us now consider the results of the most recent research, and see to what conclusions they lead. It is interesting to observe that many of the newer theories of fertilisation approximate very closely to Aristotle's opinion, which was that the female element supplied the material out of which the new individual was formed, whilst the male element supplied the impulse to its development. This coincidence of ideas must not, however, in any way influence us in judging these theories critically.

During the last few years more new facts have been observed, more experiments made, more theories invented and published on the problems of fertilisation and its relation to heredity, than perhaps on any other subject of scientific research.[1] We need not trouble about the purely speculative theories, but discuss only the scientific material from which the supports for the theoretical superstructure are taken. We shall consider the nature of these supports, and see how far anyone has yet succeeded in uniting them so as to give us any conception of the structure, which it will be the task of future generations to complete. But here at once we find ourselves involved in difficulties. Who is a trustworthy guide in this investigation? Who can give us information regarding the quality of the building materials and the best mode of combining them, so as to form at least the foundation of the future edifice? If we take one of the industrious workmen as our guide, there is some danger lest he show us especially the stones that he himself has hewn and fashioned, and give us a partial account of the reasons why these stones must be used in one way, and not in another. If, on the other hand, we take a number of the workers as guides, their explanations may involve contradictions which we cannot solve. If we have recourse to one of the theorising inspectors, we inevitably expose ourselves to the risk of falling too much under his influence and accepting his interpretations, to the neglect of other, no less well grounded, opinions. Where are we to find an 'impartial expert' on the subject?

Of all the recent publications in this department of research none perhaps is better calculated to give a fair objective account of it than the 'Allgemeiner Teil' (General Section) of Korschelt and Heider's 'Vergleichende Entwicklungsgeschichte der wirbellosen Tiere' ('Text-book of the Embryology of Invertebrates').[2] The authors have not only shown marvellous industry in collecting and tabulating an immense number of facts, but they have also displayed great circumspection in their critical appreciation of the various attempts to explain these facts theoretically.

[1] A list of works on this subject is given by Y. Delage, Korschelt und Heider, and E. B. Wilson.

[2] Part I, Jena, 1902; Part II, Jena, 1903. The 'General Section' has not been translated into English.

We have frequently referred also to Y. Delage's 'La structure du protoplasma et les théories sur l'hérédité et les grands problèmes de la biologie générale' (Paris, 1895). It is of great importance as enabling us to follow the questions propounded, although I cannot without reserve accept the author's own 'théorie des causes actuelles.'[1]

E. B. Wilson's book, 'The Cell in Development and Inheritance' (New York, 1902), contains a very good *résumé* of the phenomena of fertilisation and their connexion with inheritance; and on this subject I can cordially recommend Oskar Hertwig's 'Allgemeine Biologie,' Jena, 1906, chapters 11–13. Much has been done by E. Strasburger[2] and J. Reinke[3] to facilitate a comparison of the results obtained by zoologists with the analogous phenomena observed by botanists.

I propose to discuss the points of the subject in the following order:—

1. What are the problems to be solved?
2. How do the maturation-divisions of the germ-cells differ from the ordinary processes of indirect division of the nucleus?
3. What is the normal process of fertilisation in an animal egg, as a result of the union of the egg-cell and sperm-cell?
4. In what relation do the phenomena of superfetation in

[1] A later edition of the same work was published in Paris, 1903, entitled: *L'Hérédité et les grands problèmes de la biologie générale.* A review of the theories of fertilisation, mixed with a good deal of the hypothetical element, was given by Delage in his address 'Les théories de la fécondation,' delivered at the Fifth International Zoological Congress in Berlin (August 1901) and printed in the *Verhandlungen* of the same Congress at Jena, 1902 (pp. 121–140). Cf. also a lecture delivered by Delage in Paris on April 10, 1905, on 'Les problèmes de la biologie' (*Bull. de l'Instit. général psychologique*, V, 1905, No. 3, pp. 215–236). In an oration at the seventy-third meeting of German naturalists and physicians in September 1901, entitled 'Das Problem der Befruchtung' (Jena, 1902), Boveri expounded chiefly his own views on the subject. At the thirteenth annual meeting of the German Zoological Society in June 1903, he read a paper on the constitution of the chromatin nuclear substance ('Über die Konstitution der chromatischen Kernsubstanz,' *Verhandl.* pp. 10–33), in which he developed his views regarding the individuality of the chromosomes. In the course of this chapter we shall have occasion to refer to the works of several other scientists. L. Kathariner contributed a good review of the attempts to solve the problem of heredity to *Natur und Offenbarung*, 1903, pp. 513, &c.

[2] 'Histologische Beiträge,' No. 6: *Über Reduktionsteilung, Spindelbildung, Centrosomen und Cilienbildner im Pflanzenreich*, Jena, 1900.

[3] *Einleitung in die theoretische Biologie*, chapter 34, 'Morphologie der Befruchtung.'

animals stand to those of double fructification in plants?

5. What are the points of resemblance between the fertilising processes of multicellular animals and plants and the phenomena of conjugation observed in unicellular organisms?
6. What light is thrown on the problem of fertilisation by the facts of natural parthenogenesis?
7. Experiments in artificial parthenogenesis.
8. Attempts to fertilise non-nucleated fragments of eggs.
9. What conclusions may be deduced from this series of phenomena with regard to fertilisation in general, and our knowledge of the material bearers of heredity?

1. Problems to be solved

What is it that enables living organisms to propagate their species? The power of propagation depends upon the possession of germ-plasm, which is the means of preservation of species. In unicellular organisms the germ-plasm is contained in the cell that constitutes the body; but in multicellular animals and plants there are distinct germ-cells, out of which the body of the new individual is formed. The plasm of these cells, called by Nägeli *idioplasm* and by Weismann *germ-plasm*, is therefore the actual bearer of the phenomena of heredity. Weismann has based upon this fact his well-known theory of the continuity of germ-plasm.[1] He believes that within the tiny mass of organic substance in the germ-cell, and especially within its nucleus, are contained the material constituents for the formation of new individuals, and that these constituents are transmitted from generation to generation. He calls these constituents *idants*, *ids*, *determinants* and *biophors*, according to their size; biophors regularly arranged compose determinants, these form ids (which contain all the primary constituents necessary to the production of an individual), and the ids finally combine to make up idants. This speculation of Weismann's, according to which germ-plasm is in some degree an extremely delicate, artificial

[1] Weismann has given a detailed account of his theory in his lectures on the evolution theory, 17th lecture (Vol. I, pp. 345, &c., Eng. trans.).

sort of mosaic, is the foundation of his *Preformation theory.*[1] Opposed to this theory are the epigenetic views of O. Hertwig, Y. Delage, Hans Driesch and others,[2] who believe the development of the embryo to be determined, not by material determining constituents, but by dynamic causes, such as definite chemical and physical properties of the germ-plasm.[3]

J. Reinke has combined with this theory that of *Dominants*, which, after the fashion of teleological entelechies, direct and control the activity of the mechanical energies.[4] Driesch inclines to a similar opinion, as he upholds the autonomy of the vital processes, and thinks they cannot be accounted for by mechanical causes.[5] All these theories, which I cannot now discuss in greater detail, have been advanced as supplying answers to one and the same question: 'How can we explain the morphological processes, which present themselves to our consideration, when we observe the phenomena of fertilisation and heredity in the germ-plasm?'

A second very interesting question is: 'In the case of the higher animals and plants, which require the action of both sexes for their propagation, why is the ovum or the sperm-cell alone insufficient for embryonic development? Why is fertilisation necessary to the development of the ovum? Is the union of the two germ-cells, which takes place at fertilisation, essential to the beginning of embryonic development, or is the object of it to secure, by means of bisexual propagation (which Weismann calls *amphimixis*), the advantages of a twofold inheritance, and a mixture of the qualities of both parents? Finally, what are the real bearers of heredity in the germ-cells? May we

[1] *Preformation*, because, according to it, every part of the future individual is formed beforehand, or rather determined beforehand, by means of most minute determining constituents in the germ-cell.

[2] *Epigenesis* = development through new formations; according to these theories the various processes of development in the embryo depend upon new formations, produced by the joint action of external stimuli and internal dynamic factors.

[3] The problem of determination, i.e. the question whether preformation or epigenesis lies at the root of organic development, is obviously not limited to the beginning of the development of the germ, but covers the whole course of ontogeny (individual development). Cf. Korschelt and Heider, *Lehrbuch der vergleichenden Entwicklungsgeschichte der wirbellosen Tiere*, Part I, pp. 81–160. The problem of determination will be dealt with more fully in Chapter VIII, 'The Problem of Life.'

[4] Reinke, *Die Welt als Tat*, Berlin, 1903, pp. 275–292; also 'Die Dominantenlehre,' in *Natur und Schule*, 1903, Parts 6 and 7.

[5] Driesch, *Die organischen Regulationen*, Leipzig, 1901.

regard the chromosomes of the nucleus as such, and with what justification?'

We will now try to examine these questions more closely from the standpoint of the morphological processes in the germ-cells, as revealed by the microscope. Even if we fail to arrive at any final explanation, it is nevertheless important to see how far scientific research on this subject has advanced. We must begin with the phenomena of maturation in the germ-cells.

2. The Maturation-divisions of the Germ-cells

Both the ovum and the spermatozoon must, before becoming capable of fertilisation, undergo two divisions, which are known as maturation-divisions. Let us consider first those of the ovum.

As Y. Delage rightly remarks, what we generally call a mature egg, is really the grandmother of the egg-cell. At that stage the egg is termed a primary oocyte; after the first maturation-division it becomes a secondary oocyte, and after the second division it is an egg capable of fertilisation. This process of twofold division differs entirely in many respects from the usual form of division of cell and nucleus, as described in the preceding chapter. As a rule, the division of a mother-cell produces two daughter-cells of equal size, and, when they subdivide, four granddaughter-cells, all of the same size, are formed; but the two maturation-divisions of the egg-cell result in the formation of one large cell, which is the ovum proper, and of two, or strictly speaking three,[1] diminutive cells or portions of cells, called polar bodies. In the ordinary course of indirect cell-division a period of rest intervenes between two divisions, during which period the nucleus resumes its normal shape; but there is no resting stage between the two maturation-divisions; the second generally takes place immediately after the first, and for this reason the separation of the polar bodies from the ovum has been termed 'precipitate cell-division.' Finally, in the normal form of

[1] The first polar body often divides again immediately after its separation from the ovum, so that, when the second polar body is formed, there are in all three minute bodies present besides the ovum.

karyokinesis, the original number of chromosomes persists in the daughter-cells; in maturation-division of the germ-cell, it is a remarkable fact, that, after the separation of the polar bodies, the nucleus of the mature germ-cell contains only half the number of chromosomes that occur in the somatic cells of the same individual, and at the same time the amount of chromatin originally in the nucleus is generally reduced to a quarter. This reduction, but more particularly that in the number of chromosomes, leads us to speak of the processes of reduction, which, as will be seen later, appear to be of very great significance in the problem of fertilisation.

Like the egg-cell, the sperm-cell undergoes a twofold division in the course of maturation. The primary spermatocyte by indirect karyokinesis gives rise to two secondary spermatocytes, and each of these divides into two spermatids or ripe sperm-cells, so that in this case, too, the primary spermatocyte has four descendants. But whereas the four descendants of the primary oocyte are of unequal size and value, and only one, the ripe ovum itself, is concerned with fertilisation, those of the primary spermatocyte are, as a rule, all four of equal size, each able to fertilise an ovum.[1]

It is a most important fact that, at the completion of the processes of maturation, the number of chromosomes in both sperm and egg-cells is reduced, so that the mature cell contains only half the number that are present in the somatic cells of the same individual and of the same species. The bearing of this fact upon fertilisation will be shown later.[2]

[1] I say 'as a rule,' because Meves believes that he has recently observed a formation of polar bodies during the maturation-divisions of sperm-cells. Cf. F. Meves, 'Richtungskörper in der Spermatogenese' (*Mitteil. d. Vereins Schleswig-Holsteiner Ärzte*, XI, 1903, No. 6); 'Über Richtungskörperbildung im Hoden von Hymenopteren' (*Anatom. Anzeiger*, XXIV, 1903, pp. 29, &c.).

[2] I may incidentally remark that during the maturation-divisions of the sperm-cells of many animals, and especially of many insects, the presence of accessory or heterotropic chromosomes has been observed, the use of which has not hitherto been satisfactorily explained. See Korschelt und Heider, *Lehrbuch der vergl. Entwicklungsgeschichte*, &c., 601. R. de Sinéty, S.J., has traced the history of these accessory chromosomes very carefully in his *Recherches sur la biologie et l'anatomie des Phasmes*, Lierre, 1901; and so has Sutton, an American scientist, in his study of a grasshopper (*Brachystola magna*). Montgomery gives the accessory chromosomes, discovered by him in Hemiptera, the name of *heterochromosomes*. See also Stevens, 'Studies in Spermatogenesis, with especial reference to the accessory chromosome' (Carnegie Institution, Washington, September 1905). E. B. Wilson has recently published some important articles on the various forms of chromosomes occurring in Hemiptera, dividing them into *idiochromosomes* (of which there

Very various opinions exist as to the time and manner in which the reduction in the number of chromosomes takes place; this may partly be accounted for by the fact that different scientists have chosen different objects for observation. We must content ourselves with a condensed summary of the facts, based chiefly upon Korschelt and Heider (pp. 572, &c.).[1]

We must, in theory, distinguish two forms of maturation-division of germ-cells, viz. those called by Weismann 'equation' or equal division, and reducing division. The former follows the ordinary laws of karyokinesis, in which each chromosome of the mother-nucleus splits lengthwise, thus enabling each daughter-nucleus to have the same number of chromosomes as there were in the mother-nucleus, whence this kind of division is called *equal*. Reducing division is altogether different. When it takes place, whole chromosomes are distributed to the daughter-nuclei, so that there is a reduction in the original number of chromosomes, each daughter-nucleus having only half as many as the mother-nucleus.

When the two successive divisions of the germ-cell are both equal, the whole maturation-division is called *eumitotic*, because it follows the normal type of mitosis.[2] If, on the other hand, at least one of the two divisions is a reducing division, the whole process of maturation-division is called by Korschelt and Heider *pseudomitotic*, and we may accept this name. Three varieties of pseudomitotic division must be distinguished. The reducing division may follow the equal division, and then we have a case of *post-reduction division*; or the reducing division may precede the equal division, and then we have a case of *pre-reduction division*; or both

are various sizes) and *heterotropic* chromosomes, and discussing their biological functions. ('Studies on Chromosomes,' in the *Journal of Experimental Zoology*, II, Nos. 3 and 4, III, No. 1). In the last section of this chapter we shall refer again to the accessory chromosomes.

[1] In one of his recent works, 'Über die Konstitution der chromatischen Kernsubstanz,' in the *Verhandl. der Deutschen Zoolog. Gesellschaft* for 1903, Boveri describes the statement of the reduction problem given by these two authors as a 'model.' Cf. also O. Hertwig, *Allgemeine Biologie*, 1906, p. 282, etc.

[2] I cannot here discuss the varieties of eumitotic division known as *homœotypic* and *heterotypic*. In the former a real separation of the two halves of the split chromosome takes place, in the latter they remain connected by their ends, so that the two half-loops form a ring. Such chromosomes are termed 'heterotypic.'

divisions may be reducing, and the process may be called one of double reducing, or a *bireduction division*.[1]

These various kinds of maturation-division have a direct bearing upon the problem when, and how, the original number of chromosomes in the somatic cells is reduced to half that number in the egg and sperm-cells at the conclusion of the process of maturation.

In eumitotic maturation-division, the reduction does not take place during the divisions, but precedes them. The primary oocytes and spermatocytes have in this case the reduced number of chromosomes, before they begin to divide further. We know absolutely nothing as to the manner in which this reduction is effected, and very little as to the time when it takes place. In many plants and animals it seems to occur very early, during generations of cells preceding the formation of germ-cells.[2]

In pseudomitotic maturation-division, the chromatin reduction takes place automatically by means of one or both processes of division, but the manner in which it is effected is still very obscure, and various authors do not agree in their interpretation of their microscopical observations.

The actual results obtained stand in the following relation to the theoretical kinds of maturation-division that have been described above. The eumitotic type—in which both maturation-divisions are produced by longitudinal splitting of the chromosomes, so that no reduction in the number of chromosomes is caused actually by the divisions—seems to occur very frequently in both animals and plants. Some authors are inclined to think that this type might prove to be universal, if we could explain, in accordance with it, the microscopical observations that have hitherto been interpreted in the pseudomitotic sense.

Boveri, whose brilliant research work on *Ascaris* and other creatures has caused the eumitotic maturation-division to be known also as the 'Boveri type of division,' emphatically

[1] I have ventured to coin this word to designate the double reducing division, forming it on the analogy of the other names given to division.

[2] Cf. Wilson, *The Cell*, pp. 272, &c., also Strasburger, *Über Reduktionsteilung, Spindelbildung*, &c., Jena, 1900, pp. 81, &c. Strasburger does not call the reduced number of chromosomes in the germ-cells *reduced*, but *original*. This may possibly be correct phylogenetically, but it can scarcely be justified ontogenetically, at least in the case of multicellular animals.

maintains that the reduction in the number of chromosomes does not take place during the maturation-divisions, nor is it due to them, but precedes them, inasmuch as in the primary oocytes and spermatocytes the number of chromosomes is always half that of the chromosomes in the somatic cells of the same individual. The *Ascaris megalocephala* var. *bivalens*, chosen by Boveri for investigation, has two chromosomes in each of its primary germinal vesicles, each consisting of four grains of chromatin,[1] which Boveri believes to have been formed by a double longitudinal division of the original chromosome. This division is prepared in the nucleus of the primary germ-cells, and is effected by the two maturation-divisions, so that finally the mature ovum and spermatozoon contain each two chromosomes in their nucleus, i.e. the same number as before, whilst the somatic cells contain four.

The eumitotic type of maturation-division of the germ-cells has been described by many zoologists; by O. Hertwig and A. Brauer (in *Ascaris*), by Meves, McGregor, Janssens, Eisen, Carnoy and Lebrun (in Amphibia), Ebner and von Lenhossek (in the rat), de Sinéty (in Orthoptera), &c. Many eminent botanists, too, and especially Strasburger, with whom Guignard, Motier and Juel agree, concur in believing the maturation-divisions of plants to be of the eumitotic type, as they take place by a twofold longitudinal splitting of the chromosomes, and these writers are of opinion that the reduction in the number of chromosomes is effected before the maturation-divisions, viz. in the embryo-sac, or at the formation of the pollen.

Pseudomitotic maturation-division has hitherto been observed chiefly in Arthropods.

Post-reduction division, in which the first of the two maturation-divisions is equal, and the second reducing, is

[1] It would perhaps be well for this reason to adopt the number 8 for the chromosomes of the nucleus of the primary germ-cell, as Kathariner has done in his article in *Natur und Offenbarung*, 1903, pp. 524, 527. The adoption of this number would, however, lead to the following difficulties. First, in *Ascaris megalocephala* var. *bivalens*, the primary germ cells would contain twice as many chromosomes as the somatic cells. Secondly, the twofold maturation-division would result, not in halving, but in quartering the original number of chromosomes. I prefer, therefore, to follow Boveri, and regard the two groups of four grains as only two chromosomes, this number being half that of the chromosomes in the somatic cells, which is therefore already reduced

known also as the Weismann type, as Weismann laid great stress upon it, although he did so chiefly for theoretical reasons connected with his theory of heredity. At the maturation of the eggs of the Copepods among Crustacea, Rückert and V. Haecker observed twelve tetrads (groups of four), which, they believed, split longitudinally at the first division, and transversely at the second, which would then be a reducing division in Weismann's sense.

Vom Rath described similar phenomena occurring at the maturation of the egg of the mole-cricket (*Gryllotalpa*), but, according to Korschelt and Heider (p. 586), it is still uncertain whether the second division in this case is really a reducing division. With regard to many other insects also in the last few years the post-reduction division has been frequently called in question, and it must be observed that the interpretation of the second division as a reducing division is still a moot point; for instance, the same microscopical observations of the maturation of the sperm-cell in Orthoptera led McClung in 1900 [1] to declare the division to be reducing, and de Sinéty (1901 and 1902) to pronounce it to be a double longitudinal splitting of the eumitotic type.

The kind of reducing division that I have termed pre-reduction, in which the reducing precedes the equal division, has been described as occurring both in spermatogenesis and oögenesis of animals of widely different types. It was discovered by Korschelt, who observed it at the maturation of the egg of the annelid *Orphryotrocha puerilis*, and has been called after him the Korschelt type. Henking and Paulmier say that this kind of maturation-division occurs in many species of Hemiptera, and Montgomery has traced it in other Hemiptera and in the very obscure *Peripatus*. On the other hand, Gross [2] declares not the first, but the second, division to be reducing in the maturation of the sperm-cells of the *Syromastes marginatus*, so that this bug would seem to supply an instance of post-reduction rather than of pre-reduction division.

[1] See also McClung's more recent work, 'The Spermatocyte divisions of the Locustidae' (*Kansas Univ. Science Bullet.*, I, 1902, No. 8, pp. 185-231, with four plates).

[2] 'Ein Beitrag zur Spermatogenese der Hemipteren' (*Verhandl. der Deutschen Zoolog. Gesellsch.*, 1904, pp. 180-190).

E. B. Wilson's latest investigations regarding the maturation-divisions of germinal vesicles among Hemiptera [1] seem to show that the question of longitudinal or transverse divisions has lost its primary importance, because the chromosomes separating at the reducing division were originally distinct, and were only temporarily united during an intermediate synapsis stage.[2]

Montgomery and several other authors ascribe particular importance to the copulation of chromosomes during synapsis as facilitating the interchange of qualities between the chromosomes of the male and female parents respectively.[3]

Lastly, bireduction division, in which both maturation-divisions of the germ-cells are reducing, has been described by Julin as occurring at the maturation of the egg of an Ascidian (*Styelopsis*), and by Wilcox at that of the spermatozoon of a grasshopper (*Caloptenus*), &c. The remark that the interpretation to be assigned to the microscopical observations is by no means certain, applies to this kind of division even more than to the others.

Some idea of the difficulties which the student engaged in this department of research has to encounter, may be formed from the fact that the chief supporters of the various division theories have repeatedly changed their minds, and have assigned to their observations now one interpretation and now another. I may refer particularly to Boveri and Strasburger in this respect.

As we have seen (p. 112), Boveri first described the eumitotic type of maturation-division, which is called by his name, and in which both divisions are equal and longitudinal, the reduction in the number of chromosomes having taken place before the division ; in 1903,[4] however, he acknowledged that in a number of instances an actual reducing division takes place, 'though not precisely in Weismann's sense.' Now he thinks that only the

[1] 'Studies on Chromosomes' (*Journal of Experimental Zoology*, II, III, 1905, 1906). Cf. also p. 110, note 2.

[2] On the subject of this stage see Pantel and de Sinéty, 'Les cellules de la lignée mâle chez le *Notonecta glauca*' (*La Cellule*, XXIII, 1906, fasc. I, pp. 89–303), pp. 111, &c.

[3] See O. Hertwig, *Allgemeine Biologie*, pp. 291, 292.

[4] Boveri, 'Über die Konstitution der chromatischen Kernsubstanz' (*Verhandl. der Deutschen Zoolog. Gesellsch.*, 1903, pp. 10–32), p. 27.

first division is longitudinal, and he believes the second to be transverse, effecting a reduction in the number of chromosomes. If this is true, we have post-reduction division, approximating to the Weismann type.

In 1904 Strasburger,[1] the botanist, abandoned his earlier opinions regarding the eumitotic type of maturation. His most recent investigations of the pollen-mother-cells of *Galtonia* show the first of the two maturation-divisions of the chromosomes to be transverse, resulting in a reduction of their number; the second, on the contrary, appears to be a longitudinal or equal division. In 1904, therefore, Strasburger, it would seem, upheld, instead of the eumitotic type, the pseudomitotic, in the form of a pre-reduction division, corresponding to the Korschelt type. But we should have almost as much justification for speaking of post-reduction in this case; for, as Strasburger expressly states, the longitudinal division, which is actually the second in order of occurrence, is anticipated by a longitudinal splitting of the chromosomes, which precedes the first transverse division. In 1905, however, Strasburger returned to his earlier opinion regarding the eumitotic type of maturation-divisions,[2] and he now again maintains that both divisions are longitudinal and equal, and that the real reduction in the number of chromosomes precedes them. He agrees, therefore, now with Abbé V. Grégoire, who expressed similar views in 1905.[3]

The theory of eumitotic maturation-division seems, therefore, to have triumphed over that of pseudomitotic.[4] Whether in the chromatin skein or spireme, formed before the maturation-divisions take place, the individual chromosomes are joined longitudinally or by their apex, is a question raised by Boveri in 1903, and discussed by Grégoire, Strasburger, Schreiner[5]

[1] Strasburger, 'Über Reduktionsteilung' (*Sitzungsber. der Berl. Akademie der Wissensch.*, XIV, 1904, pp. 587–614).

[2] Strasburger, 'Typische und allotypische Kernteilung' (*Jahrb. für wissenschaftl. Botanik*, XLII, 1905, Part I, pp. 1–71).

[3] V. Grégoire, 'Les résultats acquis sur les cinèses de maturation dans les deux règnes': I. mémoire: Revue critique de la littérature (*La Cellule*, XXII, 1905, fasc. 2, pp. 221–374).

[4] Cf. J. Maréchal, 'Über die morphologische Entwicklung der Chromosomen im Selachierei und Teleostierei' (*Anatom. Anzeiger*, XXV, 1904, pp. 383–398 and XXVI, 1905, pp. 641–652).

[5] A. and K. E. Schreiner, 'Neue Studien über die Chromatinreifung der Geschlechtszellen' (*Archives de Biologie*, XXII, 1906, fasc. I, pp. 1–69).

and Bonnevie,[1] but we cannot consider it fully now. The first view is probably the correct one. I may remark incidentally that almost all the recent results of the examination of chromosomes tend to confirm Boveri's theory of their 'individuality.' But I shall recur to this theory in the ninth section of this chapter.

J. Gross[2] has recently summed up the results of his investigations into the maturation-divisions of the germ-cells in the following sentence: 'The most important results of cytological research into the problem of reduction in the last few years seem to me to be two: it has been demonstrated that a real, qualitative reduction actually takes place, and it has been found that a conjugation of the chromosomes of both parents as a rule precedes the maturation-divisions.'

I have already dwelt too long upon the various theories connected with the maturation of the germ-cells. The accompanying diagrams will enable the reader to form some idea of the maturation of the egg-cell and of the formation of the polar bodies; they represent the particular kind of division that I have termed post-reduction. It must, however, be observed that these are merely diagrams, and do not represent the actual process; they have been designed to show, in the simplest way possible, the first division as equal, and the second as reducing.

Let us assume the primary oocyte to have four chromosomes in its nucleus before the process of division begins. The first stage in the process is that the germ-nucleus or vesicle moves towards the periphery of the cell (fig. 17). Then the chromosomes of the nucleus arrange themselves in the manner described in Chapter V (p. 94), so as to form an equatorial plate or crown in the middle of an achromatic nuclear spindle (fig. 18); they split longitudinally, and the daughter-chromosomes withdraw to the poles of the nuclear spindle (fig. 19). This first nuclear division is an equation or equal division of the ordinary kind, not a reducing division. The upper group of four chromosomes with the centrosome of the egg-cell

[1] 'Untersuchungen über Keimzellen: I. Beobachtungen an den Keimzellen von *Enteroxenos Oestergreni*' (*Jenaische Zeitschr. für Naturwissensch.*, XLI, 1906, part 2, pp. 229-428).

[2] 'Über einige Beziehungen zwischen Vererbung und Variation' (*Biolog. Zentralblatt*, 1906, Nos. 13-15, &c., p. 396).

belonging to them is now forced against the periphery of the cell, until it finally passes out of the cell, surrounded by a small quantity of protoplasm (fig. 20). This forms the first polar body (r^1 in fig. 20). Meantime, a fresh nuclear spindle forms immediately round the four chromosomes left in the egg-nucleus (fig. 20); but this time there is no longitudinal

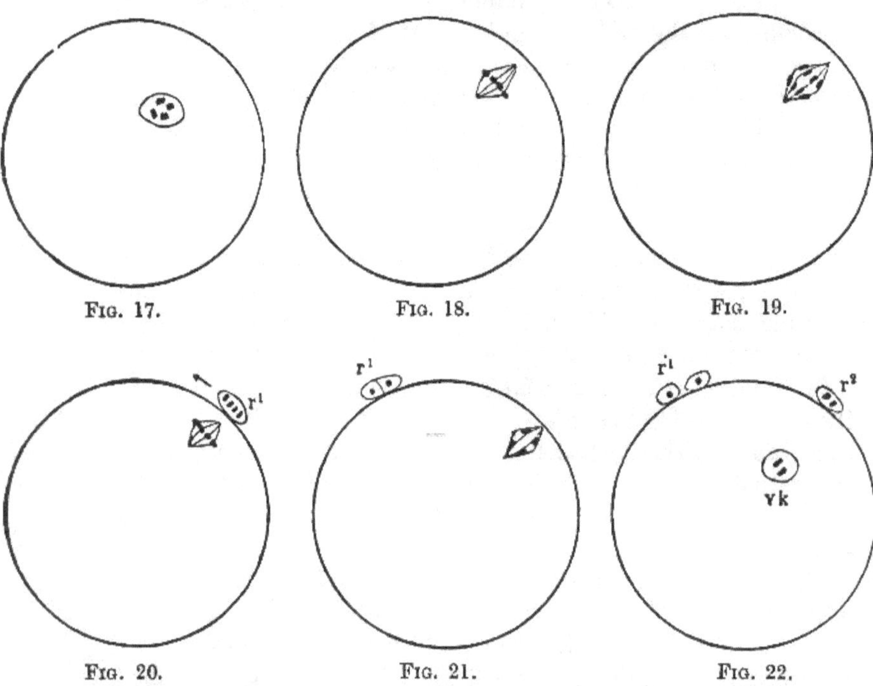

Fig. 17. Fig. 18. Fig. 19.

Fig. 20. Fig. 21. Fig. 22.

Figs. 17-22.—Diagrams representing the maturation-divisions of the egg-cell. r^1= first polar body; r^2=second polar body; *vk*=female pronucleus.

splitting of the chromosomes. They arrange themselves in pairs (fig. 21); the upper pair approach the periphery of the cell, and are expelled from it with a particle of protoplasm, and so form the second polar body (r_2 in fig. 22). This second division was reducing, for the nucleus of the egg-cell, which now resumes its original shape, and at this stage is called the female pronucleus (*vk* in fig. 22), now has only two chromosomes instead of four. If, in the meantime, the first polar body has again divided (r^1 in figs. 21 and 22), the result of the two maturation-divisions of the egg-cell has been the production of one large and three small cells, of which only the first, the egg-cell prepared for fertilisation, is of interest for us.[1]

3. The Normal Process of Fertilising in an Animal Ovum

(See Plate I)

Let us now turn to the process of fertilisation in its normal form in animal ova, as microscopical research has revealed it to us. O. Hertwig was the first to succeed, in 1875, in lifting the veil that for so many thousands of years had rested over these phenomena. In the course of observations on the eggs of the sea-urchin (*Echinus*), he saw that during fertilisation a thread-like sperm-cell passes into the ovum; the head of the sperm-cell changes into a so-called male pronucleus, and unites with the nucleus of the ovum, or female pronucleus. This union of nuclei results in the normal process of fertilisation, for it gives rise to the *cleavage-nucleus* of the fertilised ovum, which at once begins to divide by means of the nuclear spindle of the cleavage-nucleus, so forming the first pair of cleavage-corpuscles, or blastomeres, from whose further divisions all the tissues and organs of the new individual are produced.

At first sight the process of fertilisation thus described seems very simple, but it becomes very complex by reason of the vast varieties in its details, in the case of different plants and animals. Moreover, very various opinions still prevail as to the parts played by the cell-nucleus, the centrosome, and the egg-plasm respectively in the work of fertilisation. Korschelt and Heider devote over one hundred pages to a description of these phenomena in their 'Vergleichende Entwicklungsgeschichte der wirbellosen Tiere' (Allgemeiner Teil, pp. 628, &c.). I must obviously limit myself to what is absolutely necessary

[1] For the subsequent history of the polar bodies (*globules polaires*) and their importance, see Korschelt and Heider, *Lehrbuch der vergl. Entwicklungsgesch.*, pp. 549, &c. They discuss Petrunkewitsch's theory that the polar bodies continue to exist and supply the material for the germinal glands of the future embryo. But nothing is known with certainty on the subject.

in order to enable my readers to form some idea of the essential processes of fertilisation and heredity.

Although the ovum of the *Echinus* measures only $\frac{1}{10}$ mm. in diameter, it is, like all other ova, of enormous size in comparison with the spermatozoon—and this is especially true in the case of eggs containing much yolk. Such eggs have stored up in their egg-plasm a considerable quantity of nutritive matter, which is used in the development of the future embryo. The sperm-cells, on the contrary, are some of the smallest cells occurring in living organisms,[1] for their sole task is to penetrate the ovum and fertilise it. For this reason the protoplasm that constitutes the cell-body is generally only a thread-like flagellum, which serves as an organ of locomotion, and the thickened head is the nucleus of the sperm-cell; between head and tail is the so-called middle-piece containing the centrosome of the sperm-cell.

In spite of the extraordinary difference in size and shape between the ovum and the spermatozoon, their nuclei are so far of absolutely equal value, for they contain the same number of chromosomes. Both the male and the female pronuclei contain half the number of chromosomes found in the somatic cells of the same species. This fact, to which I referred in speaking of the maturation-divisions of the germ-cells, is of great importance in our consideration of fertilisation and heredity.

The union of the male and female pronuclei to form the cleavage-nucleus of the fertilised ovum does not necessarily involve a real fusion of the nuclei; on the contrary, in many cases the nuclei with their chromosomes remain distinct from one another, though they take up their positions close together, so as to form a common cleavage-spindle. We may follow Korschelt and Heider (p. 682) in distinguishing two chief types of fertilisation. The first is the so-called *Echinus*-type, deriving its name from the sea-urchin (*Echinus*), in which it was first observed and described by O. Hertwig (1875–1878). In this type the two pronuclei actually fuse together to form one resting cleavage-nucleus, which does not begin to divide until the fusion is complete. It should be noticed, however, that

[1] In mammals they often measure (without the tail filament) only 0·003 mm. See R. Hertwig, *Lehrbuch der Zoologie*, 1905, p. 49 (Eng. trans. p. 60).

the chromosomes of the two pronuclei do not fuse together, but come into close juxtaposition. The second type is the *Ascaris*-type, deriving its name from the maw-worm of the horse (*Ascaris megalocephala*), in which it was observed by E. van Beneden in 1883;[1] in it the two pronuclei remain independent, but take up their position close together, so as to produce the first cleavage-spindle in common. Having produced it, they break up, and distribute their chromosomes by longitudinal division to the two daughter-nuclei. Many instances of both types of union occur in the animal kingdom, in very various families and classes, and also in closely related species; in fact Boveri (1890) and Klinckowström (1897) have found them even within one and the same species.

I have chosen the second type to illustrate the normal phenomena of fertilisation, because it has the advantage of showing more clearly how the paternal and maternal chromosomes are evenly distributed at the cleavage of the fertilised ovum. In a lecture on the subject of fertilisation ('Das Problem der Befruchtigung,' Jena, 1902), Boveri sketched the process on the lines of the *Ascaris*-type, illustrating it by diagrams, which are reproduced on Plate I, figs. 1–7.[2]

The egg-nucleus is coloured *blue* and the sperm-nucleus *red*, in order to make it easy to distinguish the two nuclei and the chromosomes of the cleavage-spindle proceeding from them.

The nucleus of the mature egg-cell, which after the maturation-divisions is called the female pronucleus, moves from the excentric position, occupied during the formation of the polar bodies, back into the centre of the cell (Plate I, fig. 1). Meantime a spermatozoon has made its way into the ovum (at the top of fig. 1).[3] Only its head and middle-piece, however,

[1] This type was perhaps observed by O. Hertwig between 1875 and 1878 as occurring in *Mitrocoma* and *Aequorea* (Korschelt and Heider, p. 681).

[2] I say 'on the lines of the *Ascaris*-type,' because in many details this sketch is at variance with actual observations made by E. van Beneden, O. Hertwig, Carnoy, Boveri, &c., on *Ascaris megalocephala* var. *bivalens*. It should be noticed particularly that in *Ascaris* the spermatozoon does not lose a tail, but the whole sperm-cell, which in this case is conical, passes into the egg-plasm. Cf. also E. Korschelt, 'Über Morphologie und Genese abweichend gestalteter Spermatozoen' (*Verhandl. der Deutschen Zoolog. Gesellsch.*, 1906, pp. 73–82).

[3] Circumstances vary greatly in different cases. In some animals the maturation-divisions of the egg precede the entrance of the spermatozoon, in others they are simultaneous with or subsequent to it. Cf. Korschelt and Heider, pp. 630–632.

really enter it ; the tail filament, representing the protoplasmic body of the sperm-cell, is generally thrown off, or it is quickly resolved in the protoplasm of the egg-cell. The head and middle-piece of the spermatozoon rotate through 180°, so that the middle-piece, which was previously behind the sperm-head, is now in front of it ; the spermato-centrosome, or centrosome of the sperm-cell, contained in the middle-piece, now becomes visible, and sends out a ring of protoplasmic rays (fig. 2), the so-called ' sperm-aster,' which is here represented as small, although it often stretches over the greater part of the egg. A very remarkable transformation of the sperm-head now begins. It swells up—in consequence, as Y. Delage thinks, of taking in water from the egg-plasm—and, as it swells, it reveals its nuclear character by forming a chromatin framework (Plate I, figs. 3 and 4), until finally it appears as a male pronucleus (fig. 5), exactly equivalent to the female. Meantime the spermato-centrosome has undergone a series of further modifications. It divides (Plate I, fig. 3) ; the two half-centrosomes take up a position on either side of the two nuclei (fig. 4) and develop their astrospheres (fig. 5). The chromatin substance of the two pronuclei, now in close proximity, next proceeds to transform its chromatin framework, in readiness for the first cleavage of the egg-cell. Each pronucleus develops the same number of chromatin loops, which usually resemble one another exactly in size and shape. In the diagram (fig. 6), which might be taken as representing the fertilisation of the maw-worm of the horse, *Ascaris megalocephala* var. *bivalens*, each pronucleus contains two chromatin loops or chromosomes, i.e. half the number contained by the somatic cells of the same animal. The cleavage-spindle is next formed ; it gives rise to the first division of the fertilised egg-cell, and so to the first stage in the development of the future embryo.

Each of the two chromosomes in the parent nuclei splits lengthwise into two parts, which arrange themselves in the middle of the nuclear spindle formed by the centrosomes (fig. 7). Then the four daughter-chromosomes on the left, two being paternal and two maternal in origin, move to the left pole of the spindle ; the corresponding four on the right move to the right pole of the spindle, and at the two poles they give rise to the two daughter-nuclei of the first cleavage-cells (*blastomeres*) of the embryo. Thus each of the first two daughter-cells contains four chromosomes in its nucleus, two from the father and two from the mother. Hence it comes about that each of the cells in the embryo, which are produced by continued indirect karyokinesis from the fertilised ovum, contains an equal number of paternal and maternal chromosomes, and the total number is equal to that of the chromosomes in the somatic cells of the parents, and double that contained in either the male or female pronucleus. It would seem, therefore, that by this process a precisely equivalent transmission of the nuclear elements of both parents is secured to their offspring.

We must here refer to an observation, made originally by Boveri in 1887 [1] and confirmed by subsequent study of *Ascaris megalocephala*, which, whilst, to some extent, modifying the account just given, lends it additional weight in its bearing upon the question of transmission. In *Ascaris*, in all the cleavages from the two-cell stage onwards, the cells of the germinal area of the embryo present characteristics in their nuclei and processes of karyokinesis distinguishing them from the somatic cells of the same embryo. Only the cleavage-granules destined to give rise to the germ-cells preserve the original chromosomes, which they receive from the fertilised egg-cell, in unaltered form ; the cleavage-granules destined to produce the somatic cells, as soon as they begin to divide, reject the thickened ends of the chromosomes, and the rest of the chromatin loop breaks up into a number of smaller pieces, that subsequently reappear. Boveri called this phenomenon 'chromatin diminution,' and it seems to show that only in the germ-areas is the continuity of the germ-plasm fully maintained, whilst many divergencies may occur in the tracts of somatic cells.[2]

It is a fact that individuals, born of the same parents, differ to a certain extent both from their parents and from one another, and it is no less true that the qualities of grand-parents or of their collateral relatives, latent in the generation

[1] Cf. Korschelt and Heider, pp. 151, 152.

[2] For further evidence in support of this theory, see Boveri, 'Über die Konstitution der chromatischen Kernsubstanz,' pp. 18–20 (*Verhandl. der Deutschen Zoolog. Gesellsch.*, Würzburg, 1903, pp. 10–33). Cf. also O. Hertwig, *Allgem. Biologie*, 1906, pp. 199–201).

next in succession, reappear suddenly in the grandchildren. Boveri's microscopical observations, to which we have referred, may be taken as corroborating the theory that the chromatin elements of the nucleus are the means of transmitting hereditary properties. There is, therefore, actual evidence in support of the theory held by Roux, Strasburger, O. and R. Hertwig, Weismann, Kölliker, Boveri, &c., that in the chromosomes of

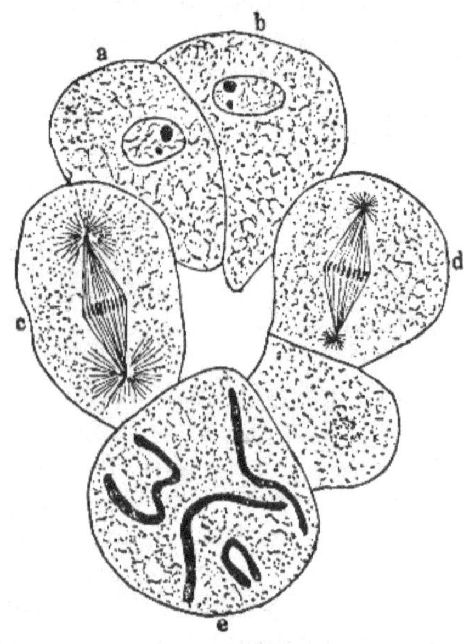

FIG. 23.—Transverse section of the blastula stage of an embryo of *Ascaris megalocephala* var. *bivalens*.

the nucleus we may discover the real substance of heredity, which Nägeli calls *idioplasm*.

In order to illustrate the differentiation of the germ-cell area from the somatic-cell area in the case of *Ascaris megalocephala* var. *bivalens*, I give, in fig. 23, an exact microscopical reproduction of a transverse section of the embryo of this creature at the blastula stage.[1]

[1] The figure is taken from a long series of sections, stained with Heidenhain's iron-haematoxylin, showing the maturation-divisions and the processes of fertilisation and development in *Ascaris megalocephala*. The series was prepared by my colleague, K. Frank, S.J., under Heider's direction. In the original the centrosomes at the two ends of the cleavage-spindle in cells *c* and *d* can be seen more plainly than in the reproduction; they seem to be little circular formations marked off from the surrounding plasmic rays.

The two uppermost cells, *a* and *b*, are two somatic cells with resting nuclei, in each of which two dark spots, nucleoli, can be plainly seen. The two middle cells, *c* and *d*, are likewise two somatic cells, but they are still in the act of mitosis; the fine chromatin rods, still grouped about the equatorial plate in the centre of the plainly visible achromatic nuclear spindle, are actually in process of division. Also the centrosomes with their astrospheres at the two poles of the spindle are shown very beautifully. Hence this illustration serves to supplement the formal diagrammatic representation given in Chapter V of the process of indirect nuclear division (see p. 95). The lowest cell, *e*, with its four large chromatin-loops, represents, according to Boveri, one of the germ-cells in the embryo. There is a great difference between the chromosomes in it and those in the somatic cells, and the fact that the future germ-cells contain much more chromatin than the somatic cells, is an argument in favour of the theory that the chromosomes of the nucleus are the bearers of heredity. We do not yet know how the normal number of four chromosomes, which subsequently are present in the somatic cells of *Ascaris*, arises out of the numerous chromatin rods of the somatic cells *c* and *d*.

Let us now refer again to the account already given of the process of fertilisation in the *Ascaris*-type. This, and the *Echinus*-type, which differs from it by the formation of one cleavage-nucleus, both show us that, in the first place, fertilisation leads to the beginning of the embryonic development of a new individual, because it causes the cells to divide; in the second place, it restores the normal number of chromosomes for all the somatic cells of the new individual; and lastly it distributes to every cell of the embryo, as an inheritance, an equal number of chromosomes derived from each parent.

The last two facts taken in conjunction show the bearing of fertilisation upon heredity; the first shows its bearing upon germinal development.

As I shall have to discuss the theoretical value of these phenomena at the close of this chapter, it must suffice for the present thus briefly to indicate the twofold object of fertilisation.

Before passing on to other points connected with the problem of fertilisation, I must once more refer to the normal process as already described and as illustrated by Boveri's diagrams (Plate I, figs. 1–7). We may ask: 'What is it in this case that gives rise to the formation of the cleavage-spindle, and thus to the first division of the ovum, which constitutes the starting point in the development of the embryo?' The impulse proceeds from the male centrosome, which penetrates into the ovum with the middle-piece of the spermatozoon. In the course of the preceding maturation-divisions the centrosome of the egg-cell either is lost or degenerates, and consequently, in spite of possessing a great quantity of nutritive plasm, the egg-cell is incapable of further division, for, in losing its centrosome, it has lost its kinoplasm, as Strasburger calls it, the active motorplasm in the cell. It requires, therefore, a new 'organ of division' before it can proceed to embryonic development, and this organ of division is, in normal fertilisation, the centrosome of the sperm-nucleus. Its division gives rise to the two centrosomes (Plate I, figs. 2–6) which form the poles of the first cleavage-spindle (Plate I, fig. 7) and cause the chromatin loops of the united male and female pronuclei to be distributed evenly between the first two cleavage-nuclei of the fertilised ovum.

This account of the process of fertilisation was first given by Boveri in 1887;[1] according to it, the impulse giving rise to embryonic development is not supplied by the union of the two pronuclei, but is the primary object of the fertilisation caused by the introduction of the sperm-centrosome into the ovum. The union of the pronuclei is the secondary object, and produces the transmission of the qualities of both parents to the offspring, but, according to this view, it is only a result of the action of the male centrosome upon the protoplasm of the female egg-cell.

As Boveri himself is careful to state,[2] this account of the process of fertilisation is not universal in its application; it cannot be applied to *all* forms of fertilisation in animals and plants, but only to those of most multicellular animals;[3] for

[1] 'Über den Anteil des Spermatozoons an der Teilung des Eis' (*Sitzungsbericht der Gesellsch. für Morphol. u. Phys.*, Munich, III).

[2] *Das Problem der Befruchtung*, pp. 23, &c.

[3] According to Wheeler the centrosome of the ovum remains in *Myzostoma*, and forms the poles of the cleavage-spindle. Cf. Korschelt and Heider, p. 657.

hitherto no centrosome has been observed at the fertilisation of the higher kinds of plants,[1] nor at the conjugation of unicellular animals.

In natural parthenogenesis the development of the ovum takes place without fertilisation by a male germ-cell, and so no spermato-centrosome occurs, therefore it is not essential to give rise to the embryonic development of the egg. Recent experiments in artificial parthenogenesis have succeeded, by means of various mechanical, thermal, chemical or other stimuli, in causing centrosomes to form, and the subsequent cell-division to take place, in the unfertilised eggs of animals, in which, under normal circumstances, the male centrosome supplies the cell with the means of division. We must therefore be careful, even in the normal fertilisation of animal ova, not to ascribe to the spermato-centrosome too much influence in setting up embryonic development in the ovum.

We can thus appreciate the reasons which led so great an authority on the problem of fertilisation as R. Hertwig to content himself with the simple statement that 'the essential feature of fertilisation consists in the union of egg- and sperm-nuclei' (Lehrbuch der Zoologie,' p. 124 : Eng. trans. p. 149).

4. The Phenomena of Superfecundation among Animals and Double-fertilisation in Plants

Under normal conditions during the process of fertilisation only one sperm-cell penetrates an animal ovum, although there may be hundreds in its immediate neighbourhood. In many eggs this is secured by the construction of the enclosing membrane, which allows spermatozoa to enter at one point only. In the case of eggs with no such point of entrance (micropyle) the same result is attained in another way—a vitelline membrane forms immediately after the entrance of one spermatozoon, excluding all others. If the reacting power of the egg be weakened by means of strychnine, or other poison, so that it admits several spermatozoa, a normal development never results ; the numerous centrosomes carried into the egg give rise to the formation of karyokinetic figures with several poles, or of very large nuclei which divide irregularly and lead to an

[1] Cf. Chapter V, p. 99.

abnormal process of cleavage and to the speedy death of the embryo. Hence Boveri was right in stating emphatically in 1902 that the entrance of two spermatozoa ruins a perfectly normal egg. The explanation of this fact is that the introduction of several centres of division into the egg hinders its normal development.

In many animals, however, exceptional cases have been observed when several sperm-cells have entered one egg under normal conditions. (Gérard, 1901.) But, when this occurs, only one sperm-nucleus unites with the egg-nucleus, and the rest are absorbed by the egg-plasm. In 1902 Boveri [1] observed these processes in sea-urchins' eggs, fertilised with two spermatozoa, and he applied the results of his observations very ingeniously to his investigations into the nature of the nucleus and the importance of the chromosomes.

We must distinguish the above-mentioned pathological superfecundation from what is called physiological polyspermy, which recent research has proved to occur in many kinds of animals. In this case also only one sperm-nucleus unites with the egg-nucleus to form the first cleavage-spindle, but, as Rückert, Oppel, Samassa (1895), and Nicolas (1900) have observed, especially in the eggs of Selachii and reptiles, only a few of the other nuclei perish—many of them are transformed into the so-called *merocytes* or yolk-nuclei of the embryo ; not much is known with certainty about their subsequent fate, but they are supposed to be connected with the vegetative functions of the egg, and to expedite the division of the abundant vitelline substance.

Closely related to physiological polyspermy among animals is double-fertilisation, an interesting phenomenon occurring in Angiosperms among the higher plants. A good deal of light has been thrown on this subject and on its biological significance by Nawaschin (1898), Guignard (1899 and 1901), and Strasburger (1900).[2]

In this process two sperm-nuclei penetrate into the embryo-

[1] 'Über mehrpolige Mitosen als Mittel zur Analyse des Zellkerns' (*Verhandl. der physikalisch-medizinischen Gesellsch.*, Würzburg, XXXV, pp. 67–90).

[2] For a good summary of works published before 1900, and dealing with the phenomena of double-fertilisation, see G. Richen, S.J., in *Natur und Offenbarung*, 1900, pp. 561, &c. Cf. also Korschelt and Heider, *Lehrbuch der vergl. Entwicklungsgeschichte*, p. 696.

sac, one of which unites as the male pronucleus with the egg-nucleus, thus forming the cleavage-nucleus of the mother-cell of the embryo. The other amalgamates with the secondary nucleus of the embryo-sac (formed by the union of the two polar cells), or in some cases with one of the polar cells before their union, and thus produces the nucleus of the mother-cell of the endosperm, which has to supply nourishment to the embryo. It is a remarkable fact that one of the two sperm-nuclei has a *generative*, and the other a *vegetative* function to discharge.

This double fertilisation in Angiosperms is of importance in explaining some mysterious phenomena in heredity, the so-called *xenia*. J. Reinke says on this subject:[1] 'It was known from earlier observations that if ripe heads of white- or yellow-grained maize (*Zea Mays*) were dusted with pollen from the blue- or brown-seeded variety, blue or brown seeds might occur, or the yellow seeds might be speckled with blue or brown spots. Focke gave the name of *xenia* to this phenomenon. It became easy of explanation after the discovery of double-fertilisation, and de Vries and Correns have proved that when maize is dusted with the pollen of another variety, not only the embryo, but also the endosperm, shows hybrid properties.'

A remarkable contrast to normal polyspermy is displayed by the specific polyembryony of certain parasitic Hymenoptera. According to Silvestri,[2] from one single egg of *Litomastix truncatellus* are produced about a thousand sexed and some hundreds of sexless larvae. One spermatozoon suffices to bring about this extraordinary productiveness in the fertilised egg, and even the unfertilised eggs, which need no spermatozoon, show the same complicated result of their parthenogenetic development. We have here one of the strangest riddles of life, that seems to be in direct conflict with the theory of the individuality of the chromosomes, but future generations may succeed in solving it.

[1] *Einleitung in die theoretische Biologie*, p. 440.

[2] 'Un nuovo interessantissimo caso di germinogonia (*Poliembrionia specifica*), &c.' (*Rendiconti della R. Accademia dei Lincei, Classe d. scienze fisiche*, &c., XIV, 1905, pp. 534–542); 'Contribuzione alla conoscenza biologica degli Imenotteri Parassiti,' I. 'Biologia del *Litomastix truncatellus*,' Portici, 1906 (*Estr. d. Annali della R. Scuola Sup. d'Agricoltura di Portici*, VI).

K

5. Conjugation in Unicellular Organisms and its Bearing upon the Problem of Fertilisation

In order to understand the importance of the union of germ-cells in the normal processes of fertilisation in higher plants and animals, we shall do well to compare them with similar processes in the lowest forms of organic life. Let us begin with the conjugation of Infusoria.

The Ciliata have two nuclei, both containing chromatin, but one—the *macronucleus*—is larger than the other—the *micronucleus*. As Bütschli showed, only the micronucleus takes an active part in conjugation, so that it may be called the sexual nucleus. The macronucleus disappears before conjugation; its activity is limited, therefore, to the period between two acts of conjugation, when the ordinary vital functions are performed, and it may be called the assimilation nucleus, which controls the processes of feeding and movement.

The multiplication of these tiny Ciliata takes place as a rule by simple division, so that one mother-cell splits into two daughter-cells. This process begins with indirect division of the micronucleus, which forms a spindle; it is only later that the macronucleus divides directly by way of elongation and constriction, and then the cell-body divides. The micronucleus reveals its character as the real sexual nucleus even at this period, but it does so more clearly in the course of conjugation.

The power possessed by Infusoria of multiplying by division is not unlimited; the periods of division are interrupted from time to time by the sexual phenomena of conjugation, by means of which, as in the processes of fertilisation amongst higher animals, a reorganisation of the living substance is effected.[1] According to R. Hertwig and Maupas the conjugation of Ciliata (e.g. in *Paramaecium*) takes place in the following way.[2]

[1] See R. Hertwig, 'Über Wesen und Bedeutung der Befruchtung' (*Sitzungsberichte der Akad. der Wissenschaften*, Munich, XXXII, 1902, pp. 57–73).

[2] R. Hertwig, 'Über Befruchtung und Konjugation' (*Verhandl. der Deutschen Zoolog. Gesellsch.*, 1892, pp. 95–112); also *Lehrbuch der Zoologie*, 1905, p. 182 (Eng. trans. p. 206); Maupas, 'Recherches expérimentales sur la multiplication des Infusoires ciliés' (*Archives de Zoologie expérimentale et générale*, VI, pp. 165–277); see also Weismann, *Evolution Theory*, Vol. I, pp. 319, &c., with fig. 85 (Eng. trans.); O. Hertwig, *Allgemeine Biologie*, 1906, pp. 294, &c.

Two individuals take up a position close to one another, and whilst the macronucleus breaks up, the micronucleus becomes active. In each individual it becomes spindle-shaped, and then divides twice in succession, so that each creature now possesses four spindles. Of these, three, which are called secondary spindles, gradually degenerate, thus recalling the polar bodies expelled from the egg-cell. The chief or primary spindle remains, and again divides into two, one of which, called the female spindle, remains in each individual, whilst the other, called the male spindle, passes into the adjacent animal, and fuses with its female spindle. The result of their union is to produce in each animal a single new division-spindle, which gives rise to the copulation-nucleus, and its development completes the conjugation. The copulation-nucleus corresponds to the cleavage-nucleus of the fertilised ovum; when it divides it forms the macronucleus and the micronucleus of the regenerated individual, which now moves away from its neighbour.

We cannot here discuss in detail all the differences between the phenomena of conjugation and the processes of fertilisation. A comparison of them shows them to be identical in principle. The conjugation of two Infusorians aims at forming in both individuals a new copulation-nucleus, which is made up of the chromosomes of the micronucleus of each in equal proportions. It is, therefore, a cross fertilisation, agreeing in its essential points with the processes of fertilisation in multicellular animals and plants, and showing that the laws, to which we have seen that they conform, are applicable also to unicellular organisms. It may be mentioned further that in many Cryptogams (*Fucus*, *Peronospora*) the phenomena of conjugation still more closely resemble the processes of fertilisation in higher organisms.

In the phenomena of conjugation in unicellular animals and plants, we can actually trace the stages of a gradual approximation to the differentiation of male and female germ-cells, which finds its complete expression in the fertilisation of higher animals and plants.[1] The two specimens of

[1] On this subject see also Y. Delage, 'Les théories de la fécondation,' 1902, pp. 122, 123 (*Verhandl. des V. internat. Zoologenkongresses*, pp. 121-140). The bearing of this series upon the history of evolution is, however, as Delage

Paramaecium, whose conjugation has just been described, were exactly similar to one another both before and after their conjugation. The same may be said of the daughter-individuals, formed by the subsequent division of the regenerated specimens; each can in its turn enter into conjugation with another of its own kind. There is, therefore, no difference at all in the sex of the cells uniting in conjugation. We might say the same of the *Noctiluca miliaris*, that causes the phosphorescence of the sea,[1] and of many other Infusorians. If, on the other hand, we consider another Infusorian, *Vorticella nebulifera*, we find a remarkable difference in the conjugating individuals; one of them, the macrogonidium, is larger and represents the egg-cell, whilst the other, the microgonidium, is smaller, and represents the spermatozoon. In one plant, *Fucus platycarpus*, belonging to a low Order, we find a still more complete sexual differentiation of the conjugating individuals; round one relatively enormous spherical egg-cell swarm numerous diminutive spermatozoa destined to fertilise it.

We can trace a distinct advance towards sexual differentiation in the case of those Infusorians, which form what are called colonies, consisting of groups of cells, each being a separate individual.[2]

In *Pandorina morum* sixteen unicellular individuals unite to form a colony, and, at the time of sexual reproduction, change into the same number of daughter-colonies of cells, all resembling one another, which swarm out of the mother-colony and unite permanently in twos by way of conjugation. In another flagellate Infusorian, *Eudorina elegans*, which also forms colonies, at the time of conjugation two kinds of daughter-colonies are produced, distinguishable as male and female.

rightly remarks, quite hypothetical. Cf. also O. Hertwig, pp. 304, &c., where he discusses the original forms of sexual generation and the first appearance of differences of sex.

[1] In *Noctiluca* fertilisation follows conjugation after a long or short interval, and multiplication takes place by a budding process and the formation of swarm spores. Cf. O. Hertwig, *Allgemeine Biologie*, p. 304.

[2] The family of Volvocineae, to which belong the species mentioned here, *Pandorina*, *Eudorina*, and *Volvox*, enjoys the honour of being claimed both by zoologists and by botanists. The former class it among Flagellata, the latter among the Green Algae. Cf. R. Hertwig, *Lehrbuch der Zoologie*, 1905, p. 171 (Eng. trans. pp. 201, 202); Strasburger, *Lehrbuch der Botanik*, 1904, p. 283 (Eng. trans. 1908, p. 355).

The female colonies have sixteen fairly large daughter-cells of the ordinary shape, and the male thirty-two much smaller cells resembling spermatozoa and called *zoosperms*, whilst the female daughter-cells are called *oosperms*. The zoosperms swarm out and penetrate the female daughter-colonies, fusing in conjugation with their oosperms.

A still higher degree of differentiation in the cells and in the processes of conjugation is shown by the well-known *Volvox globator*, which is also one of the Infusorians forming colonies. In one of these colonies there are three kinds of cells, viz. somatic or body-cells, which remain unchanged, and sexual cells of two distinct shapes, which are formed only at the time of conjugation. Some of them then become large and round, and correspond to egg-cells, whilst others change into thread-like zoosperms, which develop in clusters, then swarm out and fertilise the oosperms. As real somatic cells are developed in the *Volvox* colonies, and serve to unite the whole colony, and perform the functions of nourishment and growth for it as a whole, we are justified in regarding *Volvox* as a single animal or plant consisting of body-cells and of two kinds of germ-cells.[1] This is the link connecting the unicellular animals (Protozoa), and the phenomena of their conjugation, with the multicellular animals (Metazoa) and the processes of their fertilisation.

In other Protozoa, especially in the malaria parasites belonging to the Haemosporidae, the development of which has been studied chiefly by Grassi,[2] and in the allied order Coccididae, examined at an earlier date by Schaudinn,[3] there are two periods of reproduction, recurring alternately. The one is sexless, but in the other there are present individuals differentiated in sex, the so-called *macrogametes* and *microgametes*, which unite in conjugation.[4]

[1] Cf. also M. Hartmann, 'Die Fortpflanzungsweise der Organismen, Neubenennung und Einteilung derselben, erläutert an Protozoen, Volvocineen und Dicyemiden' (*Biolog. Zentralblatt*, 1904, No. 1, pp. 18-32; No. 2, pp. 33-61), p. 38.

[2] Cf. Grassi's address at the Fifth International Zoological Congress, 'Das Malariaproblem vom Zoolog. Standpunkt' (*Verhandl. des Kongresses*, 1902, pp. 99-114).

[3] 'Über den Generationswechsel der Coccidien und die neuere Malariaforschung' (*Sitzungsberichte der Gesellsch. naturforsch. Freunde*, Berlin, 1899, No. 7, pp. 159-78); 'Der Generationswechsel der Koccidien und Hämosporidien. Zusammenfassende Übersicht' (*Zoolog. Zentralblatt*, V, 1899, No. 22, pp. 765-783).

[4] Cf. M. Hartmann, as above.

In the *Proceedings of the German Zoological Society* for 1905 (*Verhandl. der Deutschen Zoolog. Gesellschaft*, pp. 16–35 with Plate I) Fritz Schaudinn has given an excellent summary of recent investigations on fertilisation among Protozoa. It appears from this work that 'all forms of coitus known to occur among other living organisms, both animals and plants, take place also among Protozoa.' A tabular survey of these various forms of coitus is given on pp. 20 and 21, for which Schaudinn is indebted to Stempell.[1]

I cannot do more than outline briefly the processes of conjugation in the lower organisms. They show an extraordinary variety of forms, and are in many respects instructive for us when we study the problem of fertilisation. They teach us that the difference in the germ-cells of higher animals and plants is designed to render possible the union of two cells belonging to different individuals, in order to effect the reorganisation of the vital process of the species. The greater the difference in form between the two cells, the more perfect is their physiological division of labour; inasmuch as the egg-cell stores up nourishment for the development of the embryo, and the sperm-cell acquires the greatest possible agility, in order to be able to enter the egg-cell and stimulate it to development; and the more perfectly these ends are to be attained, the higher is the degree of differentiation in the problem of fertilisation.

The feature common to all phenomena of fertilisation is the union of the nuclei of the two cells, whether the latter resemble one another or not. We cannot call the part taken by the centrosomes essential in the conjugation of lower animals, for in most of them, e.g. in Ciliata, the centrosomes seem to be absent or only temporary. Genuine centrosomes have certainly been observed in *Noctiluca*, one of the Cystoflagellata, and also in *Actinosphaerium*, one of the Rhizopods.[2]

We may perhaps conclude that among higher animals also the centrosome of the spermatozoon, as an 'organ of division,' is only an instrument for effecting the nuclear union of the two

[1] *Vegetatives Leben und Geschlechtsakt.* (Reprinted from articles contributed by the Naturwissenschaftl. Verein in Griefswald, XXXVI, 1904.)

[2] Cf. Wilson, *The Cell*, pp. 227, 228; R. Hertwig, *Lehrbuch der Zoologie*, 1905, p. 160 (Eng. trans. p. 190).

germ-cells, and that therefore the union of the male and female pronuclei is the essential point in fertilisation, and through the chromosomes of these pronuclei the properties of both parents are transmitted to their offspring.

6. Natural Parthenogenesis

In considering the phenomena of fertilisation and conjugation (§§ 3–5) we have found each process to culminate invariably in the union of the nuclei of two cells. We have now to refer to those cases in which there is no union of nuclei, and yet at least the beginning of embryonic development occurs in the egg or in the ovary. A study of these cases will help us to arrive at a general understanding of the problem of fertilisation and heredity.

In the first place we must deal with *natural parthenogenesis*,[1] which occurs in many animals and plants, and consists of the development of the egg under natural conditions without fertilisation by a sperm-cell. We are here concerned chiefly with animal eggs, and we find parthenogenetic development occurring especially in Rotatoria among worms, in Phyllopoda and Ostracoda among Crustacea, and in many butterflies, (parthenogenesis among Psychidae was discovered by Karl von Siebold in 1848), in plant-lice and their relations, in the praying-crickets, gall-flies, saw-flies, wasps, bees and ants. In considering the morphological processes during the maturation and development of the eggs of these creatures,[2] we have to distinguish two cases, viz. that in which parthenogenesis takes place regularly in definite generations, and is *obligatory*; and that in which it occurs only incidentally, and is *facultative*. It is true that in the first case parthenogenetic

[1] Under this heading we may include *paedogenesis*, in which parthenogenetic reproduction is accomplished by animals still in the larval stage of growth, for instance in Aphididae and in certain Diptera (*Miastor* and *Chironomus*); The remarkable phenomena of *polyembryony* is connected with paedogenesis; in the above-mentioned Diptera, in one larva numerous small larvae develop, and in the same way in some parasitic wasps (in *Encyrtus* and *Polygnotus* according to Maréchal, and in *Litomastix* according to Silvestri) a number of embryos develop in one egg (see p. 129). Polyembryony may therefore be described as a form of parthenogenesis in the egg; especially when it occurs in unfertilised eggs, as it does in *Litomastix*.

[2] Cf. Korschelt and Heider, *Lehrbuch der vergl. Entwicklungsgesch.*, pp. 613–622.

development is generally, at least in animals, not the exclusive mode of reproduction, as, at definite intervals in the series of parthenogenetic generations, they are replaced by sexual generation (Heterogony). The tendency to parthenogenesis is, however, stronger than when it is merely facultative.

A study of the maturation of the eggs of animals with obligatory parthenogenesis shows that as a rule only *one* polar body is formed,[1] but that *two* are present in those generations of the same species in which the eggs require fertilisation by means of spermatozoa. In these generations the normal number of chromosomes in the cleavage-spindle of the egg has subsequently to be restored by means of the male pronucleus, therefore the number is first halved by a reduction within the egg, and made up again in the course of fertilisation.[2]

We can, therefore, understand why no reduction takes place, and why consequently no second polar body is formed, in eggs that develop parthenogenetically without fertilisation. That this is the case has been proved from the examination of parthenogenetic eggs of various classes of animals by Blochmann, Weismann, Ishikawa, Erlanger, Lauterborn, Lenssen, and Woltereck. Their observations, and especially those made by Woltereck on the eggs of a Crustacean (*Cypris*), render it probable that no reduction in the number of chromosomes takes place during the maturation of these eggs, but that the original number (twelve in *Cypris*) remains unaltered until the cleavage-spindle is formed, which constitutes the first stage in embryonic development.

According to O. Hertwig, A. Brauer, Viguier, &c., there are other cases in which a second polar body is formed also in eggs that develop parthenogenetically, but its formation is incomplete, as the second polar body remains within the egg and is eventually reunited with the nucleus. Boveri thought that the second polar body might replace the spermatozoon, and that in this case parthenogenesis was the result of self-fertilisation on the part of the egg. He assumed that the polar body served, instead of the sperm-nucleus, to restore the normal number of chromosomes for the first cleavage-spindle

[1] This has been confirmed recently by J. P. Stschelkanovzew's examination of plant-lice. Cf. his article 'Über die Eireifung bei viviparen Aphiden' (*Biolog. Zentralblatt*, 1904, No. 3, pp. 104–112).

[2] Cf. pp. 110 and 120.

of the egg. According to Brauer there are two types of development in the parthenogenetic eggs of *Artemia*. In one type the second polar body is formed, but united again with the egg-nucleus; in the other type it is not formed at all. Brauer states that in the first type the number of chromosomes in the cleavage-spindle of the egg is 168 (the normal number for this species); in the second type it is only 84 (half the normal number), but, as no division takes place, the chromosomes have a double value.

The maturation of the egg of the parasitic *Litomastix truncatellus*, as observed by Silvestri in 1905, is remarkably interesting (see p. 129, note 2). The process is the same in the parthenogenetic as in the fertilised egg. In both cases two polar bodies (*globuli polari*) are formed, and remain in the front part of the egg. The first polar body divides, but its two halves unite with one another and with the second polar body to form a nucleus, which Silvestri calls from its origin a polar nucleus.

In many insects however, especially in such as have only facultative parthenogenesis, e.g. in *Liparis*, *Bombyx* and *Leucoma* among butterflies, and in the honey-bees and many ants (*Lasius*) among Hymenoptera, the maturation-divisions of the parthenogenetic egg result in the complete formation and separation of two polar bodies. At Weismann's suggestion, Dr. Petrunkewitsch[1] made a very careful examination of the unfertilised eggs of the bee, from which drones are hatched, and showed this quite conclusively. We can, perhaps, account for the formation of two polar bodies by assuming that, in these insects, fertilisation is the normal condition; where it does not take place, the egg makes the same preparations for it as when it does. But in many gall-flies (*Rhodites*) parthenogenesis is obligatory, and yet two fully developed polar bodies are formed, neither of which reunites with the egg. It is a remarkable fact that when two such polar bodies have been cast out of the egg, and when the accompanying karyokineses have reduced the number of chromosomes in the egg by a half, the normal number nevertheless recurs in the cleavage-spindle.

[1] 'Die Richtungskörper und ihr Schicksal im befruchteten und unbefruchteten Bienenei' (*Zoolog. Jahrbücher, Abteilung für Anatomie u. Ontogenie*, XIV, 1901).

Petrunkewitsch observed this phenomenon in the eggs of the bee, but was unable to account for it.

Morphological processes closely resembling parthenogenesis in the animal kingdom occur in the parthenogenetical development of many plants. In 1900 Juel observed[1] that in *Antennaria alpina* the egg developing parthenogenetically in the embryo-sac shows no reduction in the number of chromosomes; and in 1901 the same thing was observed by Murbeck[2] in the varieties of *Alchimilla* that develop parthenogenetically. In 1905 E. Strasburger devoted much attention to the study of the propagation of the *Eualchimillae,* and came to the conclusion that in the seeds of these plants the development of the mother-cell of the embryo-sac and of the embryo takes place without fertilisation. In this case there is no reduction in the original number of chromosomes, which remains constant as in the somatic cells of the plant. Strasburger prefers to call this process *apogamy*, or sexless propagation, rather than parthenogenesis, or unisexual propagation, because it takes place on vegetative and not sexual lines. Winkler, on the other hand, retains the name 'parthenogenesis,' but calls it in this case *somatic*, as opposed to the true *generative* parthenogenesis.[3]

In one of the Algae (*Ectocarpus siliculosus*) an extraordinary phenomenon has been observed. Not only the female germ-cell can develop parthenogenetically under certain circumstances, but the male cell may also do so;[4] in this case, however, the difference in size between the two is not great, and the male plant, corresponding with the smaller size of the zoosperm, tends to be poorly developed. This is the only case, occurring under natural conditions, of male parthenogenesis or *arrhenogenesis*.

There are many obscure points in natural parthenogenesis, as we have shown. Only one fact can be stated with certainty,

[1] 'Vergleichende Untersuchungen über typische und parthenogenetische Fortpflanzung bei der Gattung *Antennaria*' (*Svenska Vetenskaps Akad. Handl.* XXXIII, 1900, n. 5).

[2] 'Parthenogenetische Embryobildung in der Gattung *Alchimilla*' (*Lunds Univers. Arsskrift*, XXXVI, 1901, n. 2).

[3] Cf. Strasburger, 'Die Apogamie der Eualchimillen und allgemeine Gesichtspunkte, die sich daraus ergeben' (*Jahrbücher für wissenschaftl. Botanik*, LXI, 1905, pp. 88–164). Cf. also the article in the *Naturwissenschaftliche Rundschau*, XX, 1905, No. 27, pp. 342–344.

[4] Weismann, *Lectures on the Evolution Theory*, I, 334.

viz. that, in a good many kinds of animals and plants, the egg-nucleus alone is able to begin the embryonic development of the egg. Therefore the union of the nuclei of two cells, the male and female germ-cells, is not absolutely and universally essential to the beginning of embryonic development, even in those organisms which possess both kinds of germ-cells. If nevertheless, in normal fertilisation, the union of the nuclei of the two germ-cells is regularly the culminating point of the whole process, its object is not merely to stimulate the ovum to embryonic development, but, over and above this, its object is chiefly to secure the benefits of *amphimixis*, i.e. the transmission of the combined properties of both parents to their offspring, and this is brought about by the union of the paternal and maternal nuclear elements in the cleavage-spindle of the fertilised ovum. We must not, however, undervalue the first object in normal fertilisation. It cannot be denied that a renewal of the capability of development of the species, a 'reorganisation of the living substance,' is connected with the union of the germ-cells, and therefore it is still very doubtful whether an unlimited propagation by parthenogenesis would be possible, at least in the animal kingdom.[1]

7. Artificial Parthenogenesis

Let us now turn to experiments in artificial parthenogenesis.[2] Tichomirow discovered in 1886 [3] that in the eggs of the silk-moth, which otherwise require fertilisation, parthenogenesis may be produced by rubbing them between cloths. The same result was obtained by Tichomirow both in 1886 and in 1902 by dipping the eggs into concentrated sulphuric and

[1] In one Crustacean (*Cypris reptans*) Weismann states that he observed uninterrupted parthenogenesis (*Zoolog. Anzeiger*, XXVIII, 1904, p. 39). It seems to be possible also in some grasshoppers which are all females (de Sinéty, *Recherches sur les phasmes*, 1901, pp. 13, &c.). H. Schmitz has made the same observation in *Dixippus morosus*, a tropical praying-cricket ('*Dixippus morosus*,' in *Natur und Offenbarung*, 1906, Part 7, pp. 385–407, 402, &c.).

[2] A summary of these experiments is given by Korschelt and Heider, *Lehrbuch der vergl. Entwicklungsgesch.*, pp. 623, &c., 663 &c.; by Boveri, *Das Problem der Befruchtung*, pp. 39, &c.; by Y. Delage, *Les théories de la fécondation*, pp. 135, &c.; by Kathariner, *Das Problem der Befruchtung*, pp. 518, &c.; by O. Hertwig, *Allgemeine Biologie*, pp. 326, &c.

[3] 'Die künstliche Parthenogenese bei Insekten' (*Archiv f. Anatomie u. Physiologie*, Supplement, 1886).

muriatic acid. In 1887 O. and R. Hertwig [1] found that unfertilised eggs of sea-urchins could develop under the influence of external stimulus, and R. Hertwig continued these experiments in 1888 and 1896, and in a work [2] published in the latter year he describes the processes of division in the egg-nucleus which result from placing the unfertilised egg of a sea-urchin in a weak solution of strychnine. Many experiments in artificial parthenogenesis have been made in the last few years by American naturalists, Th. Morgan, Jacques Loeb, E. B. Wilson, and A. B. Mathews, and also by scientists of other nationalities, such as Y. Delage, Giard, Bataillon, Henneguy, Herbst, Winkler, Prowazek, Kostanecki, Boveri, Wasilieff, Schücking, Petrunkewitsch, &c.[3]

Unfertilised eggs of very various animals (Echinoderms, Medusae, Molluscs, Annelids, insects and fishes) were chosen and exposed to chemical, physical, and mechanical stimuli of many different kinds. Solutions of various poisons, narcotics and salts, such as strychnine, nicotine, hyoscyamine, ether. alcohol, chloroform, calcium chloride, magnesium chloride, diphtheria serum, a solution of cane sugar, urea, and sperm extract—all proved efficacious in setting up the processes of development; and similar results were obtained by concentrating the sea-water containing the eggs, by dipping them in warm sea-water and by applying galvanic currents and mechanical vibration. Jacques Loeb's experiments were the most successful. He was able to cause the unfertilised eggs of all kinds of Echinoderms and Annelids to form larvae, and by subjecting those of sea-urchins to the action of chloride of magnesium for two or three hours he made them develop as

[1] 'Über den Befruchtungs- und Teilungsvorgang des tierischen Eis unter dem Einflusse äusserer Agentien' (*Jenaische Zeitschr. für Naturwissenschaft*, XX).

[2] *Über die Entwicklung des unbefruchteten Seeigeleis, Festschrift für C. Gegenbaur*, Leipzig, 1896.

[3] Korschelt and Heider give a list of books dealing with the subject, pp. 733, &c. They do not, however, mention those of the last four authors named above: Boveri, 'Zellenstudien,' 1902, Part 4, p. 9; Wasilieff, 'Über künstliche Parthenogenesis des Seeigeleis' (*Biolog. Zentralblatt*, XXII, 1902, No. 24, pp. 758–772); A. Schücking, 'Zur Physiologie der Befruchtung, Parthenogenese und Entwicklung' (*Archiv für die ges. Physiologie*, XCVII, 1903); A. Petrunkewitsch, 'Künstliche Parthenogenese' (*Zoolog. Jahrbücher*, Supplem. VII, 1904, 'Festschrift für Weismann,' pp. 77–138). Cf. also a review of the last-mentioned article in the *Naturwissenschaftliche Rundschau*, 1904, No. 35, pp. 444, &c.

far as the blastula stage, and finally even as far as that of the Pluteus larva. These larvae remained alive for as long as ten days, but were unable to form any calcareous skeleton, although they developed this power when carbonate of calcium was added to the sea-water. Loeb succeeded also in inducing the eggs of an Annelid (*Chaetopterus*) actually to reach the stage of forming the trochophore larva.[1] These careful and ingenious experiments seem to have resulted in the discovery of a magic wand, capable of awakening the life dormant in the unfertilised animal ovum; and apparently they afford a brilliant confirmation of Aristotle's opinion, for he believed the ovum to contain the essentials of each animal species, and the spermatozoon merely to have the effect of stimulating the ovum to develop. Before we assent to these conclusions, we must examine the results of these experiments somewhat more closely.

The forms resulting from artificially produced parthenogenesis differ in many respects from the normal, as Kathariner already partially pointed out in 'Natur und Offenbarung,' 1903, p. 518. Their cleavage-globules have less power of resistance; they show a tendency to fall to pieces, and dwarf larvae develop from the fragments, or else several cleavage-globules unite and give rise to gigantic and monstrous embryos. In the sea-urchin larvae produced parthenogenetically, irregularities in the formation of the skeleton are apt to occur, and all these artificially developed forms seem to lack some directive power, which is supplied by normal fertilisation and results in development on definite lines. The Pluteus and trochophore larvae, produced by Loeb's experiments, are the highest achievements of artificial parthenogenesis, but it is doubtful whether they were really capable of continued existence and of developing from the stage of larvae to that of adults, for hitherto no one has succeeded in breeding even the natural larvae of these species in a laboratory. In any

[1] Loeb, 'On the nature of the process of fertilisation and the artificial production of normal larvae (Plutei) from the unfertilised eggs of the sea-urchin' (*American Journal of Physiology*, III, 1899); 'On the artificial production of normal larvae from the unfertilised eggs of the sea-urchin' (1900); 'Further experiments on artificial parthenogenesis' (IV, 1900); 'Experiments on artificial parthenogenesis in Annelids (*Chaetopterus*) and the nature of the process of fertilisation' (IV, 1901).

case, although in a few successful instances larvae were actually formed, there were many less successful, or even quite unsuccessful, attempts at artificial parthenogenesis, in which the cleavage process, artificially induced, ceased even earlier.

An attempt to account for these variations has been made by Boveri ('Das Problem der Befruchtung,' pp. 39, &c.) in his criticism of Morgan and Wilson's experiments. He points out that, when an ovum is fertilised, only one radiation sphere is formed at the head of the spermatozoon that has entered. The division of this radiation sphere gives rise to the two astrospheres which are the poles of the first nuclear spindle of the ovum. (Cf. p. 122 and Plate I, figs. 1–7.) According to the observations of the two American writers, however, artificial parthenogenesis of the same eggs, under the influence of Loeb's reagents, results in the formation of a fluctuating, but often considerable, number of radiation-spheres, each of which has a newly formed centrosome as its centre. Boveri believes that regular cleavage of the ovum can occur only in the exceptional case that only two really active radiation-spheres develop and take up their positions at opposite poles of the egg-nucleus; under all other circumstances the numerous division-centres, having no orderly arrangement, act as they do in pathological polyspermy, and give rise to an irregular mass of cells, which speedily dies. Therefore Boveri still regards the introduction of the spermatozoön into the ovum as supplying the directive quality, which, in normal fertilisation, secures the formation of a regular cleavage-spindle with two poles. It is comparatively of less importance whether the spermatozoön actually brings its own centrosome with it into the ovum, or whether, through the chemical and physical action of the sperm-nucleus, the egg-protoplasm becomes capable of forming a new centrosome for itself, which then takes up a position in front of the sperm-nucleus, and by dividing forms the poles of the cleavage-spindle. The attempts at artificial parthenogenesis seem to me to support the theory of the new formation of centrosomes in the ovum; and these experiments have in some degree caused me to modify the account that I previously gave (see p. 125) of the significance of the normal process of fertilisation, in giving which I was guided by Boveri's diagrams. (Plate I, figs. 1–7.)

Another remark must be made on the subject. Morgan,[1] and still more emphatically Wilson,[2] declare that they have observed the new formation of centrosomes as centres of the radiation spheres in sea-urchins' eggs parthenogenetically developed by the application of chloride of magnesium, and Wilson describes the new formation of centrosomes in non-nucleated fragments of an egg.[3] Wasilieff, on the other hand,[4] in his corresponding experiments with strychnine, nicotine and hyoscyamine, found that the first divisions took place without the formation of centrosomes, which, if they appeared at all, did so only in the later stages of cleavage, and were then formed of the nuclear substance of the cells. The occurrence of true centrosomes in non-nucleated fragments of an egg is questioned also by Petrunkewitsch.[5]

Should the observations of Wasilieff and Petrunkewitsch find confirmation, we shall have greater reason for regarding the centrosomes, not as a permanent formation, but as only a temporary biomechanical means of assisting the process of cell-division. (Cf. Chapter V, pp. 98, &c.) If this be so, we must consider the appearance of a centrosome beside the sperm-nucleus in normal fertilisation of the animal ovum, not as the *cause* of cell-division, but as a *consequence* of the beginning of the process. We shall then have to agree with Oskar Hertwig's older theory of nuclear fertilisation, and say, that in normal fertilisation also, the entrance of the sperm-nucleus into the ovum and its union with the female pronucleus constitute the real elements of fertilisation.

The question of chromatin-reduction is another point connected with artificial parthenogenesis on which opinions are divided. The eggs used in the experiments, to which I have referred, were such as had undergone their maturation-divisions, and so we must assume that the nucleus of each contained only half the number of chromosomes peculiar to the

[1] 'The production of artificial astrospheres' (*Archiv für Entwicklungs-mechanik*, III, 1896).

[2] 'Experimental studies in cytology,' I. 'Artificial parthenogenesis in sea-urchin eggs' (*Ibid.* XII, 1901).

[3] 'Cytasteren und Centrosomen bei künstlicher Parthenogenese' (*Zoolog. Anzieger*, XXVI, 1904, pp. 8–12).

[4] 'Über künstliche Parthenogenesis des Seeigeleis' (*Biolog. Zentralblatt*, 1902, pp. 758–772).

[5] 'Künstliche Parthenogenese' (*Zoolog. Jahrbücher*, Supplem., VII, 1904, 77–138).

species. Wilson states expressly that he found eighteen and not thirty-six chromosomes in the cleavage-cells of the sea-urchins' eggs undergoing parthenogenetic development. Y. Delage, however, says that in his experiments on the same eggs he found the normal number of chromosomes to be restored. Boveri argues[1] that eighteen, which Delage apparently took to be the normal number, is really the reduced number for that species, for his own observations and those of R. Hertwig both show thirty-six to be the normal. We must probably assume that, when eggs develop by artificial parthenogenesis, half the normal number of chromosomes suffices for the cleavage-nucleus of the developing ovum. Petrunkewitsch has gone so far as to state (1904) one essential difference between artificial and natural parthenogenesis to be that, in the former, the reduced number of chromosomes invariably remains.

We may now turn to the more general conclusions formed by various students, as resulting from the experiments in artificial parthenogenesis.

Loeb thinks he is justified by his experiments (see p. 140) in concluding that the ova of many, and perhaps of all, animals have a certain tendency to develop parthenogenetically, but as a rule this development is so slow that the ovum perishes before it attains to any advanced stage of cleavage. Artificial stimuli, such as salt solutions, &c., by hastening the development, enable the ovum to attain its end parthenogenetically. Korschelt and Heider, on the contrary,[2] and R. Hertwig[3] incline to the far more moderate opinion that the chemical and physical stimuli are able to set up in the mature, but still unfertilised, ovum that reciprocal action of the parts (and especially of the cytoplasm and nucleus) which is indispensable to embryonic development, and which under normal conditions results only from fertilisation. Boveri[4] thinks that the

[1] 'Über mehrpolige Mitosen als Mittel zur Analyse des Zellkerns' (*Verhandl. der physikal.-mediz. Gesellsch.*, Würzburg, XXXV, 1902).

[2] *Lehrbuch der vergl. Entwicklungsgesch.*, p. 624; cf. also *ibid.* pp. 65–67.

[3] 'Über Korrelation von Zell- und Kerngrösse und ihre Bedeutung für die geschlechtliche Differenzierung und die Teilung der Zelle' (*Biolog. Zentralblatt*, 1903, Nos. 2 and 3); also 'Über das Wechselverhältnis von Kern und Protoplasma,' Munich, 1903. (Reprinted from the *Münchener Medizin. Wochenschrift*, I.)

[4] *Das Problem der Befruchtung*, pp. 22–23, 39, &c.

phenomena observed in artificial parthenogenesis afford a confirmation of his theory of fertilisation, according to which the mature ovum resembles a complete piece of mechanism, still at rest, and needing only to be wound up, in order to begin to work. The key to it is, in normal fertilisation, the centrosome of the spermatozoon; but in artificial parthenogenesis it consists of some chemical or physical agents; which affect the egg-plasm in a way similar to the action of the centrosome under ordinary circumstances. As early as 1886 Tichomirow put forward the theory that the egg-cell responded to all exterior action—no matter of what kind—invariably in the same way, peculiar to itself, viz. by development; just as the cells of the optic nerves react invariably through their susceptibility to light, and the cells of the muscular fibres contract under external stimulus. This idea was borrowed from Johannes Müller's law of specific energies of the senses. The same view has been recently formulated by Y. Delage in the following terms:[1] 'The mature but still unfertilised ovum is in a condition of unstable equilibrium; any stimulus, destroying the equilibrium, gives rise to development.'

Loeb goes perhaps rather too far when he says that all animal ova have an original tendency to parthenogenetic development, for the results of experiments show that artificial parthenogenesis seldom attains the normal end, and that the cleavage stages thus produced cease, as a rule, without developing to a larva. Moreover, at the present time most zoologists agree in regarding natural parthenogenesis, where it actually occurs among animals, not as the original mode of development, but as a later simplification of the original mode; they believe propagation by fertilisation to be the normal condition.

We must therefore not overestimate the capacity of many eggs to develop without fertilisation under artificial stimulus; but, on the other hand, we must not underestimate it, for, taken in conjunction with natural parthenogenesis, it proves plainly enough that under certain circumstances one nucleus alone, viz. the egg-nucleus, suffices to begin embryonic development. The chief object, then, of the union of two different nuclei in normal fertilisation is not merely to stimulate the ovum

[1] *Les théories de la fécondation*, p. 138.

L

to develop, but rather to secure the benefits of amphimixis, i.e of transmitting to the offspring the properties of both parents, and this is effected by the union, in the cleavage-spindle of the ovum, of the nuclear elements of the male and female pronuclei. I shall recur to this subject at the end of the present chapter.

The other object of fertilisation, viz. to stimulate the ovum to develop, can be attained by very various means without fertilisation, as the experiments in artificial parthenogenesis prove.[1]

As Delage puts it the mature egg really gives us the impression of being in a state of unstable equilibrium ; anything that disturbs that equilibrium suffices to cause the egg to develop.

Closely akin to this idea is the further suggestion that, in normal fertilisation also, there may be certain chemico-physical processes which result in the development of the egg. Thus we arrive at the physical and chemical theories of fertilisation, which have been propounded in the last few years. They are still hardly ripe for discussion, and consist chiefly of rather vague speculations, so we may limit ourselves to what is absolutely necessary in dealing with them.

The question to be answered is : 'In normal fertilisation. what does the spermatozoon bring into the ovum to render it capable of development ?' The answer given by Boveri's morphological theory is : 'In its centrosome the spermatozoon imports a new division-centre into the ovum.' The physical and chemical theories, however, reply: 'The spermatozoon produces in the ovum certain physical and chemical changes which result in the process of division.'

The two classes of theories are not necessarily antagonistic, but are complementary. Carnoy and Bütschli had already suggested that the centrosomes stimulate the cell to divide, by exerting some chemical influence on the protoplasm, and Boveri himself expressed an idea, which Wilson subsequently elaborated, that possibly some chemical substance,

[1] I must remind the reader here, as I did on p. 141, that this object is only imperfectly attained by artificial parthenogenesis. We must therefore assume that a particular kind of 'reorganisation of the vital substance' is connected with natural fertilisation, and especially with the union of the nuclei.

stimulating the ovum to develop, is brought into it by the spermatozoon.[1]

The morphological theory only shows itself really antagonistic to the chemico-physical theory, when there is a choice between one or other of them, as being exclusively valid; J. Loeb seems inclined to adopt the chemico-physical theory, in spite of the obscurity in which it is still involved. There is a wide divergency of opinions regarding the nature of the chemical and physical processes which underlie fertilisation. Loeb, the chief champion of the new theory, originally thought that electrolysis might account for it, and that new metallic ions were brought by the spermatozoon into the ovum. Subsequently, he came to the conclusion that some alteration in the osmotic conditions of the ovum was effected by the action of the spermatozoon. In 1900, Wilson suggested that the middle-piece of the spermatozoon, which contains its centrosome, might be the bearer of a specific chemical substance stimulating the ovum to development, quite apart from the sperm-nucleus. Finally, Yves Delage has set up a still simpler hypothesis of chemical and physical fertilisation;[2] he thinks that the ovum becomes capable of fertilisation in consequence of the breaking up of the nuclear membrane during the maturation-divisions, and the distribution of the nuclear fluid to the protoplasm of the ovum. The head of the spermatozoon penetrating the ovum becomes the male pronucleus, and grows by taking up water from the egg-plasm, thus depriving it of some of its fluid. In this dehydration of the ovum by the sperm-nucleus Delage thinks he has discovered the chemico-physical cause of the beginning of the dividing process in the ovum. He does not, however, exclude the specific action of salts, metallic ions, &c., which may be contained in the sperm-nucleus.

Loeb considered that his experiments in artificial parthenogenesis had transferred the problem of fertilisation from the domain of morphology into that of chemico-physical science.

[1] Cf. Korschelt and Heider, *Lehrbuch der vergl. Entwicklungsgesch.*, pp. 663, &c., and Wilson, *The Cell*, pp. 354, &c. The phenomena of natural parthenogenesis are against these theories, as in that case there is no spermatozoon, nor any special chemical stimulus, present.

[2] On this theory and those akin to it, see Y. Delage, *Les théories de la fécondation*, pp. 135, &c.

L 2

Y. Delage seems to share this opinion, and Max Verworn has long desired to replace the morphological theory of fertilisation by a physiological one. Quite recently B. Hatschek too has brought forward a new 'Hypothesis of organic inheritance' ('Hypothese der organischen Vererbung,' Leipzig, 1905) based upon a physiological and chemical foundation. I agree with Boveri [1] in thinking that this bold speculation is still far from having a basis of ascertained scientific facts. After showing what a vast number of distinct morphological problems are involved in the fertilisation, cleavage, and embryonic development of the ovum, with regard to the physical and chemical factors of which we still know nothing at all, Boveri continues: 'As we have said, a transference of the problem of fertilisation into the domain of physico-chemical science would involve the assumption that the process of cell-division has been traced back to physical and chemical factors. How far we really are from having accomplished this is known to everyone who has studied the question; and it is scarcely possible at the present time to speculate as to how deeply we may eventually penetrate into the mystery.'

The problem of fertilisation and heredity is, at any rate, no merely morphological problem; on the contrary, its physiological aspect is the chief point, as enabling us to understand the morphological processes, but the morphological and physiological aspects must be taken in conjunction to support and complete one another.

My opinion regarding the importance of artificial parthenogenesis as bearing upon the problem of fertilisation may be expressed thus: These ingenious experiments have proved that the problem of fertilisation must not be studied, as has been done hitherto, exclusively by morphological methods, but also by completely new methods belonging to physico-chemical science. Only in this way shall we arrive at a satisfactory insight into the true nature of the fertilisation and cleavage of the ovum, and the embryonic development that follows these processes. For the present we have no certain information, but only bold speculations, as to the physico-chemical factors engaged in these processes, nor do we know how they co-operate mechanically and teleologically to accom-

[1] *Das Problem der Befruchtung*, p. 47.

plish them. The naturalists who fancy that they have at last succeeded in giving a purely physico-chemical explanation to life itself are doomed to disappointment.

8. The Fertilisation of Non-nucleated Egg-fragments (Merogony)

There still remains one class of phenomena which we must consider shortly, because it throws some light on the problem of fertilisation, namely, artificial fertilisation of non-nucleated fragments of ovum, called by Y. Delage *merogony*.[1] The first experiments, now become classical, in this subject were begun in 1887 by O. and R. Hertwig, and continued by Boveri in 1889 and 1895. They resulted in the surprising discovery that non-nucleated fragments of sea-urchins' ova could, if fertilised, develop to the larval stage. Others have subsequently confirmed this discovery by means of experiments on the eggs of sea-urchins and other animals; we may mention particularly Morgan (1895), Ziegler (1896 and 1898), and Delage (1898, 1899 and 1901). Similar experiments were made by Rawitz in 1901 on the immature eggs of holothurians, the nucleus of which is unimportant and in course of time disappears, so that they may be regarded as non-nucleate. In 1896–8 H. E. Ziegler made some experiments at artificially constricting sea-urchins' eggs, and his results are not without bearing on the question.[2]

Experiments in merogony have been made with plants also, and I may draw attention particularly to those undertaken in 1901 by Hans Winkler on the eggs of a seaweed (*Cystosira*).[3] Let us now examine some of the above-mentioned experiments more closely.

Oskar and Richard Hertwig succeeded in proving[4] conclusively that, if sea-urchins' eggs are broken by shaking, fragments containing no nucleus may be fertilised by the

[1] Cf. Korschelt and Heider, *Lehrbuch der vergl. Entwicklungsgesch.*, pp. 149–151 and 625–626.

[2] A full list of the works to which I have referred will be found in Korschelt and Heider, pp. 733–750.

[3] H. Winkler, 'Über Merogonie und Befruchtung' (*Jahrbücher für wissenschaftl. Botanik*, XXXVI, pp. 753–775).

[4] 'Über Befruchtungs- und Teilungsvorgänge des tierischen Eis' (*Jenaische Zeitschrift für Naturwissenschaft*, XX, 1887).

entrance of a spermatozoon. In Boveri's experiments, such non-nucleated fragments of the ovum, after fertilisation with one spermatozoon of the same species, developed and actually reached the stage of the Pluteus larva, thus showing such ova to be capable of normal development. In this way Boveri obtained dwarf larvae, larger or smaller according to the size of the fragment of ovum.

The experiments made by Hertwig and Boveri prove that under certain conditions the sperm-nucleus alone, without the egg-nucleus, suffices for the fertilisation and development of the animal ovum, in exactly the same way as, in parthenogenesis, the egg-nucleus suffices without the sperm-nucleus. Giard called this phenomenon simply *male parthenogenesis*, as in this case the sperm-nucleus receives from the non-nucleate egg-cell the cytoplasm necessary for its development. The same idea had been expressed somewhat differently by M. Verworn in 1891, and in 1901 Rawitz invented the name *ephebogenesis* to designate the process.

The embryos of the non-nucleated eggs of sea-urchins only reached the stage of cleavage into sixteen cells in Morgan's experiments,[1] but he was able to demonstrate that their nuclei contained only half the normal number of chromosomes (eleven instead of twenty-two) belonging to that species. It is easy to see why this is so, for the sperm-nuclei, which fertilised the fragments of egg, contained the reduced number. This fact therefore agrees with similar phenomena observed in artificial parthenogenesis (see p. 144), and shows that sometimes half the normal number of chromosomes suffices for the embryonic development of the egg. Whether these chromosomes are paternal or maternal in origin is immaterial for the purpose of embryonic development, but not for that of heredity, as Boveri's next experiments show with a degree of probability almost amounting to certainty.[2]

He began by crossing two distinct varieties of sea-urchin, and fertilised ova of *Sphaerechinus granularis* with spermatozoa of *Echinus microtuberculatus*. The Pluteus larvae of these two species can easily be distinguished—those of *Echinus* have

[1] 'The fertilisation of non-nucleated fragments of Echinoderm eggs' (*Archiv für Entwicklungsmechanik*, II, 1895).

[2] 'Ein geschlechtlich erzeugter Organismus ohne mütterliche Eigenschaften' (*Sitzungsberichte der Gesellschaft für Morph. und Phys.*, Munich, 1889).

a much more slender shape and a different formation of the calcareous skeleton. Boveri succeeded in showing that the result of crossing these two species was to produce hybrid larvae (fig. 26) occupying a position midway between the two larvae of pure breed (figs. 24 and 25) and displaying a mixture of the peculiarities in shape of both parents.

Boveri next proceeded to fertilise ova of *Sphaerechinus*, partially broken by shaking, with spermatozoa of *Echinus*. Of the larvae produced by the fragments, some showed the hybrid type, and Boveri assumed that they developed either from uninjured ova, or from fragments containing part of the

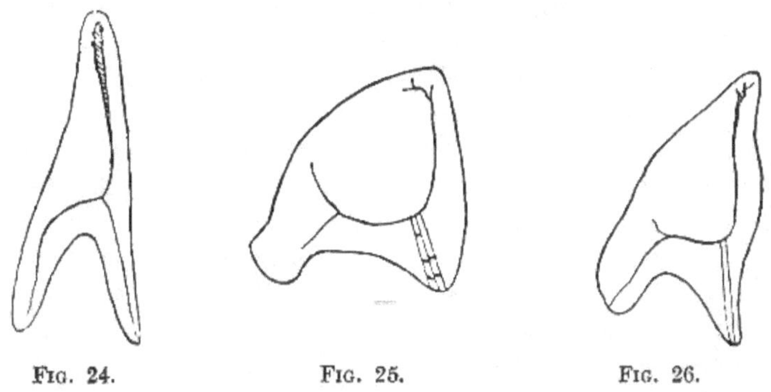

FIG. 24. FIG. 25. FIG. 26.

FIGS. 24–26.—Side view of Pluteus larvae: FIG. 24 of *Echinus*, FIG. 25 of *Sphaerechinus*, FIG. 26 of the hybrid of *Sphaerechinus* ♀ and *Echinus* ♂.

From Korschelt and Heider, according to Boveri's diagram.

egg-nucleus, into which a spermatozoon of the other species had found its way. Other larvae were particularly small, but displayed the pure *Echinus*-type (fig. 24). Boveri calls these dwarf Plutei, and believes them to have developed from non-nucleated fragments of *Sphaerechinus* ova, and therefore to represent altogether the paternal *Echinus*-type because the sperm-nucleus fertilising them belonged to this latter species. According to Boveri's view, these dwarf Plutei are organisms without any maternal characteristics, and, if this view is the true one, we have here a proof that the cell-nucleus does not merely determine the shape of the embryo, but is the real bearer of heredity, for only the cell-nucleus on the father's side, and not the egg-plasm on the mother's side, stamped upon the embryo its specific characteristics as a pure *Echinus* larva.

Boveri's explanation is rendered more probable by the fact that the dwarf Plutei of the *Echinus* type possessed an unmistakably smaller nucleus than larvae of the same size of the hybrid type. This difference in the size of the nucleus is quite intelligible if we may assume that in the former case the cell-nucleus was formed from only one pronucleus, and so contained only half the amount of chromatin, whereas in the second case the nucleus was produced by the union of two pronuclei.

Boveri assumed, therefore, that the dwarf larvae of pure *Echinus*-type, produced in the course of his experiments at cross-breeding, really developed from non-nucleated fragments of ovum, and consequently were organisms devoid of any maternal characteristics. Morgan, Seeliger, Driesch and Delage have brought forward a number of objections to this theory, but Boveri adheres to it even in his most recent works. Yves Delage himself classes Boveri's experiments among what he calls *expériences décisives*, as furnishing evidence of great weight in the solution of the scientific problem under discussion. In fact, when we take into consideration, firstly, that non-nucleated fragments of sea-urchins' eggs can be fertilised, and, secondly, that Boveri fertilised them with spermatozoa of another species, we can hardly avoid agreeing with him in regarding the dwarf larvae, which display only paternal characteristics, as the products of non-nucleated ova, deriving from the father's side alone their nucleus, and consequently the substance which bears heredity.

Quite recently E. Godlewski has made experiments [1] at cross-breeding between sea-urchins (Echinidae) and sea-lilies (Crinoidea), by fertilising the eggs of the former with spermatozoa of the latter, and the results which he obtained are exactly the reverse of Boveri's. All the hybrid larvae displayed purely maternal, and no paternal characteristics, even in cases where a non-nucleated fragment of *Echinus* ovum was fertilised with an *Antedon* spermatozoon. Godlewski argues from this that Boveri's whole morphological theory of heredity is untenable,

[1] 'Untersuchungen über die Bastardierung der Echiniden- und Crinoidenfamilie' (*Archiv für Entwicklungsmechanik*, XX, 1906, pp. 579, &c.).

and Verworn's physiological theory must be substituted for it; and that not the chromosomes, but the egg-plasm, constitute the vehicle of transmission. Such far-reaching conclusions need confirmation from other experiments before they can be accepted, for the bulk of the evidence afforded by biology seems to show decisively that the chromosomes of the nucleus are the material bearers of heredity. The physiological fact that the chromosomes of the nucleus and the protoplasm of the egg act reciprocally upon one another, is of course included as a fully recognised condition.

The successful attempts made by Boveri and others to fertilise non-nucleated fragments of ova show that under certain circumstances the sperm-nucleus alone suffices for the development of the egg. But this statement does not imply that it is the sperm-nucleus itself which gives rise to the process of development: it may be the sperm-centrosome which penetrates into the egg with the nucleus. An observation made by Boveri in 1887 on the subject of 'partial fertilisation' suggests that this may be the case. He saw a spermatozoon enter a sea-urchin's egg. Its nucleus remained near the periphery of the egg, whilst the centrosome alone with its amphiaster approached the egg-nucleus, and thereupon the first cleavage-division of the egg-nucleus took place. The sperm-nucleus united with one of the daughter-nuclei of the egg. Wilson, too, considers that [1] this observation affords a beautiful illustration of Boveri's theory that it is the centrosome of the sperm-nucleus which supplies the normal stimulus to division on the part of the ovum.

Further light is thrown upon this interesting question by the experiments made by H. E. Ziegler in 1896 and 1898 on sea-urchins' eggs, which he fertilised artificially and then divided by constricting them with fine threads.[2]

In every case in which the egg was so divided that the sperm-nucleus, with its centrosome and centrosphere, was contained in one-half of the egg, and the egg-nucleus in the other half, the former half divided in the ordinary manner, whereas an aster was formed near the egg-nucleus, and all

[1] *The Cell*, p. 190.

[2] Cf. H. E. Ziegler, 'Experimentelle Studien über die Zellteilung: I. Die Zerschnürung der Seeigeleier: II. Furchung ohne Chromosomen' (*Archiv für Entwicklungsmechanik*, VI, 1898, Part 2, pp. 249–293).

preparations were made for cell-division, which, however, never actually took place. These experiments seem to show again that, in normal fertilisation, it is the sperm-centrosome that renders the egg-nucleus capable of active division. In some experiments made in 1897 and 1901, Boveri broke up some sea-urchins' eggs after fertilisation, and found asters, leading in some cases to cell-divisions, also in fragments containing only egg-nucleus, and no particle of the sperm-nucleus or its centrosome. Wilson, Winkler, and others are inclined to explain this last phenomenon by assuming that, as soon as the spermatozoon enters the egg, its centrosome sets up a kind of fermentation [1] in the whole egg-plasm, so that even the parts remote from the centrosome are stimulated to division. This explanation would bring us back to the chemical side of the problem of fertilisation, and, as was said on p. 148, we cannot do more at present than advance some vague speculations on the subject.

The experiments in merogony suggest this question: Is it possible that the sperm-centrosome alone, without the sperm-nucleus and without the egg-nucleus, has the power of setting up a regular process of division and so of beginning embryonic development in the fragments of ovum?

In 1897 Boveri made an experiment [2] and fertilised some non-nucleated fragments of *Echinus* eggs with spermatozoa of another species (*Strongylocentrotus*). It happened that the whole nuclear substance of both nuclei passed into one half of the egg, and the centrosome alone into the other. The former half divided in the regular way, but in the other a series of divisions took place in the centrosomes and attraction spheres, but no cell-division occurred. This observation led Boveri to conclude that, at any rate for sea-urchins, at least *one* nucleus is indispensable for cell-division. H. E. Ziegler, however, believes that he succeeded in 1898 in effecting a 'cleavage without chromosomes.' In an egg of *Echinus microtuberculatus*, fertilised with spermatozoa of the same species, at the first division the entire nuclear substance of both the sexual nuclei passed into one of the cells formed by division,

[1] Cf. Korschelt and Heider, *Lehrbuch der vergl. Entwicklungsgesch.*, pp. 663–665.

[2] 'Zur Physiologie der Kern- und Zellteilung' (*Sitzungsberichte d. physik.-mediz. Gesellsch.*, Würzburg).

whilst a centrosome with its centrosphere was left in the other. The cell containing the nuclei divided with perfect regularity, but also in the non-nucleated cell a series of cleavages took place in the cell-body; they were, however, incomplete and irregular. It is unfortunate that in this interesting experiment Ziegler did not use nuclear stains, but only treatment with acetates, to prove that there was really no chromatin present in the division-cell that apparently contained no chromosomes. This flaw has left the matter still doubtful.

My own opinion is that, in these instances of merogony also, the centrosome is a biomechanical instrument for assisting nuclear division, but is not an independent division-organ of the cell. It is true that the experiments described above confirm Boveri's opinion (cf. p. 126), that in the case of most animal ova the centrosome of the spermatozoon gives the immediate impulse to cell-division in the normal course of fertilisation, but it is not absolutely indispensable to the beginning of the process of embryonic development. This is proved by the phenomena of natural and artificial parthenogenesis (see pp. 135 and 139), where no male centrosome can possibly be present. Moreover, many circumstances to which I have referred (see p. 143) suggest the idea that centrosomes are not permanent organs in the cell, but are formed afresh in the egg-plasm as occasion requires.

9. General Review of the Subject of Fertilisation and Conclusions

(See Plate II)

We have now completed our examination of the relations in which cell-division stands to the problems of fertilisation and heredity. The facts to be taken into account are so numerous and of so many kinds, and the interpretations put upon them are so varied, that it is naturally no easy task to draw from them any clear and definite conclusions. We might almost say that we cannot see the wood because of the trees in it! And yet the wood is one whole, composed of the trees which various naturalists have laboriously planted and cultivated. And there are some paths through it, though they are still footways and not carriage drives, for the wood is still wild, and not a park.

Let us try now to follow these paths by surveying the facts once more and seeing in what respects they conform to general laws. We must be on our guard against adopting the methods of those theorists who simply cast aside and reject all that does not coincide with their subjective ideas.

Both the male and the female germ-cells prepare for their union in the process of fertilisation by a double maturation-division. These preparatory divisions cause a reduction in the number of chromosomes (if it has not taken place before), so that the cells contain only half the normal number contained in the somatic cells of the same species. The act of fertilisation restores the number to the normal, as the chromosomes of the male and female pronuclei meet in the cleavage-spindle of the ovum, and by splitting lengthwise furnish an equal number of paternal and maternal chromosomes for the daughter-nuclei of the ovum in process of cleavage.

Normal fertilisation has as its essential feature the union of two germ-cells, one being male and the other female, and the union is more especially a union of their nuclei. E. B. Wilson sums up this result on p. 230 of his excellent work 'The Cell' (1902) in the following words: 'We thus find the essential fact of fertilisation and sexual reproduction to be a union of equivalent nuclei; and to this all other processes are tributary.' This is true both of the animal and of the vegetable kingdom. With reference to the latter Wilson says (p. 216): 'The essential fact is everywhere a union of two germ-nuclei—a process agreeing fundamentally with that observed in animals.' Richard Hertwig uses similar language in the seventh edition of his 'Lehrbuch der Zoologie,' 1905, p. 124 (Eng. trans. p. 149): 'Since not until this point (i.e. the union of the sexual nuclei) is fertilisation complete, we arrive at the fundamentally important proposition that the essential feature of fertilisation consists in the union of egg- and sperm- nuclei.'

Nuclear union can, however, assume various forms. It may—as in the *Echinus*-type—lead to the formation of a resting cleavage-nucleus, in which the chromosomes of the two pronuclei are already brought into contact, or—as in the *Ascaris*-type—the two pronuclei may remain apart, so that their chromosomes are not grouped in a common division-figure until the cleavage-spindle is formed. Moreover, the part played by the centrosomes in the processes of fertilisation varies. In normal fertilisation of the animal ovum, the male centrosome acts as an organ of division, inducing the formation of the cleavage-spindle, but no centrosomes have been observed in the fertilisation of the higher orders of plants. In many animal ova (e.g. *Myzostoma*, according to Wheeler) the place of the sperm-centrosome as an organ of division seems to be taken by the oocentrosome. Finally, in physiological super-fecundation among animals and in double-fertilisation among angiosperms in the vegetable kingdom, not only *one* sperm-nucleus, but two or more, are concerned in the process of fertilisation, although only one, which unites with the egg-nucleus, has a distinctly generative function, the duty assigned to the others being rather of a vegetative character, and consisting of the formation of nourishment for the embryo.

So far we have spoken only of the usual case in which two nuclei, the male and female pronuclei, carry on the fertilising process in the ovum. Analogous to this are the phenomena of conjugation which occur in unicellular organisms. But in artificial fertilisation of non-nucleated fragments of ovum, only the sperm-nucleus is concerned, and in animal eggs this is generally accompanied by a sperm-centrosome.

In parthenogenetic development of the ovum there is no fertilisation by a spermatozoon, but the process is carried on by the egg-nucleus alone; in natural parthenogenesis it is assisted by the oocentrosome, and in artificial parthenogenesis by centrosomes newly formed in the egg-plasm by means of exterior agents. Wasilieff considers that even these centrosomes may be absent. The centrosome alone, without either egg- or sperm-nucleus, seems to be able to begin the process of cell-division, but not to succeed in carrying it through.

Let us now sum up the results of these observations and experiments.[1] It seems safe to infer from them that the nucleus of the germ-cell is of primary importance in normal fertilisation, as well as in artificial fertilisation of non-nucleated fragments of ova, and in parthenogenesis. Opinions are still divided

[1] Cf. on this subject Korschelt and Heider, *Lehrbuch der vergl. Entwicklungsgesch.*, pp. 697–706 ('Wesen und Bedeutung der Befruchtung').

as to the centrosomes, whether they originate in the achromatic nuclear substance or in the egg-plasm; they seem to me to be of secondary importance as merely assisting the division of nucleus and cell. That the egg-plasm is an essential factor in the processes of fertilisation and development is proved beyond question, especially by the phenomena of artificial parthenogenesis, which gave rise to the modern chemico-physical theories regarding fertilisation.

What, then, is the answer to the question raised by Aristotle, and repeated from age to age in the course of the dispute between ovulists and animalculists: 'Is the essence of the animal and vegetable species contained in the egg-cell or in the sperm-cell?'[1]

Many facts, and especially the phenomena of natural and artificial parthenogenesis (see pp. 135, 139, &c.) seem to support Aristotle's opinion that the material required to form the new individual is all contained in the egg-cell, and that the sperm-cell only supplies the stimulus causing this material to develop.[2]

In a modernised form this opinion is revived in Boveri's theory of fertilisation, which regards the ovum as a complete piece of clockwork, lacking only the mainspring, or rather, lacking only the key to wind up the mainspring. This key is the sperm-centrosome, that sets in action the dividing process of the ovum. The same fundamental idea is present in Delage's chemico-physical theory of fertilisation, according to which the mature but unfertilised egg-cell is in a state of unstable equilibrium; this equilibrium is disturbed by a reduction in the water of the egg-plasm, caused by the entrance of the spermatozoon, and the ovum is thus stimulated to independent development.

Other considerations of no less weight are directly opposed to the theory that the egg-cell alone contains the essence of the new individual. The experiments in artificial impregnation of non-nucleated fragments of ova, and especially the results obtained by Boveri (see p. 150), show that the sperm-nucleus alone—just as in parthenogenesis, the egg-nucleus alone—in conjunction with the egg-plasm, is able to cause the egg to

[1] Cf. p. 104. See also O. Hertwig, *Allgemeine Biologie*, p. 352.
[2] Cf. Aristotle, '*De animalium generatione*,' cap. 2 (*Aristotelis opera omnia*, ed. Didot, III, 320). Aristotle does not of course speak of the elements of reproduction as cellular, for he had no knowledge of cells at all.

produce a new individual of the species concerned. The embryonic material for the formation of the new individual must therefore be contained as completely in the nuclear substance of the spermatozoon as in that of the ovum. The nuclei of both germ-cells have then with regard to the development of the embryo the same *prospective potency*, as Driesch calls it.

Let us now turn to a third series of phenomena, viz. to the facts of normal fertilisation, which are of great importance for our purpose (cf. pp. 119, &c.). We have already seen that the process of fertilisation culminates in the union of the nuclei of the two germ-cells, and that the originally insignificant sperm-nucleus finally becomes exactly equivalent to the egg-nucleus in size and shape and in number of chromosomes. The sperm-nucleus supplies for the development of the new individual exactly the same amount of chromatin nuclear substance as the egg-nucleus ; the nuclear substance of the cleavage-spindle of the embryo represents the sum of that contained in the nuclei of the ovum and spermatozoon ; the essence of the animal or vegetable species, as propagated by normal fertilisation is therefore first contained in the sum of the chromatin nuclear substance of the male and female pronuclei, and the essence of normal fertilisation culminates therefore in the union of the chromosomes of both to form one new cell-nucleus.[1] In his 'Allgemeine Biologie' (1906), p. 301, O. Hertwig states his conclusions in the following words : 'The nuclear substances supplied in exactly equal quantities by two distinct individuals are the especially active materials, the union of which is the chief object of the act of fertilisation ; they are the real materials of fertilisation.' [2]

We cannot avoid asking further questions : What is the object of this union of paternal and maternal nuclear elements in the normal course of fertilisation ? Is it not altogether superfluous, if what is essential to the species is contained

[1] This explains why the number of chromosomes in the somatic cells of animals and plants that are propagated by sexual reproduction is always *even*. Cf. Chapter V, p. 92.

[2] A detailed proof that the nucleus is the physical basis of inheritance is given by Hertwig in the thirteenth chapter of the same work, pp. 354–363. His proof depends upon four kinds of evidence, which agree on the whole with those that I have adduced.

completely either in the egg-cell alone, or in the sperm-cell alone? What is the use of the vast difference between the ovum and the spermatozoon in the higher organisms, where the former is very large and richly provided with nutritive plasm, and the latter is diminutive and consists of a thread of cytoplasm by way of tail, a head containing a nucleus, and a middle-piece? What is the use of the complicated maturation-divisions, by which the egg-cell and the sperm-cell prepare for their future union in the process of fertilisation? What is the good of all these complicated arrangements? Are they not perfectly aimless?

It is true that the two kinds of germ-cells are in their origin essentially alike. This is proved, on the one hand, by the embryonic development of the individual, in which the egg- and sperm-cells proceed from similar germinal *Anlagen* and are differentiated only at subsequent stages of development. It is proved, on the other hand, by the phenomena of conjugation in unicellular organisms, in which *isogamy*, i.e. the union of two similar germ-cells, represents theoretically and practically the first condition of propagation by germ-cells (see p. 132 on *Pandorina morum*). Nevertheless, the differentiation of the male and female germ-cells in the organic kingdom, and their union in the normal course of fertilisation, are processes of the highest teleological significance.

In order to see this more clearly, we must follow Boveri, Weismann, R. Hertwig, Y. Delage, &c., in recognising a two-fold object in fertilisation. (1) It aims at inciting to develop a new individual, and (2) it aims at transmitting the combined properties of both parents to this individual.

1. The first of these two aims can be realised both among animals and plants by other means besides fertilisation. We have seen this in the case of Infusorians and other unicellular organisms, which increase either by simply splitting in two, or by breaking up a colony of cells into single cells. Although with them from time to time periods of conjugation have to intervene between the periods of non-sexual or agamous multiplication, R. Hertwig's recent observations seem to show that there is no direct connexion between conjugation and the multiplication of individuals by division. In multicellular animals and plants, which are propagated by gemmation, we noticed that

the new individuals come into existence independently of any process of fertilisation. This is seen still more plainly in the case of plants that can be propagated indefinitely by means of cuttings and tubers, without weakening their growth, such as the grape-vine and the potato. The absolutely sexless propagation of *Laminaria* and other plants bears witness to the same fact, and natural parthenogenesis in animals and plants shows that the development of a new individual from an egg is not necessarily connected with its fertilisation.

In spite of all this, however, it cannot be denied that where the normal process of fertilisation is the rule, it is of great, even of essential, importance in realising the first of the two aims of fertilisation, viz. in stimulating the formation of a new individual.

According to Bütschli, the organic substance requires a periodical rejuvenescence of its vital powers. The capacity for growth and multiplication of cells is gradually weakened and exhausted as life goes on, and eventually death from senile decay must follow. But in order that the species may not perish with the individual, it is necessary that certain cells, viz. the germ-cells, of one individual should unite with those of another in the process of fertilisation, that thereby their vital force may be regenerated and renewed. There is certainly much truth in this theory, although it has been vigorously contested by Weismann in his 'Lectures on the Evolution Theory' (vol. i. pp. 325–328, English translation). His germ-plasm theory leads Weismann to regard the germ-cells as 'potentially immortal,' and so he thinks there can be, in connexion with them, no suggestion of senile decay calling for rejuvenescence. But even Weismann does not venture to deny that a strengthening of the metabolism or constitution of the germ-cells is connected with fertilisation, and this differs very little from an actual rejuvenescence of their vital force. This is the reason why R. Hertwig[1] has recently adopted

[1] 'Über Wesen und Bedeutung der Befruchtung' (*Sitzungsberichte der Akad. der Wissenschaften*, Munich, XXXII, 1902, pp. 57–73); 'Über Korrelation von Zell- und Kerngrösse und ihre Bedeutung für die geschlechtliche Differenzierung und die Teilung der Zelle' (*Biolog. Zentralblatt*, 1903, Nos. 2 and 3); 'Über das Wechselverhältnis von Kern und Protoplasma,' Munich, 1903 (reprinted from the *Münchener Medizin. Wochenschrift*, I); 'Über das Problem der sexuellen Differenzierung' (*Verhandl. der Deutschen Zoolog. Gesellsch.*, 1905, pp. 186–214).

M

Bütschli's theory under a somewhat modified form, and with fresh evidence in support of it. Hertwig sees in the conjugation processes of unicellular organisms, and in the phenomena of fertilisation in multicellular, an important reorganisation of their organic substance, and he lays particular stress upon the restoration, by these means, of that relation between nucleus and cytoplasm in the cell which is best adapted for carrying on the vital functions.

Fr. Schaudinn's opinion approximates closely to Hertwig's.[1] He thinks that the object of fertilisation is to restore the proper equilibrium between the vegetative and animal properties of the organism; he regards the egg-cell as the principal bearer of the vegetative, and the sperm-cell as that of the animal properties, because in the former the cytoplasm, and in the latter the nucleus, predominates.

Whilst R. Hertwig and Fr. Schaudinn in their theories emphasise particularly the physiological interaction of the various constituents of the cell, A. Bühler[2] has based a rejuvenescence theory of his own upon the chemical nature of the metabolism in the living cells. He sums it up in the following words: 'I have therefore arrived at the conclusion that, through the act of fertilisation, something is again imparted to the new organism which the old organism gradually lost in life and through the processes of life, until its eventual death; this something being a molecular constitution of its parts, rendering them capable of metabolism, and so fit to underlie all the vital processes.'

From what has been already said on the subject it is quite clear that there are very various views regarding the rejuvenescence of the capacity of the germ-cells to develop, especially in normal fertilisation. Let us therefore return to the consideration of some of these theories.

According to Boveri and Strasburger, the centrosome of the spermatozoon supplies the egg-cell with fresh kinoplasm, whilst the trophoplasm of the egg-cell assists the sperm-nucleus to develop. According to Y. Delage, the sperm-

[1] 'Neue Forschungen über die Befruchtung bei Protozoen' (*Verhandl. der Deutschen Zoolog. Gesellsch.*, 1905, pp. 16–35, and especially p. 33).

[2] 'Alter und Tod; eine Theorie der Befruchtung' (*Biolog. Zentralblatt*, XXIV, 1904, Nos. 2, 3 and 4).

nucleus renews the developing capacity of the egg-cell, by taking away water from the egg-plasm, whilst the sperm-nucleus grows into the male pronucleus precisely by absorbing this water.

We must not overlook the fact that a rejuvenescence of the developing capacity is probably connected with the maturation-divisions of the germ-cells, but it presupposes the reunion of the reduced and consequently rejuvenated nuclear substance of both cells in the process of fertilisation, for the formation of a new and particularly vigorous nucleus. If this union is not effected, both cells generally perish, and no further development results.

This seems to point to the fact that the nuclear union of the two germ-cells in fertilisation must have some other, higher purpose than the mere renewal of vital capacity in the single germ-cells, for the differentiation of the germ-cells into egg- and sperm-cells, and the physiological division of labour connected with this differentiation, and the maturation-divisions of both germ-cells all result in this—the egg-cell alone and the sperm-cell alone are made incapable of further independent development; the new life of the embryo has to proceed from their union. The object of this union is the second of the objects of fertilisation that we have already mentioned, viz. the transmission to the offspring of the combined properties of both parents.[1]

2. This aim can in fact be attained only by fertilisation, in the case of the higher organisms, or by the corresponding processes of conjugation, in that of the lower organisms. In agamous propagation the properties of one individual only can be transmitted to its offspring, and the same is true of unisexual propagation. The egg-cell that develops parthenogenetically can transmit only maternal qualities to the new creature, and in the same way, if we accept Boveri's observations on this subject as evidence, when non-nucleated egg-fragments are fertilised, only the paternal sperm-nucleus is the bearer of heredity. But in normal fertilisation, on the contrary, both parents' properties, united or blended,

[1] I need hardly point out that heredity is not in itself part of fertilisation, for this is plain from the cases of non-sexual or unisexual propagation. Cf. Reinke, *Einleitung in die theoretische Biologie*, pp. 413, 414.

are transmitted to their offspring. This end is served by all the morphological and physiological processes in the germ-cells that prepare them for fertilisation, or take place during it.

All modern cytologists are agreed in regarding the reduction in the number of chromosomes in the mature germ-cells, the restoration of the original number by the union of the chromosomes in the male and female pronuclei, and the even distribution of the paternal and maternal chromosomes at the cleavage of the fertilised ovum, as constituting a process of great regulative importance, to which we must ascribe an eminently final teleological significance, as do E. B. Wilson[1] and J. Reinke.[2] Let us begin by forming a clear idea of the process of reduction by which the number of chromosomes in the germ-cells is reduced to half.[3]

The reason for this is given by Weismann[4] and by Oskar Hertwig[5] and others; it is a process to prevent a summation of the hereditary substances. Let n represent the number of chromosomes constantly present in the somatic cells of any definite species of animal or plant; if no reduction took place before fertilisation, the fertilised ovum and the somatic cells developed from it in the next generation would each contain $2n$ chromosomes, and the number would go on increasing for ever in geometrical progression. As the chromosomes of each species have a definite maximum size, fluctuating it is true within certain limits, it follows that in course of time either all the somatic cells would consist exclusively of chromosomes, or the size of the cells, and consequently of the body of the individual, would attain such huge dimensions, that there would be no room for them in the world. Both conclusions are obviously absurd and quite chimerical. In the first case we should have creatures more preposterously constructed than the fabulous Hydra, which consisted entirely of heads. In the second case we should have giants whose heads would touch the moon. Therefore, some kind of regular

[1] *The Cell*, 1902, chapter v, pp. 233–288.
[2] *Einleitung in die theoretische Biologie*, p. 442.
[3] See pp. 110, &c.; cf. also Korschelt and Heider, *Lehrbuch der vergl. Entwicklungsgesch.*, pp. 606–713 ('Wesen und Bedeutung der Chromatinreduktion').
[4] *Lectures on the Evolution Theory*, I, pp. 303, &c., Eng. trans.
[5] *Die Zelle und die Gewebe*, I, Jena, 1892; II, Jena, 1898, chapter 9.

reduction in the number of chromosomes in the germ-cells may be described as absolutely necessary.

But it would be quite possible for the numerical reduction, accompanied by a corresponding quantitative diminution in the amount of chromatin, to be effected in some other way than that in which it actually occurs in the reduction processes preparatory to fertilisation. It might take place after fertilisation by means of some regulative process, causing some of the chromosomes to dissolve and be incorporated with the protoplasm of the cell. This consideration has led Weismann, O. Hertwig and others to conjecture that the processes of reduction aim at the elimination of important factors in organisation; Weismann goes so far as to think that by the reduction of chromosomes definite 'ancestral plasms' are eliminated from the parental germ-cells. In other words, according to these authors, whose views are now almost universally accepted, *numerical* and *quantitative* reduction of the chromatin is connected with a *qualitative* reduction.[1] There is, however, great diversity of opinion as to the way in which this is effected and its real significance. There are even a few naturalists who, like Yves Delage,[2] absolutely question the expediency and the necessity of any such qualitative reduction of the chromatin.

In spite of all these and many other difficulties and objections, we cannot avoid regarding as of great teleological importance the fact that, before normal fertilisation, the number of chromosomes in the germ-cells is regularly reduced to half, and then is brought up again to the normal by means of fertilisation. The maturation processes in the germ-cells take place unmistakably in view of subsequent fertilisation. Independently of it, they would be perfectly aimless, if not actually harmful, because they render the egg-cell incapable of further division, and so condemn it to death, if no fertilisation follows. This is still more true of the spermatozoon, which in its whole structure is simply designed to be able to fertilise an egg-cell. There must be some deeply significant purpose hidden under these phenomena, and it is this: The union of the nuclear substances of the egg- and sperm-cell

[1] Cf. Korschelt and Heider, pp. 149, 712, &c.
[2] *Les théories de la fécondation*, p. 131.

renders possible the transmission to the offspring of the properties of both parents. The transmission of the combined properties is effected in a very sure and simple way by the reduction in the number of chromosomes in the two pronuclei, by the union of the pronuclei in the process of fertilisation, and by the regular distribution of equal numbers of paternal and maternal chromosomes to the daughter-nuclei of the dividing egg-cell.

As a matter of fact, both among animals and plants, the force of heredity is as strong on the father's as on the mother's side, although the sperm-cell often contains only one-thousandth or one-hundred-thousandth part of the living protoplasm contained by the egg-cell.[1] This can be explained only by assuming the nuclei of the two germ-cells, and especially the chromosomes in the nuclei, to be the chief material bearers of hereditary properties. Oskar Hertwig in 1898[2] pronounced this to be his opinion, but as far back as 1884 he and Strasburger declared the nuclear substance to be what Nägeli called *Idioplasm*. Boveri, too, says very aptly on this subject:[3] 'However widely the male and female germ-cells may differ, they resemble one another in one point, viz. their nuclear substance. The full-grown sperm-nucleus is indistinguishable from the egg-nucleus, the paternal and maternal nuclear elements are absolutely alike in size, shape, and number. All imaginable care is shown in effecting their distribution in equal proportions to the daughter-cells, and, as we may assume, to all the cells of the embryo.[4] In these paternal and maternal nuclear elements must reside the directing forces, which stamp upon the new organism not only the characteristics of its species, but also the individual qualities of both parents combined. This combination of the nuclear elements as means of transmitting qualities would seem to be the object of all copulation from that of the lowest Infusorians to that of mankind.'

In our task of considering the problems of fertilisation and

[1] In one sea-urchin, *Toxopneustes*, the bulk of the spermatozoon is between $\frac{1}{400000}$ and $\frac{1}{800000}$ the volume of the ovum (Wilson, *The Cell*, p. 134).

[2] *Die Zelle und die Gewebe*, II. 232, &c.

[3] *Das Problem der Befruchtung*, p. 35. Cf. also O. Hertwig, *Allgemeine Biologie*, pp. 354, &c.

[4] This applies especially to the cells in the germinal tract of the embryo. Some deviation from this law may occur in the somatic cells, as part of the chromatin loops is thrown off. Cf. pp. 123, &c., and p. 169.

heredity, we have here arrived at one important result, which we can regard as fairly certain: Fertilisation consists essentially in the nuclear union of two germ-cells, and through this nuclear union the parental characteristics are transmitted to the offspring. The chromosomes of the cell-nucleus are shown in this process to be the immediate material bearers of heredity in the organic world.

It is important once more to draw attention to the fact that, in the nuclear union that takes place in fertilisation, the chromosomes of the two pronuclei retain their individuality. Whether—as in the *Echinus*-type [1]—the male and female pronuclei coalesce and form one common, resting cleavage-nucleus, or whether—as in the *Ascaris*-type—the two pronuclei remain distinct until they break up in forming the first cleavage-spindle of the fertilised ovum: in both cases alike the paternal and maternal chromosomes remain separate, divide themselves independently, and distribute their longitudinal segments equally between the two daughter-nuclei of the first cleavage stage of the ovum. This independent action on the part of the chromatin derived from father and mother respectively may, as V. Haecker [2] has shown, be traced in favourable cases from the nucleus of the fertilised ovum, through numerous generations of cells, to the nuclei of the germ-cells in the embryo resulting from this fertilisation. This independence of the chromatin elements is what Boveri calls 'the individuality of the chromosomes'; to some extent it stamps these morphological constituents of the cell as being the visible bearers of heredity.

Boveri's well-established hypothesis of the individuality of the chromosomes [3] has been accepted in the last few years,

[1] See pp. 120 and 156 for the difference between these two types of fertilisation.

[2] 'Über die Autonomie der väterlichen und mütterlichen Kernsubstanz vom Ei bis zu den Fortpflanzungszellen' (*Anatomischer Anzeiger*, XX, 1902). Rabl, Boveri, and Rückert have made similar observations. Cf. Wilson, *The Cell*, p. 208; O. Hertwig, *Allgemeine Biologie*, pp. 289, &c.

[3] On the subject of Boveri's theory of the individuality of the chromosomes, see his lecture on the problem of fertilisation (*Das Problem der Befruchtung*, Jena, 1902), and also the following works by the same author: 'Über mehrpolige Mitosen als Mittel zur Analyse des Zellkerns' (*Verhandl. der physikal.-medizin. Gesellschaft.*, Würzburg, XXXV, 1902, pp. 67–90); 'Über die Konstitution der chromatischen Kernsubstanz' (*Verhandl. der Deutschen Zoolog. Gesellsch.*, 1903, pp. 10–33); 'Ergebnisse über die Konstitution der chromatischen Substanz des Zellkerns,' Jena, 1904. In the last-named work Boveri

not only by most zoologists, but also by eminent botanists, such as E. Strasburger[1] and J. Reinke;[2] others, however, such as Yves Delage, have opposed it, whilst it has been only partially adopted by E. B. Wilson ('The Cell,' pp. 294–301)[3] and Oskar Hertwig ('Allgemeine Biologie,' 1906, pp. 205–208). Should it be fully confirmed, our comprehension of the material basis of heredity would undoubtedly be facilitated. According to Boveri, the chromosomes during karyokinesis are in a state of rest, and in this condition they have clearly defined shapes and are strongly susceptible to nuclear stains, which render them visible in fixed numbers. When the fresh nuclei of the daughter-cells are formed, the chromosomes in them revert from a state of rest to one of activity, in which they control all the vital functions of the cell. Their free ends approach one another, unite and become matted together by means of amœboid processes, so that they form a coil of chromatin thread or a chromatin network. It is not until the next division of the cell that the chromosomes reappear in the same form, number, and order as before, in fact, in the same 'individuality'; they are again separate, just as the oxygen and hydrogen which make up water are given off again when the water is resolved into its chemical constituents. In the case of chemical compounds, we may assume persistence in the elements of which they are composed, and, in exactly the same way, we may assume a similar latent persistence of

formulates his theory most precisely. A good account of the development of the theory of individuality up to 1900 is given by Wilson, *The Cell*, pp. 294–301. Fresh confirmation of it, in a department where it was formerly contested, is added by J. Maréchal, 'Über die morphologische Entwicklung der Chromosomen im Keimbläschen des Selachiereis' (*Anatomischer Anzeiger*, XXV, 1904, Nos. 16 and 17, pp. 383–398) and 'Über die morphologische Entwicklung der Chromosomen im Teleostierei' (*ibid.* XXVI, 1905, No. 24, pp. 641–652). That the chromosomes are not to be regarded literally as individuals is obvious, and Boveri himself does not mean this; he considers only that they are clearly defined parts of the cell, capable of independent division and redintegration.

[1] 'Über Reduktionsteilung' (*Sitzungsber. der Berl. Akad. der Wissensch.*, XIV, 1904, pp. 587–614). Cf. p. 116.

[2] *Philosophie der Botanik*, 1905, pp. 60, 69, 70, 143.

[3] Wilson is of opinion that not the chromosomes themselves, but only their constituents, the chromomeres, remain as constant elements through all changes in the nucleus. It is, therefore, to the chromomeres that we ought to ascribe 'individuality,' rather than, as Boveri does, to the chromosomes. For our purpose it is, however, a matter of indifference whether the chromosomes or the chromomeres should eventually be proved to possess individuality.

the chromosomes as the material bearers of the laws of organic development during the whole life of the cell.

On p. 166 I quoted Boveri's statement to the effect that in the cleavage-divisions of the fertilised ovum the chromosomes of the cleavage-spindle are distributed in equal proportions to all the cells of the embryo. The exceptional cases, to which I have already referred shortly (pp. 123, &c.), confirm Boveri's opinion that the chromosomes possess a certain individual independence. At the cleavage of the ovum of *Ascaris megalocephala* var. *bivalens*, from the two-cell stage onwards, the four chromosomes of the cleavage-spindle remain only in those daughter-cells which are to supply the germ-cells of the embryo, whereas they undergo a striking modification in those daughter-cells which are to produce the somatic cells. In these the ends of the chromosomes are cast off and lost, and the remaining middle-piece breaks up into a number of little rods (see p. 124, fig. 23). In subsequent divisions, giving rise to somatic cells, the chromosomes always appear in this form and order, but in the cells of the germinal area the original number, form, and arrangement of the chromosomes are preserved, until finally, before the maturation-divisions of the germ-cells, the ordinary chromatin reduction occurs, and the number of chromosomes is reduced to half, and then is brought back to the normal by fertilisation. In biological language this morphological result is stated thus: The chromosomes of the germinal areas represent in an unbroken series the bearers of heredity for the species in question. In a series of observations made in 1901 on a water-beetle, *Dytiscus*, Giardina[1] has described processes which show a difference in the nuclei of sexual cells and somatic cells. Here, too, in the former chromatin elements remained constant, which were lost in the latter.

Some recently discovered facts pointing to a qualitative difference in the chromosomes of one and the same nucleus are very significant.[2] It is enough for the present to say that in the spermatogenesis of various insects (especially in bugs, beetles, and grasshoppers) a so-called superfluous or accessory

[1] 'Origine dell' oocite e delle cellule nutrici nel Dytiscus' (*Internat. Monatschr. für Anatomie und Physiologie*, XVIII, 1901).

[2] Boveri gives a short summary of them. 'Über die Konstitution der chromatischen Kernsubstanz,' pp. 20–26.

chromosome occurs,[1] which at the last maturation-division passes undivided into one of the two sperm-cells, whilst the other receives one chromosome less. Montgomery calls these accessory chromosomes *heterochromosomes* ; he has observed them in the spermatogenesis of spiders.[2]

Sutton noticed, in the spermatogenesis of a grasshopper—*Brachystola magna*—(1900 and 1902), that in the secondary spermatogonia (descendants of the male germ-cells) not only did the extra chromosome appear regularly for nine generations of cells, but the other chromosomes of the same cells fell into two groups of different sizes, and always occurred in pairs. Quite recently E. B. Wilson has made a very careful study of the qualitative differences of the chromosomes in the germ-cells of bugs (Hemiptera), and their biological functions.[3] He distinguishes normal chromosomes, or idiochromosomes, from abnormal or heterotropic (accessory) chromosomes. The idiochromosomes are of two sizes, which he calls respectively macrochromosomes and microchromosomes ; they occur either in pairs or singly. In the egg-cells the mature ova invariably contain half the normal number of chromosomes, but among the sperm-cells there are three different types, with chromosomes varying in quality or quantity. Wilson attempts to account for the sex differences in Hemiptera as depending upon the different combinations of these male chromosomes with the female.

When we consider that Mendel's Law of Hybridisation,[4]

[1] For a fuller account of it see Wilson, *The Cell*, pp. 271, 272 ; Korschelt and Heider, *Vergleichende Entwicklungsgesch. der wirbellosen Tiere*, Allgem. Teil, pp. 599–601 ; R. de Sinéty, *Recherches sur les Phasmes*, 1901, pp. 123–126 ; Sutton, 'The Spermatogonial Divisions in *Brachystola magna*' (*Kansas Quarterly Journal*, 1900 and 1902) ; J. Pantel and R. de Sinéty, 'Les cellules de la lignée mâle chez le *Notonecta glauca*' (*La Cellule*, XXIII, 1906, fasc. I. pp. 89–303, 138, &c., 245). See also the works mentioned on p. 110, note 2.

[2] T. H. Montgomery, 'Spermatogenesis of *Syrbula* and *Lycosa*, with general remarks on the reduction of Chromosomes and on Heterochromosomes' (*Proceedings Acad. Nat. Science*, Philadelphia, LVII, 1905, pp. 161–205). Montgomery classes as heterochromosomes all those that differ from the normal in size or structure. Cf. on this subject a review in the *Naturwissenschaftliche Rundschau*, 1906, No. 4, p. 44.

[3] 'Studies on Chromosomes,' I, II, and III (*Journal of Experimental Zoology*, 1905, Nos. 3 and 4 ; III, 1906, No. 1).

[4] Gregor Mendel (1822–84) was abbot of the Augustinian monastery in Brünn. Cf. C. Correns, 'Gr. Mendels Briefe an C. Nägeli,' 1867–73 (*Abteil. der mathemat.-physikal. Klasse der Kgl. Sächsischen Gesellsch. der Wissenschaften* XXIX, 3, 1905). Mendel's laws of segregation are dealt with very fully

which is to a great extent confirmed by the phenomena of hybrid fertilisation, may have a quite simple morphological basis, if we accept Boveri's theory of the individuality of chromosomes (as Boveri himself was the first to show),[1] we can scarcely refrain from ascribing to the chromosomes a certain individual independence, in virtue of which they become the material bearers of heredity. At the seventy-seventh meeting of German naturalists and physicians at Meran in September 1905, the interesting connexion existing between the individuality of the chromosomes and Mendel's Law was discussed by C. Correns from the botanical point of view,[2] and by C. Heider from the zoological and cytological.[3] I must limit myself here to a very brief account of the matter.

Mendel's *Law of Hybridisation,* which has recently attracted so much attention, comprises three rules: the rule of dominance, the rule of segregation, and the rule of independence of characters. According to the rule of dominance, when two sub-species (e.g. red and white peas) are crossed, the hybrid offspring of the first generation resemble one parent (the white pea) in every respect, and the characteristics of the other parent (the red pea) do not show themselves. The character that appears in the first hybrid generation is called the *dominant,* and the contrasted character that disappears is called the *recessive.* According to the rule of segregation, if the breeding of these hybrids be continued, the contrasted characters of both parents are again distinguished or segregated, and in such a way that half the germ-cells of the hybrid tend to give rise to the character of one parent, and the other half to the character of the other parent. According to the rule of independence of characters, the various individual characters,

by de Vries, in the second volume of his *Mutationstheorie,* 1903, and by Lotsy in his *Vorlesungen über Deszendenztheorien,* 1906, Lecture 8. They have been applied to cross-breeding among silkworms by K. Toyama, 'Mendel's laws of heredity as applied to the silkworm crosses' (*Biolog. Zentralblatt,* XXVI, 1906, Nos. 11 and 12). Cf. also J. Gross, 'Über einige Beziehungen zwischen Vererbung und Variation' (*ibid.* Nos. 13–18). According to Gross (p. 414) no typical instances of Mendelism occur when species are crossed.

[1] *Über die Konstitution der chromat. Kernsubstanz,* pp. 32–33. J. Reinke too thinks (*Einleitung in die theoretische Biologie,* p. 539) that Mendel's law supports the theory that the chromosomes are the chief bearers of heredity.

[2] 'Über Vererbungsgesetze' (*Verhandl. der 77 Versammlung deutscher Naturf. und Ärzte,* Leipzig, 1906, Part I, pp. 201–221).

[3] 'Vererbung und Chromosomen' (*ibid.* pp. 222–244).

which distinguished the parents of the first hybrid, appear quite independently of one another, when cross-breeding is continued.

When two sub-species of the same species are crossed, and the characters of the offspring follow Mendel's laws, they are said to ' mendelise.' Mendel formulated his law in consequence of experimental observations on hybridisation, and quite apart from microscopic research. Working, however, on other lines, cytologists have found three important principles, which lead them to regard the chromosomes as the material bearers of heredity and to ascribe to them a certain individual independence. Firstly, each germ-cell receives exactly half the normal number of chromosomes, and of those which it contains, half are paternal and half maternal. Secondly, each germ-cell receives the total number of chromosomes necessary to normal development, these chromosomes being parental in origin, but qualitatively different. Thirdly, these chromosomes may meet in the germ-cells of the offspring in very various combinations (arranged mostly in tetrads or groups of four), and there they form regularly fresh combinations in their maturation-divisions and fertilisation.

If we may assume that qualitatively different chromosomes are the bearers of definite hereditary qualities, these three principles will enable us easily to explain, not only Mendel's three rules, but also most of the other phenomena of variation and heredity.

Of the numerous instances quoted by Correns and Heider in the above-mentioned lectures, I may give one by way of illustration.

A red and a white specimen of *Mirabilis Jalapa* were crossed. The hybrids of the first generation all bore pink blossoms,[1] those of the second generation were partly white, partly red, partly pink, in the ratio 1 : 1 : 2 ; so that the pink blossoms were twice as numerous as either the white or the red. Let us assume (see Plate II) that the tendency to produce red blossoms is represented by a definite chromosome

[1] According to Mendel's rule of dominance the red colour of one parent ought to have been the dominant, but this was not the case. The rule of dominance, therefore, is not illustrated by this example, and it is more difficult to account for it by the chromosome theory than for the rule of segregation. On this subject see p. 173, note 2, of Gross's work.

A, and the tendency to produce white blossoms by a definite and qualitatively different chromosome *a*. The red variety of *Mirabilis Jalapa* has among its chromosomes only A, the white variety only *a*, as influencing the colour of the blossom. The first hybrid generation receives in its fertilised egg-cell and in all the somatic cells the combination A+*a*, i.e. all its blossoms are pink. At the maturation-divisions of the germ-cells of this first hybrid generation a separation of the A+*a* pair of chromosomes takes place, and by the reduction processes half of all the mature germ-cells receive chromosome A, and the other half chromosome *a*. What is the result to the second generation, produced by the union of these germ-cells in twos? In the somatic cells the chromosomes will be thus combined, A+A, A+*a*, *a*+A, *a*+*a*, and each combination will probably occur the same number of times; in other words, in this generation there will be pink blossoms as well as pure red and pure white, but the pink will be about twice as numerous, which was actually found to be the case. Plate II at the end of the book illustrates this relation of the chromosome theory to the phenomena of hybridisation. The diagrams were used in Heider's lecture.

We have now learnt to regard the mixture of qualities as the chief aim of fertilisation, in which the combined properties of both parents are transmitted to their offspring, and we have seen further that the chief part in this transmission is played by the chromosomes of the cell-nucleus. The next question we must answer is this: What is the object of this blending of qualities? Why is it of so much importance to the maintenance of organic species that Nature has taken great pains to secure it, by means of these complicated and regular arrangements?

The opinions held on this subject are to some extent contradictory. We may safely take it for granted that the rejuvenating or regenerating effect, ascribed by Bütschli, R. Hertwig, A. Bühler and others to the process of fertilisation, is due, at least in part, to this blending of qualities. But I have already referred to their theories (pp. 162, &c.), and so we need now only answer the question: What is the significance of blending qualities for the race development of different species? Does it act in a conservative or in a liberal sense?

Does it promote permanence of species or does it supply the means of altering them ?[1]

Charles Darwin, Spencer, Romanes, Hatschek, O. Hertwig and others have regarded this blending of parental qualities effected by fertilisation as a means of compensating for individual fluctuations; they are therefore of opinion that this union of qualities preserves the purity of the race, and so makes for permanence.

According to these authors it would be possible for a new variety, race, or species to arise only if the possibility of breeding with individuals of the same species were restricted, by either exterior or interior circumstances, to definite and limited groups of individuals, which then had the power to propagate and intensify their peculiarities. On this idea are based Wagner's theory of migration, Romanes' theory of physiological selection, Gulick's theory of segregation, &c.

August Weismann's view is, however, directly opposed to all these.[2] He thinks that amphimixis, i.e. the mixing of qualities resulting from fertilisation, is the chief means of modifying species. It gives rise to fresh combinations of the nuclear elements, and to corresponding new variations in the hereditary qualities of the offspring. These variations offer a wide field for natural selection, which 'breeds' from them new races and species.

At first sight this theory is very attractive. Let us assume that the male and female pronuclei of the germ-cells of some organic species possess each eight chromosomes before their union in the process of fertilisation, and that these sixteen chromosomes differ qualitatively from one another. In the cleavage-spindle of the fertilised ovum they may be paired in no less than sixty-four different ways, and so may produce sixty-four descendants, all differing qualitatively from one another and from their parents. Now, as a matter of fact, in most species of plants and animals the number of chromosomes is far higher than sixteen,[3] and therefore the possible number of variations due to fertilisation is correspondingly higher. It appears to be true that by blending qualities a very vast

[1] See Korschelt and Heider, *Lehrbuch der vergl. Entwicklungsgesch.*, pp. 702, &c.
[2] *Lectures on the Evolution Theory*, I, pp. 331, &c.; II, pp. 192–237 (Eng. trans.).
[3] See Chapter V. pp. 92 and 93.

field is opened to natural selection. Boveri agrees with Weismann to a certain extent,[1] and thinks that the mixture of qualities, which is the chief object of fertilisation, is one means, and even one of the most efficacious means, whereby organic species have developed from the simplest Protozoa to the highest animals and plants.

My own opinion nevertheless is, that the amphimixis resulting from fertilisation may not be of such importance to the evolution theory as Weismann believes.[2] I need not now lay much stress on the many objections to it that can be raised. For instance, it is quite common to find the number of chromosomes differing greatly in closely connected species of animals and plants—e.g. in the *Ascaris* class of worms—whilst forms as far removed from one another as the frog, the salamander, the mouse, the salmon, a crab (*Branchipus*), a bug (*Pyrrhocoris*) and the lily all have the same number of chromosomes, viz. twenty-four. Some experimental evidence is needed to show that the variability of the species is directly connected with, and dependent upon, the number of its chromosomes. Weismann anticipated these difficulties by suggesting, in his theory of determinants, that only the larger complexes of bearers of heredity (the ids) correspond to the chromosomes; each of these is built up of a great number of smaller bearers of heredity (determinants), which are equivalent to the chromomeres or smallest grains of chromatin in the chromosomes, and are able to vary independently of one another. As very little is actually known of the finer structure of the chromosomes,[3] these theoretical speculations cannot be tested by means of microscopical research.

There are, however, other objections to Weismann's theory of the importance of amphimixis, and they are, perhaps, of greater weight. We must notice at the outset that indiscriminate cross-breeding between individuals of the same species

[1] *Das Problem der Befruchtung*, pp. 36–38.

[2] We must be careful to distinguish amphimixis in Weismann's sense, in which it refers to the blending of qualities of individuals belonging to the same species, from the other use of the word, in which it refers to sexual cross-breeding between individuals of different species. I shall discuss the latter kind of amphimixis, as bearing upon the Evolution theory, in Chapter IX, 'Theory of permanence or theory of descent.'

[3] Wilson, *The Cell*, pp. 301, 302.

can never lead to a new permanent variety, as the average will always recur. Moreover, a completely new quality in the offspring can never be produced by a mere combination of qualities present in the parents. It is therefore difficult to see how a mixture of qualities can ever give rise to new species, families, classes, &c., in which some new organ or system of organs is frequently the distinguishing characteristic. Natural selection is, according to Weismann, the sole directive element in the evolution of a race, but all that it can do is to make choice out of the variations furnished by amphimixis, and to preserve the individuals best capable of existence, and therefore Weismann's whole theory of evolution seems unsatisfactory; mere amphimixis and selection could never have produced the present system of animals and plants from extremely simple primitive organisms.

Since 1895 Weismann has very ingeniously tried to meet this objection by bringing forward his theory of germinal selection as a new factor in evolution. He now no longer regards the determinants of hereditary qualities in the nuclear substance of the germ-cells as invariable, but is of opinion that they 'are continually oscillating hither and thither in response to very minute nutritive changes, and are readily compelled to variation in a definite direction, which may ultimately lead to considerable variations in the structure of the species, if they are favoured by personal selection, or at least if they are not suppressed by it as prejudicial.' [1]

He goes so far as to speak of 'vital affinities,' [2] i.e. of definite interior forces uniting the determinants into ids, and the biophors into determinants. It is undoubtedly a very interesting concession on Weismann's part, when he says: [3] 'In all vital units there are forces at work which we do not yet know clearly, which bind the parts of each unit to one another in a particular order and relation.' Weismann seems here to acknowledge that it is impossible ever to understand a development of the organic world, with definite arrangement, and consequently ordered in conformity to law, unless there are interior laws governing that development. If—as Weismann

[1] Weismann, *Evolution Theory*, II, p. 196, Eng. trans.
[2] *Ibid.* I, p. 374; II, p. 36.
[3] *Ibid.* II, p. 35.

suggests in these quotations—there is a connexion, tending to some aim, between the material bearers of heredity among themselves and the influences of the outer world, so that the former are modified by the latter and directed into new channels of development, he seems also to grant that there is a teleological element in the constitution of these material bearers of heredity, to which they own their capacity to adapt themselves to new circumstances by corresponding changes in their constitution, and thereby to effect a regular development of the organic species.

This teleological element, which I have described as the interior laws governing the development of organisms, is no 'mystical, intangible thing' hovering vaguely in the air, as some of my opponents have imagined. It is the original chemico-physical and morphological constitution belonging to the first bearers of the hereditary qualities of the race, at least in its material aspect. If we wish to explain the phenomena of heredity, we must consider in this material constitution not only the morphological character of the smallest and most elementary parts of living substance, that make transmission of qualities possible, but also their dynamic and physiological action.[1]

It cannot be denied that we need moreover some formal principle to explain these laws of evolution. J. Reinke, the well-known botanist, has lately acknowledged this, by declaring the chromosomes of the nucleus to be the chief agents, in all probability, in transmitting specific dominants.[2]

Hans Driesch,[3] one of the best and most thorough students of organic development, seems to hold a very similar opinion, for he says that the processes of organic development require,

[1] On this subject see J. Reinke, *Philosophie der Botanik*, Leipzig, 1905, p. 106; O. Hertwig, *Allgemeine Biologie*, 1906, chapter xii, 'Die Physiologie des Befruchtungsprozesses.' From what is stated above and also from what follows, it is plain that Gemelli, in his Italian translation of the last edition of this work (Wasmann-Gemelli. *La Biologia moderna*, 1906, pp. 218–221), completely misunderstands me, if he thinks that I regard the chromosomes as a transmitting substance in a purely morphological sense.

[2] Cf. J. Reinke, *Einleitung in die theoretische Biologie*, 1901, p. 455; see also pp. 386–408 and especially p. 396. Cf. further his *Philosophie der Botanik*, 1905, pp. 53, &c., pp. 71, &c.

[3] Of his works cf. especially the following: *Die organischen Regulationen; Vorbereitungen zu einer Theorie des Lebens*, Leipzig, 1901; *Die Seele als elementarer Naturfaktor*, Leipzig, 1903; *Kritisches und Polemisches* (*Biolog. Zentralblatt*, 1902, Nos. 5, 6, 14, 15; 1903, Nos. 21, 22, 23).

N

as an indispensable directive power, a teleological formal principle which may be compared with the entelechies. If this is true for the development of the individual, we may regard it as still more necessary for the hypothetical development of the race. I shall recur to this topic at the close of Chapter VIII ('The Problem of Life'), and in the course of Chapter IX ('Thoughts on the Theory of Evolution').

From the evidence given in the present chapter it appears that we may, with great probability, regard the chromosomes of the nuclei in the germ-cells as the chief material bearers of heredity.[1]

We have now obtained a scientific foundation for the interior laws of development, which are the necessary premiss for the hypothesis of a race evolution of organic species. I shall have to deal with this hypothesis in a subsequent chapter: 'Thoughts on the Theory of Evolution.' For the present I will only draw the reader's attention to the fact that all the results of modern biological research, in this department as in others, increase our appreciation of the Creator's wisdom and power, and show us in what a simple and yet wonderfully regular way the transmission of the parents' qualities to their descendants is effected, by means of most diminutive material portions of the germ substance.

[1] Further information of great interest, and tending to confirm this theory, may be found in C. Correns' lecture *Über Vererbung* (On Heredity) and C. Heider's *Vererbung und Chromosomen* (Heredity and Chromosomes). These lectures, to which I have referred on p. 171, were delivered in September 1905, at the seventy-seventh meeting of German naturalists at Meran.

CHAPTER VII

THE CELL AND SPONTANEOUS GENERATION

I HAVE already shown (Chapter III, pp. 55, 65, &c.) that the cell is not a simple entity, but a compound formation of very delicate and artistic structure, as recent research has proved. We have also considered the life of the cell (Chapter IV) and convinced ourselves of the great and universal importance of the nucleus in every function of cellular life, but especially in cell-division and in the processes of fertilisation (Chapters V and VI). We have now sufficient material at our disposal to enable us to answer with assurance the question propounded long ago: 'Is the cell the ultimate unit of organic life, or is it merely an aggregation of still more elementary units?' The solution of this problem will help us to form a really scientific opinion on spontaneous generation or *generatio aequivoca,* for almost all attempts to disprove the unity of the cell have been motived by a desire to make the origin of organic life in the world more intelligible by the assumption of spontaneous generation.

1. The Cell as the Ultimate Unit in Organic Life

The question of the unity of the cell resolves itself into two other questions, which we shall answer each in turn. The first is: 'Are there really in nature organic entities of a still lower organisation than the cell?' The second is: 'Do the morphologically different elements of the cell form together one biologically indivisible unit, or can they be divided into subordinate biological units?' On the answers which facts supply to these questions, depends our acceptance of the various theories which represent the cell as a mere aggregation of lower units, or our rejection of the same as fictions. What does recent research tell us as to the existence of living entities of still lower organisation than the cell? It has really answered this question plainly enough already; it has shown us that the cell-nucleus is also the principle of organisation for the living cell, directing its most important vital activities, and, by means of heredity, maintaining the continuity of organic life. Consequently we should expect to find no organism with a protoplasmic body containing no nucleus, and none with a nucleus that is not inserted, or meant to be inserted, in a protoplasmic body.

This does not, however, prove that in all organisms the cell-nucleus must be developed in equal perfection. On the contrary, the graduated perfection of organic beings may extend also to the organisation of the cells, and we need not be surprised to find, even among the lowest living creatures, some in which the nucleus is not formed into one morphological whole, but is scattered in little grains of chromatin (chromidia) about the protoplasm of the cell. As we shall see directly, this occurs, apparently at least, in many Bacteria. In Chapter III, p. 49, I pointed out that the nucleus was essential to the existence of the cell, either in a complete and centralised form, or in a diffused and incomplete one. This latter statement need not surprise us, as we have seen, in Chapters V and VI, that during indirect cell-division the distinct nucleus ceases for a time to exist as such, because the nuclear membrane breaks up and the chromatin framework of the nucleus divides into small pieces, viz. the chromosomes, and is only reorganised in the newly formed nuclei of the daughter-cells. The sharply defined form of the nucleus is not therefore essential to the cell, although the presence of the nuclear substance is essential.

Attempts have been made to demonstrate the existence of really non-nucleate primitive organisms, or at least to assert the possibility of their existence. Let us examine them in order and test their value.

For a short time it was believed that the long-sought organic matter, devoid of all structure, which Ernst Haeckel announced as the Promised Land of Darwinism, had really been discovered. The discovery was made whilst the North Atlantic cable was being laid in 1857. Huxley subsequently described this primitive matter as consisting of little organic masses, without nucleus and without any structure, found at the bottom of the ocean, and named by him *Bathybius Haeckelii,* after the famous prophet of Darwinism. But the godfather himself was obliged later on to declare this hopeful scion of the Evolution Theory to be a changeling, foisted upon him by an impish trick of bad luck. He had to withdraw his discovery, and acknowledge that there had been a mistake about the *Bathybius.* It was nothing but a deposit formed accidentally in a test-tube filled with alcohol. Bessels, the explorer of the North Pole, afterwards thought that he had rediscovered the primitive organism, which he called *Protobathybius;* but in spite of the amœboid movements that he said he observed, the *Protobathybius* has not yet been admitted to the rank of a living creature; at best it appears to be a deposit of organic substance which has formed at the bottom of the sea from the remains of plancton organisms. Haeckel's own creations, the ostensibly non-nucleated *Monera,* still demand consideration. Haeckel classed together as Monera, the lowest division of Protozoa, all those that he thought contained no nucleus. Their number seemed at first to be legion, and to justify the hopes set upon them by the advocates of the Evolution Theory. But as our microscopes and our methods of research were improved, they melted away like snowflakes in the sunshine. Apochromatic objectives and modern staining methods have revealed the hitherto obscure nucleus in almost all Protozoa, and all possessors of a nucleus were at once banished from the class of Monera, which grew smaller and smaller. The day is not far distant when the last Moneron will share the fate of the last of the Mohicans. On this subject we may refer to R. Hertwig, an eminent zoologist and a favourite pupil of Haeckel's. In the seventh edition of his 'Lehrbuch der Zoologie' (1905, p. 159), he writes as follows: 'The most important feature in the Monera is said to be the lack of a nucleus. Like every negative characteristic, this is somewhat unsatisfactory. In many cases it is difficult to recognise nuclei, especially when the protoplasm is abundant and filled with chromatin granules, and thus it may happen that animals are described as devoid of nucleus, simply because the existing nucleus has been overlooked. For this reason the number of "Monera" was at one time very large; it has diminished, as improved technical methods have revealed nuclei, and so it is not only possible, but even probable, that, in the few forms still reckoned as Monera, the nuclei have only escaped notice; perhaps their functions are discharged by chromidia.'

Unicellular animals without a nucleus have therefore no longer any scientific justification for existence; and no one can refer to them as affording evidence of there being living creatures of a still lower degree of organisation than cells possess. It may, however, be asked: Can the long-sought non-nucleated forms be discovered amongst the lowest plants?

Botanists are still not agreed as to the presence of a genuine cell-nucleus in Bacteria and Cyanophyceae, to which the Oscillaria also belong.[1]

Bütschli thought that he had discovered in Bacteria a very large nucleus, not clearly marked off from the layer of cytoplasm, but Fischer contradicted this statement. Arthur Meyer ('Flora,' 1899, pp. 428, &c.) believed that several little nuclei could be traced in the cells of some Bacteria. Fritz Schaudinn

[1] For the bibliography of this subject, see Strasburger, *Lehrbuch der Botanik*, sixth edition, 1904; Bütschli, *Weitere Ausführungen über den Bau der Cyanophyceen und Bakterien*, Leipzig, 1896; Fischer, *Untersuchungen über den Bau der Cyanophyceen und Bakterien*, Jena, 1897; G. Schlater, 'Zur Biologie der Bakterien: Was sind Bakterien?' (*Biolog. Zentralblatt*, 1897, pp. 833, &c.); J. Reinke, *Einleitung in die theoretische Biologie*, Berlin, 1901, chapter 25, pp. 256, &c.; R. Hertwig, 'Die Protozoen und die Zellentheorie,' *Archiv für Protistenkunde*, I, 1902, pp. 1–40); Fr. Schaudinn, 'Beiträge zur Kenntnis der Bakterien und verwandter Organismen' (*Archiv für Protistenkunde*, I, 1902, pp. 306, &c.; II, 1903, pp. 421, &c.).

has discovered quite recently that in the case of *Bacillus Bütschlii,* a large parasitical fission fungus found in the intestine of the cockroach, *Periplaneta orientalis,* a genuine nucleus appears temporarily during the formation of spores, although otherwise the nuclear substance is dispersed in the cell. R. Hertwig's investigations into Bacteria and Oscillaria have led him to conclude that these organisms ought to be regarded as cells without a clearly differentiated nucleus, but having the nuclear substance distributed among the protoplasm. He gives the name *chromidia* to the little particles of chromatin in Bacteria, corresponding to the chromosomes and their constituents, the chromomeres, in true nuclei.

J. Reinke does not venture to express a general opinion as to the non-nucleate character of Cyanophyceae and Bacteria, but he considers that the cell of *Beggiatoa,* a tiny, thread-like Bacterium, is non-nucleate to this extent, that it does not contain any distinct nucleus, in the sense in which the higher plants and animals contain nuclei.

In the sixth edition of his 'Lehrbuch der Botanik,' p. 46, Strasburger says: 'The two most essential constituents of the protoplasm (i.e. of the living cell) are the nucleus and the cytoplasm, and the vital functions of the cell depend upon the interaction between them. But in the lowest plants, Cyanophyceae and Bacteria, the existence of a nucleus is still uncertain.' On p. 270 of the same book, Schenk, in writing of Bacteria, remarks: 'In the protoplast there are one or more granular structures called chromatin-bodies, which may be deeply coloured by stains, and are regarded as nuclei by various authors. Hitherto no one has succeeded in demonstrating undoubted karyokinesis in them, and therefore the presence of nuclei (in Bacteria) is still not established.' On p. 274 Schenk remarks with reference to the Cyanophyceae: 'Within the coloured zone (of the protoplast) lies the colourless central body, which perhaps corresponds to a nucleus. However, the structure and division-figures characterising typical nuclei have not been observed with any degree of certainty.'

F. G. Kohl on the other hand, in a recently published work,[1] declares with assurance that the central body in the

[1] 'Über die Organisation und die Physiologie der Cyanophyceenzelle und die mitotische Teilung ihres Kerns' (mit 10 Tafeln), Jena, 1903.

Cyanophyceae is a true nucleus, and he proves such to be the case from the processes of mitotic division that occur. Orville P. Phillips[1] has come to the same conclusion, and thinks that the Cyanophyceae can no longer be regarded as devoid of nucleus.

The existence of true nuclei in Bacteria has lately been asserted also by R. Raymann and R. Kruis,[2] and by F. Vejdowsky.[3]

Even if we are obliged to regard the question of the non-nucleate character of Bacteria and other diminutive representatives of the lowest vegetable orders as to some extent still doubtful, we can at least learn from the investigations made on the subject, that the nuclear substance is present in them, although it is broken up into little chromatin granules or chromidia. They possess, therefore, what Wilson calls a scattered or distributed nucleus (' The Cell,' p. 40), and they ought not to be considered simply non-nucleate, although they seem to form a kind of transition to those cells which contain a fully developed nucleus. That the chromidia in Protozoa are the biological equivalents of nuclei and only represent a particular condition of nuclear configuration has been conclusively proved lately by Fritz Schaudinn.[4]

Oskar Hertwig, one of the greatest biologists of the present day, has declared it to be his opinion that really non-nucleated organisms do not exist (' Allgemeine Biologie,' 1906, pp. 44, 45). No actual facts can be brought forward in support of them, only ' various theoretical considerations ' of a purely speculative character; as R. Hertwig expresses it (' Lehrbuch der Zoologie,' 7th ed. p. 159) : ' It is easier to imagine that, in spontaneous generation, those organisms first came into being which consisted of only one kind of substance, than those in which nucleus and protoplasm were already distinguished.'

[1] ' Vergleichende Untersuchung der Cytologie und der Bewegungen der Cyanophyceen ' (*Contributions from the Botanical Laboratory, University of Pennsylvania*. II, 1904, pp. 237–306).

[2] ' Über die Kerne der Bakterien ' (*Bullet. International de l'Acad. des Sciences de Bohême*, VIII, 1903).

[3] ' Über den Kern der Bakterien und seine Teilung ' (*Zentralblatt für Bakteriologie*, XI, 1904, 2nd Part, pp. 481–496). Cf. the review in the *Naturwissenschaftliche Rundschau*, XIX, 1904, No. 29, pp. 366–369.

[4] ' Neuere Forschungen über die Befruchtung bei den Protozoen ' (*Verhandl. der Deutschen Zoolog. Gesellsch.*, 1905, pp. 16–25 and Plate I. See particularly pp. 3–6).

We cannot, therefore, name any independent unicellular organism having either a cell-body without a nucleus, or a nucleus without a cell-body. Is it possible that these forms, so eagerly sought under Haeckel's name *cytodes* by the upholders of the theory of spontaneous generation, may occur within the tissues of multicellular animals and plants? If they did occur, it would prove nothing in support of the theory of spontaneous generation, for once-living cells can degenerate and lose their nucleus, whilst cells still in process of formation may have a nucleus before the layer of protoplasm belonging to it can be traced.[1] But in these cases we should have to deal with the products of living, nucleated cells; not with a spontaneous coming into existence of non-nucleated cell-bodies, or of bodiless nuclei, out of still unorganised primitive matter. Let us examine the facts rather more closely.

The young red blood-corpuscles of vertebrates have a nucleus, which multiplies itself by direct division, and so causes an increase in the number of red blood-corpuscles, as we have already stated (Chapter V, pp. 86 and 87). The old red blood-corpuscles lose their nuclei and become enucleate, but they have ceased to be living cells, and are only the remains of cells once alive, which still for a time are of use to the organism as bearers of the oxygen loosely attached to their hæmoglobin, but soon they are dismissed from service, and the white blood-corpuscles come and devour them. The existence of red blood-corpuscles without nuclei, accepted by most authors,[2] is of no use as evidence that there can be *living* cells without a nucleus, and that the nucleus is not, therefore, indispensable to the life of the cell. Just as a living cell must have a nucleus or its equivalent, so a living nucleus must have a protoplasm body, if it is to continue in existence. It is true that there are cells in which the volume of the nucleus is far greater than that of the cell body. Spermatozoa belong to this class; they often have an enormous head consisting

[1] I observed instances of this when I was preparing the series of sections of *Lomechusa* larvae. They occurred during the formation of new œnocytes in the hypodermic region.

[2] I say 'by most authors' for some maintain that they have observed nuclei even in old red blood-corpuscles. Cf. M. Duval, *Précis d'Histologie*, pp. 50, 614, &c.

of the nucleus of the sperm-cell, whilst the thin threadlike tail and probably also the middle-piece, connecting it with the head, are the protoplasmic elements of the cell; but no sooner has the spermatozoon lost its tail in the process of fertilisation, than its existence as a cell is over; its nucleus perishes, unless it can unite with a female pronucleus to form the cleavage-nucleus of the fertilised ovum (cf. Chapter VI, pp. 119, &c.).

We come now to the reverse case, in which new nuclei are formed apparently without a cell-body. In the history of the genesis of cells, these phenomena play an important part, as we shall see later on. This is the so-called *free nuclear formation*, which is supposed to lead to *free cellular formation*. These formations were called *free*, because the new nuclei were not formed by division from an old nucleus, nor the new cells by division from an old cell, but both were supposed to originate in an indifferent mass of protoplasm called *blastem*, a product of the mother-cells in the same organism. Such a mode of forming fresh nuclei, destined to become the centres of fresh cells, even if it really existed, would have had nothing to do with spontaneous generation, and it had no real existence at all. The theory of free nuclear formation was, as we shall see, to all intents and purposes dead at the end of the nineteenth century, and in the twentieth no one can have recourse to it to support any favourite theory.

Let us now sum up shortly the results of these investigations. They amount to this: There are no living organisms simpler in organisation than the cell.

We can now approach the question: 'Is the cell the ultimate unit of organic life, or is it composed of still lower and more elementary units?'

According to the laws of logic, we ought to describe as the lowest unit of life only that part of a morphologically complex living creature, which, at least under certain conditions, is actually capable of an independent existence. Otherwise it is no longer a biological unit, but only a part of a biological unit. Now we have just shown that no organism is actually of lower organisation than the cell, therefore the cell is actually the lowest and ultimate unit in organic life.

We have seen moreover, in the previous sections, that within the cell the nucleus and the protoplasm of the cell-body, as well as the morphologically distinguishable elements of these two chief parts of the cell, are in no sense independent of one another, but are closely connected, so as to make up one cell, capable of life, to which they belong partly as essential, partly as integral portions. The nucleus is in a certain degree the material principle of organisation in the cell, controlling its activities, but the protoplasm is indispensable to its life. It is true that the chromosomes of the nucleus take the leading part in the processes of cell-division, fertilisation, and transmission of qualities, and possess some amount of individuality (see pp. 167, &c.), as they always appear at the cell-divisions in definite shape and number, and within these limits have an independent power to propagate themselves and develop by means of segmentation and growth; but still no chromosome can exist and become a nucleus without its corresponding particle of protoplasm. And what does this show? That the chromosomes are not lower biological units within the cell, but they are merely essential morphological and physiological constituents of the cell. What is true of the chromosomes, applies also to the centrosomes and to all the other less important morphological elements of the cell. None of them is capable of independent existence apart from the cell; they are, consequently, only parts of the cell, not lower and more elementary units out of which the cell is composed as a secondary formation.

The cell, therefore, from the biological point of view, represents an indivisible unit, although it is composed morphologically of many different parts, whose various functions co-operate in the one biological process of life. The life of a multicellular animal or plant is one biological whole, in which the various organs, tissues, and cells, with their respective functions, all unite and work together in conformity to law, and the discovery of the intercellular bridges connecting the various cells in the body of an animal or plant has furnished a histological explanation of this fact,[1] and in just the same

[1] See Wilson, *The Cell*, 1902, pp. 59, 60. An excellent account of the biological unity of the whole process of growth and development in the living organism is given by the same author, pp. 58, 59, and 393, &c. According to him (p. 59) cells are 'local centres of a formative power pervading the growing mass as a whole.' O. Hertwig too, in his *Allgemeine Biologie*, 1906, chapter xiv, has done much to remove the obscurity prevailing on the subject of 'Individuality,' although I am unable to agree with him on all points, e.g. in his conception of personality, pp. 378 and 383.

way the life of a unicellular organism is an individual biological unit, in spite of the fact that the cell is composed of various parts with various functions. The impossibility of maintaining the opinion that multicellular organisms are mere aggregations of cells, has been brought out very clearly by O. Whitman in an article 'On the inadequacy of the cell-theory of development' (*Wood's Hall Biological Lectures*, 1893).

The cell-bridges forming protoplasmic connexions between the cells of the organism may, according to Hammar,[1] be recognised even between the cleavage-globules of the first divisions of the fertilised ovum. In his 'Allgemeine Biologie' (1906, chap. xiv), Oskar Hertwig stoutly upholds the individual unity of the multicellular organism. He distinguishes clearly (p. 371) two different conceptions of individuality, viz. the physiological and the morphological individual. The former is 'an independent living being,' and it is to this alone that the idea of individuality strictly speaking applies. The latter is 'a formal unit, which resembles a physiological individual morphologically, i.e. in appearance, structure, and composition, but not in the physiological sense, for it is not an independent living being, but is taken as a dependent part into another higher physiological individuality, or, in other words, is adopted as an anatomical element of the same.'

The idea of organic individuality has in recent times often been transferred from unicellular organisms to every single cell of a multicellular organism, so that each cell in the body of an animal or plant has been wrongly raised to the dignity of an 'individual,' although it is not one at all physiologically, i.e. it is not an independent individual, from the biological point of view, but only a part of an individual.

In just the same way, in the lowest histological unit, viz. in the morphological individual represented by the cell, the part has often been confused with the whole, and attempts have been made to prove, from the composition of the cell,

[1] 'Über eine allgemein vorkommende Protoplasmaverbindung zwischen den Blastomeren' (*Archiv für mikroskopische Anatomie*, XLIX, 1897); 'Ist die Verbindung zwischen den Blastomeren wirklich protoplasmatisch und primär?' (*ibid.* LV, 1900). Cf. also Korschelt and Heider, *Lehrbuch der vergl. Entwicklungsgesch.*, Jena, 1902, Allgem. Teil, Part I, pp. 159, 160. On the subject of intercellular bridges, see also O. Hertwig, *Allgemeine Biologie*, pp. 400–406.

that there must be organic units of a lower order than the cell. This line of argument is quite wrong, and we must clearly understand that we may regard as the lowest units of organic life only those parts of organisms which, at least under definite conditions—such as occur among unicellular animals and plants—are capable of independent existence. To call the parts of these units ' subordinate units ' is most deceptive, for they are not units at all, but only parts of units. All the arguments adduced by Altmann, Schlater, and other modern writers against regarding the cell as the final biological unit are based upon this quibble. Flemming has shown this very clearly with regard to Altmann, and says [1] that evidence is still inadequate to prove that Altmann's granula are really elementary organic units or *bioblasts,* inasmuch as the chief point in it is absent, viz. conclusive proof that one of his famous granula is capable of exercising its elementary vital functions outside the cell. We arrive therefore at the same result as Oskar Hertwig in his ' Allgemeine Biologie ' (1906, p. 375), where he declares cells to be the elementary units in the whole organic world.[2]

If we wish to find a justification in fact for speaking of ' lower elementary units ' of living substance, we can do so only in the sense in which Sachs spoke of *energids.* An energid is a particle of nuclear substance with a definite amount of protoplasm belonging to it and subject to its control. In this way it would be possible to avoid the difficulties that seem to prevent our giving the same account of cells with one nucleus and of those with more than one. A cell with more than one nucleus would be made up of a number of energids not so completely distinct from one another as to be called separate cells. A cell with one nucleus would be one fully developed energid. The acceptance of this idea would obviously not affect our opinion of the essential unity of the cell. We may even imagine, as Lotsy does,[3] that the first living beings were *monoenergids,* i.e. very simply organised cells, consisting each of a single energid. These might swim about freely, but we cannot

[1] Cf. W. Flemming, ' Über Zellstrukturen ' (*Naturwissenschaftliche Rundschau,* XIV, 1899, No. 35, p. 444).

[2] To understand his meaning more clearly, see also chapter xvii, pp. 424, &c., of the same work.

[3] *Biolog. Zentralblatt,* 1905, No. 4, p. 97.

possibly imagine biophors or other 'lower elementary units' to have swum about, because they, as far as they have any real existence, are only parts of an energid, and not creatures capable of independent life.

Thus we arrive again at the conclusion: The cell (or energid) is actually the lowest unit in organic life. Therefore the alleged 'lower elementary units' of the upholders of the Theory of Descent are nothing but fictions. It is a matter of complete indifference for this subject whether the formations in question can be seen under the microscope, as definite morphological elements of the cell, or whether they exist solely as figments of the imagination in the brain of some philosophising naturalist, for their interpretation as elementary *units* is in both cases equally imaginary, although they may retain their significance as more or less hypothetical elementary *parts* of the living substance.

I should stray too far were I to attempt to give my readers anything like a complete account of the many various theories in which these elementary units are concerned. The names given to these units by those who believed they had discovered them are very numerous. In 1864 Herbert Spencer began the list by calling them *physiological units*; Darwin called them *gemmules*, Erlsberg and Ernst Haeckel *plastidules*, Nägeli *micellae*, Detmer *Lebenseinheiten* or vital units, Hugo de Vries *pangens*, Verworn *biogens*, and Weismann *biophors*, which by combining make up the units next above them or *determinants*, which in their turn compose *ids* and ids *idants*. (Cf. Chapter VI, pp. 107, &c. and pp. 175, &c.) W. Roux called his elementary units *metastructural parts*, Wiesner *plasomes*, W. Haacke *gemmae*, which he imagines as rhomboid crystals lying side by side to form magnetic columns or gemmaria.[1]

L. Zehnder[2] conceives of the elementary units of life as annular hollow cylinders, formed of organic molecules, and he calls them *fistellae*. Oskar Hertwig calls his units *bioblasts*,[3]

[1] For a criticism of Haacke's fantastic 'Doctrine of Creation,' see my article, 'Zur neueren Geschichte der Entwicklungslehre in Deutschland: Eine Antwort auf W. Haacke's *Schöpfung des Menschen*,' Münster, 1896 (*Natur und Offenbarung*, XLII).

[2] *Die Entstehung des Lebens aus mechanischen Grundlagen entwickelt*, I, Freiburg i. B., 1899, pp. 50-52.

[3] *Allgemeine Biologie*, 1906, pp. 52, &c.

Simroth[1] *biocrystals*, and Altmann *granula*, *bioblasts* or *autoblasts*—granula, inasmuch as they are visible under the microscope as very fine grains; bioblasts, inasmuch as they represent the hypothetical elementary units of the life of the cell; and autoblasts, inasmuch as they are said to be capable of a free existence outside the cell. It is a pity that neither Altmann himself nor any of his followers, among whom Gustav Schlater is conspicuous for his energy,[2] have succeeded in demonstrating the existence of granula as bioblasts and autoblasts.

I am far from denying that the above-mentioned theories contain many ideas that are both accurate and fruitful for the philosophy of life. (Cf. O. Hertwig, 'Allgemeine Biologie,' chapter xxxi.)

Richard Hertwig has drawn attention[3] to the fact that according to most recent research, the chromatin of the cell-nucleus really possesses the properties which Nägeli required theoretically for his *idioplasm* as the material substance of heredity (1884). This hypothetical substance in the first place must not only be organised at the time of fertilisation, but it must have possessed its organisation beforehand, and have constantly preserved it; secondly, it must be present in the egg- and sperm-cell in equal quantities; and thirdly, it must occur in all cells in a state of living metamorphosis, and influence their vital processes. The chromosomes of the nucleus possess all these properties, as I have shown plainly in my account of the processes of cell division and fertilisation (Chapter V, pp. 123, &c. and pp. 165, &c.). That the chromatin of the cell-nucleus is a real idioplasm, a real physical basis of inheritance, we must acknowledge to be extremely probable; but, on the other hand, it is wrong to follow Nägeli in regarding the single particles of chromatin, *micellae*, as he calls them, as elementary vital units; for, from their very nature, the chromosomes can only be parts of the nucleus of a living cell, with which the substance of inheritance is necessarily connected. A living

[1] 'Bemerkungen zu einer Theorie des Lebens' (*Verhandl. der Deutschen Zoolog. Gesellsch*, 1905, pp. 214–232).

[2] Cf. his articles: 'Der gegenwärtige Stand der Zellenlehre' (*Biolog. Zentralblatt*, XIX, 1899, Nos. 20–24); 'Monoblasta—Polyblasta—Polycellularia' (*ibid.* XX, 1900, No. 15).

[3] 'Über Befruchtung und Konjugation (*Verhandl. der Deutschen Zoolog. Gesellsch.*, 1892, p. 101).

chromosome apart from a corresponding particle of living protoplasm is an impossibility.

I will gladly acknowledge that many of these theories of heredity display a marvellous wealth of ingenuity and intellectual effort. This is particularly true of Weismann's Germ-plasm theory, especially in the form of the Theory of Determinants, in which he stated it in his lectures on the Evolution Theory in 1892. It aims at explaining the nature of the germ-plasm, and of all the phenomena of heredity, by reference to particular structures and particular distribution of even the smallest material parts of the germ-plasm. As a general theory, however, it proves to be untenable.[1]

It seeks in a one-sided way to account for the development of the individual out of the preformed structure of most minute material particles of germ, and finally it is reduced to the necessity of assuming the existence of 'vital affinities' between these minute particles, and this necessity reveals the inadequacy of the ingeniously thought-out mosaic theory. I should prefer to accept Oskar Hertwig's Theory of Biogenesis ('Allgemeine Biologie,' chap. xxii, &c., and especially pp. 635, &c.) which, in a successful and logical manner, connects the principle of preformation with that of epigenesis. It too regards the chromosomes as the material bearers of heredity, but takes into account also the dynamic and physiological force of their interaction in the vital unity of the whole process of development. If we therefore consider O. Hertwig's hypothetical *bioblasts* to be elementary particles, and not elementary units of living substance, the theory of biogenesis, as a working hypothesis, is of assistance to us in trying to solve the problem of evolution. O. Hertwig himself frequently emphasises the facts that a cell containing a nucleus is the lowest morphological unit in organic life, and that the cells in multicellular organisms unite to form a true, physiological, living unit. On p. 569 he sums up his opinion as to the causes of development as follows: 'Continuity in development is not attained by means of the *emboîtement* of miniature creatures, nor by the

[1] For a criticism of it, see Y. Delage, *La structure du protoplasma et les théories sur l'hérédité*, pp. 196, &c., 512, &c., 667, &c.; also J. Reinke, *Philosophie der Botanik*, 1905, pp. 63, 64. O. Hertwig, *op. cit.*, 1906, pp. 361, 452, &c., 620, 633, &c. Cf. also Chapter VI, pp. 174, &c.

secretion of an unorganised formative material endowed with a *nisus formativus*, nor by a substance composed of tiny germs, and so to some extent representing an extract of the body, but rather by the cell, a living elementary organism, which by its multiplication and combinations gives rise to all forms of vegetable and animal life. Continuity of organic development and of organic life depends therefore on the principle *omnis cellula ex cellula*.'

Zoological and botanical research, whilst it has enlarged our knowledge, has tended more and more to prove the non-existence, among unicellular organisms, of any that really consists of a simple lump of plasm, such as the theorists are so anxious to discover. Fritz Schaudinn, who is one of our best authorities on Protozoa, gave an address on 'Recent Research into Fertilisation among Protozoa' ('Neuere Forschungen über die Befruchtung bei Protozoen') at a meeting of the German Zoological Society at Breslau, on June 14, 1905, and the opinion, which he expressed in the following resigned terms, must be valuable. He said: 'As in the class of Flagellata, universally regarded as one of the lowest groups of Protozoa, the study of the problem of fertilisation alone shows the finer structures of the cell to be almost as highly differentiated and complicated as in the highest organisms, the discovery among Protozoa of our day of that tiny drop of simple plasm, whence the animal cell is supposed to have originated, may present some difficulties.'

2. Spontaneous Generation of Organisms

The question as to the lowest actual units of organic life is closely connected with the question whether spontaneous generation is possible.

The Monists assure us that it is undoubtedly possible, because it must have taken place; organic life exists now in the world, and yet there was a time when it did not exist, as the world was still in a state of molten heat. Therefore there must have been an epoch when, under particularly favourable chemico-physical conditions, the first primordial plasm or plasms were produced from inorganic combinations of carbon. The assumption of spontaneous generation is therefore an

O

indispensable postulate of science, according to Monism. M. Verworn, the eminent physiologist,[1] argues in the following way in favour of spontaneous generation: 'Living substance is actually a part of the matter composing our world. The combination of this matter to form a living substance was as much a necessary result of the evolution of the world as the formation of water, viz. a necessary result of the gradual cooling of those masses which made up the crust of the earth. In the same way the chemical, physical, and morphological properties of living substance, as we know it, are the inevitable consequence of the working of our present exterior conditions of life upon the interior conditions of earlier living substance. Interior and exterior conditions of life stand in inseparable interaction, and the expression of it is life.' Thus the assumption of spontaneous generation is scientifically irrefutable!

What are we to say in answer to this demand made upon us in the name of science? I am quite ready to admit that the first organisms were made of inorganic matter, for, if they were not, they would have to be created out of nothing, which I am by no means inclined to believe. But the theory of spontaneous generation requires inorganic matter to have produced the first organisms by itself and out of its own resources. The latter assumption cannot be a 'postulate of science,' because, as I shall show, it plainly contradicts actual facts. If I were to maintain, on the contrary, that the first living beings were brought forth from matter still not organised,[2] under the action of a higher power proceeding from the Creator of matter, I should have given up the idea of spontaneous generation, and have replaced it by that of creation in the wider sense. I say 'creation in the wider sense,' because the matter out of which the organisms were formed already existed, and the creative action was limited to the organisation of this matter. It is quite indifferent to our question how we imagine this organisation to have taken place, whether it was by an *eductio formarum e potentia*

[1] *Allgemeine Physiologie*, 1901, pp. 333, &c.

[2] The antithesis is between organised and not organised, not between organic and inorganic, for many organic substances, i.e. such as under natural conditions are formed only in living organisms, can be made artificially in chemical laboratories.

materiae, or by some other method ; nor do we know when the first organisation of matter occurred.[1]

It is obvious that the material basis for the origin of the first forms of life must be supplied by definite arrangements of atoms and the physical and chemical laws governing them ; but this no more proves spontaneous generation to have taken place than does the fact that also at the present time the phenomena of life rest on a chemico-physical foundation.

The problem with which we are now concerned is therefore the following : ' What are we to think of the theory of spontaneous generation, which requires lifeless matter of itself to have produced the first living organisms ? ' We must examine the scientific character of this spontaneous generation more closely.[2] We may disregard those rash and untenable theories which, like Ernst Haeckel's carbon theory, aim at giving a direct account of spontaneous generation. It is impossible not to be amazed at the audacity with which these hypotheses are published as being the results of scientific research. For instance, in 1892, Schaaffhausen seriously asserted that water, air, and various mineral substances had united directly under the influence of light and heat, and had produced a colourless *Protococcus*, which afterwards turned into the *Protococcus viridis*. Yves Delage remarks somewhat sarcastically :[3] ' If the matter is so simple, why does not the author produce a few specimens of this *protococcus* in his laboratory ? We would gladly supply him with the necessary chlorophyll.' Still more fantastic is Haeckel's discovery

[1] Hamann (*Darwinismus und Entwicklungslehre*, 1892, p. 58) and Fechner assume that matter was originally in a ' cosmo-organic ' state, subject to the laws of neither organic nor inorganic nature, but this hardly seems to be a tenable hypothesis, for the chemico-physical laws governing the atoms and molecules in matter can scarcely have differed from those that now govern inorganic matter, and, in the same way, the mechanical laws governing the movement of atoms, molecules, and masses must have been identical with the present laws. It follows that primitive matter in itself must be judged according to the laws of the present inorganic world, and so the ability to produce organisms spontaneously cannot have belonged to its essence.

[2] On the differences between living creatures and lifeless matter see also L. Dressel, *Der belebte und der unbelebte Stoff*, Freiburg i. B., 1883 I cannot here discuss the other reasons for declaring the theory to be philosophically untenable. Stölzle remarks very justly (*A. v. Kolliker's 'Stellung zur Deszendenzlehre*,' 1901, p. 14) that as an explanation the theory of spontaneous generation is worthless, if for no other reason, because it attempts to explain the unknown, not by the known, but by another unknown.

[3] *La structure du protoplasma et les théories sur l'hérédité*, p. 402.

of an organic primitive pulp to which he gave the classical name of *Autoplasson,* or self-forming substance. We have already seen how badly *Bathybius Haeckelii* has fared, which was supposed to be the first real representative of this pulp. On a level with Haeckel's autoplasson is the plastic primary substance discovered in 1874 by an Italian named Maggi, who called it *Glia,* and declared it to be the starting-point of the development of the organic world. It does not altogether savour of genuine science.

Thoughtful naturalists cannot regard as serious such clumsy attempts to solve the most delicate problems; it is obvious that they are doomed to be failures. The chemical composition of nucleinic acid,[1] which is present chiefly in the chromatin (nuclein) of the nucleus, and is therefore intimately connected with the problem of heredity, defies all the attempts made hitherto, and likely to be made in future, by the upholders of the carbon theory to explain its chemical formula $C_{26}H_{49}N_9P_3O_{22}$. That it is a hopeless task to seek the origin of life directly from inorganic matter is acknowledged frankly by most naturalists. If theories, such as Haeckel's carbon theory, are still brought forward, it is not for the benefit of really scientific circles, but that the so-called 'general' readers may be disposed thereby to accept a realistic and monistic view of life.

I have, of course, no intention of condemning the ingenious attempts, which chemists are making with ever-increasing success, to produce organic matter artificially in their laboratories. By means of unwearied industry, Emil Fischer and other eminent workers in this department of research have advanced steadily towards mastering the chemical construction of a molecule of albumen,[2] and, perhaps, erelong the

[1] For a detailed account of the chemistry of the nucleus see Dr. Hans Malfatti, *Zur Chemie des Zellkerns*: reprinted from the *Berichte des naturwissenschaftlich-medizin. Vereins in Innsbruck* (XXII, 1891–2). Cf. also Hofmeister, 'Über den Bau des Eiweissmoleküls' (*Verhandl. der 74 Versammlung Deutscher Naturforscher zu Karlsbad*, 1902, communicated to the *Naturwissenschaftliche Rundschau*, 1902, No. 42). Also Wilson, *The Cell*, pp. 41, 330, &c.; O. Hertwig, *Allgemeine Biologie*, 1906, pp. 29, &c.

[2] Cf. Karl Kautzsch, 'Über das Eiweiss, insbesondere die neuesten Forschungen auf dem Gebiete der Eiweisschemie' (*Natur und Schule*, V, 1905, pp. 195–208); G. v. Bunge, *Lehrbuch der Physiologie des Menschen*, II, 1905, pp. 55–70; Fr. Samuely, 'Die neueren Forschungen auf dem Gebiete der Eiweisschemie und ihre Bedeutung für die Physiologie' (*Biolog. Zentralblatt*, 1906, Nos. 11, 12, 13–15); O. Hertwig, *Allgemeine Biologie*, chapters ii, iii.

artificial synthesis of the simplest forms of albumen will be accomplished by these indefatigable students. But this would prove nothing about spontaneous generation. The albumen molecules, with their highly complicated chemical composition, are the constituents of living creatures, but even in the smallest cell these constituents are *alive*, and no astute human intelligence will ever succeed in breathing the breath of life, capacity for growth and propagation, into one of these artificially prepared, lifeless molecules of albumen, and still less can chance ever have been in a position to form molecules of albumen by itself. Oskar Hertwig remarks very aptly in his 'Allgemeine Biologie' (1906, p. 19): 'Even if chemistry in course of time were able to produce artificially by synthesis all existing forms of albumen—to undertake to form a protoplasmic body would still resemble Wagner's attempt to crystallise out a homunculus in a test-tube.'

Modern physics will in vain strive to do what organic chemistry fails to accomplish. It is not long since people believed that the discovery of radium had removed the hindrance which had frustrated all previous attempts to produce life.[1]

On June 30, 1905, John Butler Burke, of the Cavendish Laboratory in Cambridge, startled the scientific world by announcing that, with the help of radium, he had succeeded in producing from sterilised bouillon a substance that showed certain signs of life: the first living albumen body appeared to have been born artificially! But it was unhappily a miscarriage. Sir William Ramsay, the famous physicist and investigator of the properties of radium, soon explained what Burke had observed, and accounted for it in a very simple way. The powdered radium, which Burke had strewn upon the bouillon, produced in it chemical changes. The emanation of the radium decomposes the water in the bouillon into oxygen and hydrogen, and has at the same time the peculiarity of coagulating albumen. Consequently this emanation could not fail to form, in any watery fluid containing albumen, little bubbles of gas surrounded by a covering of coagulated albumen. As more gas is produced, these bubbles increase and occupy more space, so as to present the appearance of a very small, growing organism. In reality, therefore, this

[1] 'Das Radium und die Urzeugung' (*Gaea*, XLII, 1906, Part I, pp. 34–36).

alleged living creature was nothing but a lifeless covering of albumen filled with gas! This explains a phenomenon observed by Burke, viz. that the new-born organism melted away in the water, for the water gradually removed the gelatine from the 'cell-walls,' and they returned to lifeless non-existence.

We cannot waste time here on the refutation of the various old and new theories of spontaneous generation; we will rather turn our attention to the attempts made by scientific men to present the problem of spontaneous generation in a 'more comprehensible or more acceptable' form. To this category belong the theories that have devised the simplest possible elementary units of life, in order by their means to bridge over the chasm between the atoms and molecules of the inorganic world and the simplest forms of life; or, if the chasm cannot be actually bridged, they aim at diminishing its width to such an extent that a bold 'stroke of genius' may help them over it. To leap from inorganic matter, or even from artificially produced organic combinations, to the living cell is a very hazardous proceeding, which even the most daring advocate of the theory of evolution would hesitate to venture upon. Therefore there is only one way of getting over the difficulty. The chasm must be crossed, not at one bound, but by degrees, and so intermediate halting-places are necessary. These hypothetical intermediate stations are called 'simpler elementary units of life'; they are used to make up the phylogeny of the cell by means of the assertion that nature has taken these steps before us, in order to produce the first cell out of inorganic matter. In this way the theory of spontaneous generation is supposed to be made more acceptable from the scientific point of view.

The statement just given is not an invention of my own, it is only a short summary of the way in which Gustav Schlater in the *Biologisches Zentralblatt* for 1899 (pp. 729, &c.) tries to give a phylogenetic value to Altmann's granular theory. Schlater thinks that Altmann's newly discovered elementary units are of great importance, chiefly because they bring us nearer to a comprehension of spontaneous generation. He says on this subject (p. 732): 'Although at the present time we are naturally not yet in a position to fix the moment when, through a complicated molecule of albumen, the first ray of life flashed, which changed the dead molecule into a living organism, or, let us say, into an autoblast; nevertheless such a change is much more within our comprehension than such a gigantic transition in evolution, as that from a dead molecule of albumen to a complicated organism like the cell.'

There must have been a flash somewhere for life to have begun at all; even Schlater acknowledges this. But it is eventually a matter of perfect indifference whether the flash was at the spontaneous generation of an autoblast or of a cell; the flash of the first spark of life in lifeless matter is as inexplicable in the one case as in the other. Schlater might have saved himself the trouble of writing over a hundred pages in support of bioblasts and autoblasts, for by so doing he has quite gratuitously brought himself into conflict with scientific facts, which know nothing of autoblasts, i.e. of Altmann's granules with a free and independent existence, but recognise cells as the lowest units of organic life. He has brought himself needlessly into conflict with scientific laws of thought, which forbid us to regard Altmann's granules as bioblasts, i.e. as real elementary units of life, because they are actually only biologically dependent parts of the real biological units, viz. the cells. So Schlater's whole argument misses its point. He has not succeeded in establishing the existence of elementary units, having a lower degree of organisation than the cell; nor has he succeeded in explaining the origin of life, even by assuming the existence of these units. The *summa summarum* is in his case another unmistakable breakdown of the theory of spontaneous generation.

Therefore in 1899 the theory did not fare better than in the previous contests that it had had to undergo. It has always suffered defeat, and as scientific research advances, it withdraws into obscurity. It may, perhaps, be interesting to give my readers a short sketch in broad outlines of this retreat of the theory of spontaneous generation.

There was a time when *generatio aequivoca* or *spontanea* was regarded not only as possible, but as of actual occurrence. This was during the so-called 'dark ages' and the still darker mediæval period. At that time men believed that the origin of organic beings was influenced to a great extent by the stars. I am not referring to the dreams of astrologers, but to the Aristotelian theory of the formation of new organic beings from decaying substances, the cause of which was supposed to be a mysterious power proceeding from the heavenly bodies. This ancient theory of spontaneous generation is far less contrary to common sense than the modern theory, and considering the state of scientific knowledge at the time was far more pardonable. It was taken up in very various ways by the naturalists, poets, and quacks of the period. As an example I may refer to Vergil's 'Georgics,' where there is a recipe for making bees. A dead ox is to be laid out, beaten vigorously, and left to decompose in its hide, until the bees develop in its body. Vergil did not draw upon his imagination when he gave this recipe; it is based upon real observations wrongly interpreted. There are some robber flies that resemble bees very closely, belonging to the genus *Eristalis*, the larvae of which develop in decomposing matter. It might easily escape the notice of a casual observer that the old flies had already laid their eggs there. Even the famous ant-stone, *lapis myrmecias*, which was supposed to grow in ants' nests and to combine the nature of the ant with that of a precious jewel, able to cure various ailments among mankind, is no mere fiction. The story originated in the discovery in ants' nests of the cocoons of the rose chafer (*Cetonia floricola*) which, when the beetle has developed, really contain a living jewel of a golden or emerald green colour, in a covering of the size of a pigeon's egg, formed of earth.[1]

As methods of observation improved in modern times, the theory of spontaneous generation gradually lost favour. As early as the seventeenth and eighteenth centuries it was challenged by naturalists, such as Redi, Malpighi, Swammerdam and Réaumur, and was pushed into the background, although in the nineteenth century it had some champions who defended it theoretically. In the middle of the nineteenth century much was done to overthrow it by von Siebold and R. Leuckart in the department of parasites, by Ehrenberg in that of Infusoria, and by de Bary, and especially by Pasteur in that of Bacteria. Thus modern scientific research has removed one support after another from the theory of spontaneous genera-

[1] Cf. Lochner v. Hummelstein. 'Lapis myrmecias falsus, cantharidibus gravidus' (*Ephem. Ac. Nat. Curios* 1687, Observ. ccxv, 436–441).

tion, until now nothing is left of it—except that it is 'a postulate of science.'

As early as 1651 an Englishman, William Harvey, formulated the famous principle *Omne vivum ex ovo*, in his work 'De generatione animalium.' In this form the dictum is not universally true, for the unicellular organisms multiply themselves not by eggs, but by cell-division or gemmation, which is, however, only a special form of cell-division (see p. 86). Therefore Harvey's saying must be amended and receive the form: *Omne vivum ex vivo*. It was not until two hundred years later that Rudolf Virchow, the founder of cellular pathology, in 1858 set the modern axiom of biology, *Omnis cellula ex cellula*, beside Harvey's dictum.

The theory of spontaneous generation found for a time its last refuge in just that cellular theory which subsequently dealt it its death-blow. In order to account for the origin of the cell, Schwann propounded his *Cytoblastema* theory, according to which cell-formation took place by way of a sort of crystallising process in matter still unorganised. The first deposit from the primitive matter or cytoblastema was, according to Schwann, the nucleolus of the cell, round which a membrane formed; between the nucleolus and the membrane a fluid penetrated by endosmosis, so forming the cell-nucleus; round this again there was a second membrane, and by endosmosis more fluid made its way between this membrane and the nucleus, so that finally the membrane enclosed the cell, having in its centre the nucleus with the nucleolus. Schwann imagined the cell to have been formed in this way spontaneously out of unorganised matter by *generatio aequivoca*. It was a most ingenious idea, but it did not correspond with facts, and it soon had to be given up.

The somewhat later *blastem* theory advanced by Charles Robin, a French scientist, has this advantage over Schwann's cytoblastema theory, that it does not assume the formation of cells out of unorganised matter. Robin's blastems, which give rise to new cells, are the product of previously existing cells of the same organism. It is, therefore, not correct here to speak of a *generatio aequivoca*. Robin's theory was nearer to the process that really goes on in cell-formation in another respect also, for he thought that the nucleus of the new cell was formed before the nucleolus. Round the nucleus a layer of protoplasm took up its position and finally surrounded itself with a membrane. This account of the genesis of the cell also failed to agree with ascertained facts. It is true that for a considerable time it found much support in the embryonic development of insects. Hugo von Mohl had proved that free cell-formation did not occur among plants, and Albert von Kölliker had proved its non-occurrence among animals; it had long been established that among higher animals the blastoderm of the embryo had its origin in continued cell-division from the cleavage-nucleus produced by the union of the egg- and sperm-nuclei, and yet for some time it seemed that among insects there was free cell-formation in Robin's sense. In 1864, in his classical studies on the development of Diptera, August Weismann still felt bound to uphold this theory of free cell-formation, as he could not perceive any processes of cell-division in the formation of the blastoderm in these insects. As recently as 1888 Henking[1] thought that he had found that the nuclei of the blastoderm in the egg of *Musca* were not formed by division from the cleavage-nucleus, but by free nuclear formation in the isolated particles of plasm dispersed among the masses of yolk.

On this subject Korschelt and Heider remark in their excellent 'Lehrbuch der vergleichenden Entwicklungsgeschichte der wirbellosen Tiere' (special section, part 2, Jena, 1892, p. 764), that this opinion seems to be quite untenable. In those insect eggs which are so extraordinarily rich in nutritive yolk (deuteroplasm) as are the eggs of flies, the processes of cell-division are very apt to escape observation under the microscope. In other insect eggs that contain less yolk (such as those of the plant-louse, gall-gnat and gall-fly), these processes have undoubtedly been observed, and we must take the latter, rather than the former, as illustrating the normal course of blastoderm formation in the eggs of insects. Thus the last support of free cellular formation has been removed, and we now have a general law that, not only does every new cell arise out of a previously existing cell, but each new nucleus out of a previously existing nucleus.

[1] 'Die ersten Entwicklungsvorgänge im Fliegenei und freie Kernbildung' (*Zeitschrift für wissenschaftliche Zoologie*, XLVI).

Walter Flemming in 1882 added the third dictum, *Omnis nucleus ex nucleo,* to the two biological axioms laid down by Harvey and Virchow respectively. As Boveri's theory of the individuality of the chromosomes (see p. 167) is constantly receiving fresh confirmation, we must add yet a fourth dictum, dating from 1903, viz.: *Omne chromosoma e chromosomate.* In it the antagonism shown by modern biology to the theory of spontaneous generation has reached its climax. The four axioms—*Omne vivum ex vivo, Omnis cellula ex cellula, Omnis nucleus ex nucleo, Omne chromosoma e chromosomate*— have destroyed the theory as far as modern naturalists are concerned. It can continue to exist only outside the sphere of scientific thought.

Very descriptive of the scientific weakness of the theory of spontaneous generation are the following remarks which occur in the famous biologist, Oskar Hertwig's 'Allgemeine Biologie' (1906, p. 263): 'Considering the state of natural science at this time, there seems but little prospect that any one engaged in scientific research will succeed in artificially producing even the simplest living organism from lifeless material. He has certainly no more hope of success than Wagner in Goethe's "Faust" had in his attempts to brew a homunculus in a retort.'

J. Reinke, the distinguished botanist, has expressed himself much more sharply still on the subject of the theory of spontaneous generation, in many places in his works.[1]

It is, therefore, an absolutely necessary consequence that organic life on earth did not begin by way of spontaneous generation, and that it is altogether unscientific to represent this theory as a postulate of science, in spite of its being quite untenable. Our modern evolutionists above all others lay great stress upon the fact that the laws of nature now existing

[1] See his book *Die Welt als Tat* (Berlin, 1899), the third edition of which appeared in 1903. In it J. Reinke devotes a chapter, almost thirty pages in length, to proving the impossibility of spontaneous generation, and he deduces from this argument the conclusion that we shall never be able to account for the origin of organic life unless we accept the creation. In 1905 a fourth edition was published. Cf. also J. Reinke's *Einleitung in die theoretische Biologie*, 1901, pp. 555–562, and his treatise 'Der Ursprung des Lebens auf der Erde' (*Türmer Jahrbuch* 1903); also his inaugural oration at the International Botanical Congress in Vienna, June 12, 1905, 'Hypothesen, Voraussetzungen, Probleme in der Biologie' (*Biolog. Zentralblatt*, XXV, 1905, No. 13, pp. 433–446), pp. 442, 443. He has an excellent refutation of the hypothesis of spontaneous generation in his last book, *Philosophie der Botanik* (Leipzig, 1905), chapter xii, On the Origin of Life.

must have existed from the beginning, and that we must regard them as a safe standard, applicable also to the most remote history of the animal and vegetable world, if we wish to solve the problem of descent scientifically. It is quite in vain that they appeal to the 'uniform causal connexion of natural phenomena' to support the theory of spontaneous generation. J. Reinke says very aptly ('Einleitung in die theoretische Biologie,' p. 558): 'I am of opinion . . . that the assumption of spontaneous generation in past ages agreed no more with our ideas of causality than a hypothesis that a million years ago water flowed uphill of its own accord would agree with them.' And in another place he says ('Philosophie der Botanik,' p. 188): 'Just as at no stage of the earth's cooling was it possible for two lines to form a triangle, so was it never possible for an organism of the most primitive kind to be produced by the forces and combinations of inorganic matter.' There is therefore, as Reinke rightly points out, scarcely a greater incongruity possible, than for one and the same man to reject spontaneous generation, as a thoughtful naturalist, and in the same breath to declare it to be a postulate of science, when he speaks as a philosophical thinker. What is a 'postulate of science?' This name can properly be given only to a truth that proceeds logically from facts, and never to a hypothesis that is in antagonism to them.

From this point of view, what true postulate of science is there to account for the first origin of organic life?

Life cannot always have existed on our earth; modern cosmogony leaves us no room to doubt this, for it teaches us that the earth was once in a condition of molten heat. How, then, did the first organisms come into being?

It is an unprofitable amusement to fancy, with Thomson and Helmholtz, that they were brought by meteors from other planets, or, with H. E. Richter and Arrhenius, that they fell upon the earth as cosmic dust,[1] for life must have had a beginning

[1] In his *Einleitung in die theoretische Biologie*, p. 559, Reinke says: 'Men like Lord Kelvin (Thomson) and Helmholtz would not have devised their hypothesis of the advent of primitive cells from other planets, if they had not regarded spontaneous generation as lost beyond all hope of recovery.' It should be noticed that Thomson has repeatedly and decidedly said that we must assume the existence of a Creator. Cf. Karl Kneller, S.J., *Das Christentum und die Vertreter der neueren Naturwissenschaft*, Freiburg i. B., 1903, pp. 28–30, and *The American Quarterly Review*, XXVIII, 1903, p. 603.

on the planets of other solar systems also, since they too are subject to the same cosmogonic laws.

Therefore, how did the first organisms come into being? Every effect must have an adequate cause. Inorganic matter cannot have been this cause, for science teaches us this when she declares spontaneous generation to be contradictory to facts. But at that time there was still nothing in the world but inorganic matter and its laws. Therefore there must have been some cause extraneous to this world, which brought forth the first organisms out of matter. This cause, extraneous to the world, and differing substantially from it, in spite of its omnipresence in it, is an intelligent cause, and is the personal Creator, so often denied and feared by modern monism.

Monism, in its desire to get rid more easily of the theistic conception of God, has caricatured it, until finally the Creator has been represented as a 'gaseous vertebrate' (Haeckel) bearing alarming testimony to its discoverer's want of philosophical knowledge. The new idea of God invented by monism, and set up in place of the personal Creator, is nothing but a fantastic sort of idol draped in a covering of theism to hide its atheistic nakedness. Everything acceptable in the monistic idea is borrowed from theism, the omnipresence of God in nature, His action in all creation, &c. But what is peculiar to monism, and marks it off from theism, is the theory of the substantial identity of God and the world, which is nonsense from the philosophical point of view. A god identical with the world, and developing himself through it, is not an infinitely perfect being, having the reason of his existence always in himself, but he is a mass of imperfections and contradictions. Any thoughtful student of nature must be able to see this for himself.

We may therefore close our examination of the theory of spontaneous generation with the following statement: Organic life has not always existed in our world, nor can it have originated by itself from inorganic matter. Natural science brings us thus far; and natural philosophy leads us on to the further irrefutable conclusion:—It follows that some cause superior to the world produced the first organisms from lifeless matter. When and how this took place is perfectly indifferent, as far as the necessity of this conclusion is concerned.

Even if we did not need to assume the existence of any special vital principle, and if the living atoms differed from inorganic matter only by being in a state of movement peculiar to themselves, we could still not dispense with the Creator to create primitive matter, and to impart to those atoms their state of movement, in order thereby to make them the constituents of the first living creatures. But we are still more forcibly constrained to acknowledge the existence of a personal Creator by the fact that modern science proves, more and more clearly, that all vital processes are subject to their own particular law, and we are thus compelled to accept the entelechies, or formal principles, which raise the laws governing inorganic matter to a higher, vital conformity to law in the case of living creatures.

Thus the acceptance of a personal Creator is seen to be a real 'postulate of science.' For, as J. Reinke rightly points out: 'If we assume at all that living creatures once were formed of inorganic matter, as far as I can see, the theory of creation is the only one which satisfies the demands of logic and causality, and so satisfies those of reasonable scientific research.'[1]

[1] *Einleitung in die theoretische Biologie*, p. 559. See also the quotations from Charles Darwin and Lyell on the indispensability of a creation in Chapter IX, at the end of § 6.

CHAPTER VIII

THE PROBLEM OF LIFE

INTRODUCTION AND SURVEY OF THE VARIOUS TYPES OF CLEAVAGE

LIFE is for the student of nature a fact which he must take as his starting point for the further investigation of the phenomena of life. All attempts to account for the origin of life from inorganic matter by way of spontaneous generation have failed, as they contradict what modern cytology teaches. This has been shown clearly in Chapter VII. Organic chemistry may make a bold and triumphant advance by means of the laborious and ingenious experiments, by which she examines the elementary composition of living organisms and the chemical processes of their metabolism. She may succeed eventually in producing synthetically a highly complex molecule of albumen in her test-tubes, but one thing will always be wanting to the artificial product, viz. life.

The laws of inorganic matter apply also to living creatures, but in their case the laws are subordinate to a higher unity, which brings their activities into that wonderful harmony, tending to fulfil a purpose, that we call a vital process.

Even in the simplest unicellular organisms, Amœbae and Bacteria, we encounter the mysterious problem of life. We meet with it in a more astonishing form in the fertilised egg-cell, out of which a multicellular plant or animal is produced by a long series of cell-divisions. In Chapter VI we have traced the microscopical processes that go on within the germ-cells, before their union in the fertilised ovum. Now let us consider the following deeply important questions concerning the continuation of the same great problem of life :—

1. How does the organism, in its individual ontogeny, develop from the egg-cell into a perfect animal or plant?
2. How have the organisms on our earth been evolved, each according to its kind, from the first appearance of life in the world to our present Fauna and Flora?

In this chapter we can deal only with the first of these questions, leaving the other for subsequent discussion.

It will conduce to a better understanding of the following arguments if we begin by studying the chief kinds of cleavage in animal ova.[1]

After fertilisation is effected, the egg-cell divides rapidly into 2, 4, 8, 16, &c., cells, which become smaller as the process of cleavage continues. These cells are called *cleavage-spheres* or *blastomeres*. We speak of each egg as having an *animal* and a *vegetative pole,* inasmuch as the substance at one pole serves chiefly to form the animal organs or nervous system, and that at the other pole serves chiefly to form the vegetative organs or digestive tract.

Different types of cleavage processes are distinguished; the peculiarities of which depend upon the quantity of food-yolk or deuteroplasm in the egg, and upon its position.

The cleavage of the egg is *total* or *partial,* according as the whole substance of the egg, or only part of it, undergoes the process of cleavage. It is total in eggs poor in yolk, partial in those rich in yolk, as the yolk impedes cleavage. In total cleavage the whole substance of the egg is used to build up the embryo, and therefore eggs that show this type of cleavage are called *holoblastic,* whilst those with partial cleavage are called *meroblastic.* In holoblastic eggs with total cleavage,

[1] For further details see R. Hertwig, *Lehrbuch der Zoologie,* 1905, pp 125, &c. (Eng. trans. pp. 151, &c.)

it is either *equal* or *unequal*, according as the cleavage-spheres are equal or unequal in size ; this depends upon the quantity of yolk in the egg.

In meroblastic eggs with partial cleavage, it is either *discoidal* or *superficial*. This distinction depends upon the position of the yolk in the egg. If the yolk is accumulated about the vegetative pole, the cleavage is limited to the animal pole (discoidal cleavage) ; if the yolk lies in the centre of the egg, only the surface of the egg shows a thin layer of cleavage cells surrounding the unsegmented central mass (superficial cleavage).

The eggs which have their yolk more or less concentrated at the vegetative pole are called *telolecithal* ; those with a mass of yolk in the centre are called *centrolecithal*.

Superficial cleavage occurs among arthropoda and especially among insects. Discoidal cleavage occurs in birds and in most of the other vertebrates, among molluscs, also in cuttle-fish, and in some Arthropoda and Tunicata. Equal and unequal cleavage, however, may appear in all kinds of multi-cellular animals.

The account just given of the different types of cleavage does not depend immediately upon the question whether preformation or epigenesis controls the cleavage of the egg. We shall have to study the behaviour of animal eggs towards these two factors in development in the third part of this chapter.

1. The Problem of Determination and its History

It was chiefly through Karl Ernst von Baer (1791–1876) that the study of the individual development of animals became a special branch of zoology, to which the name ontogeny has been given. There is an analogous branch of botany, dealing with the individual development of plants.[1]

Both confront us with the same old and yet ever new questions with which from remote antiquity the minds of ordinary men have busied themselves, no less than the inquiring spirit of the scholar. Why are children like their parents ?

[1] See the general sketch of the departments of biological science, Chapter I, pp. 3, &c.

P

Why does an oak always grow out of an acorn, and why is a chicken always hatched out of a hen's egg? Whence comes the specific conformity to law in accordance with which, from the fertilised egg of any given species, there is invariably produced a being similar to that which gave life to the egg-cell? What is the influence determining the germ of the new individual to follow one line of development rather than another? Moreover, are the laws controlling this development purely mechanical, or are there also vital laws, essentially superior to what goes on in inanimate nature?

These are undoubtedly very interesting and important questions, having a bearing not only upon biological research, which is seeking to solve the problem of life by way of natural science, but also upon philosophy, which is striving to penetrate into the essential nature of life by means of the phenomena of life.

We stand therefore face to face with the problem of determination, i.e. with the question: 'What are the causes controlling embryonic development?' Regarded from afar this problem may seem to the layman to resemble a porcupine bristling with all manner of technical difficulties, so that an ordinary intellect can scarcely venture to approach it. Let me, however, see if I cannot succeed in inducing this porcupine of the problem of determination to lay down his prickles, and show himself to my readers in a harmless form, presenting no particular difficulties to a man of average intelligence.

To begin with, I must follow Oskar Hertwig [1] in pointing out that a one-sided view of the subject cannot fail to be a false one. Many internal and external causes co-operate in the development of organic beings, and they do so in such a way that the internal causes are invariably the foundation for the action of the external factors.

The problem that we have to discuss is closely connected with the subject of Chapter VI, viz. the relation of the processes of cell-division to the problems of fertilisation and heredity. We came to the conclusion that the chromatin constituents of the nuclei of the germ-cells, that is to say their chromosomes, might with great probability be regarded as the chief material bearers of the phenomena of heredity, and

[1] *Allgemeine Biologie*, pp. 132, &c., and 138, &c.

consequently also as the chief bearers of the laws governing the particular development of each kind of animal and plant.

Yet in making this statement, we have alluded to only one side of the problem of organic development, viz. to that which is the subject of microscopical cytology. We now encounter a series of other questions which are of great interest as affecting the problem of life:—Does the development of the fertilised ovum depend upon a self-differentiation, controlled exclusively by the interior factors already present in the egg, or does it depend upon a differentiation controlled chiefly by exterior causes? Must we uphold the theory of preformation, which assumes that there is in the egg a foreshadowing of the whole future being, or the theory of epigenesis, which asserts that the organs of the embryo are formed afresh in the course of its development? The so-called problem of determination is comprised in the answers to these questions. It will be well to show shortly what success has hitherto attended the attempts made to solve it. Incidentally we shall have to be careful to ascertain whether the individual development of organic beings is controlled by some special laws of life, as vitalism asserts, or whether it can be satisfactorily explained, as the mechanics theory maintains, by merely chemico-physical causes.

The branch of biology that deals with experimental research into the laws and causes of organic formation is known as the physiology of development. Wilhelm Roux, the principal founder of this branch of science, called it 'mechanics of development.' But as the mechanical explanation of the processes under consideration is only a part of the problem, we agree with Hans Driesch, who has done excellent work in this department of research, that it is better to adopt the name physiology of development.[1]

[1] Among the publications bearing on this subject I may mention particularly *Das Archiv für Entwicklungsmechanik der Organismen*, edited by W. Roux in Halle a. S. Also W. Roux, 'Einleitung zu den Beiträgen zur Entwicklungsmechanik des Embryo' (*Zeitschrift für Biologie*, XXI, 1885); *Die Entwicklungsmechanik der Organismen, eine anatomische Wissenschaft der Zukunft*, Vienna, 1890; *Die Entwicklungsmechanik, ein neuer Zweig der biologischen Wissenschaft*, Leipzig, 1905. E. Pflüger, 'Über den Einfluss der Schwerkraft auf die Teilung der Zellen und auf die Entwicklung des Embryo' (*Archiv für die gesamte Physiologie*, XXXII, 1883); 'Beiträge zur Entwicklungsmechanik des Embryo': No. 1. 'Zur Orientierung über einige Probleme der embryonalen Entwicklung' (*Zeitschrift für Biologie*, XXI, 1885); 'Über die Bestimmung der Hauptrichtungen des Froschembryo im Ei und über die

P 2

It may appear to some readers that these questions have already been answered satisfactorily by the results previously described of microscopic morphology. Among the higher organisms at least, under normal circumstances, the development of a new individual can result only from fertilisation, which consists essentially in the union of the nuclei of the ovum and spermatozoon, as we saw at the end of Chapter VI (pp. 156, &c.). As the chromosomes of the nuclei of the germ-cells are the bearers of heredity, visible under the microscope and passing in definite number and order from the parents to the children, and as (according to Boveri's theory of the individuality of the chromosomes) they preserve some amount of independence during the whole process of development, it may seem a superfluous question to ask whether the development of the fertilised ovum depends upon preformation or epigenesis, upon an independent or a dependent differentiation. Has not

erste Teilung des Froscheis' (*Breslauer ärztliche Zeitschrift*, 1885). O. Hertwig, 'Über den Wert der ersten Furchungszellen für die Organbildung des Embryo' (*Archiv für mikroskopische Anatomie*, XLII, 1893); *Zeit- und Streitfragen der Biologie*, I, Jena, 1894; *Präformation oder Epigenese?* II, 1897; *Mechanik und Biologie; Die Zelle und die Gewebe*, II, Jena, 1898; *Allgemeine Biologie*, Jena, 1906 (especially recommended). A. Weismann, *Das Keimplasma*, Jena, 1892; *Vorträge über Deszendenztheorie*, Jena, 1902 (*Lectures on the Theory of Evolution*, Eng. trans.). E. B. Wilson, '*Amphioxus* and the Mosaic Theory of Development' (*Journal of Morphology*, VIII, 1893). H. E. Crampton, 'Experimental Studies of Gastropod Development' (*Archiv für Entwicklungsmechanik*, III, 1896). C. O. Whitman, 'Evolution and Epigenesis' (*Wood's Hall Biological Lectures*, 1894). Hans Driesch, *Analytische Theorie der organischen Entwicklung*, Leipzig, 1894; *Die organischen Regulationen*, Leipzig, 1901; 'Kritisches und Polemisches' (*Biolog. Zentralblatt*, XXII, 1902, Nos. 5, 6, 14, 15; XXIII, 1903, Nos. 21–23); 'Ergebnisse der neueren Lebensforschung' (*Politisch-Anthropologische Revue*, II, 1903, part 10). O. Herbst, *Formative Reize in der tierischen Ontogenesis*, Leipzig, 1901. Th. H. Morgan, *Regeneration*, New York and London, 1901. O. L. Zur Strassen, 'Über das Wesen der tierischen Formbildung' (*Verhandl. der Deutschen Zoolog. Gesellsch.*, 1898, pp. 142–156). K. Heider, 'Das Determinationsproblem' (*Verhandl. der Deutschen Zoolog. Gesellsch.*, 1900, pp. 45–97). L. Kathariner, 'Über die bedingte Unabhängigkeit der Entwicklung des polar differenzierten Eis von der Schwerkraft' (*Archiv für Entwicklungsmechanik*, XII, 1901, part 4); 'Weitere Versuche über die Selbstdifferenzierung des Froscheis' (*Ibid.* XIV, 1902, parts 1 and 2); 'Schwerkraftwirkung oder Selbstdifferenzierung?' (*Ibid.* XVIII, 1904, part 3, pp. 404–414). An excellent general account of the problem of Determination is given by Korschelt and Heider in their *Lehrbuch der Entwicklungsgeschichte der wirbellosen Tiere*, Allgem. Teil, I, Jena, 1902, § 1, cf. especially chapter ii, 'Das Determinationsproblem' (pp. 81–150). In the same book will be found a list of all the literature on the subject up to the year 1902; for works published since that date see O. Hertwig, *Allgemeine Biologie*. Of botanical works dealing with embryology I may mention particularly: W. Pfeffer, *Pflanzenphysiologie*, I, Leipzig, 1897; II, first part, 1901. Also G. Klebs, 'Über Probleme der Entwicklung' (*Biolog. Zentralblatt*, XXXIV, 1904, Nos. 8, 9, 14, 15, 16, &c.).

this question been already answered in what has gone before, and have we not already decided in favour of preformation, and of independent differentiation?

The matter is not so simple as it appears. Even if we assume that the chromosomes of the nuclei of the germ-cells are the chief material bearers of heredity, passing on from one generation to another, we still have to solve the problem of the development of the organism from the fertilised ovum. This difficult question still remains: 'What causes the groups of cells, formed out of one egg-cell by cleavage-division, to differ from one another more and more, both morphologically and physiologically, as the development of the embryo proceeds? How is it that these groups of cells develop into the various tissues and organs of one and the same individual?' In other words: 'What causes underlie the process of harmonious differentiation, by means of which the wonderful and complicated structure of the complete organism with all its manifold parts is produced from the apparently simple ovum?'

The physiology of development, which we now have to study, approaches this problem on lines quite unlike those followed by microscopical anatomy. The latter has recourse to modern methods of staining and cutting sections, and examines the tissues and cells of animals under the strongest microscopes, and strives to trace all the morphological changes in the nucleus and cytoplasm of the cells, but the former proceeds by way of actual experiment. It takes, for instance, the living ovum of a frog, subjects it to all possible kinds of artificial treatment, to pressure, twisting, division or partial destruction of its cleavage-spheres, and then observes how the embryo develops under these conditions. From these observations it draws its conclusions regarding the laws and causes of the embryonic development of living creatures.

It proceeds also to study the course of regeneration in the living organism by similar methods. It tries experimentally, in the case of a creature that has reached an advanced stage of development, how far, and in what way, the faculty is retained of forming afresh lost organs and tissues. The experiments made by G. Wolff and others with a view to determining the power of regeneration in the lens of the eye of a salamander have become particularly famous.[1]

Before we discuss the results of modern research in embryology, we must refer shortly to the previous history of the problem of determination.[2]

The question whether the future individual is contained in the egg, and, if so, under what form, has aroused the interest of students in all ages, although until recent times there has been very little certain knowledge upon which to found any theory. In the seventeenth and eighteenth centuries the most eminent scientists, such as Swammerdam, Malpighi, Leeuwenhoek, Haller, Bonnet and Spallanzani declared themselves to be in favour of the preformation theory, then known as the doctrine of evolution, or unfolding.[3]

They observed the development of the butterfly in the pupa, and the blossom in the bud, and laid down the dictum: 'Evolution is merely the unfolding of parts already present in the egg- or sperm-cell, but imperceptible to us by reason of their diminutive size and transparency.' It is true that we can trace in the pupa all the organs of the future butterfly, and in the ripe bud all the parts of the future flower, but when this theory of unfolding is applied to the embryonic development of living creatures, it leads to very peculiar results. According to it, in the first ovum of each species[4] all the individuals of all the succeeding generations must have been contained in infinite numbers and in infinitely diminutive size. For instance, the ova of the first cat must have contained extremely small editions of all the future cats that would ever be born to the

[1] G. Wolff, *Entwicklungsphysiologische Studien*, I. 1895; *Die Regeneration der Urodelenlinse*. Cf. also Part II, 1901, and Part III, 1905, of the same series of studies in the *Archiv für Entwicklungsmechanik*. Hans Spemann, 'Über Linsenbildung nach experimenteller Entfernung der primären Linsenbildungszellen' (*Zoolog. Anzeiger*, XXVIII, 1905, No. 11, pp. 419–432). A list of the other works on this subject by Barfurth, Colucci, Fischel, Herbst, Lewis, Menčl, E. Müller, Schaper and Spemann will be found in Spemann, p. 432. Cf. also O. Hertwig, *Allgemeine Biologie*, pp. 546, &c.

[2] Cf. O. Hertwig, *Allgemeine Biologie*, pp. 350, &c.

[3] At the present day we generally speak of the theory of evolution with reference to the evolution of the species, not with reference to that of the individual. In order to avoid confusion, I have used the expression 'theory of preformation' to designate the theory of evolution in the earlier sense.

[4] Or in the first spermatozoon, for, according to the theory of the animalculists, it was not the egg-cell, but the sperm-cell, which transmitted hereditary qualities. See p. 104 and p. 158.

end of the world. This has also been termed the theory of *emboîtement*.

In 1759 Kaspar Friedrich Wolff in his 'Theoria generationis' for the first time opposed the old theory of preformation, and by so doing became the founder of the theory of epigenesis. After a careful examination of the development of a chicken, he came to the conclusion that the egg was only a mass of unorganised matter, which was gradually organised in the course of the development of the embryo. Wolff's opinion is right to this extent, that the organs of the embryo are really formed anew, because the fertilised egg (as was recognised only in the nineteenth century) still has the character of a simple cell, and so cannot consist of organs. But Wolff was wrong in thinking the egg a mere mass of unorganised matter, for modern microscopical research has revealed to us the wonderfully delicate structure of the egg-cell and its nucleus, and has shown us the chromosomes, which, being definite parts of the nucleus, are the material bearers of heredity, and are distributed with such marvellous exactitude among the cleavage-cells of the egg as it develops. I will not, however, at this point anticipate the historical development of the problem of determination.

As the study of embryology advanced in the first half of the nineteenth century, the theory of epigenesis found increasing favour, and soon became predominant.

In 1853, Rudolf Leuckart, a famous zoologist, wrote in his article on procreation: 'Our knowledge of the development of the embryo and of the formation of the procreative substance admits of only one interpretation, viz. in the sense of epigenesis—there can be no further doubt on the subject; the embryo is the product of a new formation in connexion with the procreative substance.'

As late as the year 1872, Ernst Haeckel in his 'Anthropogeny' described the human embryo in the so-called monerula stage[1] as a 'completely homogeneous, structureless mass,'

[1] We owe the 'discovery' of this stage in the embryonic development of man to a mistake on Haeckel's part. He believed, though wrongly, that the germinal vesicle of the embryo broke up as soon as embryonic development began. According to Haeckel's fanciful anthropogeny, the monerula stage in the human germ is a lineal repetition of the monera stage of our most remote ancestors. As a matter of fact, not only is this monera stage existent only

as a 'simple lump of primitive matter.' Haeckel must certainly have studied the human embryo through very cloudy glasses, if in the year 1872 he was still able to see so little of its finer histological structure, although Goette fared no better in 1875, when he studied the egg of the toad, and declared it to be an unorganised lifeless mass, produced by a transformation of one or more germ-cells.

The theory of epigenesis, however, was not destined to stand its ground much longer. As microscopes became more perfect, both the ovum and the spermatozoon were seen to contain elements of very complicated composition, which had to prepare, by a special process of maturation, for the union of their nuclear substances, effected by fertilisation. At once the breath of popular favour veered round to the preformation theory, although it was no longer the old theory of *emboîtement*, but assumed an entirely new form.

In 1874 Wilhelm His [1] propounded the theory of there being germ regions or local areas for the formation of organs in the individual development of vertebrates.[2] According to this theory definite tracts in the fertilised ovum are, in virtue of some special interior tendency or *Anlage*, destined to form definite organs in the embryo. At the same time he submitted Haeckel's fantastic ideas on human embryology to a most destructive criticism in his article. The new theory of germ-regions for the formation of organs found support in observations made on many other animals, and it was discovered that even in the ovum the so-called primordial axis gave rise to an animal and a vegetative pole, determining the direction in which the future embryo was to develop. Embryology had therefore again taken an appreciable turn in the direction of the preformation theory.

But in 1883 there was an apparent reversion to epigenesis, in consequence of the experiments made by Edward Pflüger, with a view to determining the influence of gravitation upon

in the imagination, but so is also the ontogenetic monerula stage in the development of the human embryo. For a criticism of Haeckel's pedigree of man see Chapter XI.

[1] *Unsere Körperform und das physiologische Problem ihrer Entstehung*, Leipzig, 1874.

[2] Wilson suggests 'Germinal Localisation' as a name for this theory.—*Translator's Note.*

the development of frogs' eggs. To these experiments we owe Pflüger's principle of the isotropy of the egg-plasm, according to which all the protoplasmic constituents of the egg are collectively of equal value with regard to the formation of the organs in the embryo. Pflüger put frogs' eggs in what he called a position of constraint, so that the egg was prevented from turning round in its gelatinous envelope, owing to defective swelling of the latter. Under normal circumstances the animal half of the frog's egg, which consists of lighter substances and contains black pigment, always is uppermost, whilst the pale yellow vegetative pole is underneath. If, however, the egg is prevented from turning, the axis of the egg can be made to form any desired angle with the vertical. Even in this case the first cleavage-plane of the egg as it develops will always be vertical. This might lead us to believe that gravitation alone determined the arrangement of the parts of the embryo, and that it was a matter of indifference which part of the egg lay above or below at the beginning of cleavage.

The conclusions, which Pflüger deduced from this fact in favour of the isotropy of egg-plasm, proved, however, not to be tenable. Wilhelm Roux and Oskar Hertwig soon suggested that the dependence of the evolution of the frog's egg upon gravitation was only a consequence of the unequal specific gravity of its parts. In the eggs placed in abnormal positions the egg envelope was prevented from turning, but the rearrangement of the substances within the egg was unaffected. Born proved this by experiments of his own.

In order to disprove Pflüger's theory of the importance of gravitation in directing the development of the embryo, Roux placed some frogs' eggs, already developing, on a disc that rotated vertically, so that their position with regard to gravitation was constantly changing. In spite of this, their development was normal both as to time and manner. Yet, as Kathariner has recently pointed out,[1] in his clinostatic experiments Roux had replaced the force of gravitation by

[1] Über die bedingte Unabhängigkeit des polar differenzierten Eis von der Schwerkraft' (*Archiv für Entwicklungsmechanik*, XII, 1901, Part 4, pp. 597–609); 'Weitere Versuche über die Selbstdifferenzierung des Froscheis' (*ibid.* XIV, 1902, Parts 1 and 2, pp. 289–299); 'Schwerkraftwirkung oder Selbstdifferenzierung?' (*ibid.* XVIII, 1904, Part 3, pp. 404–414).

another force, viz. the centrifugal; and consequently it was still not certain that the development of the egg was completely independent of an external directive force.

In order to settle this point, Kathariner had recourse to another method. He kept the fertilised frogs' eggs in constant rotation by means of a stream of water. Even then they developed in a perfectly normal way, although somewhat more slowly than usual. These experiments have proved conclusively that the reasons for the specific development of a frog's egg into a frog are in the egg itself, and cannot be found in any external influences. The development of the egg depends on self-differentiation, as Roux declared. We must regard as disproved, once for all, the theory which Pflüger enunciated as follows, in support of epigenesis: 'I am of opinion that the fertilised ovum no more bears an essential relation to the subsequent organisation of an animal, than the snowflakes do to the size and shape of the avalanche to which they contribute: the fact that out of a germ the same thing is always produced is due to its being always subjected to the same external conditions.'

2. More Detailed Discussion of the Problem of Determination

When we find scientific men like Oskar Hertwig,[1] who are not far from being vitalists, still feeling bound to ascribe to external factors, such as heat, the rank of causes of specific development, we must believe that this is due to a confusion of the general conditions of development with its particular causes. We have many external means of accelerating or retarding development, and of making it follow a normal or an abnormal course, but we are never able to alter the laws of specific development, for instance in the frog's egg. If, therefore, such an egg invariably produces a frog, it does so through some self-differentiation in the fertilised ovum.

If we regard the egg with its capacity for development as a *whole*, the question whether preformation or epigenesis controls its action is therefore already answered in favour of

[1] *Die Zelle und die Gewebe*, II, 1898. Cf. my remarks on O. Hertwig's opinions on p. 220.

preformation; there are in the egg some dormant tendencies which underlie its specific development. But this is not a complete solution of the problem of determination.

We have to answer another and a much more difficult question: 'In what relation do the individual parts of the fertilised ovum stand to one another? Is their development fully independent, based on self-differentiation, or is it in a state of regular dependence upon the other parts of the egg, and based, therefore, on a dependent differentiation?'

I have already discussed Pflüger's theory of the isotropy of egg-plasm, according to which all parts of the egg are quite uniform in material and in their influence on the development of the various organs of the embryo (see p. 217). This theory must be given up, for, as Roux pointed out, even before cleavage begins, the median plane of the future embryo is determined by the position of the cleavage-nucleus in copulation, i.e. by the course taken by the male pronucleus in order to unite with the female pronucleus, and so form the cleavage nucleus of the fertilised egg. Recent microscopical research has revealed the regular distribution of the chromatin of the cleavage-nucleus to the daughter-cells of the egg, and this distribution introduces the development of the embryo. We must therefore ascribe to the chromosomes of the nuclei an important part in determining the formation of the organs in the embryo. This consideration gives support to Roux and Weismann's theory of nuclear regions for the formation of organs. Here too, therefore, the theory of preformation seems to prevail over epigenesis.

In fact, epigenesis seems almost hopelessly weak as a theory, if we take into account only those epigenetic opinions which are based on mechanics, and aim at accounting for the whole development of the embryo merely by the attraction and pressure of the cleavage-spheres. But the chief supporters of epigenesis—men like Oskar Hertwig and Hans Driesch—are by no means adherents of the theory of mechanism in the ordinary sense of the word. Oskar Hertwig's views on the subject of organic development have much in common with vitalism; he has expressed them in his earlier works, but a concise statement of them may be found in his 'Allgemeine Biologie,' 1906 which is practically a new edition of his previous textbook 'Die Zelle und die Gewebe' ('The Cell and the Tissues') published in 1898.

In discussing the various internal and external causes of development (pp. 132–140), he says that both factors must co-operate in every process of development; but, as he thinks the internal causes (or tendencies to development) always form the basis for the action of the external influences, it is impossible to say that he gives a purely mechanical explanation of the process of development. On the contrary (pp. 141, &c.), he expressly emphasises the 'very important differences existing between machines and organisms, between what is mechanical and what is organic.' In his 'Allgemeine Biologie' he devotes only two chapters (xx and xxi) to the external factors of organic development, but no less than four chapters (xxii–xxv) to the internal factors, and ascribes to them the chief importance, especially in the case of animals (p. 508). He expresses himself as a vitalist in speaking of the various stages of the process of development, and says (p. 519): 'The form at any given moment appears to be in many respects a function of the growth of the organic substance; its persistence is subject to definite conditions; and as they change in consequence of advancing growth, they effect a modification, adapted to the purpose in view, in the form of the substance, which is capable of reacting under their influence.'

At the close of this chapter I shall recur to Oskar Hertwig's attitude towards vitalism. In 1898 he felt bound to ascribe to external mechanical causes [1] a direct formative influence upon the process of development in many cases, but in 1906 he modified this opinion considerably. His earlier views were challenged by O. L. Zur Strassen in a lecture delivered on June 3, 1898, at the eighth meeting of the German Zoological Association at Heidelberg.[2]

According to O. Hertwig, the division of the fertilised ovum into cells of equal size and similar structure is effected by the vitelline contents of the cells and the external shape of the cleavage-spheres (blastomeres). He thinks that the delicate mechanism of mitotic karyokinesis, in which the egg changes into the groups of cells in the embryo, is the cause of cell-division as such, but not of the differentiation of these cells

[1] *Die Zelle und die Gewebe*, II.
[2] 'Über das Wesen der tierischen Formbildung' (*Verhandl.*, pp. 142–156).

to form organs and tissues, although the two processes are connected. Hertwig attempts to account for unequal cell-division by means of the mechanical influence of the yolk contained in the egg, which, he thinks, causes the daughter-cells to be of different sizes. If more deuteroplasm is accumulated at one pole of the egg than at the other, the nucleus of the egg-cell is, according to Hertwig, mechanically pushed to the opposite pole, and the result is the division of the egg into two cleavage-spheres of unequal size.

Reasonable as this may sound, the rule still does not universally hold good, and there is not a purely mechanical regularity in the process of cell-division. There are, for instance, as Zur Strassen points out, a number of cases (e.g. in the cleavage of the egg of the maw-worm, *Ascaris*) where the actual process is the direct reverse of that required by Hertwig's 'law.' In this particular egg, when the first cleavage-spindle is formed, the upper part of the plasm is pale in colour and poor in yolk; whilst the lower part is rich in yolk. Nevertheless, after the cleavage the upper daughter-cell is the larger, and the lower is the smaller, in spite of its abundance of yolk.

O. Hertwig attempted to give a very simple account of the uneven rate of division of the cleavage-spheres by means of the mechanical action of the yolk. He thought that cells containing much yolk divided more slowly than those containing less, because the yolk offered an external resistance to the cleavage processes of the protoplasm. But here, too, there are facts in direct opposition to Hertwig's mechanical law. According to Jennings, in the development of the Rotifer *Asplanchna* and of many other species, the larger cells, that are rich in yolk, have a decided tendency to divide more quickly than the smaller cells, that are poor in yolk.

Purely mechanical factors must by their very nature always act in the same way, and these 'exceptions' to Hertwig's mechanical laws show that the laws, even where they are apparently observed, are not purely mechanical, but a vital conformity to law underlies them, controlling and regulating the action of the mechanical factors.

Of still greater importance for the decision of the question whether the development of the organism can be accounted for on purely mechanical grounds, is the regular direction in which the cells of the embryo divide, for all growth in a definite direction is accompanied by a corresponding formation of the nuclear figures in the processes of mitotic division, and therefore the series of cleavage stages in the developing embryo is based primarily upon that definite direction of division. If it were possible to find a purely mechanical principle to account for this, it would go far towards enabling us to explain the processes of development on mechanical lines. Oskar Hertwig thought that he had discovered a principle of this kind, and enunciated the following 'law' regarding it: 'The division-spindle of the cell is, in the case of non-spherical cells, placed in the direction of the largest mass of protoplasm, i.e. in the longest axis of the cell.'

From the purely mechanical point of view this is quite natural, and there are in fact many cases of agreement with this law—but there are, on the other hand, a great many other facts that contradict it.

As Zur Strassen points out, it is easy to bring forward an overpowering number of instances in which the division-spindle does not follow the longest axis of the cell, which would be a convenient and natural arrangement from the mechanical point of view, but it follows a shorter axis, often the shortest possible, so that it seems to challenge the greatest pressure instead of avoiding it, as it should do, if Hertwig's mechanical theory were correct. This occurs in all cylindrical epithelia and also in very many of the early blastula stages of various organisms.[1]

With regard to the cleavage stages of the embryo, it has been conclusively shown by Jennings in the case of a Rotifer, *Asplanchna*, by Conklin in the case of a snail, *Crepidula*, by Bergh in various Crustacea, and by Sobotta in the lancet fish, *Amphioxus*, that there is no such thing as a direct influence of the shape of the cell upon the direction of the spindle that is easily explicable on mechanical lines. There is therefore no justification for Hertwig's 'mechanical law,' as stated above.

[1] By the *blastula stage* we understand the first development of the embryo, in which the ectoderm is formed as a hollow sphere consisting of one layer of cells. The next is the *gastrula stage*, in which, by means of invagination of part of the blastula, the intestine is formed and the entoderm begins to grow. Between ectoderm and entoderm there is formed subsequently a third layer of cells, called the mesoderm.

Still less is there any justification for a theory propounded by J. Loeb, an American. He thinks that the regular interaction of the parts of the embryo depends upon the mechanical pressure exercised upon one another by the crowded cleavage-spheres, forcing them by merely external means to assume a definite geometrical form. Such crude attempts at explaining facts on mechanical lines are almost as unsuccessful in embryology as in animal psychology.[1]

Zur Strassen has arrived at the following conclusion :—'That the cell in its living plasm contains mechanisms enabling

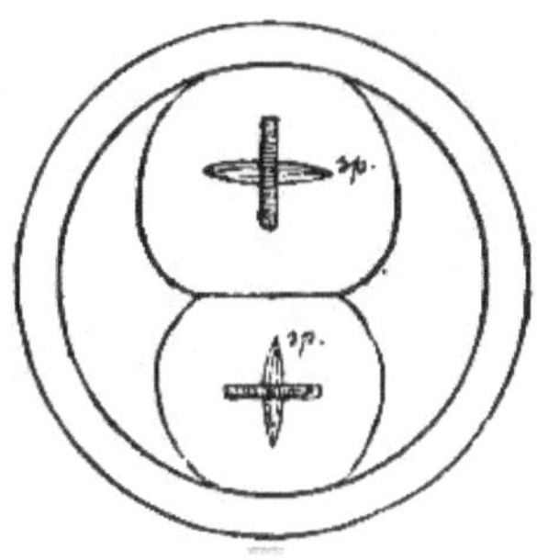

FIG. 27.
sp = spindle.

it independently to discover and adopt a definite direction in division, corresponding to the aim of its ontogeny.'

He proved this by experimenting with the eggs of the maw-worm, *Ascaris*. The second cleavage-division affords a classical instance of the formation of the spindle (*sp*) in the shortest axis of the cell (fig. 27).

If there were only purely mechanical causes forcing the protoplasm to set the spindle in this position, it ought to be easy to induce the lower cell, which is subject to greater pressure than the upper (see fig. 27), to develop its spindle on its longest axis, when the pressure is removed. In order to effect this, Zur Strassen rolled the eggs to and fro under a glass until they were no longer spherical, but of a long oval

[1] On the latter see the author's article 'Zur mechanischen Instinkttheorie' (*Stimmen aus Maria-Laach*, LX, 1901, parts 2 and 3). Also *Instinkt und Intelligenz im Tierreich*, 1905, chapter viii. A criticism of Loeb's chemico-physical theory of fertilisation may be found on pp. 147, &c.

shape, and thus the two cleavage-cells had room enough to develop their spindles in the longest diameter. But they did not do so; in the lower cell also the spindle retained its normal position, although it was in the shortest axis of the cell. Similar observations were made by Zur Strassen at the two-celled stage of the giant eggs of *Ascaris*, which have a long, oval shape, and their cleavage-spheres are so far from being subject to any mechanical pressure that they float freely within the

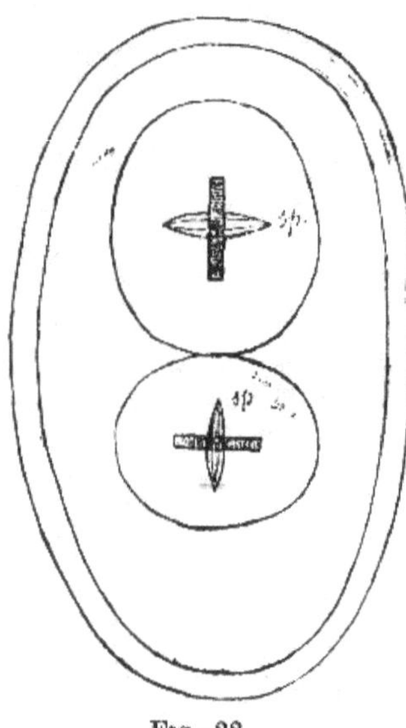

Fig. 28.

covering of the egg, and touch one another at one point only. Yet even in this case the two cells developed their spindles in the shortest axis (fig. 28).

These experiments in embryology lead us chiefly to the negative result, that the mechanical laws laid down by Hertwig, Loeb, and others are inaccurate, and supply no causal explanation of the processes we are discussing. Zur Strassen thought that his experiments justified the positive conclusion: 'That the cell, when ready to divide, contains most delicate mechanisms which determine the moment when mitosis shall take place, the direction of the spindles, and the comparative size of the products. This really seems as if the cleavage-cell possesses an unerring instinct directing the process of cleavage.'

Therefore not only do the causes determining the specific development reside in the egg itself, but the interaction of the various parts of the egg, as it develops, is controlled by a teleological law, which directs the mechanical factors towards the aim of the embryonic development.

This has brought us at least somewhat nearer to a solution of the problem of determination, but we have still not decided whether preformation or epigenesis underlies the whole process of development. Weismann, the extreme supporter of the theory of preformation, says that ontogeny can be explained only by evolution, and not by epigenesis.[1]

Oskar Hertwig, on the contrary, asserts:[2] 'The development of a living creature is by no means a piece of mosaic work, but all the various parts develop always in relation to one another, or the development of any one part is always dependent upon the development of the whole.'

Here, as in every case where scientists hold different opinions, we must put the question in a clear and definite form, in order that we may know what each of these theories involves.

We shall therefore ask with Korschelt and Heider:[3] 'Are there present in the egg, when it begins to develop, any special, independent *Anlagen* or fundaments, which develop quite apart from the other portions of the egg and become definite formations in the embryo? And, if there are such *Anlagen*, how have they come into existence? Can other *Anlagen* of a similar kind arise later?

'Or: Do the various formations in the embryo never develop independently? Are they always dependent upon the other parts of it? In this case we should have to acknowledge the existence of a constant, mysterious influence exercised by the whole upon its several parts.

'Or: Do both methods of formation, the dependent and the independent, participate in the development of the embryo? and, if so, to what extent?'

In the first case, if preformation alone controls development,

[1] *Das Keimplasma*, Jena, 1892, p. 184. In his recent lectures on the Evolution Theory, 1902, Weismann still maintains a decidedly preformistic attitude, although he concedes a great deal more to epigenesis than he did previously.

[2] *Ältere und neuere Entwicklungstheorien*, Berlin, 1892, p. 29. Cf. also his *Allgemeine Biologie*, p. 632.

[3] *Lehrbuch der vergl. Entwicklungsgesch.*, Part I, Jena, 1902, pp. 93–94.

the development not merely of the egg as a whole, but of each separate organ in the future creature would depend upon self-differentiation ; it would be mosaic work, and nothing else.

In the second case, if epigenesis alone controls development, the whole ontogeny of the organism would be based upon dependent differentiation, upon which the idea of the whole would be impressed.

In the third case, we should have to trace development partly to preformation and partly to epigenesis, working together harmoniously to produce the due result. We might then follow Driesch in describing the ontogeny of the individual as an epigenetic evolution. As we shall see presently, this third alternative is the best, and comes closest to the truth.

The well-known saying, 'What suits one does not suit another,' is applicable not only to the circumstances of human life, but to the phenomena occurring in the development of living beings. In different kinds of eggs, and in different stages of the development of one and the same organism, intrinsic and dependent differentiation act very variously. We must therefore follow Korschelt and Heider, and examine the individual cases and the embryological experiments of modern research. Before doing so, however, I ought to explain some expressions introduced mostly by Hans Driesch, the most consistent advocate of epigenesis. In spite of their learned sound they are all quite simple.

Driesch distinguishes the *prospective value* and the *prospective potency* of a cell or a cleavage-segment, in the course of the development of an individual organism. By prospective *value* he understands the *real* destiny of the cell, by prospective *potency* its *possible* destiny. We may therefore call prospective value also destiny in development, and prospective potency possibility in development. We shall understand the distinction better, if we consider something analogous in human life. Let us imagine a boy with an *Anlage* for being a tinker. If the circumstances of his life permit, and he really becomes a tinker, it was his prospective value to be a tinker. But the prospective potency of the same boy was plainly far wider ; according to his natural disposition he might eventually become a knife-grinder or a schoolmaster, a gunner or an author. Now, the prospective potency of a cell comprises all that it is possible for it to develop into, or the sum of the dispositions that it contains, of which, however, only one or very few can ever be set in action in the process of development; these latter represent the prospective value of the cell and its descendants. According to Brauer, any cleavage-sphere of the freshwater polypus *Hydra* has the power to produce ectoderm and entoderm cells. But the ectoderm cells of later stages in the development of the same animal have lost the power to produce entoderm cells. Thus in course of ontogeny (or the development of the individual) the prospective potency of the cells of *Hydra* suffers limitation. In general we may lay down this principle: The prospective potency of a cell is more limited in higher organisms than in lower, and in the more advanced stages of ontogeny than in the earlier; it may even cease to exist, and we have an instance of this in the cornified cells of our skin.

Whoever accepts the theory of prospective potency has practically recognised the truth of epigenesis, for whenever we speak of the possibility of development, we mean that cells, or groups of cells, which were originally designed to make up some definite formation, may, under certain circumstances, take another direction and serve another end. This process of transformation has been called redifferentiation or redetermination. In such processes the influence of the whole in some mysterious way is brought to bear upon the parts of the organism, and through this influence they co-operate, so as to develop a creature capable of life. All processes of development that have this character are known as *regulatory*, or as *organic regulations*, these being the names used by Driesch.[1]

Closely connected with Driesch's theory of prospective potency or possibility of development in cells is his other idea of the *equipotential system*. Such a system is formed by a group of cells, each of which possesses the same potency. Driesch subdivides these systems into *determined equipotential systems* and *undetermined* or *harmonious equipotential systems*. In the former, the number of things that can possibly be made from the group of cells under consideration is strictly limited.

[1] I need not discuss the further distinction, also due to Driesch, between primary and secondary regulations, primary and secondary prospective potencies, &c.

Q 2

For instance, from any transverse section of a willow branch either a shoot or a root may be formed, but the prospective potencies of the cells of the piece of willow are limited to these two things. But in the harmonious equipotential systems any one element can assume any part, and so the number of possible developments is very great. Each portion of such a system can likewise accomplish a whole complicated process of formation; which form it will assume depends upon the position borne by the part with regard to the whole, for all parts are harmoniously subordinated to the whole, whence the system has its name of 'harmonious equipotential.' Thus, for instance, each of the cells of the thirty-two cell cleavage stage in the egg of the sea-urchin is not only able to form the $\frac{1}{32}$ part of the embryo, which it is its proper function to form, but, if the 32 cells are artificially separated from one another, each of them is capable of developing into a very small, but still complete, sea-urchin larva.

3. Embryological Experiments on the Eggs of Various Kinds of Animals and their Results

The scale seems now to be turning again in the direction of epigenesis, but before pronouncing a final decision, and deducing conclusions for or against the theories of mechanism and vitalism respectively, we must briefly consider the various groups of animals on which embryological experiments have chiefly been made.

We must mention first the experiments on the eggs of Amphibia, begun by W. Roux in 1883. With a heated needle he pricked one of the first pair of cleavage-spheres of a frog's egg, and so killed it. The other half, that remained uninjured, developed exactly as if the destroyed portion had remained alive, but, as the latter was incapable of development, the result of the experiment was the production of a half-embryo (*hemiembryo lateralis*), i.e. a future frog cut in two lengthwise. Roux succeeded also in destroying a cleavage-sphere at the four-cell stage, and then a three-quarter embryo was produced.

These results justified the conclusion that under ordinary circumstances the two cleavage-spheres of the two-cell stage of development in the embryo frog contain the rudiments of the right and left half of the future frog respectively, and these rudiments have the power to develop independently of one another. In the same way, each quarter at the four-cell stage seemed able to produce a corresponding quarter of a frog, without being affected by the remaining three quarters.

Roux formulated his results as follows: 'Normal development is from the outset a system of definitely directed processes; it is intimately connected with the chief directions in which the embryo develops, so that the first four cleavage-cells do not merely each occupy the position of a definite quarter of the embryo, but are capable of producing each its proper quarter independently.' 'The development of the frog gastrula and of the embryo resulting immediately from it, is, from the second cleavage onwards, a mosaic, made up of at least four vertical pieces developing independently.'

The development of the frog's egg appeared, therefore, to obey the laws of preformation and intrinsic differentiation, not those of epigenesis and dependent differentiation, but obviously it was not permissible to regard this result as applicable generally to the ontogeny of other organisms. Even in the case of the frog, Roux observed subsequently that his half-embryos afterwards grew into complete ones, as the missing half of the body was supplied by the existent half, by means of the materials from the cleavage-sphere which was injured by the operation. A process of *redifferentiation* set in, changing the half into a whole embryo—a *regulation* which unmistakably aimed at the production of a complete creature, capable of life. All the theories of preformation and mechanism fail to account for this phenomenon.

Oskar Hertwig repeated Roux's experiments on frogs' eggs, but came to quite different results. He observed that whenever he destroyed one of the first pair of cleavage-spheres, with one solitary exception, the uninjured half did not produce a half-embryo, but a complete embryo of half the normal size. Here, therefore, we find no trace of mosaic work, but only confirmation of the laws of dependent differentiation, which is dominated by the idea of the whole.

It was reserved for O. Schulze and Th. Morgan to give by their experiments a satisfactory explanation of the apparent discrepancy between the results at which Roux and Hertwig had arrived, whilst employing the same methods on the same object.

Whenever Morgan left the frogs' eggs after the operation in their natural position, i.e. with their black (animal) pole upwards, the uninjured halves invariably produced half-embryos. When he turned them round, so that the white (vegetative) pole was uppermost, as a rule complete embryos of half the normal size were developed. In the former case, the original arrangement of the egg-substance was retained in the uninjured blastomere, which continued its ordinary course of development, and only turned into a complete embryo by later redifferentiation. In the latter case, on the contrary, turning the egg round altered the arrangement of its contents in a way which led directly to a regulation of the development in accordance with the design of the whole. In neither case can we dispense with a principle regulating embryonic development.

From the above-mentioned embryological experiments, and from others of a similar nature, we may conclude that under normal circumstances the first two cleavage-cells in the frog's egg possess a different prospective value, inasmuch as they form each one symmetrical half of the embryo. But their prospective potency is identical, and equivalent to that of the egg before cleavage, for each half can produce a whole embryo. The same is true of the four blastomeres at the four-cell cleavage stage of the frog's egg. Each is under normal circumstances designed to give rise only to a definite quarter of a frog, but if they are separated, each can produce a complete, though very diminutive creature. At later periods of embryonic development, however, from the eight-cell stage onwards, the cleavage-cells are not any longer all of the same value. At this stage the four cells of the animal half of the ovum can produce only organs of the animal sphere, and those of the vegetative half only organs of the vegetative sphere. The prospective potency of the cleavage-cells of the Amphibian egg becomes more limited and restricted as development proceeds.

We come now to experiments on the eggs of Echinoderms. In these, as in the eggs of Amphibia, the chief axes of the embryo are probably determined before the beginning of the cleavage process, although we do not know with certainty on what material and structural circumstances this preformation depends. In the Amphibian egg the different colouring of the two poles indicates an animal and a vegetative half of the egg, but in the Echinoderm egg no such difference in the egg-substance is perceptible.

Among the eggs of Echinoderms, those of the sea-urchin are particularly well suited for embryological experiments, and are often chosen for the purpose. In them it is possible to separate the blastomeres of the egg undergoing cleavage, not only by means of needles or by shaking the vessel of water containing the eggs, but the blastomeres can be isolated much more satisfactorily, as Curt Herbst was the first to discover, if the eggs are put into water containing no lime. The absence of lime alone suffices to induce the blastomeres to develop in isolation; in fact, at somewhat advanced stages in the development of the embryo, it is only necessary to put it into water containing no lime, in order to separate the cells from one another.

The capacity for regulation, or power of redifferentiation in the cleavage-spheres, is possessed by sea-urchins' eggs in a very unusual degree, and has led to true triumphs for the theory of epigenesis. In the eggs of Amphibia only the first four cleavage-cells of the embryo, if separated from one another, are capable of producing a fresh, complete embryo ; but in sea-urchins' eggs this power lasts as far as the blastula stage, which, according to Hans Driesch's very careful calculation, consists of 808 cells. Each of these 808 cells is equivalent to all the rest, as far as its power of development is concerned. Driesch used a fine pair of scissors to cut up some sea-urchin embryos at the blastula stage. He cut them in all directions, haphazard, and first the raw edges drew together and closed the wounds, then the piece cut off became a little round blastula, which followed the normal course of development and finally produced a perfect, though small, larva (*Pluteus*) of the sea-urchin. If the blastula had been left untouched, and had followed the usual course of development, the cells situated where the incisions were made would have occupied quite a different position in the embryo, and would have served to form quite different tissues ; for instance, they might have formed the intestine and not the outer skeleton of the body. Driesch's experiments have proved, therefore, that in the sea-urchin blastula all the cells are still equivalent to one another with regard to their power of development; each of them can occupy any position and discharge any function in the formation of the organism. All the cells of the *Echinus* blastula are alike in their prospective potency, and what each cell becomes, i.e. its prospective value, is determined by its position in the whole blastula, which is itself already determined by the direction of its axes. Driesch has, as a result of his experiments, enunciated the statement: 'The prospective value of the cell is a function of its position.'

The *Echinus* blastula is a beautiful instance of a harmonious, equipotential system, in which each part is able to take the place of any other part, or to become a complete embryo. Just as the soul of man is wholly in every part of his body, and wholly in the entire body, so is the power of organic development in this case present wholly in every part of the embryo and wholly in the entire embryo. Without a principle regulating its development and controlling the mechanical factors, this wonderful unity in multiplicity would be inconceivable. Only vitalism can offer any satisfactory explanation of this phenomenon; mechanics cannot account for it.

The further the development of the organs has advanced in the *Echinus* larva, the less is the power of redifferentiation possessed by the individual cells. In this case too, as in that of the development of the embryo frog, the prospective potency of each cell is diminished as growth goes on, although in the sea-urchin it remains unrestricted until the blastula stage is reached. Driesch remarks that the organs in their original *Anlage* or disposition are without exception the result of dependent differentiation in the widest sense, but in their development they show intrinsic or self-differentiation in the literal sense of the word. It seems, then, that here too epigenesis must be reconciled with preformation, if we are to give any complete account of the process of development.

Let us now refer shortly to experiments on the ova of other classes of animals.

In the ova of Hydromedusae (Polypi and Medusae) the cleavage-spheres, when isolated, behave as do those in the ovum of the *Triton* among Amphibia. A cleavage-sphere after isolation becomes round, and forms a diminutive whole, continuing its cleavage-divisions and resulting finally in the formation of a very small, but otherwise normal larva. Zoja bred perfect Hydroid polypi from isolated blastomeres of the two- and four-cell stages, but only larvae (*Planulae*), from those of the eight- and sixteen-cell stages, and these larvae had no power of further development. Therefore, we have here another instance of restriction of the prospective potency in the cleavage-cells of the embryo, proportionate to the advance in its development.

A comparison between these embryological experiments and others, made on the eggs of Ctenophora with tentacles, will show what great diversities can exist in the laws governing the development of closely related groups of animals. In the ova of the Ctenophora a limitation of the prospective potency of the individual blastomeres sets in very early, so that we are reminded of the mosaic theory. The first experiments were made by Karl Chun, who succeeded in shaking apart the two blastomeres resulting from the first cleavage of the ovum of tentacular Ctenophores and in breeding from them two half-larvae, each possessing four ribs instead of eight (the normal number), and having only half the usual number of other organs too. Subsequent research has confirmed Chun's observations on all essential points, and we may say that in Ctenophores the first two cleavage-spheres of the fertilised ovum have each a clearly defined prospective potency.; each can produce only half a normal organism, whilst among the true Medusae belonging to the same subdivision of the animal kingdom, each cell at the sixteen-cell stage is still capable of producing a complete little larva. The development of the first pair of blastomeres in the ovum of a Ctenophore is a genuine mosaic, which depends on self-differentiation, each half of the ovum being quite independent of the other half. The same is true of the formation of the fourth and eighth parts of the embryo, which are produced by subsequent cleavage-divisions. Not until the ectoderm has grown over the embryo is any co-operation and reciprocal action perceptible between the fourth and eighth parts.

The development of the ribs in the embryo of a Ctenophore is peculiarly interesting. All who have made experiments on the fertilised ovum of Ctenophore agree in believing that it can produce eight ribs and no more. As the process of cleavage goes on, the possibility of producing them is so far localised, that to each eighth is assigned the task of forming one rib. As the *Anlagen* for the ribs arise from the little cleavage - spheres, or micromeres, of the embryo, which differentiate themselves from the large cleavage-spheres, or macromeres, at the sixteen-cell stage, we must say that each of the eight micromeres possesses the *Anlage* to form one rib, and its development is therefore a real intrinsic differentiation.

Although there is no connexion between Molluscs and Ctenophores, their eggs behave in the same way during the process of cleavage. Isolated blastomeres continue to divide as if they were still in union with the whole, and show consequently partial cleavage. It is true that the half- or quarter-embryos thus produced do not correspond exactly to a half or a quarter of the organism under observation, but they become so far complete as to be capable of life, the ectoderm covers them abundantly, and there are some attempts at forming the velum of the normal larva. But it has proved impossible to breed these creatures any further; they died in every case at this point. The development of the Mollusc ovum depends, therefore, essentially upon self-differentiation of the individual blastomeres, and can be described as a mosaic. An equally pronounced mosaic character is displayed by the cleavage process of the ovum of Annelida and Nematoda.

Chabry's experiments on the eggs of Ascidia seemed also to support the mosaic theory and preformation, for by separating the first two cleavage-spheres half-larvae were produced, but subsequent experiments made by Driesch and Crampton have shown that these eggs resemble in this respect those of many of the Echinoderms, for instance those of a sea-urchin (*Sphaerechinus*). Interference with the cleavage-cells and their isolation cause at first a defective cleavage, producing only part of an embryo, but subsequently readjustment sets in, and the part develops to a whole, so that finally complete blastulae, gastrulae, and larvae are formed, but of reduced size.

In the eggs of Ctenophora, Molluscs and many worms there is only a very slight power of readjustment, and their development appears as a mosaic work, but the eggs of the bony fishes (Teleostei) and those of the famous *Amphioxus* in regulatory power resemble those of the Echinoderms. There is, however, in the eggs of the *Amphioxus* a certain tendency to defective cleavage, i.e. to the formation of imperfect embryos, and there is also a very rapid diminution in the power of redifferentiation as the process of cleavage goes on. In spite of this, however, at least in the early stages of cleavage, dependent differentiation is far more apparent than independent.

4. Conclusions

We have now completed our survey of the embryological development of the eggs of various kinds of animals, and we may pass on to the conclusions to be deduced from it. It will tend to brevity and clearness if I present them in the form of questions.

First: 'Is the ontogeny of the organism based upon independent or dependent differentiation, on preformation or epigenesis?'

If we regard the fertilised ovum as a whole, then its embryonic development from beginning to end is based upon independent differentiation, and consequently upon preformation. But if, on the contrary, we take into account the relations to one another of the individual parts of the egg and of the embryo to be produced from it, the answer to the question is:—Development is based partly on intrinsic, and partly on extrinsic or dependent differentiation. Viewed as a whole, the process of development appears to be an epigenetic evolution. Considered in detail, in the ontogeny of living organisms dependent and independent differentiation act in many respects conjointly, but in many other respects quite distinctly, not only in the eggs of various animals, but in the stages of development in the same embryo. Sometimes the development of the parts of the embryo resembles a mosaic work, in which each part takes its form irrespective of the other parts, as in the Ctenophores. Sometimes it is more like a harmonious equipotential system, in which each part is able to exchange its rôle with every other part, or even to undertake the duty of the whole, as in the sea-urchin blastula. In both cases, however, the regular course of the various phases in development is controlled by the idea of the whole that is to be produced, although in the latter case the idea is certainly clearer and more definite than in the former.

We have seen that at the beginning of embryonic development, the cleavage-cells of the embryo generally display a far greater power of readjustment or redifferentiation than they do later, and thus the prospective potency of the individual cells is diminished the further the organs of the new creature develop. From this point of view, development begins with dependent differentiation and ends with intrinsic differentiation of the various parts of the embryo.

Second: 'What connexion is there between the nuclear substance of the egg-cell and the development of the embryo?'

This difficult question has already been discussed from the standpoint of microscopical morphology in Chapter VI; we must now refer to it shortly on its embryological side. On this subject there are two opinions current, in direct antagonism to each other. According to one, supported chiefly by Wilhelm Roux and August Weismann, the chromatin nuclear substance of the fertilised ovum and the cleavage-cells formed from it exercises a controlling and regulating influence over the processes of development. By means of what Weismann calls *erbungleiche Teilung*, or differential division, the chromosomes of the cell-nuclei, which are the material bearers of heredity, are distributed in different ways to the different cells of the organism that is to be produced, and thus they determine the character of the future tissues and organs. The other theory, however, which is upheld chiefly by Oskar Hertwig [1] and Hans Driesch, denies both the existence and the necessity of any differential division of the chromosomes. It recognises the facts that they are to be regarded as material bearers of heredity, and that they possess a certain amount of individual independence, but it does not ascribe to them so great a determining importance in the processes of development as the former theory assigns to them.

Both theories find support in significant facts, although there are other facts which can hardly be reconciled with them.

The theory of differential division stands, perhaps, in more logical connexion with the processes of karyokinesis that

[1] *Allgemeine Biologie*, pp. 356, &c., 454, &c.

have been observed under the microscope as taking place during fertilisation. These show us not merely the regular distribution of the chromatin substance of the nuclei of the germ-cells to the daughter-cells of the embryo, but also a division, which at least in many cases seems to be differential, as the future germ-cells and the future somatic cells receive remarkably unequal amounts of chromatin. Boveri and other scientists have shown this to occur in the egg of the maw-worm, *Ascaris megalocephala* var. *bivalens*, and Giardina has observed it in that of the water-beetle, *Dytiscus*.[1]

The theory of differential division may find support also in the embryological phenomena already described, in which the development of the embryo is controlled chiefly by the self-differentiation of its various parts, and therefore represents a mosaic, as, for instance, in the Ctenophores. Moreover the fact that, as the development of the embryo advances, the prospective potency of its cells diminishes and becomes more limited, can easily be explained by the theory of differential division.

But against this theory and in favour of *erbgleiche Teilung*, or integral division, there are many other facts in embryology which have been carefully observed and are of no less significance, the chief of them being that the single cells of the embryo may form an equipotential system, the component parts of which may be set to discharge the functions of any other parts or even of the whole. When the sea-urchin egg is in course of cleavage, each part of the blastula, cut haphazard in any direction, is capable of becoming a complete blastula able to develop further. This fact would seem to justify the conclusion that the nuclear substances of the single cells in the embryo are absolutely equivalent to one another, and that consequently no differential division can have taken place at the cleavage of the ovum. Against the theory of differential division is the further fact that the development of the special *Anlagen* for the future organs in the embryo is based chiefly upon dependent differentiation, whilst self-differentiation asserts itself more in subsequent stages. It appears, therefore, that, if we leave out of consideration the very early differentiation between germ-cells and somatic cells,

[1] Cf. p. 122 and fig. 23, p. 124; also p. 169.

as a rule only an integral division of the bearers of heredity takes place at the beginning of embryonic development. It is possible that future research will show us how to reconcile these two theories of integral and differential division, but at present they are involved in many difficulties, and it is not easy to view them impartially.

Of far greater importance than this purely technical question is another, which is concerned with the philosophical solution of the problem of life, and must therefore be discussed more fully.

Third: 'Do mechanical causes suffice to afford a satisfactory explanation of the processes of development, or must we accept a special "vital" law to account for them,—a law governing the chemico-physical factors of development, and directing them to the formation of an organism capable of living?' In other words: 'In attempting to offer a philosophical account of the phenomena of embryonic development must we profess ourselves adherents of the "machine theory" or of vitalism?'

Vitalism is as old as natural philosophy itself. It is well known that the scholastic philosophers adopted special formal principles (entelechies) as the actual essential forms of living matter, in order to account for the phenomena of life.

This is the earliest kind of vitalism, but, at the beginning of the nineteenth century, it had been more or less forgotten in scientific circles. Liebig and other chemists thought that they must assume the existence of a special kind of vital force working in living organisms, over and above mechanical forces. Towards the end of the century neovitalism entered upon a new stage, approximating to the vitalism of the old philosophers. Two of the chief advocates of neovitalism, J. Reinke, the botanist, and Hans Driesch, the zoologist, do not regard the principle of life as a *causa efficiens* of the vital processes, but as an internal formal principle of the living organism. We shall recur to this topic later (cf. p. 243).

The machine theory was the outcome of the great success with which the mechanical view of nature was applied to physics and chemistry in the nineteenth century, but, when it is closely examined, it is found to be based upon a one-sided overvaluation of the importance of mechanics in explaining natural phenomena, and it cannot hold its own against a thorough criticism. It still has many adherents, for old prejudices die hard. Professor Otto Bütschli defended it against the supporters of neovitalism at the fifth international Zoological Congress at Berlin, and read a long paper entitled 'Mechanismus und Vitalismus' on August 16, 1901.[1] In this paper Bütschli remarks: 'The machine theory regards it as possible, though for the moment only to a very limited extent, to account for the forms and phenomena of life on the lines of complex physico-chemical conditions. Vitalism, on the contrary, denies this possibility. The vitalist is convinced that the physico-chemical action of inorganic nature is not sufficient to account for organic life, that an altogether peculiar action, unknown to inorganic nature, must exist in the world of organic life.' Bütschli states the question clearly and accurately, but unfortunately we cannot say as much for his arguments in favour of the machine theory. I listened to what he said with attention, and read a report of it afterwards still more attentively, but I discovered only one real piece of evidence in favour of the machine theory as an explanation of life, and this one piece of evidence occurred in the closing words of his discourse: 'Of all the phenomena of life we can understand only what admits of a physico-chemical explanation.'

Professor Bütschli will, I hope, forgive me for saying that this kind of evidence seems to me quite unintelligible. If it were accurate, the thoughts of the speaker would be pronounced unintelligible for himself as well as for his hearers and readers. According to his own opinion, his thoughts undoubtedly belong to the category of phenomena of life. He ought, therefore, first to give us a physico-chemical explanation of his own process of thought, before he calls upon us to understand his defence of the machine theory!

Bütschli was certainly arguing in a circle, and thus his arguments had no logical force. He confused the ideas of 'to understand' and 'to give a physico-chemical explanation,' and regarded them as synonymous, but I must protest against being required to accept this. Either he assumed that the phenomena of life, considered scientifically, admitted only of a physico-chemical explanation—which was exactly what he

[1] See *Verhandlungen*, pp. 212–235.

undertook to prove—or he did not assume it, and then he has simply not given us the evidence to prove that the phenomena of life have no special vital laws governing them, over and above what is physical and chemical. It is time for people to give up attempting to combat the vitalist theory with such threadbare arguments.

In the interests of modern biology I must enter a further protest against Bütschli's entirely ungrounded assertion, that we can understand only what admits of chemico-physical explanation, and can understand it only as far as it can be explained on these lines. If this were true, the scientific value of the greatest biological triumphs of the present day would be absolutely nothing. Are we in a position to give a physico-chemical explanation of the processes of indirect karyokinesis, of fertilisation, and of ontogeny? Are they therefore simply unintelligible to us? No, they are not; for we understand these phenomena chiefly by considering their purpose and not their mechanical cause. Just as we can understand why a key of a particular shape can turn in a lock, without needing to know by what mechanical process the key and the lock have been made, so we can grasp the significance in fertilisation and development of the processes involved in karyokinesis, although we do not know their chemico-physical causes. The assertion that the scientific intelligibility of a biological process is limited by the knowledge we possess of its physico-chemical causes, is therefore false and misleading, as well as materialistic. A reasonable explanation of biological phenomena cannot be given, unless they are observed from both the teleological and the causal, mechanical points of view, since both are worthy of equal consideration.[1]

An opinion identical with my own was expressed by L. Rhumbler in an address delivered at the seventy-sixth meeting of German naturalists and physicians at Breslau: 'The mechanical processes of the cell do not exhaust the powers of a living cell, but concern it only on its physico-mechanical side.'[2]

Other advocates of the machine theory have not been

[1] On this subject see also J. Reinke, *Philosophie der Botanik*, 1905, chapter iii, 'Kausalität und Finalität'; also 'Neovitalismus und Finalität in der Biologie' (*Biolog. Zentralblatt*, 1904, Nos. 18 and 19, pp. 577–601).

[2] *Naturwissenschaftliche Rundschau*, 1904, Nos. 42 and 43, p. 549.

much more successful in adducing satisfactory evidence to support it. Max Verworn, a famous physiologist, writes as follows in the introduction to his 'Zeitschrift für allgemeine Physiologie' (Vol. I), when attacking neovitalism and defending the machine theory: 'The principles of action must be the same everywhere, as long as we move in a material world.'

But why? Can this be decided at all *a priori*? Must not the question, whether the principles underlying inorganic and organic action are identical or not, be answered by experience? Experience tells us that the vital processes are of such a kind as not to admit of any purely mechanical explanation. Therefore a vitalist is justified in saying: 'The vital processes are governed by laws of their own, which are superior to chemico-physical activity.' By his method of defending the machine theory Verworn has really cut away the ground from under his own feet. He asserts that purely mechanical principles must be equally applicable to living and to lifeless bodies, and he goes on to prove the truth of this assertion by saying that 'physiology can never be anything but physics and chemistry, i.e. the mechanics of the living body.' Therefore physiology, as a special branch of biology, is quite superfluous; we may quietly let it drop, and incorporate it with physics and chemistry—though perhaps Verworn, being one of our most eminent physiologists, will hardly agree to this.

If physiology were to be nothing more than applied physics and chemistry; if the whole scientific value of physiology were to be measured by its success in tracing all living action back to chemico-physical causes, then indeed modern physiology with its imposing achievements would be in a sad plight. G. von Bunge says in his famous manual of human physiology ('Lehrbuch der Physiologie des Menschen,' II, 1905, 3): 'The opponents of vitalism and adherents of the mechanical explanation of life are accustomed to justify their views by maintaining that, the further physiology advances, the more successful are they in referring to physical and chemical laws those phenomena which used to be ascribed to some mystical vital force; it is therefore now only a matter of time, and eventually the whole vital process will appear to be a complicated set of movements, governed solely by the forces of inanimate nature. It seems to me, however, that the history of physiology teaches us the exact opposite, and I maintain that the supporters of the machine theory are wrong. The more thoroughness, acumen, and impartiality we bring to bear upon our examination of the phenomena of life, the more do we perceive that processes, for which we had thought it possible to account by means of physics and chemistry, are of a far more complex character, and for the present defy every attempt to explain them in a mechanical sense.' Bunge had previously declared that the machine theory of the present day would inevitably drive us towards the vitalism of the future, and he was quite right. Oskar Hertwig uses similar language in his 'Allgemeine Biologie' (1906), p. 551, where he says: 'The development of the eye, the ear, and the larynx, as well as of the bones, has hitherto not been explained on mechanical lines, in fact, we may say the same of every process of development; for everywhere we meet with a factor outside the scope of mechanical knowledge, although it is the most important of all, and this factor is the activity of the cell-organism.'

'But,' say the champions of the machine theory, 'vitalism directly contradicts the universally recognised law of mechanical energy. If there were a special vital activity, it would violate the law of the conservation of a constant amount of energy in the universe—and therefore we cannot accept the theory of vitalism.' What answer can we give to this argument?

The law of energy in its original form is a purely mechanical law, and can therefore apply only to the operation of mechanical factors. It is applicable to psychical and vital factors only in so far as they make use of mechanical agencies in doing their own work, and no further. Whoever has recourse to the law of energy in order to prove a psychical or vital action impossible, is either silently assuming that all action in the universe must be essentially mechanical,—and then he is taking for granted what it was his business to prove—or his whole line of proof is useless.

The assumption of a special vital action would be really contradictory to the law of energy only if the operation of the vital principle either increased or diminished the fixed amount of mechanical energy; but this is a complete misrepresentation of true vitalism. We need no old-fashioned 'vital force' acting like a *deus ex machina*, pushing and pulling and interfering with mechanical factors, but we require a vital principle, which as *causa formalis* enables the atoms and molecules of the living body to accomplish their chemico-physical tasks with a definite vital aim. All the mechanical work performed may be put down exclusively to the chemico-physical factors, and not to the vital principle, therefore it is impossible for the latter to violate the law of the conservation of energy.

The only correct view of the laws of life, which constitute the essential difference between living organisms and inorganic natural bodies, was stated centuries ago by the Aristotelian philosophers (see p. 238), and has recently been adopted by eminent naturalists of our own day.[1] Especial mention must be made of Hans Driesch,[2] a great embryologist, who has declared himself a supporter of the 'Autonomy of the Vital Processes,' and has lately expressly described the vital or formal principle, as one corresponding to Aristotle's entelechies.

J. Reinke, the well-known botanist, speaks of *dominants*, which are closely akin to the idea of entelechies.[3] These statements may suffice to weaken the objections raised against vitalism by the upholders of the machine theory, and, on the other hand, to give a correct idea of what vitalism really is.

If we are now asked the question whether the assumption of a special vital law, controlling the chemico-physical agencies, is absolutely necessary, in order to supply a reasonable explanation of the embryological processes described in this section, we may answer shortly: 'The assumption of a vital principle is absolutely necessary in order to account for the phenomena of development.'

I have already alluded to the inadequacy of the attempts made by J. Loeb and others to explain the cleavage process of

[1] On this subject see Hans Malfatti, 'Über die Chemie des Lebens' (*Die Kultur*, 1905, Part I, pp. 41–49).

[2] *Ergebnisse der neueren Lebensforschung*, 14; see also by the same author, *Organische Regulationen*, Leipzig, 1901, and *Die Seele als elementarer Naturfaktor*, Leipzig, 1903.

[3] *Die Welt als Tat*, Berlin, 1903, pp. 275–292; *Einleitung in die theoretische Biologie*, Berlin, 1901, chapters 19 and 20. 'Die Dominantenlehre' (*Natur und Schule*, 1903, Parts 6 and 7). See also Reinke's more recent work, 'Der Neovitalismus und die Finalität in der Biologie' (*Biolog. Zentralblatt*, XXIV, 1904, Nos. 18 and 19, pp. 577-601); also *Philosophie der Botanik*, 1905, chapter iv.

R 2

the ovum on purely mechanical lines (see p. 222), I have referred to dependent differentiation and to redifferentiation or readjustment as facts supporting the theory of epigenesis, and have shown in several places (pp. 229, 230, &c., and 235), that we can account for these facts only if the whole process of development is dominated by the idea of the whole that is to be produced—a form of expression frequently used by Korschelt and Heider in their excellent 'Lehrbuch der vergleichenden Entwicklungsgeschichte.'

We cannot dispense with a teleological interpretation of the processes of development; they are absolutely incomprehensible, unless we assume the existence of a formal principle controlling the mechanical agencies, and directing them to the aim of producing an organism capable of life.

But is it altogether impossible to regard the fertilised ovum from the point of view of the preformation theory, as a wonderfully delicate and complicated machine, set in motion by purely mechanical agencies and effecting the regular construction of the organism in the process of development? This machine theory of life was once upheld by Hans Driesch, but he has recently subjected it to a very searching criticism and condemned it as quite untenable. In his 'Ergebnisse der neueren Lebensforschung' (p. 15), he writes: 'Eggs are the result of an extremely complicated formative process; therefore each egg might be considered as a very complex piece of machinery, though so small as to be invisible to the naked eye. Now in the course of the ontogeny of an individual, all the eggs have been formed from one cell, by division. How can a complex piece of machinery go on dividing and yet remain complete? It is impossible, and therefore, in this department also, the machine theory breaks down.'

In fact a machine, at once so delicate and so ingeniously constructed, able spontaneously to divide itself a hundred times, and yet to preserve in all its parts the power to become a complete machine again automatically, would be so wonderful a piece of mechanism as to be absolutely inconceivable.

The machine theory of life breaks down in the equipotential systems (see p. 227) no less than in the development of the ovum. Let us refer to a statement made on p. 231 with regard to the blastula of the sea-urchin egg. Such a blastula may be cut up in any direction, and each piece will grow into a complete blastula; in fact every one of the 808 cells forming the blastula is capable of exchanging its original function with any other cell of the same blastula. Now imagine a machine consisting of 808 parts; hack the machine to pieces, and see if each single piece is able 'by means of physico-chemical factors' to complete itself automatically, and produce a whole machine able to work. A machine, capable of doing this, is again something absolutely inconceivable.

I may quote from Driesch [1] another classical instance showing that the machine theory of life is absolutely untenable. He made a series of experiments on an Ascidian, *Clavellina lepadiformis*, a rather highly organised creature, which he describes as follows: '*Clavellina* is about an inch long, and its body consists of three chief parts; at the top is an extremely large, basket-shaped branchial sac, with openings for water to flow in and out; in the middle is a slender portion of the body, which contains the stomodæum and proctodæum, and behind it we see the intestinal sac, containing the stomach, intestine, heart, organs of propagation, &c.

'If a *Clavellina* is cut in two, across the narrow part of its body, so that the branchial and the intestinal sacs are separated, each of these two parts is able in three or four days to grow into a complete animal, as, by means of regeneration from the wounded surface, the branchial sac supplies itself with an intestinal sac, and the intestinal sac with a branchial sac. But the branchial sacs of *Clavellina* do not, when isolated, always behave in the way just described. About half of them, and especially those belonging to small specimens, arrive at the formation of a new whole, but by a totally different method. They do not begin by producing any new formation at all, but they undergo a complete transformation. The organisation of the branchial sac, its ciliated stigmata, apertures, &c., gradually vanish, and after five or six days it is no longer possible to trace any organisation at all, the creatures look like uniform white balls; in fact, when I first saw these shapeless

[1] 'Studien über das Regulationsvermögen der Organismen': 6. 'Die Restitutionen der *Clavellina lepadiformis*' (*Archiv f. Entwicklungsmechanik*, XIV, 1902, Parts 1 and 2, pp. 247–287); see also *Ergebnisse der neueren Lebensforschung*, pp. 10–12.

masses before me, I thought they were dying, if not actually dead. But such is not the case. They may remain for as long as two or three weeks in this shapeless condition ; then, one day, they begin to show signs of life and to stretch, and in two or three more days they are again complete Ascidians, with branchial sac, intestinal sac, &c. They are absolutely new creatures, having no part in common with the original, but made of the same material. Their branchial sacs are not the old ones that were cut off, but are much smaller, with fewer channels, and fewer and smaller apertures.

'The organisation of the isolated branchial sac seems to have been reduced to undifferentiated material, out of which, as in embryonic development, a complete little organism has been formed. Sections made by the microtome through the balls undergoing retrogressive transformation show that the change of differentiated into undifferentiated substance had gone very far. We now come to the most important point in the results of our experiments on isolated branchial sacs of *Clavellina*. Not only is the isolated branchial sac itself able to become a little Ascidian by means of retrogressive transformation and regeneration, but it may be cut in half in any direction, so as to form an upper and a lower, or a front and a back half, and each half still possesses the power to undergo retrogressive transformation, and to develop into a little Ascidian, complete in every detail of its organisation. This is undoubtedly an extremely strange phenomenon in organic formation.'

So far I have quoted from Driesch. Let us now compare the capacity of reformation possessed by the branchial sacs or portions of them, undergoing retrogressive transformation, with the favourite example of a machine of very complex structure, such as the upholders of the machine theory regard as essentially equivalent to a living organism. Let us imagine that we break the machine in pieces, and choose one piece which we break again, for closer observation. After a few days this piece falls into a confused mass of fragments, so that nothing of the original parts of the machine can be recognised. It remains in this condition for some weeks, and then suddenly begins to move, the various bits of iron come together quite spontaneously and form, not the original piece of the machine which gave rise to the mass of fragments, but a new and complete little machine, constructed on the same lines as the old one. Any one would say that nothing short of witchcraft could accomplish this, and it is a fact that a *Clavellina*, acting in accordance with the machine theory of life, would never naturally succeed in performing such a feat. We declare, therefore, that the machine theory, which, in spite of the accomplishment of such wonders, persists in regarding the *Clavellina* as a mere machine, makes large demands upon our credulity. But as we are convinced that natural causes, and not magic arts, underlie the marvels of development, we come to this conclusion: Vitalism is the only philosophical theory of life that is in accordance with reason, for it does not regard the living organism as a mere machine, but it knows how to find the architect residing in it!

'In the smallest cell we have all the problems of life before us.' These words of Bunge's[1] have found abundant confirmation in the preceding pages. A diminutive egg-cell, once fertilised, contains already the design of the whole complex organism which is to proceed from it, and it contains it in a way that defies all purely mechanical explanation. The study of ontogeny has brought us to the same conclusions as those which we expressed at the end of Chapter VI (pp. 177, &c.), although by another road, that, namely, of modern embryology. In Chapter VI, the results of microscopical study of the phenomena of fertilisation and heredity led us to assume the existence of internal laws of development, controlling the maturation-divisions of the germ-cells and their union in the course of fertilisation, and directing these processes to a definite end. We found that the chromosomes should probably be regarded as the chief material bearers of heredity, but their morphological function was by no means a satisfactory explanation of the real problem of development. Even if the supporters of the chromosome theory really succeeded, by means of most accurate microscopical observations, in showing conclusively that their theory agreed with the results of embryological physiology; even if they were able to express the amazing processes of regeneration in *Clavellina* by

[1] *Lehrbuch der Physiologie des Menschen*, II, 11.

a complicated formula of chromosomes (which would have to surpass in ingenuity the System of the Universe, the outcome of Laplace's giant intellect)—they would still not have solved the mystery of life, as it is presented to us by the problem of ontogeny. The external aspect of the problem, and no other, can be dealt with by means of microscopical observation, and by considering the morphological peculiarities of chromosomes of definite shape, dividing in definite ways, and distributing themselves in definite numbers to the various cells of the new organism—we have still not touched the other side of these embryological processes, which is concerned with their interior dynamics. The physiological part played in the maturation and fertilisation of the germ-cells, and in the subsequent cleavage-divisions of the embryo, by the chromosomes, as bearers of heredity, upon one another and upon the cell-plasm, goes far beyond the scope of the most subtle machine theory, and reaches far into the domain of the mysterious conformity to vital laws that manifests itself in living creatures. In studying the processes both of fertilisation and of development, we must necessarily assume the existence of some inner causes working harmoniously to one common end, and thus only shall we understand the physiological importance of the chromosomes. If, on the one hand, these material parts, visible only under the microscope, are really the smallest wheels, setting the wonderful clockwork of life in action from generation to generation, and if the movements of these wheels are due immediately to some still unknown chemico-physical laws acting upon the molecules of albumen and nuclein in the cells, we must remember that, on the other hand, they are *living* wheels, and it is only from their uniform action, which has the whole vital process as its aim, that the chromosome theory of the future will ever be able to supply a really satisfactory explanation of the phenomena of life. This uniform action, however, must have a uniform interior cause, and this we perceive in the vital principle of the organism to which I have already alluded.

In Chapter VII we considered a number of facts, that led us to accept this immanent teleological principle, whilst they revealed the impossibility of spontaneous generation. Now that we have surveyed the results of modern embryology, the acceptance of this same principle has been shown to be necessary in a far higher degree.

The vital principle, that controls what goes on in a diminutive fertilised ovum, is at the same time the architect, directing the course of the whole resulting process of development, and bringing it to completion by means of the mechanical agencies that are subordinate to him. But this little architect is not himself an intelligent being ; he has power to act in the various cells and in the whole organism, and to direct all to their aim, but he does so in virtue of the laws which a higher intelligence, superior to our universe, imposed upon living matter when the first organisms came into being. This higher intelligence we call a personal Creator. The necessity for assuming the existence of this first cause for all conformity to law in organic life would remain undiminished, if the machine theorists succeeded in accounting for all the vital processes without a vital principle. Only an architect of infinite intelligence could possibly construct a machine capable of developing, growing, and propagating itself for millions of years by means of purely mechanical agencies. The reasons for regarding the machine theory of life as untenable are therefore not theological, but scientific. Unicellular living creatures and the fertilised ovum and the organism proceeding from it, all have in themselves the vital principles, which uniformly direct the action of the chemico-physical forces of the single atoms towards the higher aim of life.

Our praise is due, not to these diminutive, unconscious architects, but to the eternal creative Spirit that has connected them with matter.

CHAPTER IX

THOUGHTS ON EVOLUTION [1]

---

[1] An article published in the *Biologisches Zentralblatt* for 1891 (Nos. 22, 23), dealing with the evolution of the varieties of *Dinarda*, gave rise to a number of unfair remarks upon my attitude towards the theory of evolution. I thought it possible to show that the varieties of the *Dinarda* beetle, living among our

1. The Problem of Phylogeny

The ontogeny of organisms, which we discussed in the previous chapter, is a direct object of scientific observation. That the seed of a rose develops into a rosebush, and a hen's

ants, were not strictly speaking species at all, but races, standing on various levels with regard to the formation of species. Further, I was able to show that the differences in our various kinds of *Dinarda* appeared to be characteristics due to adaptation of their way of life to that of the various kinds of ants who were their hosts. In this article I mentioned shortly several other facts, that I had observed in the course of my special study of the inquilines among ants and termites, and that I considered were arguments in favour of a modified theory of evolution. I remarked emphatically that I regarded the theory as justified only in so far as it is really based on ascertained facts in the case of definite series of forms; I altogether refused to accept the so-called 'Postulates,' which the monists set up in the name of the theory of evolution.

In spite of this important reservation, a reviewer in the *Schlesische Zeitung* of January 21, 1902, ventured to claim me simply as a supporter of the theory of descent. In the Supplement to the *Allgemeine Zeitung* for June 17, 1902 (No. 136), a longer article appeared by Dr. K. Escherich, entitled, 'A Jesuit as an adherent of the theory of descent.' It is true that my own opinions were reproduced in it with praiseworthy accuracy, and that attention was drawn explicitly to my not regarding as justifiable the extension of the theory of evolution to man. But the reviewer went on to express a hope that the theory would soon be accepted without reservation by me and the whole Catholic Church! I think, therefore, that I am absolutely bound in this place to state clearly what I am ready to accept in the theory of evolution, and what I reject as mere additions from Darwinian and monistic sources. Moreover, in his review Dr. Escherich spoke of me as an opponent of the other advocates of the Christian cosmogony, and especially of all other Catholic theologians, and this is certainly not the truth. It is not a dogma that every species owes its existence to a particular act of creation. More than twenty-five years ago Father Knabenbauer, S.J., contributed a very careful article on 'Glaube und Deszendenztheorie' ('Faith and the Theory of Descent') to *Stimmen aus Maria-Laach* (XIII, 1877). On p. 72 of this article he says: 'Faith does not forbid us to assume that the now existing varieties of plants and animals are derived from some few original forms.' Professor Schanz expresses similar views in his *Apologie des Christentums*, 1895, to which attention was drawn by articles in the supplement to the *Germania*, July 3, 1902, No. 150, and the *Deutsche Reichszeitung*, No. 326. More than twenty years ago, the *Stimmen aus Maria-Laach* several times contained emphatic warnings to be careful to distinguish Darwinism and the theory of evolution; although the former must be rejected, there are many facts to support the theory that organic species have developed within definite series of forms.

Extracts from Escherich's review concerning my attitude towards the theory of descent were subsequently reprinted in the *Frankfurter Zeitung* of July 18, 1902, No. 197; in the *Deutsche Zeitung*, No. 168; and in the *Bohemia* of July 20, No. 198; with the unfortunate title 'Ein Jesuit als Anhänger des Darwinismus' ('A Jesuit as an adherent of Darwinism'). In order to remove all misunderstandings that may have arisen in consequence of these newspaper reports, I intend to make a clear and detailed statement here of my opinions on the subject of evolution, which have also been expressed in a number of lectures of a popular scientific nature, delivered in various German towns and in Luxemburg since the year 1901. It was easy to foresee that the extreme Darwinists would attack my views, but I can notice only those attacks which have some foundation on facts. Further remarks on this subject will be found at the beginning of this book in the 'Few Words to my Critics,' and at the end, in the appendix containing my Innsbruck lectures.

egg into a chicken, are facts of everyday occurrence. Therefore the study of individual ontogeny, which concerns itself with the way in which the various living organisms of the present day come into being, is in its nature an empirical science. In it hypotheses and theories begin only at the point where we seek a deeper insight into the laws and causes of the actual development which we can observe.

But with the race history of organisms it is otherwise. The science dealing with this subject is generally called simply the doctrine of evolution or the theory of descent. It is not empirical, but by its very nature it is a hypothesis, which has grown into a theory by the aid of the circumstantial evidence adduced in its support. I propose to do my best to give my readers a clear idea of what it implies.

Roses and poultry have not always existed, both in fact are of very recent date; the earliest representatives of the family to which our poultry belong are found in the upper Eocene, i.e. in the Tertiary period of the earth's history. Whence came the first rose, and the first hen? Were they suddenly created, just as we know them, or were they developed from other kinds of plants and animals that lived before them? If so, how was this development or evolution effected? These questions are very simple and obvious, and yet they are of great importance in our comprehension of the vegetable and animal world about us. The Flora which *now* covers the face of the earth with leaves and blossoms, and the Fauna which *now* under various forms inhabits sea and land, are not the original occupants of our world, but late-born epigoni. They took the place of other plants and animals which lived in the same world before them, and are to some extent known to us through their fossil remains; and these earlier plants and animals had other predecessors in still more remote periods, and so we may go on, until at last we come to the first and oldest forms of animal and vegetable life on our planet. And here again the same question confronts us: 'Did the later representatives of the Flora and Fauna come into existence quite independently of the earlier ones, or are they chiefly their modified descendants?'

We know that geology divides our earth into a series of strata, formed successively one after the other, and arranged one above the other.

I. Azoic or archaic strata, containing no organic remains.

II. Palæozoic strata, containing the earliest traces of organic life—
   1. Cambrian (including Pre-Cambrian).
   2. Silurian.
   3. Devonian.
   4. Carboniferous (Coal).
   5. Permian (Dyas).

III. Mesozoic strata (the middle ages of organic life)—
   1. Triassic (red sandstone, shell lime, marl).
   2. Jurassic (black, brown, and white Jura or Lias; Middle Jurassic or Dogger; Upper Jurassic or Malm).
   3. Cretaceous (Chalk).

IV. Cænozoic strata (the modern period of organic life)—
   1. Tertiary age (Eocene, Oligocene, Miocene, Pliocene).
   2. Quaternary age (Pleistocene or Diluvium, Present or Alluvium).

Man, the highest of all created beings, appeared only in the Pleistocene period; but the history of animal and vegetable life upon earth began thousands, perhaps millions of years before man's appearance. No human eye beheld the beginning of the drama of life on our planet, no human eye watched the thousands of scenes enacted from the moment when the great drama opened, to the moment when man came forth as the last and noblest figure on the stage of life. And now he ventures boldly to look back into the past and survey the whole history of the evolution of organic life on earth. He tries to find out in what order the various forms of animals and plants have succeeded one another, from the earliest times down to the present day, and he attempts to account for this succession by tracing the later forms back to the earlier, by means of natural evolution of species, genera, families, &c.

It is therefore quite intelligible that this theory of evolution, having as its subject the conjectural race-history of the organic world, cannot be an empirical science, but bears, and must inevitably bear, a hypothetical character. But as the human spirit of research makes use of facts as a starting point for its comparisons and deductions, the theory of evolution rightly claims to be called a science, *scientia rerum ex causis*; for race-evolution, if we accept it, enables us to give a comparatively simple and natural explanation of a number of phenomena actually occurring in various departments of biology. Inasmuch as it is in a position to offer the most probable account of these facts, we must undoubtedly regard the theory of evolution as scientific, although the evidence which the scientist can use in support of the theory is almost exclusively circumstantial; and indeed we cannot expect it to be otherwise, for we are dealing with the previous history of the living organisms known to us, with a primæval period, of which at the present day we find only faint traces and fragmentary remains. Like a skilful advocate, the man of science must carefully collect his circumstantial evidence, and fit it together, so as to reconstruct from it a course of events which no one actually witnessed.

The circumstantial evidence in support of race-evolution is of many different kinds. It consists firstly of the facts of palæontology, which offers us the fossil remains of extinct animals and plants as silent witnesses to the primæval history of our present Fauna and Flora. We have also the facts of variation and mutation, which show us how the properties of still existing creatures can be modified, and new species formed. Comparative bionomics shows us how animals and plants undergo adaptation to one another, and are influenced by very various external factors, and these facts enable us to infer how the altered relations have come about. The facts of comparative morphology also, the points of likeness in interior and exterior structure that exist among members of definite families, these too are quite explicable if we may assume that they have a common descent. Lastly, there are the facts concerned in the ontogeny of the individual, which incidentally reveals to us traces of former race-evolution. In short, the various branches of zoology and botany—both empirical sciences—supply innumerable pieces of circumstantial evidence, of which the theory of descent makes use. If it does so in a critical and careful manner, we have a scientific foundation for the theory of evolution, although we have no wish to deny its hypothetical character. If, however, the circumstantial evidence is used in a superficial and fanciful way, and involves groundless generalisations and reckless jumping at conclusions, we have, instead of a scientific theory of evolution, merely a fantastic semblance of it, which is pretentious enough to put forward its arbitrary statements as historical truths.

The very subject-matter of the theory of evolution shows—and I am careful to emphasise it again—that it is indeed based upon many results of the empirical sciences, but can never be itself an empirical science, and will always remain a hypothetical explanation of observed facts, and as such it has risen to the rank of a theory. We must, however, always be careful to distinguish hypotheses and facts; and this is especially necessary, because the theory of evolution in many respects stretches beyond the domain of natural science into that of natural philosophy, and it is often difficult to define the boundaries of each. For this reason we must act cautiously with regard to the 'postulates' which so-called monism has set up in the name of the theory of evolution, for they are not based on scientific facts, but on materialistic dogmas.

Without entering upon a full account of the history of the theory of evolution, I may shortly sketch the outlines of the problem with which we are going to deal.

In order to explain the origin of the existing species of plants and animals, we have to assume one of two things. We may assume that the systematic species (e.g. lion, tiger, polar bear) are invariable—apart from the formation of varieties and breeds within the species—and that they were created originally in their present form. Or we may assume that the systematic species are variable, and constitute definite lines of descent, within which an evolution of species has taken place during the geological periods. The first of these assumptions belongs to the theory of permanence, the second to the theory of evolution or descent. In the latter we must make a further distinction between monophyletic and polyphyletic evolution. According to the monophyletic theory, all organisms have originated in one single primitive cell, or perhaps there is one pedigree for all animals and one for all plants, each having one primitive ancestor. According to the polyphyletic theory there are several pedigrees for both plants and animals, independent of one another, but each one going back to one special primitive form as its starting point.[1] In the following pages we shall see that the latter assumption alone can claim to have any positive scientific probability—and we shall see, moreover, that this assumption is perfectly reconcilable with the Christian doctrine of the Creation.

2. The Various Meanings of the Word 'Darwinism'

For over forty years a conflict has been raging in the intellectual world, which both sides have maintained with great vehemence and energy. The war-cry on one side is 'Evolution of Species,' on the other 'Permanence of Species.' No one could fail to be reminded of that other great intellectual warfare regarding the Ptolemaic and the Copernican systems, which began about three hundred and fifty years ago, and raged with varying success for over a century, until finally the latter prevailed. Perhaps the present conflict between the theories of evolution and permanence only marks a fresh stage in that great strife, and, if so, how will it finally be decided?

The contest that we have to consider was stirred up by Charles Darwin, when he published his book on the 'Origin of Species' about the middle of last century. The theories advanced by Lamarck and Geoffroy St. Hilaire at the end of the eighteenth and the beginning of the nineteenth centuries may be regarded as causing preliminary skirmishes, but Cuvier's powerful attacks soon succeeded in overthrowing the new ideas of evolution (see p. 28). It was not until the year 1859[2] that the great battle began, which has received its name from the commander-in-chief of the attacking army, Charles Darwin. The warfare with which we are now concerned centres round Darwinism, so-called.

I say, so-called Darwinism. A few words of explanation are absolutely necessary. The thick smoke of the powder, which hid the battlefield from our gaze, is gradually dispersing,

[1] It is of secondary importance to consider how many individuals there were of each primitive form. The chief point is that the *Anlage* for evolution in each primitive form differed from those of the primitive forms of other lines of descent.

[2] The first English edition of *Origin of Species* was published in November 1859, as Darwin himself stated, although 1858 is sometimes erroneously given as the date of its publication. See Francis Darwin, *Life and Letters of Charles Darwin*, I (London, 1888), p. 84.

and it is much easier now than it was twenty or thirty years ago to survey the armies on both sides and to judge of their positions, their strength and their mode of fighting, and to value rightly what they have achieved and what they still have to accomplish. It now appears that the number of scientific combatants gathered under Darwin's banner is still comparatively small. By far the greater number of supporters of what was once called Darwinism are now ranged under the standard of the theory of evolution, and no longer under that of Darwinism. These troops form the rank and file, but Ernst Haeckel is the leader of a corps of free-lances and freebooters, conspicuous for the disturbance that they cause in the name of 'Science.' [1]

Their weapons are not, however, of the best and noblest sort, and their aim is not the triumph of truth, but rather the plunder of the Christian camp, that they suspect to be situated somewhere in the rear of their opponents' position. But victory does not incline to them; with their wooden swords they bring upon themselves one defeat after another, and only succeed in hindering the triumph of the picked troops of really scientific men, who fight with better weapons on the side of the theory of evolution.

It is time, however, to explain in simple words the simile of the battle which has presented itself to our sight.

If we want to answer the question: 'What are we to think about Darwinism?' we must first of all try to grasp clearly the different senses in which this name is used.

The first and most obvious way in which the word Darwinism is used, is to designate the *theory of selection*, put forward by Charles Darwin; i.e. the special form of the theory of descent, which traces back the evolution of organic species to natural selection, as its chief, if not its only cause. Man uses his intelligence to produce artificial breeds of domestic animals, by selecting for breeding those that show the peculiarities that answer his purpose. Darwin, however, assumes the occurrence of a natural selection with no purpose at all; he thinks that, by its means, in the struggle for existence some varieties prove better able to hold their own than others, and

[1] On January 11, 1906, they founded the 'German Monistic League' (Deutscher Monistenbund) in Jena, under Haeckel's presidency.

S

their peculiarities are accentuated by transmission to following generations, whereas the varieties that are less capable of self-preservation die out. This is the fundamental idea of Darwin's theory of selection.

The word Darwinism received a second meaning when it was applied to an extension of the theory of selection to a new and, as it was called, philosophical theory of the universe. It was assumed that not only the organic species, but the whole orderly arrangement of the world, had arisen out of an originally lawless chaos by means of accidental 'Survival of the Fittest.' In Germany Ernst Haeckel has been the chief founder and champion of this Darwinian theory of the universe, and therefore it is also known as Haeckelism. It bears the misleading name of 'Realistic Monism,' but it would be better designated 'Materialistic Atheism.'

The third use of the word Darwinism proceeded from the extension to man of Darwin's theory of selection. In this sense, the theory that man is descended from beasts is called Darwinism, whether it be Vogt's theory of the descent of man from apes, or some more modern opinion of the same kind. According to this 'Darwinian' view of man, he is in both body and soul nothing but a beast, that has accidentally reached a higher point of development than his fellows. The first to deduce this conclusion from the Darwinian System was an Englishman, Huxley, in his work 'Evidence as to Man's Place in Nature' (London, 1863). He was followed by Haeckel in his 'Natürliche Schöpfungsgeschichte' (1868). It was not until 1871 that Darwin himself made up his mind to extend his theory to man in his 'Descent of Man.' This book is really the weakest of all Darwin's scientific works.

In 1887 Wiedersheim attempted to give a detailed anatomical foundation for the descent of man from apes in his book on the structure of man as evidence of his past ('Der Bau des Menschen als Zeugnis für seine Vergangenheit,' 3rd ed., Tübingen, 1902). An excellent refutation of this piece of fiction was given in 1892 by O. Hamann in an article on 'Darwinism and the Theory of Evolution' ('Darwinismus und Entwicklungslehre') (see p. 108, &c.). The weakness of the Darwinian methods of proof is thoroughly displayed by J. Ranke in his work on Man ('Der Mensch,' 2 vols.).

The fourth and last meaning attached to the name Darwinism is due to its having been applied first to a particular form of the theory of descent, and afterwards transferred to the theory of descent in general. Although this use depends upon a confusion of ideas, the name is still in popular language applied to the whole doctrine of the evolution of organic species, as opposed to the theory of permanence, which assumes that the systematic species never change, and were created originally in their present form. In this sense, therefore, every student of nature, who declares the species in any one genus of animals or plants to be related to one another, is a Darwinist, though erroneously so-called.

This last application of the name Darwinism ought to be given up, as it only leads to confusion. It is based—and I must again emphasise the fact—upon a logical blunder, for it confuses the theory of evolution as a whole with a particular form of it. This blunder was pardonable forty years ago, when Darwin's theory of evolution was the only one known, but it is pardonable no longer. At the present day it is unfair to identify the ideas conveyed by the names 'Darwinism' and 'Theory of Evolution,' and it is done only with a special intention; the adherents of Darwinism, on the one hand, have recourse to this device in order to propagate their obsolete theory in popular circles, and the opponents of the theory of evolution, on the other hand, try to annihilate every attempt to question the permanence of species, by hurling at it the epithet 'Darwinism.'

It will now be an easier task for us to answer the question: 'What are we to think about Darwinism?' We see that the question resolves itself into four.

1. What are we to think of Darwin's Theory of Selection?
2. What are we to think of the extension of Darwin's Theory of Selection, so as to make of it a realistic and monistic theory of life?
3. What are we to think of the application to man of Darwin's Theory of Selection?
4. What are we to think of the Theory of Evolution as opposed to that of Permanence?

It is the object of our present discussion to supply an answer to the last of these questions, and I can deal with the first three only briefly, for they have often been answered before, and admit also of much shorter answers than the fourth.

First.—Modern science can hardly be said to take into account Darwin's theory of selection as the exclusive form of the theory of evolution. It is full of weak spots, to which attention was drawn as early as 1874 by Albert Wigand,[1] and it is impossible any longer to avoid recognising them. In the *first* place the theory of selection is in principle not satisfactory, for natural selection may be able to destroy what is inexpedient, but not to produce what is expedient. Therefore it simply leaves to chance the origin of advantageous modifications, which lead to the formation of new species. A theory based on chance is worthless as affording an explanation of conformity to law in nature. In the *second* place, most of the variations which serve as the groundwork of classification are biologically indifferent, and do not affect the individual or the species in the struggle for existence; they can therefore not be due to natural selection in their breeding, because they present no *points d'appui* on which it can work. In the *third* place, in order to account for the formation of one new species, this theory requires innumerable, almost imperceptible variations to have existed for immense periods of time and to have been gradually accumulating and intensifying. This contradicts known facts of palæontology, for the Fauna and Flora of remote ages display a definite system of classes, orders, families, genera and species, just as do those of the present day, and not a chaos of imperceptibly slight variations, such as the theory of selection requires.

For these reasons most naturalists have by this time abandoned the theory in its exclusive form. An eminent

[1] *Der Darwinismus und die Naturforschung Newtons und Cuviers*, I. Cf. also G. Wolff, 'Beiträge zur Kritik der Darwinschen Lehre' (*Biolog. Zentralblatt*, X, 1891, Nos. 15 and 16); O. Hamann, *Entwicklungslehre und Darwinismus*, Jena, 1892, chapter ix; A. Goette, 'Über den heutigen Stand des Darwinismus' (*Die Umschau*, 1898, Part 5); Aug. Pauly, *Wahres und Falsches an Darwins Lehre*, Munich, 1902; *Lamarckismus und Darwinismus*, Munich, 1905; Max Kassowitz, 'Die Krisis des Darwinismus' (*Die Zukunft*, February 15, 1902); E. Dennert, *Am Sterbelager des Darwinismus*, Stuttgart, 1905 and 1906; H. Kranichfeld, 'Die Wahrscheinlichkeit der Erhaltung und der Kontinuität günstiger Varianten in der kritschen Periode' (*Biolog. Zentralblatt*, 1905, No. 20; 1906, No. 8); Chr. Schröder, 'Kritische Beiträge zu den strittigen biologischen Fragen der Gegenwart' (*Natur und Schule*, V, 1906, Part 6, pp. 233–247); O. Zacharias, 'Planktonforschung und Darwinismus' (*Zoolog. Anzeiger*, XXX, 1906, Nos. 11, 12, pp. 381–388).

modern zoologist, Dr. Hans Driesch, condemned it perhaps rather harshly in the *Biologisches Zentralblatt* for 1896, p. 355, when, in speaking of Darwinism, he said: 'It is a matter of history, like that other curiosity of our century, Hegel's philosophy. Both are variations on the theme "how to take in a whole generation," and neither is very likely to give ages to come a high opinion of the latter part of our century.' In the same publication for 1902, p. 182, he says: 'For men of clear intellect, Darwinism has long been dead, and the last argument brought forward in support of it[1] is scarcely more than a funeral oration in accordance with the principle *De mortuis nil nisi bonum,* and with an underlying conviction of the real weakness of the subject chosen for defence.'

Professor Oskar Hertwig, Director of the Anatomical and Biological Institute at the University of Berlin, expressed himself almost as strongly in an address delivered at the meeting of German naturalists at Aix-la-Chapelle, on September 17, 1900, on the growth of biological knowledge in the nineteenth century. He points out the necessity of distinguishing clearly between the theory of evolution and the theory of selection, and then continues (p. 15): 'They stand on a very different foundation and basis, for we might say with Huxley: "The theory of evolution would stand where it did, even if Darwin's hypothesis were blown away." In the former we have a permanent achievement of our century, based upon facts, and certainly worthy to be numbered among the chief attainments of our age.' We shall have to examine later on to what extent the theory of evolution is really based upon facts.

In one of his lectures given in April 1905, at the Berlin Singakademie, even Ernst Haeckel frankly acknowledged, in at least one passage,[2] that the theory of natural selection alone ought to be termed Darwinism in the stricter sense, and he added: 'We cannot now discuss the extent to which this theory is justified, nor how far it has been amended by other

[1] The reference is to a paper by L. Plate in the *Verhandlungen der Deutschen Zoologischen Gesellschaft* for 1899: 'Die Bedeutung und Tragweite des Darwinschen Selektionsprinzips.' The paper has since appeared in an enlarged form with title: *Über die Bedeutung des Darwinschen Selektionsprinzips und Probleme der Artbildung*, Leipzig, 1903.

[2] *Der Kampf um den Entwicklungsgedanken*, Berlin, 1905, p. 20.

newer theories, such as Weismann's Germ-plasm theory (1884) and de Vries' theory of mutation.' He did not refer to this delicate question in his later lectures. The passage is particularly noteworthy, because Haeckel, as the 'Prophet of Darwinism,' has for nearly forty years been confusing Darwinism and the theory of evolution to suit his own ends, and has extolled Darwin's theory of selection as the highest intellectual achievement of the nineteenth century, because it teaches us how to understand design in nature without recognising a wise Creator! And, after all, Haeckel himself finally acknowledges that the confusion between Darwinism and the theory of evolution is a mistake, and he can scarcely find any scientific justification for the theory of selection. I feel inclined to put on Darwin's lips the words 'Et tu, Brute,' uttered by the dying Caesar!

This confession on Haeckel's part must have been very unwelcome to those who support Darwinism from the point of view of popular science, and who try to mislead the general public by confusing it with the theory of evolution. One of them, R. H. Francé, in a work entitled 'Die Weiterentwicklung des Darwinismus' ('The further development of Darwinism'), 1904,[1] has tried to represent all the progress made by the theory of evolution since Darwin's time, and even modern vitalism itself, as a triumphant 'further development' of Darwinism, whereas in reality he is uttering a sort of funeral oration over it.

That Darwinism and the theory of evolution are two essentially different things is quite evident from the evolution theories of Mivart,[2] Wigand,[3] Kölliker,[4] Heer,[5] Nägeli,[6] Eimer,[7]

[1] *Gemeinverständliche Darwinistische Vorträge und Abhandlungen*, published by W. Breitenbach, Part 12. To show the method of proof adopted by Francé, I may mention that in the above-mentioned work (p. 24), by means of unmistakable falsification of a quotation from *Stimmen aus Maria-Laach*, he tries to make out that the Jesuit Father Wasmann is a supporter of the theory of permanence, in order thus to render 'Jesuitical science' harmless from his point of view.

[2] *The Genesis of Species*, London, 1871.

[3] *Die Genealogie der Urzellen als Lösung des Deszendenzproblems*, Brunswick, 1872.

[4] 'Allgemeine Betrachtungen zur Deszendenzlehre' (*Abhandl. der Senkenbergschen Naturforschenden Gesellsch.*, VIII, 1872, pp. 206–237).

[5] *Urwelt der Schweiz*, Zürich, 1883, chapter 18.

[6] *Mechanisch-physiologische Abstammungslehre*, Leipzig, 1884.

[7] *Die Entstehung der Arten*, I, Jena, 1888; II, Leipzig, 1897.

de Vries,[1] Gulick [2] and others, who either attack Darwin's principle of selection, or impose very strict limitations upon it.[3] Kölliker and Eimer's theories unfortunately resemble Darwinism in having a mechanical and monistic basis,[4] but they have the great merit of combating it on scientific grounds, for they admit internal causes of evolution as the chief factors in the hypothetical phylogeny of living organisms. Eimer's researches into evolution proceeding towards some definite aim (orthogenesis) were continued after his death by his pupils, Countess Maria von Linden and Dr. Fickert. It is worth noticing that E. Strasburger, the well-known botanist, who formerly upheld the theory of selection, has recently given it up very decidedly.[5] It is true that there are still at the present day in Germany some eminent zoologists, especially Professor August Weismann at Freiburg im Breisgau, who profess to defend Darwin's theory of the all-importance of natural selection,[6] but on closer examination Weismann's 'Neo-Darwinism' also appears to be gradually beating a retreat, the first stage in which is marked by W. Roux's 'Histonal Selection,' or selection of the tissues; Roux tries to supply the deficiencies of the principle of selection by transferring Darwin's personal selection to the struggle among the various parts in the living organism. When, therefore, in 1895, Weismann propounded his theory of germinal selection, as the last bulwark of the principle of selection, he acknowledged that not Darwin's natural selection, but interior causes of evolution, must be the chief factor in an orderly evolution of the organic world.[7]

[1] *Die Mutationstheorie, Versuche und Beobachtungen über die Entstehung von Arten im Pflanzenreich*, I, Brunswick, 1901; II, *ibid.*, 1903.

[2] Rev. John T. Gulick, *Evolution racial and habitudinal (Theory of Divergence)*, Washington, Carnegie Institution, 1905.

[3] In his *Konvergenz der Organismen*, Berlin, 1904, H. Friedmann has even attempted to substitute the principle of divergence for that of descent. I cannot say that I think his attempt successful; the two principles are complementary to one another, but neither can take the place of the other.

[4] With regard to Kölliker's theory see an article by Professor Stölzle, 'A. von Köllikers Stellung zur Deszendenzlehre,' Münster i. W., 1901 (*Natur und Offenbarung*, 1901). On the principles underlying Eimer's theory of orthogenesis see Wasmann, 'Die Entstehung der Arten nach Eimer' (*Natur und Offenbarung*, 1889, pp. 44, &c.).

[5] Cf. *Jahrbücher für wissenschaftliche Botanik*, 1902, pp. 518, &c.

[6] Cf. Weismann's 'Lectures on the Evolution Theory,' Eng. trans., London, 1904.

[7] See remarks in Chapter VI, p. 176.

In the scientific theory of descent, selection is now regarded as a subordinate factor of more or less importance, but it cannot take the place of the interior factors determining the evolution of the race, in fact it presupposes their existence. O. Hertwig remarks very aptly on this subject (' Allgemeine Biologie,' 1906, p. 620): ' It seems to me perfectly plain that no advantage is gained by the use of such phrases as " Struggle between the parts of an organism," " intraselection," " histological selection," " germinal selection," they do not enable us better to understand the processes of organic nature. They teach us no more about what goes on within the organism than a chemist would learn about the formation of any organic compound, if he were to content himself with using such a phrase as " the struggle of the molecules in a test-tube " for explaining some chemical process.'

Neo-Lamarckism stands in direct contrast to Weismann's Neo-Darwinism. In 1809, Jean Lamarck wrote his ' Philosophie Zoologique,' in which he traced the development of species to direct functional adaptation, viz. to the principle of the use or disuse of organs; from this followed inevitably the theory that the qualities thus acquired by the individual could be transmitted to his descendants. Charles Darwin did not by any means exclude the principle of direct adaptation and the power of transmitting acquired qualities, but he assigned to them less importance than to natural selection. Weismann, however, and the Neo-Darwinists after him, denied the possibility of direct adaptation and the transmission of acquired qualities. According to them, nothing was inherited but modifications working directly upon the germ-plasm. This view was opposed by the Neo-Lamarckians under the guidance of Herbert Spencer and K. von Nägeli, who upheld the principle of direct adaptation, and maintained that acquired qualities could be transmitted. Among the modern representatives of Neo-Lamarckism we may mention particularly two zoologists, viz. Oskar Hertwig[1] and L. Hatschek,[2] E. Koken, a palæontologist,[3] and R. von Wettstein,

[1] *Allgemeine Biologie*, Jena, 1906, esp. chapters 27–30.

[2] 'Hypothese der organischen Vererbung': an address delivered at the seventy-seventh meeting of German naturalists at Meran, Leipzig, 1905.

[3] 'Paläontologie und Deszendenzlehre' (*Verhandl. der 73 Versammlung deutscher Naturforscher zu Hamburg*, I, Leipzig, 1902, pp. 221, &c.).

a botanist.[1] As a matter of fact, both direct adaptation and selection seem to take part in the processes of evolution; the former to a greater degree than the latter, because it results from the interior laws of evolution, whilst selection only plays the negative part of eliminating the unfit. It is self-evident that only those modifications can be hereditary which in some way have stamped themselves on the germ-plasm, but how and to what extent the characteristics acquired by individuals are transmitted to the germ-plasm, is a very dark, mysterious question.[2] Oskar Hertwig in his 'Allgemeine Biologie,' p. 598, has made a suggestion which is certainly very important in connexion with the theory of evolution. He says: 'Is it not possible that, just in the same way as the multicellular organism develops by epigenesis from the egg, so, when we survey the matter from the point of view of the theory of descent, each species may develop in accordance with a permanent, regular principle of progress, not as the plaything of chance, but with the same interior necessity as, in ontogeny, the blastula must grow out of the gastrula?'

Second.—We can give a still shorter answer to the question regarding the extension of Darwin's theory of selection, so as to make of it a realistic and monistic cosmogony[3]—it is simply a mischievous act committed in the name of science.

It is mischievous philosophically, because it traces back the origin of all conformity to law in the natural order to a denial of all conformity to law as to its primary cause. It is mischievous theologically, although it vaunts itself to be

[1] *Über direkte Anpassung*, Vienna, 1902; *Der Neolamarckismus und seine Beziehungen zum Darwinismus*, Jena, 1903.

[2] In his book '*Lamarckismus und Darwinismus*, Munich, 1905, A. Pauly aims at adducing fresh psychological evidence in support of Lamarckism. His ideas on teleology are, however, mostly wrong and psychologically without foundation.

[3] The physical arguments in favour of this extension are stated in Haeckel's *Riddle of the Universe*, but they have been submitted to a very destructive criticism in a work entitled *Hegel, Haeckel, Kossuth and the Twelfth Commandment*, by O. D. Chwolson, Professor of Physics at the University of St. Petersburg, and author of a valuable textbook of Physics, that has been translated into German. We may assume that everyone knows the sharp criticisms pronounced upon Haeckel's *Riddle of the Universe* by Professor Paulsen in his *Philosophia militans*, by Professor Loofs in his *Antihaeckel*, by Professor Seeberg and others. E. Dennert's popular works, *Die Wahrheit über Ernst Haeckel und seine Welträtsel* (Halle a. S., 1904) and *Haeckels Weltanschauung*, Stuttgart, 1906, are very well worth reading.

the 'Religion of the Future,' for it alters the conception of God, the most perfect Being, and reduces it to absolutely nothing, whilst ostensibly preserving it; hence it would be more honest to call it atheism than monism. Finally Haeckel's cosmogony is mischievous socially, and constitutes one of the greatest dangers for human society, inasmuch as it proclaims the 'struggle for existence' and the accidental 'survival of the fittest' to be the only laws in the natural order, and it exalts them to be the only laws governing human society also. Haeckelism is, therefore, the support of anarchy and of social democracy, as Bebel once informed us in the German Parliament.[1]

Third.—We saw that the third use of the name Darwinism was to designate the application to man of Darwin's theory of selection.[2] If man is really nothing more than a higher animal, if God does not exist for him, nor an immortal soul, nor any retribution beyond the grave, then human society is indeed delivered over to anarchy, and the anarchists are the only sensible people. But to uphold such a doctrine in the name of science is worse than humbug, it is a grievous offence against the highest possessions of mankind.[3] Those periodicals are guilty of participation in this offence, which profess to present science in a popular form, and recklessly represent the application of Darwinism to man as justified by assured scientific results. Even men like Rudolf Virchow, who do not

[1] In his well-known speech on September 16, 1876, in which he proved the connexion between social democracy and Darwinism, that Haeckel denied, Bebel's words were: 'Gentlemen, in my opinion Professor Haeckel, the decided advocate of the Darwinian theory, because he does not understand social science, has no idea at all that Darwinism must necessarily promote socialism, and vice versa, socialism must harmonise with Darwinism, if its aims are to be correct.' Cf. also a little pamphlet, *Darwinismus und Sozialdemokratie, oder Haeckel und der Umsturz*, Berlin, 1895. It is a matter of especial psychological interest that recently even anarchists have attacked the theory of the struggle for existence. The Russian anarchist, Prince Peter Kropotkin, has done this in his book on mutual help in development, which G. Landauer translated into German, *Gegenseitige Hilfe in der Entwicklung*, Leipzig, 1904. Even to men of this type the theory of selection is beginning to seem untenable, but apparently they do not see that, by acknowledging this fact, they are undermining the foundations of their own social theories.

[2] A further discussion of this subject will be found in Chapter XI.

[3] For a scientific criticism of Darwin's theory of the descent of man, see the works of Hamann and Ranke, mentioned on p. 258; also J. Bumüller, *Mensch oder Affe?* Ravensburg, 1900; C. Gutberlet, *Der Mensch, sein Ursprung und seine Entwicklung*, Paderborn, 1903; Wilh. Schneider, *Göttliche Weltordnung und religionslose Sittlichkeit*, Paderborn, 1906.

claim to speak from the point of view of Christianity, have felt bound to protest vehemently against this mischievous doctrine.

3. The Subject of the Doctrine of Evolution as a Scientific Theory

It is high time for us to go on to the real question under discussion, and ask: 'What are we to think of the theory of evolution in itself? Have the systematic species always existed in their present forms, or are they mostly related with other species, some still existing, and others extinct, and known to us only by fossil remains dating from earlier ages of the world? Are they the result of an historical evolution of the organic world, or were they originally created in their present condition?'

In order to be able to deal with this important question objectively and impartially, it is indispensable for us to disregard altogether the misuse made of the theory of evolution by those who distort it to answer the purposes of atheistic materialism. It is much to be regretted that this misuse of it occurs. It is embodied in Haeckelism, which is by no means a feather in the cap of modern science. Nothing has more injured the reputation of the theory of descent—as the doctrine of evolution is called in scientific circles—than the fact that one section of atheists and materialists have used it as a battering-ram against Christianity; nothing has done more to vulgarise it and disfigure its scientific character than this misuse of it, which has rendered it almost unrecognisable. It is chiefly owing to this misuse, that those who profess to be Christians regard the theory of descent with so much suspicion, and think themselves bound to hold aloof from it, because they confuse the anti-Christian character thus given it with the essence of the theory of evolution. We must resolutely put aside all thoughts of this misapplication, and consider the doctrine of evolution as what it really is, viz. a scientific theory, which we may either accept or reject on its own merits.

I repeat, we have to consider the doctrine of evolution as a scientific theory, which arises out of the facts of the organic world, and seeks to offer the best and simplest natural explanation of them, in accordance with strictly logical methods of thought. We are not concerned with that pseudo-theory of descent,[1] which, starting from the *a priori* considerations of a false philosophy, takes as its fundamental axiom: 'We refuse to admit the existence of a personal Creator, and therefore, whatever exists, must have developed itself by purely mechanical means.' No less false than this fundamental principle are, of course, the various so-called postulates, which the pseudo-theory of descent is fond of stating in the name of science. In the name of true science we are forced to oppose an emphatic veto to these postulates, for the methods of *this* theory of descent are utterly antagonistic to those of true scientific procedure. We must take up, however, another attitude with regard to the question what we are to think of the theory of evolution, from the point of view of natural science. We need not feel any scruple about attempting to answer this question, for we lay down no false postulates of materialism, but we approach it taking as our starting points real facts, viz. the works of God in nature.

Why should we fear to look the truth in the face? We know with absolute certainty that one truth can never contradict another, therefore the recognition of what is really *true* in the theory of evolution can tend only to the glory of Him who is the highest and eternal Truth.[2] Let us, therefore, try to give an honest and careful answer to the question: 'What is the scientific value of the modern theory of evolution? What does it explain? How far is it necessary to a scientific comprehension of the organic world about us?'

Is the theory of descent able to account for the origin of organic creatures and of organic life on our earth? No,

[1] The advocates of Haeckelism are doing their best to identify this pseudo-theory of descent with the scientific theory of evolution. An instance of this was given by H. E. Ziegler, in an address delivered at the seventy-third meeting of German naturalists at Hamburg on September 26, 1901, and printed at Jena, 1902, with the title: *Über den derzeitigen Stand der Deszendenzlehre in der Zoologie.* It is the counterpart of Haeckel's address delivered in Cambridge in 1898: *Über unsere gegenwärtige Kenntnis vom Ursprunge des Menschen*, Bonn, 1899. Haeckel's influence on Ziegler is plainly apparent in the latter's Hamburg lecture (cf. for instance pp. 18, 19, 24, 28, 43, &c.). I think it unnecessary for this reason to criticise Ziegler's views more fully.

[2] On this subject see J. Knabenbauer, S.J., 'Glaube und Deszendenztheorie' (*Stimmen aus Maria-Laach*, XIII, 1877, pp. 71, &c.).

it cannot, for it is a theory of natural science, and natural science can tell us nothing of the source of life on our planet. It only knows the facts and the laws to be deduced from them. But, however carefully we compare these laws with one another, and however skilfully we combine them, they give us no suggestion of spontaneous generation, i.e. of the spontaneous development of living creatures from lifeless matter; on the contrary, modern biology is directly opposed to the theory of spontaneous generation (cf. Chapter VII, 'The Cell and Spontaneous Generation'). If, therefore, a modern scientist, acting not as an investigator of nature, but as a monistic 'philosopher,' appeals to natural science for evidence that the assumption of spontaneous generation is 'a postulate of science,' he is entangling himself in a very obvious contradiction. What biology actually knows is nothing but an uninterrupted series of living beings, living cells, living nuclei, which find a truthful expression in the fourfold law: *omne vivum ex vivo; omnis cellula ex cellula; omnis nucleus ex nucleo; omne chromosoma e chromosomate.* The student of nature must necessarily accept these laws as a foundation, if he wishes to trace the origin of life on earth, but they will carry him no further—they will lead him round in a circle and never let him see the beginning of the mystery. If, as a philosopher, he wishes to study the origin of life more deeply, he is forced to conclude that only some cause apart from the world could have produced the first living organism out of matter. We have already discussed this point in the section dealing with spontaneous generation (pp. 204, &c.). If the student of nature refuses to accept this conclusion, and is resolved to be content with what natural science as such can offer him, he must simply say: 'We know nothing about the origin of life.' Many naturalists of the present day have actually adopted this empirical standpoint; it was done, for instance, by Branco in the address that he delivered on the occasion of his admission to the Royal Academy of Science in Berlin ('Sitzungsberichte der Königlichen Akademie der Wissenschaften,' 1900, pp. 679–696).

What, then, are we to think of the theory of evolution? It certainly does not profess to account for the origin of organic life on earth, it has simply to accept it as a fact; and at the same time it accepts as a fact the existence of laws governing organic development. Just as philosophical examination has as its necessary foundation the fundamental principles of thought; just as no human being can think over any philosophical problem without assuming that his understanding is able to recognise the truth, that everything must have a sufficient cause, and that two contradictory propositions cannot both be true at the same time; so no student of nature can consider theories of evolution, unless he assumes at the outset as a fact the existence of laws governing organic evolution. If he refuses to admit that essentially the same laws of organic formation, which now govern the genesis of living creatures, were in force from the very beginning, he has no clue at all to his phylogenetic research; as soon as he tries to set aside this fundamental principle, his scientific investigations become mere fictions, with no basis of fact. Therefore, in considering the race-evolution of living organisms, we must never lose sight of the conclusions stated at the end of Chapters VI and VIII (pp. 176, &c., and pp. 247, &c.). In dealing with the race-evolution of the living things about us, we can far less dispense with internal laws of evolution, which are the expression not of a purely mechanical, but of a higher, vital activity, than we can dispense with them in dealing with the phenomena of fertilisation, heredity, and ontogeny.

What is, then, the real scope of the doctrine of descent, in so far as it has a scientific basis? Its task is, and can only be, to determine the sequence in which the organic forms appeared upon earth, and so to establish their relationship with one another; it has, moreover, to investigate the causes underlying the gradual modifications in organic forms. The task of the theory of descent is, in other words, to examine the actual facts and causes of the sequence of organic forms, chief amongst which are the species of the present time, being the last offshoots of one or many hypothetical pedigrees.

The theory of evolution is not, and cannot be, an empirical science (cf. p. 253 in § 1), because it is concerned with the earliest history, antecedent to that of the present organic world. By collecting traces of that evolution from the fossil records of palæontology and by comparing them with the facts of the present, it becomes a theory in natural science, aiming at offering a probable explanation of the connexion between these actual phenomena.

From what has been said of the limitations of the theory of descent, it follows that it is by no means essential for it to trace the origin of all living organisms back to one single primitive cell. Nor need it be thus restricted within the limits of the animal and vegetable kingdoms respectively, and trace all animals back to one stock, and all vegetables back to another. It is not essential to the theory of evolution to insist upon a *monophyletic* evolution; it may just as well decide in favour of a *polyphyletic* evolution, for, in examining the hypothetical race-evolution of living organisms, it is bound to conform to facts, and not to monistic postulates. As I shall show later, facts point to a polyphyletic evolution among both animals and plants. Whether a monophyletic or a polyphyletic evolution is to be accepted is therefore, for the scientific theory of descent, a question of fact and not of principle.

From this we may deduce two statements that are important in our investigation: 1st. The extreme champions of the theory of descent, who recognise only a monophyletic evolution as the real theory of descent, and reject polyphyletic evolution as being merely the theory of permanence in disguise, are influenced by monistic prejudices and not by a genuinely scientific spirit;[1] they completely misunderstand what the scientific doctrine of evolution really is. 2nd. Equally mistaken is the attitude of those opponents of the theory of descent, who try to prove that the whole doctrine of evolution has broken down, because no one has yet succeeded, and probably no one ever will succeed, in tracing back the chief types of the animal and vegetable kingdoms to one single stock. I cannot therefore concur with Fleischmann's opinions, expressed in his book 'Die Deszendenztheorie' (Leipzig, 1901). In many passages he bases his arguments against the theory of descent on the statement that the types of organisation among animals cannot phylogenetically be derived from one single type. This proves nothing but that polyphyletic evolution must be accepted rather than monophyletic; it does not

[1] I wish this remark to be taken to heart by Escherich, Forel, Haeckel, von Wagner and others, who criticised my first edition. See also 'A Few Words to my Critics,' at the beginning of this volume.

prove that an evolution of the species within definite series of forms or genera is impossible. Arguments of this kind affect only the monistic, and not the scientific, theory of descent. In general, I am unable to share Fleischmann's views, which are involved in pure empiricism and agnosticism.

4. The Theory of Evolution considered in the Light of the Copernican Theory of the Universe

'But,' some one may say, 'why do we not simply assume that the species in the world of organic life have always been what they are at the present day? Why do we want any theory of evolution at all?'

I am bound to explain this point to my readers, at least to some extent, before I go on to discuss the modern theory of descent more in detail. Three hundred and fifty years ago, when war broke out between the old Ptolemaic view of the universe and the new Copernican view, people had no conception of the distance to which they would be carried by the ideas that then took possession of the human intellect. It was not until the nineteenth century, that from the heliocentric theory of the universe inferences were made affecting the natural development of our solar system, and the whole history of the universe, of which the geological development of our earth occupies but an insignificant moment of time. And within this insignificant period (which, in comparison with the development of the whole universe, is like a second between two eternities, although according to geologists it really lasted millions of years) is another period of history preeeding that in which man appeared upon the world, and this is the history of the animal and vegetable kingdoms from the earliest palæozoic age until the present time.

The Copernican system revealed to us the earth as a mere atom in the universe, as one of the many planets attendant upon a central sphere, that we call the sun. But our sun is not the only sun; there are thousands of others, many being still far larger than it is. All the innumerable fixed stars that we see in the sky at night are so many suns, which are not, however, scattered at random in space, but form one single huge cosmic system. This system is not an unalterable mathematical formula in its various components. Astronomy teaches us that the constellations are, at different stages in their evolution, ranging from gaseous vapour to molten matter like the sun, and even to the dark planets, that are visible only by the light of others.

This is where the theories or Kant and Laplace on cosmogony find their *points d'appui* ; they strive to account for the genesis of the whole universe by one uniform law.[1] By means of the laws which now control the movements and conditions of the celestial bodies, this cosmogony seeks to ascertain how our solar system, and the cosmic system as a whole, assumed their present form. It was led to accept the existence of an original enormous sphere of gas, in which, as it gradually cooled and condensed, a rotatory movement arose, that caused the formation of the solar systems. According to the same cosmic laws, the planets subsequently separated from each sun, in order to circle round it on definite paths. And one of these planets is our earth. Many modifications have recently been introduced into the theories that are called after Kant and Laplace,[2] but it is not likely that any new theory will take its place, at least as far as its essential outlines go.

T. C. Chamberlin's 'Spiral Nebulae Theory'[3] suggests a different explanation for the origin of the planetary system of a sun, but still it presupposes the existence of the gaseous sphere.

No matter what scientific form the theories regarding cosmogony may take, their problem is always to account for the present form and arrangement of the heavenly bodies, and to explain how this form and arrangement may have been evolved by natural means.

At the present day there are probably very few who still cling to the old theory that sun, moon, earth, planets and

[1] Cf. J. Epping, S.J., *Der Kreislauf im Kosmos*, Freiburg, 1882 (Supplement to *Stimmen aus Maria-Laach*, Part 18); also an excellent work by K. Braun, S.J., *Über Kosmogonie vom Standpunkt christlicher Wissenschaft*, Münster, 1905.

[2] The theories of Kant and Laplace on cosmogony are somewhat different, and cannot be united under one name, as Stölzle, Gockel, and other recent authors have shown. See A. Gockel, *Schöpfungsgeschichtliche Theorien*. Cologne, 1907.

[3] Cf. F. R. Moulton, 'The Evolution of the Solar System' (*Astrophysical Journal*, XXII, 1905, pp. 165-181). See also the review in the *Naturwissenschaftliche Rundschau*, 1906, No 5, pp. 53-56.

T

all the fixed stars in the universe were created once for all as we now know them. Even to St. Augustine it seemed a more exalted conception, and one more in keeping with the omnipotence and wisdom of an infinite Creator, to believe that God created matter by one act of creation, and then allowed the whole universe to develop automatically by means of the laws which He imposed upon the nature of matter.

God does not interfere directly with the natural order when He can work by natural causes : this is a fundamental principle in the Christian account of nature, and was enunciated by the great theologian Suarez,[1] whilst St. Thomas Aquinas plainly suggested it long before, when he regarded it as testimony to the greatness of God's power, that His providence accomplishes its aims in nature not directly, but by means of created causes.[2]

Is it not reasonable for us to try to apply the same principle of independent evolution also to the living creatures that inhabit our globe? The obvious complement to the geological history of our world is the history of the creatures that have dwelt on it, since the time when organic life first made its appearance. In the geological arrangement of strata we see the working of natural forces influencing the formation of the earth's surface, and, in the same way, in the fossil animals and plants we see the remains of organisms that really lived at those respective epochs.[3]

Palæontology teaches us that our present species of animals and plants have not always existed. It shows us that there was a succession of different organic forms in the

[1] *De opere sex dierum*, l. 2, c. 10, n. 12. Further evidence to show that this idea was by no means strange to St. Augustine, St. Thomas, St. Bonaventure and others may be found in Father Knabenbauer's 'Glaube und Deszendenztheorie' (*Stimmen aus Maria-Laach*, XIII, 1877, pp. 75, &c.). Cf. also T. Pesch, *Philosophia naturalis*, II, pp. 241, &c., and *Die grossen Welträtsel*, II, pp. 349, &c.

[2] *Summa c. gent.*, l. 3, c. 77.

[3] The idea that fossils were originally created as such, and represent mere *lusus naturae*, is just as groundless as the other opinion, that all fossils date from the deluge. The first idea is wrong in principle, and contradicts the fundamental laws of all intelligent research; it is opposed, therefore, to the true philosophy of nature, and leads inevitably to occasionalism, and is equivalent to a complete abandonment of all hope of giving a natural account of palæontological facts. The second theory may not be intended to clash with geology and palæontology, but it is manifestly wrong in assuming that all the strata containing fossils, more than 20,000 in number, can be accounted for by the deluge.

various geological periods, and in this succession the species of animals and plants that appeared later approximated more and more closely to those of the present time, and in many cases—e.g. in the extinct connexions of the horse—the succession suggests upward lines of evolution,[1] and our present species are their latest developments.[2]

We have now to face the critical question: 'Does this gradual or more abrupt approximation of the fossil Fauna and Flora to those of the present depend upon a mere succession of forms, constantly becoming more like the present forms, or is it a real evolution, a genetic production of various systematic species from one another? Are these "evolutionary series," which lead us from fossil ancestors to now existent species, merely apparent? Do they owe their origin to the fact that, at the close of the various geological formations and groups of formations, a great catastrophe occurred, destroying all living creatures, which at the beginning of the next period were replaced, by means of a new creation, by similar creatures, for the most part somewhat more highly organised? Or are these evolutionary series real and natural, depending upon a genealogical connexion between the organisms of various periods?'

There can scarcely be any doubt as to the answer. Cuvier's theory of a catastrophe has been given up by geologists, because, when generalised, it proved to be inconsistent with facts; consequently it had to be given up by palæontologists also. In place, therefore, of the periodically repeated 'new creations,' the theory of a natural evolution of organic forms has won

[1] The hypothetical pedigree of the Equidae does not, however, form a simple line of evolution, but it has many ramifications, and since the Lower Eocene age they have developed on distinct lines in Europe and North America. Cf. Zittel, *Grundzüge der Paläontologie* (Munich and Leipzig, 1895), p. 871.

[2] I am not, however, speaking here of evolutionary series in the sense of Darwin's theory of transmutation, i.e. not of series of very small and gradual transitions, for these, if they occur at all, are an exception to the more usual transitions 'by steps,' that involve greater changes. Hilgendorf's famous *Planorbis* series has proved not to be a progressive sequence of variations, but rather a cycle of recurring variations, and it is of no use for the purposes of phylogeny. (Cf. K. Miller, 'Die Schneckenfauna des Steinheimer Obermiocäns,' in the *Jahreshefte für vaterländische Naturkunde in Württemberg*, 1900, pp. 385–406 with Plate VII.) L. Döderlein's dictum (*Zeitschrift für Morphologie und Anthropologie*, IV, 1902, Part 2, p. 408) that complete knowledge of any group of animals requires all the forms in that group to stand in unbroken sequence, is not based on fact, but is a theoretical postulate of the Darwinian theory of evolution.

acceptance,[1] in logical application of the principle that God does not interfere directly with the natural order, when He can work by natural causes.

The theory of evolution, regarded without prejudice, is then for us the latest outcome of the Copernican theory of the universe, which no one probably, at the present day, will call un-Christian.

A few instances may be added by way of illustration. If we find the Brachiopod Order *Lingula* occurring frequently in the Silurian and Devonian strata, and continuing to appear at different geological epochs in various species down to the present day, we must undoubtedly say: 'The modern species of *Lingula* are really connected with those of the Silurian age; in fact they are their modified descendants.' If in the Cambrian, the oldest strata containing any fossils, we find representatives of the family of *Nautiloidea,* various genera and species of which still exist, we must say in the same way: 'The still existing four species of *Nautilus* are modified descendants of members of the same family belonging to earlier ages of the world.' If we compare our crickets (*Phasmidae*) with those of the Carboniferous period, we shall be forced to ascribe to them not merely a theoretical, but a real relationship with the *Protophasma* and the *Titanophasma* of the coal age. If we compare our ants and *Paussidae* with those found in Baltic amber of the Tertiary period, we cannot possibly think that they are new creations, but must regard them as genuine descendants of the Tertiary forms, although differing from them partly specifically and partly generically. Any other view of the matter seems scientifically almost impossible.

If we compare fossil termites[2] with those of the present

[1] It is a remarkable fact that more than two hundred years ago, a famous Jesuit, Father Athanasius Kircher, in his book, *Arca Noe in tres libros digesta* (Amsterdam, 1675), expressed his belief that our modern species had originated by transmutation within definite series of forms. (On this subject see Daniele Rosa, 'Il Rev. Padre Kircher trasformista.' *Bolletino dei Musei di Zoologia ed Anatomia comparata d. R. Università di Torina*, XVI, No. 421, March 14, 1902.) Although Father Kircher's views were based on insufficient data, we are all the more justified in holding similar opinions, as our scientific knowledge is much greater.

[2] According to Handlirsch, remains of termites occur with certainty only after the early Tertiary period; he does not regard as termites what Heer described as such occurring in the Black Jurassic strata. His views, however, do not in any way affect the above statement respecting the connexion between our present termites and those of the Tertiary period.

day, we cannot doubt that they all form one single natural stock, continuing from the Mesozoic group of formations through the Cœnozoic to the Alluvial present. The extinct fossil genus *Clathrotermes*, of the black Lias, represents one natural stock with the fossil varieties of the genus *Calotermes*, belonging to the same period. Of this latter genus many species still exist, which differ, however, from those occurring in the lias. With regard to tho much greater variety of fossil termites of the Tertiary period, which include a great many still existent genera and one that is extinct (*Parotermes*), we cannot question the fact that they are genetically connected both with the termites of the Lias and with those of the present day, although their species are different both from the Mesozoic and the modern. We still find in Australia a curious genus of termites, *Mastotermes*, whose wing-veins, in my opinion, show that they are unmistakably connected with the Palæozoic *Blattinae* of the coal age; and this fact justifies our assuming that we have in *Mastotermes* the last living representative of the oldest and most original form of termite, which as a 'collective type'[1] has united in itself the venous systems of cockroaches and termites, that afterwards became entirely distinct. Australia is particularly rich in old forms, which occur in other parts of the world only in a few still surviving representatives, or as fossils dating from earlier ages.[2] These instances are quite enough to prove that it is hardly possible to deny the existence of a genuine race-connexion between our modern forms of animals and the extinct species of bygone ages. We may now return to our consideration of the doctrine of evolution.

Under Haeckel's guidance, the monists have misused the

[1] Forms which show the characteristics of several systematic groups are called 'collective types.' Such, for instance, is *Peripatus* among the Arthropods, which by its low organisation approaches the Annelids. According to Handlirsch (*Verhandl. d. Zool. Bot. Gesellsch.*, Vienna, 1906, Part 3, p. 91), it ought to be classed among the Annelids. Numerous collective types occur, especially among the palæozoic insects, to which Skudder gives the general name of *Palaeodictyoptera*.

[2] In support of this statement I may refer to the Monotremata and Marsupials among mammals, and to the genus *Arthropterus* in the family of *Paussidae*. Australia seems to have preserved the oldest type of the human race, for Macnamara has recently shown that the cranial formation of modern Australian and Tasmanian blacks approximates very closely to that of the fossil Neandertal man. We shall come back to Macnamara's statements in Chapter XI.

theory of evolution, and by making it serve as a weapon with which to attack the theism that they hate, they have brought it into disrepute in conservative circles; and so the idea has arisen that the theory of evolution is an absolutely atheistical device, directly opposed to Christianity. I have just shown this idea to be erroneous, and to have no foundation. If we wish successfully to combat the modern theory of descent, in so far as it has proved serviceable to atheism, we must carefully distinguish truth and falsehood in it. We shall then have no difficulty in depriving our antagonists of their weapons, and even in smiting them with the same sword with which they fancied we were already conquered. If we let ourselves be misled by the skilful tactics of our monistic opponents, and take up an attitude hostile to evolution in every form, we shall be playing into their hands and giving them an easy victory. We shall in fact be assuming the same mistaken position as the champions of the Ptolemaic system once assumed against the advocates of the Copernican theory. They were obliged to be always on the defensive, and to limit themselves to weakening this or that actual piece of evidence adduced by their opponents, as not holding good. In an intellectual conflict such a position must, in course of time, be abandoned. A succession of retreats brings the defenders on to more and more dangerous ground, and finally leads to a decisive defeat. If Christianity is not to succumb to the attacks of monism based on natural philosophy, it must determine upon bold action on the offensive; it must seize the enemies' arsenal, and by accepting without reserve whatever is right in the theory of evolution, it will turn its opponents' weapons against themselves. In such proceedings caution is always advisable. Not everything alleged by the supporters of the theory of descent to be 'based on actual facts' really deserves belief. I need only remind my readers of Haeckel's famous pedigree of man, of which one critic sarcastically remarked that it was just as worthy of credence as those of the Homeric heroes. We must examine carefully how far we can accept the ideas involved in the theory of evolution, both from the philosophical and the scientific points of view. There must be no mention of concessions. We make concessions to error only when, through cowardice or weakness, we accept what is wrong as right, or what is half true as quite true; but it is not a concession when we deprive error of the weapons that it is using in the struggle against truth.

5. Philosophical and Scientific Limitations of the Theory of Evolution

1. What are we, therefore, to think of the theory of descent in its relation to philosophy? It has already been shown that the acceptation of an evolution of the organic species is only a logical consequence of the cosmic and geological evolution. On the philosophical side it would be possible to reject the theory of descent only if it could be proved, on purely philosophical grounds, that our present species are absolutely unchangeable, and that therefore there can be no question of their having evolved from older forms. But philosophy cannot adduce any proof of this kind, because the subject does not fall within her scope. She is obliged to leave natural science to decide whether the systematic species are altogether constant magnitudes or not, and we have already seen what this decision is, and shall refer to it again later.

The fundamental principle laid down by philosophy with reference to the theory of evolution agrees perfectly with Christianity, and may be stated thus: 'It is not permissible to assume any immediate interference on the part of the Creator, where the facts can be explained by natural evolution.' In applying this principle we must be careful to distinguish the philosophical and the scientific standpoints. Many things are possible in themselves, and even probable, *a priori*, but there is no scientific proof of their occurrence. The limits assigned to us by philosophy, with regard to our acceptance of the theory of evolution, are far wider than those imposed upon us by natural research as to details. Moreover, the former are fixed and cannot be overthrown; the latter are constantly changing as our positive knowledge advances. We must, therefore, carefully distinguish between the limits set by philosophy and natural science respectively to the theory of evolution; and, in dealing with the philosophical limits, we must again distinguish between purely philosophical questions and those that are of a mixed character.

Let us first consider the philosophical limits. In one sense philosophy has only to sketch the broad outlines of the theory of evolution; it is the task of natural science to fill in the details. The first and foremost boundary, admitting of no modification whatever, is the principle that the hypothetical race-evolution of the organic species must have had an adequate first cause. This principle contains two postulates, one purely philosophical, and one partly philosophical and partly belonging to natural science. The first is: 'We must assume the existence of a personal, all-wise and all-powerful Creator as the first cause, extraneous to the world, of the whole cosmos and its laws of evolution.' The second is: 'In order to account for the origin of the first organisms, we must accept some special action, direct or indirect, on the part of the Creator upon matter.' Here we are not concerned with 'Creation,' strictly speaking, as we are in the first postulate; but only with the production of the primitive organisms from already existent inorganic matter, which had been formed by a definite act of creation at the beginning of the cosmic evolution.[1] The formation of the first living creatures followed, therefore, by an *eductio formarum e potentia materiae*, as scholastic philosophy expressed it. As intelligent beings we cannot dispense with this postulate; all the efforts of monism to set it aside are fruitless. This postulate is of a mixed character, not purely philosophical like that regarding the creation of primitive matter, for natural science proves that an essential difference exists between animate and inanimate substances, and shows us the absolute incompatibility of the laws of biology and the theory of spontaneous generation. (Cf. Chapter VII. pp. 193, &c.) Neither philosophy nor natural science can tell us how the first organisms came into being; no facts that we can observe enable us to infer anything on this subject. Nor can philosophy say how many primitive organisms were produced, and whether they differed essentially from one another or not. Yet a somewhat important limitation seems to meet us here. As sensitive life is on a higher level than vegetative, it is reasonable to suppose that the former could

[1] I wish to draw particular attention to this passage, as some of the critics of my previous edition fell into the error of regarding the creation of the first organisms as a *creatio e nihilo*.

not have evolved itself out of the latter. We must therefore assume that when organic forms first came into being, there was in all probability a differentiation among them into animals and vegetables. This postulate is of a mixed character, partly philosophical and partly scientific, and is by no means absolutely fixed. For, on the one hand, while observed facts show us the great difference between the vegetative and sensitive life of the higher animals and the merely vegetative life of the higher plants; on the other hand, they reveal to us a number of unicellular organisms, which zoologists reckon among the lower animals, and botanists among the lower plants;[1] and in their case it is impossible to say whether the sensitiveness of the protoplasm, which is a general characteristic of all living cells, amounts to real sensation or not.[2] We have also to take into consideration the movements made by certain plants in response to external stimulus, for which a purely vegetative interpretation seems inadequate,[3] although I agree with J. Reinke[4] in thinking that the so-called 'sense organs' of plants represent merely the receptive centres for physical and chemical stimuli, and we are not justified in arguing from them that plants have sense perception. We probably ought not to regard the original difference of animal and vegetable organisms as an unalterable philosophical postulate;

[1] On this subject see von Wettstein, *Handbuch der systematischen Botanik*, I, 1901, pp. 16, &c.

[2] We derive our ideas of plants and animals from the higher varieties of both, which it is perfectly easy to distinguish, but there are obviously great difficulties in applying these ideas to unicellular organisms.

[3] Cf. Haberlandt, *Die Sinnesorgane im Pflanzenreich zur Perzeption mechanischer Reize*, Leipzig, 1900; 'Die Sinnesorgane der Pflanzen' (paper read at the seventy-sixth meeting of German naturalists at Breslau, September 23, 1904, published in the *Naturwissenschaftliche Rundschau*, 1904, Nos. 45 and 46); 'Über den Begriff "Sinnesorgan" in der Tier- und Pflanzenphysiologie' (*Biologisches Zentralblatt*, 1905, No. 13, pp. 446–451); 'Die Lichtsinnesorgane der Laubblätter' (*ibid.*, No. 17, pp. 580–588). See also Fr. Noll, *Das Sinnesleben der Pflanzen*, Frankfurt a. M., 1896; F. R. Schrammen, 'Kritische Analyse von G. Th. Fechners Werk: Nanna oder über das Seelenleben der Pflanzen' (*Verhandl. des Naturhist. Vereins*, Bonn, LV, 1903, pp. 133-199). On p. 198 Schrammen seems to think that we ought to ascribe to plants a sensitive, but not an intelligent existence. This is intelligible only if he means by a sensitive existence merely susceptibility to mechanical and other stimuli, not amounting to perception. Many botanists speak of plants as sensitive to light, but the word is then used inaccurately, as it is when photographic paper is so described. It is not possible in either case to use the word 'sensitive' in its strict psychological meaning; we ought rather to say susceptible to light.

[4] *Philosophie der Botanik*, 1905, pp. 83–87 and 113.

that the whole organic world may have been evolved from one single primitive cell is not an absolute impossibility, though it is improbable. This improbability appears greater when we take into account the important physiological distinction between the two kingdoms, which O. Hertwig ('Allgemeine Biologie,' 1906, p. 602) states as follows: 'In consequence of their characteristic metabolism, the whole formation of chlorophyll-bearing plants is directed towards the exterior and is visible from the exterior, but, unlike animal organisms, plants either show no interior differentiation into organs and tissues, or they show it in a relatively limited degree.'

Philosophy can give us no information at all regarding the number of forms of plants and animals originally produced, nor can it tell us whether they were produced once for all and in one place, or on many occasions and in various places. Natural science, too, in the present state of our knowledge, can throw very scanty light upon this subject, but I shall return to this topic later; let us now consider it only from the point of view of philosophy.

Philosophy is not concerned to decide whether the animal world on the one hand, and the vegetable world on the other hand, were each descended from one primitive form (monophyletic evolution), or whether they originated simultaneously or successively from several primitive forms, independent of one another (polyphyletic evolution). Nor does philosophy tell us anything of the causes that motive race-evolution; however, the fact that, as natural science shows us, at the present time interior laws of development are the ultimate foundation of all organic genesis,[1] justifies philosophy in inferring that the race-evolution of living organisms chiefly and essentially must have been the result of interior causes of development. All the exterior causes are simply aimless, unless we presuppose the existence on the part of the organism of a corresponding interior capacity for development; and this capacity must ultimately have been implanted by the Creator in the nature of the primitive types. Therefore philosophy is justified in drawing the further inference that

[1] Some suggestions respecting the probable material bearers of these laws of development were made in Chapter VI, pp. 164, &c. Cf. also the conclusion of Chapter VIII.

those theories of descent, which attach the utmost importance to the exterior causes of development, whilst underrating the interior, must be regarded as unsatisfactory. Thus far philosophy may and must utter her decisions ; but in herself she can tell us nothing as to the character of these interior causes of development and how they co-operate with the exterior factors ; any knowledge that she possesses on these points is borrowed from natural science.

She does not inform us whether a race-evolution of the organic species ever really took place or not ; she does not tell us anything as to the number of the original primitive forms ; she teaches us nothing about the laws governing the hypothetical race-evolution of organisms, nor the order in which it took place. If she is wise, she will leave it to natural science to express an opinion on these points ;[1] but there is one thing of great importance, which she is able to tell us. Without a first cause outside the world, the existence of matter and the laws governing its development would have been impossible ; without the same first cause outside the world, the origin of living organisms from inorganic matter would have been inconceivable, and consequently a race-evolution of the organic world would have been out of the question; and, in exactly the same way, we can account for the existence of man only by assuming some special action on the part of the Creator, this special action being the creation of the human

[1] The writer of a review on the first edition (in the *Innsbrucker Zeitschr. für katholische Theologie*, 1905, p. 561), asks : 'Is there philosophically no difficulty in assuming that the sparrow and the hippopotamus have branched off from the same primitive form ? Are we not forced to believe that there is an essential difference in their inner nature, in their very soul ? ' I do not think that this question admits of a purely philosophical answer. If it were worded scientifically, it would be simply : 'Are birds and mammals to be regarded as related ? ' On examining the scientific limitations of the evolution theory we shall find that there is very little to be said in support of the common descent of all vertebrates. Moreover, as mammals appear in the Triassic, and birds only in the Jurassic strata, there are no intermediate forms between birds and mammals. It is true that in some respects our present Monotremata (Ornithorhynchus and Australian ant-eating Echidna) occupy a position midway between these two classes of vertebrates, but in other respects there are important differences. (Cf. Fleischmann, *Deszendenztheorie*, 1901, chapter i.) If birds and mammals are two branches of a common stock, which is very doubtful, they are still not directly related, but are only connected through long extinct Saurians. The question whether the sparrow and the crocodile have branched off from the same primitive form no more admits of a philosophical answer, than does the question regarding the sparrow and the hippopotamus.

soul; for God's almighty power cannot have produced the soul, which is a spirit, out of matter, as it produced the forms of plants and animals.[1]

No evolution theory is capable of bridging the gulf between the mind of man and matter, which our experience teaches us really exists. It is far greater than the gulf between inorganic matter and organised substances, or than that between vegetative and sensitive life; its width is such that it will never be bridged, just because mind and matter are diametrically opposed. Modern monism has, of course, forgotten this ancient truth, and is doing its best to ignore the essential difference between them, but it is successful neither in the mental physiology of man, nor in the comparative psychology of man and beast; between the movement of the atoms in the brain and thought, between animal instinct and human intelligence, yawns ever the old impassable gulf.[2]

Materialists are only wasting their time when they collect stone after stone and fling them into the abyss; the stones vanish like dust in a bottomless pit, and the gulf remains as wide as ever. Equally futile are all attempts to bridge it by references to the 'intelligence' of apes, or ants, or any other animal, and by depreciating to the utmost extent the

[1] In the *Biologisches Zentralblatt* for 1903, p. 602, note 1, Professor L. Plate expresses his disapproval of this sentence, and criticises it as 'curious logic,' adding: 'That God is almighty, and nevertheless is limited in His sphere of action, is a *contradictio in adiecto*.' Has my good colleague never heard that there are things which are not affected by God's omnipotence, because they contain an interior contradiction? Does he perchance fancy that God's omnipotence could make $2 \times 2 = 5$ and not $= 4$? If so, he has a more exalted idea of the divine omnipotence than all the theologians in the world put together. —*Have, pia anima!*

[2] A classical attempt to bridge this gulf was made by H. E. Ziegler in the lecture already mentioned, 'Über den derzeitigen Stand der Deszendenzlehre in der Zoologie.' On p. 28 he says: 'If the stag can be related to the roebuck, in spite of the fact that the stag has large antlers and the roebuck only small horns, so man can be related to beasts, although man has a great intellect and beasts only a small one.' This profound statement deserves to be inscribed in golden letters on the annals of comparative psychology, that generations to come may benefit by it. One might almost fancy that it was written at the time of shedding horns, when the old antlers of intellect had been cast, and the new ones had not yet grown. Not long ago H. E. Ziegler published a new treatise 'Über den Begriff des Instinktes' (*Zoolog. Jahrbücher*, Supplementary volume, VII, 1904, pp. 700–726), the historical part of which abounds in superficialities and biased misrepresentations. The author unfortunately gives a very poor account of instinct as it is usually understood in Christian philosophy. It may be interesting to compare the definition of instinct given in my book, *Instinkt und Intelligenz im Tierreich*, 1905, pp. 23, &c

mental level of the wildest savages; no success ever has followed, or ever will follow, such attempts. The essential difference between the mental life of man and the sentient existence of beasts, and the impossibility that an alleged brute ancestor of man should ever have become the first *homo sapiens* by natural evolution, are facts that cannot be set aside.[1] Therefore, it is a real 'postulate of science' to account for the mind of man by an act of creation. This involves no violation of the laws of nature; for as mind cannot be produced out of matter, it is obvious that origin by creation is, in the case of mind, the only natural mode of origin.

2. We have now completed our examination of the philosophical limits to the theory of evolution and may pass on to those assigned to it by natural science, although here, too, we must begin with a philosophical preamble.

The theory of evolution is a scientific hypothesis, and in its further development is a scientific theory. By an hypothesis is meant a proposition, the truth of which cannot be demonstrated directly by way of observation or experiment, but which follows as a reasonable deduction from facts, because it is capable of supplying a satisfactory explanation of them. Hypotheses or suppositions are indispensable in natural science; without them there is in fact no *science* in this department of knowledge, for science is *scientia rerum ex causis*; so, apart from hypotheses, we should have only a crass empiricism, contenting itself with observations, and caring nothing for the why and the wherefore of them. As our immediate perception of the things of nature around us reveals to us only their outer husk, our mind is forced to have recourse to hypotheses, in order at least to some extent to be able to penetrate into the working of the laws of nature. If various modes present themselves of explaining one and the same phenomenon or group of phenomena, the mind compares and examines them to see which agrees best with the facts that bear upon the subject, taken collectively. One is then selected as the most probable hypothesis, which the student of nature must accept, until a better is found.

[1] On this subject cf. my two works, *Instinkt und Intelligenz im Tierreich*, 1905, and *Vergleichende Studien über das Seelenleben der Ameisen und der höheren Tiere*, 1900.

As an hypothesis obtains additional probability when pieces of evidence from various sources concur to establish it, it develops into a uniform scientific structure, and ceases to be an hypothesis and becomes a theory. The nature of things requires that we can never demand such a degree of certainty for a scientific hypothesis, or even for a theory, as for a mathematical formula. Metaphysical (mathematical) certainty can never exist with regard to it, and physical certainty only seldom; as a rule it can only claim a lower or higher degree of probability. The Copernican theory supplies us with an instance how an hypothesis, originally possessing only a moderate degree of probability, may eventually rise to the rank of a theory, having so much physical certainty that at the present day no educated person doubts its accuracy. It would be unfair to demand at the outset, in order to justify the scientific existence of an hypothesis, that irrefutable evidence in support of it should be adduced. To demand this would be almost as foolish as, before partaking of any food, to require a chemical guarantee that it contains no poison.

Let us now apply these principles to the theory of evolution. The weight of the evidence in its favour is as often diminished by exaggeration of its value on the part of its champions, as by depreciation of its cumulative force on the part of its opponents.

With regard to the nature and origin of the organic species, we have to choose between two opposite theories, each of which consists of a group of connected hypotheses. Of these theories one, that of permanence, maintains the absolute invariability of the systematic species. It is of opinion that the species are perfectly unchangeable, although varieties and breeds may be formed within them; therefore it regards relationship between the species as impossible, and as equally impossible the suggestion that our present species can be the descendants of other extinct ones. Consequently it assumes so many special acts of creation to have been performed as there are distinct systematic species, and we may assume that at least 800,000 are known to exist now. But in the various geological periods, as a rule, species have followed one another,—they appear at the beginning of a period and vanish at its close; so that this theory requires the acts of creation to have been constantly repeated during the whole geological evolution of our earth. 'But why,' some one may ask, 'need we lay so extreme an interpretation upon the theory of permanence? Why do we not rather say that it requires a relative, but not an absolute, invariability of the species?' Simply because to accept a merely relative permanence of the species involves necessarily the acceptance of a relative variability. A theory of permanence, which declares the systematic species to be 'relatively variable,' regards them as variable either only within the limits of the species or beyond those limits. In the first case it asserts practically the absolute permanence of the limits of the species, and restricts the variability to the characteristic marks of the varieties and breeds within the species; in the second case, on the contrary, it ceases to be a theory of permanence, for it accepts the principle of the theory of evolution, which regards the systematic species as related by belonging to a common stock. It must not be forgotten that the historic strife between the theories of permanence and descent concerns the systematic species in natural science, not the so-called natural species. Our idea of the latter is based on natural philosophy, and has taken its present form under the influence of the theory of evolution. I shall have to recur to it in the next section of this chapter.

Our second alternative is the theory of evolution, according to which the organic species have been evolved from earlier forms belonging to previous ages. It holds that the species are relatively permanent for a definite geological period, and that palæontological research shows shorter periods of transformation to alternate with longer periods in which the organic forms do not vary.[1] We are now in one of the latter, more permanent periods, and this explains the normal persistence of our systematic species; they correspond to the conditions of life around them; but as there is only a relative, and not a fundamental difference between the characteristics of

[1] Cf. Zittel, *Grundzüge der Paläontologie*, 1903, p. 15. Attention was drawn to this phenomenon by Oswald Heer in his *Urwelt der Schweiz*, 1883, chapter xviii. What de Vries calls the 'periods of mutation,' and the periods of 'explosive' transformation of species (Koken, Standfuss), are only other names for the above-mentioned periods of change. The view which de Vries takes of his 'periods of mutation' is extremely hypothetical (*Mutationstheorie*, II, 1903, § 12, p. 697).

species and of genera in systematics, this theory extends the idea of a natural evolution also to the origin of genera. The genera of systematic classification are only groups of natural species, more closely akin to one another than to the species of other groups, although they may originally have branched off from the same stock. The theory of evolution affects families and orders in the same way, and, as far as facts allow, also the higher divisions of the animal and vegetable kingdoms. So much for the theory.

What are the limits of the theory from the point of view of natural science? How far do facts enable it to answer the three following questions, with which philosophy cannot deal? At what date did organic life begin? Must we assume the evolution of plants and animals to have been monophyletic or polyphyletic? What internal and external causes gave rise to the hypothetical race-evolution?

We know very little as yet regarding the date when living organisms first appeared upon our earth. It is certain that life was possible only after the surface of the earth had cooled down, and had formed an atmosphere about itself. The earliest organisms probably lived in the water.[1] In geological language, the date of the first appearance of organic life coincides with the end of the Azoic and the beginning of the Palæozoic age. The dividing line between these two periods in the history of our planet must probably be set further back than has hitherto been done. It is well known that geologists used to regard the Cambrian formation as the oldest stratum containing fossils. But recently Pre-Cambrian fossils have been found in North America, Great Britain, Scandinavia, Bohemia and elsewhere, so that now the Pre-Cambrian is regarded as the oldest stratum containing fossil remains of living creatures.[2] In the present state of our knowledge it is still quite impossible for us to fix the age of this stratum; very likely millions of years have passed between the time when it was formed and now.

[1] Dependent on this is the further question whether the first centres of creation were at the poles, i.e. at the ends of the shortest axis of the earth, or in the equatorial zone, at the ends of the longest axis. On the latter hypothesis see Simroth, 'Über das natürliche System der Erde' (*Verhandl. der Deutschen Zoolog. Gesellschaft*, 1902, pp. 19–42).

[2] Cf. on this subject, Credner, *Elemente der Geologie*, 1902, pp. 389–394; R. Hertwig, *Lehrbuch der Zoologie*, p. 151 (English translation, p. 180).

We do not know whether the primitive forms of all the creatures that lived later, of all classes in the animal and vegetable kingdoms, existed in the Pre-Cambrian period. Probably they did not, for, as far as we know, vertebrates appeared first in the Silurian, and flowering plants seem to be of still later origin. Whether the occurrence of any particular class of forms was really the first or not, is a point on which no final answer can be given, and therefore, from the scientific standpoint, we are still far from being able to decide whether the primitive types of the chief classes of animals and plants were produced simultaneously or in succession, nor can we say when they first appeared.

I may here give a short sketch of what palæontology teaches us regarding the sequence of plant and animal forms in the course of the earth's history. The list of the geological strata with the names of the various formations has been already given (p. 253), and I need not repeat it here.

In speaking of animals I shall follow chiefly Zittel's 'Grundzüge der Paläontologie,' and R. Hertwig's 'Lehrbuch der Zoologie.' No living organisms can be assigned with certainty to the Azoic or archaic age. The animal nature of the famous *eozoon* found in the Archæan (Laurentian) strata is, to say the least, very doubtful. The Palæozoic age supplies the earliest organisms. In the Pre-Cambrian strata of Brittany there are numerous remains of *Radiolaria,* if Barrois is correct in his interpretation of the discoveries made. The Cambrian strata contain only remains of various classes of invertebrates, amongst which Arthropods (Trilobites), Brachiopods (reckoned by Hertwig among Worms), Echinoderms and Molluscs are the chief. In the Silurian, besides the above-mentioned, occur the first vertebrates of the class of fishes, and the first insects among the Arthropods. In the Devonian there are many different kinds of fishes. In the Carboniferous begin the Amphibia, and in the Permian the reptiles. In many cases the forms of these palæozoic creatures very closely resemble those of the modern representatives of the same classes (*Nautilus, Lingula*), but as a rule they are very different (e.g. Trilobites), although frequently they are not inferior to their modern relations in their degree of organisation. The Mesozoic age is that in which reptiles reached their highest development, and the insect fauna of the Lower Jurassic or Lias is very numerous. The first mammals appear in the Triassic or earliest Mesozoic age, and in the Upper Jurassic the first birds, if we may reckon the *Archaeopteryx* as a genuine bird, in spite of its many points of resemblance to a reptile. The fauna of the Cænozoic age approaches more and more to that of the present time; in the Tertiary period the still existent orders of mammals and birds developed, and the likeness between the insects of that period and our own is still more striking. Man appeared only in the Quaternary period, on the threshold of modern times.

According to Reinke's 'Philosophie der Botanik' (pp. 182, &c.) the geological sequence of plant-forms is as follows. There are no remains at all of plants in the Pre-Cambrian and Cambrian strata; the earliest are ferns, which occur in the Silurian, at the same time as the first land animals (insects). Of other Cryptogams, the chalk-algae also occur in the Silurian, the flint-algae in the Carboniferous strata, and they form enormous deposits in the Chalk and Tertiary strata. Ferns, shave-grasses, and Lycopodia reached the highest point of their development in the Coal age, and had then in some ways a more perfect organisation than at the present time. There are no fossils that can serve as links connecting the Algae and the mosses, or the mosses and the ferns.

The Gymnosperms were the first Phanerogams to make their appearance. The earliest of them are the Cordaitae, relations of the Cycadaceae, which appear first in the Devonian, reach their highest point in the Carboniferous, and vanish in the Permian. The first undoubted remains of Cycadaceae occur in the Permian, as well as the first Ginkgos and Conifers. In the Mesozoic age, in the Triassic, Jurassic and Cretaceous periods, the three above-mentioned families of Gymnosperms developed still further, and in the Tertiary strata occur only such kinds as are still known. The earliest Angiosperms, both monocotyledons and dicotyledons, appear suddenly in a great variety of forms in the Upper Chalk, and are unconnected with the Gymnosperms that preceded them. During the Tertiary period more and more representatives occur of still existent families, genera and species of Gymnosperms, and their frequency increases in the more recent strata.

What information as to the hypothetical history of the primitive forms in the organic world is given us by palæontology in its two branches, palæozoology and palæophytology? It tells us nothing certain as to the date of the appearance of the first living organisms or as to their structure, for those organisms alone could be preserved as fossils which were solid enough to make impressions or hollows in the stone; all soft protoplasmic formations must have perished and left no trace. Moreover, it gives us only faint suggestions, though they are extremely valuable, as to the order in which the chief classes of animals and plants appeared upon earth, but it affords certain evidence that the Fauna and Flora of former ages gradually approximated more and more to those of the present time. Numberless families and genera of ancient animals and plants have become extinct, some long ago, some more lately, leaving no descendants; but on the other hand very many seem to have been really the ancestors of our present Fauna and Flora, in spite of the inevitable gaps in the palæontological records, and in spite of the uncertainty still attaching to the interpretation to be put upon many palæontological discoveries.[1]

Let us now turn to the second question and ask: 'Are we to assume that the evolution of animals and plants was monophyletic or polyphyletic?' There is no trace of any scientific evidence to show that the two organic kingdoms were descended from one common primitive cell. It is true that now every multicellular organism in its ontogeny proceeds from a unicellular stage, and among unicellular organisms there are many of which it is impossible to decide whether they are plants or animals; but it is a very bold speculation to conclude from these considerations that all organisms are descended from a common ancestral cell. We are quite ignorant too as to whether we must assume the vegetable kingdom and the animal kingdom respectively to have had a monophyletic or polyphyletic evolution. This alone is certain; there is no evidence at all in support of a monophyletic phylogeny.

All honest supporters of the theory of evolution, who

[1] In his book on the theory of descent (*Die Deszendenztheorie*) Fleischmann has emphasised these two points as detrimental to the theory of evolution, but he has exaggerated their importance. Cf. the discussion in *Stimmen aus Maria-Laach*, LXII, 1902, Part I, pp. 116, &c.

U 2

pay due attention to facts, acknowledge further that the grounds for assuming the existence of a real relationship between the forms in question become more scanty when the higher divisions of the system are considered. For the species of one genus these grounds often amount to great and even irrefutable probability,[1] and the same may be said in not a few cases of the genera of one family, and occasionally for the families of one order, but it can seldom be maintained of the orders of one class. The evidence afforded by natural science for the theory of common descent becomes steadily weaker the higher we ascend in the system, and it becomes weaker, too, the deeper we go into the palæontological history of our earth in order to seek the common ancestors of the subsequently distinct, systematic divisions.

In the latest (7th) edition (1905, p. 152) of his 'Lehrbuch der Zoologie' R. Hertwig gives the chief natural groups of the animal kingdom as seven in number (Protozoa, Coelenterata, Worms, Echinodermata, Mollusca, Arthropoda, Vertebrata); C. Claus reckons nine, and the number is variously given by other zoologists; but the evidence in support of the theory that these groups are of common origin is so weak that we must describe it as improbable rather than probable, in the present state of our knowledge. The truth of this statement becomes apparent if the different hypotheses be compared; for instance, those put forward to account for the descent of Vertebrata or of Arthropoda from other groups of animals; with regard to these hypotheses we might almost say: *Quot capita, tot sensus.* When the opinions of scientists diverge so greatly on one and the same point, we may safely conclude that nothing certain is known about it. Whether we accept seven or seventeen, or any other number, as that of the chief types of the animal kingdom, it is always impossible to assign to them a monophyletic descent from a common primitive form. This has been thoroughly proved by Hamann ('Entwicklungslehre und Darwinismus,' 1892), and by Fleischmann ('Die Deszendenztheorie,' 1901); recently even Theodor Boveri expressed the same opinion in his rectorial address on May 11, 1906 ('Die Organismen als historische Wesen,' Würzburg, 1906, pp. 7 and 51).

[1] Instances of this will be given in Chapter X. See also pp. 276, &c.

The same holds good with regard to the chief classes among plants; R. von Wettstein thinks that we must distinguish seven, all independent of one another ('Handbuch der systematischen Botanik,' I, 1901, p. 16).

In fact, among modern zoologists and botanists, and still more among palæontologists,[1] the number is ever increasing of those who think that the evolution of both animals and plants was polyphyletic, and who regard the monophyletic hypothesis as merely a pretty fancy on the part of the supporters of the theory of descent in its crude form—a fancy that they cannot hope to prove true, for comparative morphology and ontogeny of living organisms, as well as the discoveries made by palæontology, all alike render it more and more improbable that anyone will ever succeed in establishing a monophyletic evolution of either the animal or the vegetable kingdom on a scientific basis. It becomes more and more probable that a monophyletic evolution does not correspond at all with facts.

No serious student is at present able to tell us with certainty how many independent lines of descent, or series of evolution, we must assume to exist among animals and plants respectively. This is due partly to the fact that the answer to this question depends greatly upon the subjective ideas of each individual, but the chief reason for it lies in the significant circumstance that a final answer will be possible only when we have a perfect knowledge of both the present and the fossil organic world. At the present day we are at an immense distance from possessing such knowledge, and therefore we do not know how many original acts of creation must be assumed, in order to account for the existence of the living organisms in the world. Koken says on this subject (1902, p. 218): 'All the great Phyla go back, sharply distinguished, to the Cambrian period, and we have no records at all of those periods when they might have been connected, or when they branched off from a common stock.' Steinmann (1899,

[1] Cf. on this subject E. Koken, *Die Vorwelt und ihre Entwicklungsgeschichte*, Leipzig, 1893; 'Paläontologie und Deszendenzlehre,' address given at the seventy-third meeting of German naturalists at Hamburg, on September 26. 1901 (*Verhandl.* I, Leipzig, 1902, pp. 212-228. Reprinted Jena, 1902). G. Steinmann, *Die Erdgeschichtsforschung während der letzten vier Jahrzehnte* (Freiburg i. B., 1899); *Paläontologie und Abstammungslehre am Ende des Jahrhunderts* (ibid., 1899).

p. 33) goes so far as to believe that men will never attain to this knowledge: 'I feel certain that the oldest representatives of animals and plants of every kind will for ever remain unknown to us; all trace of them has probably vanished, owing to the great changes undergone by the oldest strata.'

We still do not know, and probably we shall never know, under what form we are to imagine the hypothetical primitive types of the various series of evolution; whether we are to think of them as very simple cells, having however an already definite tendency or *Anlage* to evolution; or as phylembryos, or as further differentiated forms, displaying the exterior characteristics of the various types in the shape of definite morphological designs. Nor can we state anything as to the appearance of these primitive types; we do not know whether they all appeared at the same time, or in succession, nor when they were produced.

We come now to the third question: 'What does natural science tell us of the interior and exterior causes of the hypothetical race-evolution?' Here we are still more completely in the dark. Leaving aside those prejudiced persons who are blindly in love with their own theory—the theory of selection, or orthogenesis, or whatever it is—and fancy that it explains everything (although, as a matter of fact, it explains very little), we may frankly acknowledge that our knowledge of the real causes of the race-evolution of the organic species is still in its infancy. One thing alone seems to be fairly certain: Numerous interior and exterior factors must be regarded as the causes of the race-evolution, and the part played by these factors with respect to various series in evolution differs greatly as to the extent both of their participation and co-operation.[1]

Just as, in the development of the individual organism, preformation and epigenesis work together in accord,[2] and definite interior tendencies are regularly modified by exterior influences, so, as we may suppose, is it in the race-evolution of living organisms. In general we must follow Nägeli in distinguishing, in the case of organic species, characteristics due

[1] Some instances taken from zoology will be found in Chapter X.

[2] See Chapter VIII, p. 225, and p. 235. Also O. Hertwig, *Allgemeine Biologie*, 1906, pp. 132, &c., pp. 139, &c.

to organisation from those due to adaptation. The former, which determine the degree of organisation, must primarily be referred to the interior causes of evolution, whilst the latter are connected with the influence of the exterior causes. The active parts taken by both series of causes are more or less mixed, and the interior causes are always the foundation, acted upon by the exterior (e.g. nutrition, temperature, light, &c.), which affect evolution by means of various attendant stimuli.[1]

I cannot at present discuss this topic further. I have considered both the philosophical and the scientific limitations of the theory of evolution, and, as I believe, have dealt impartially with both philosophy and science. We must not undervalue, but neither must we overvalue, the achievements of the theory of evolution hitherto. Centuries will pass before it succeeds in establishing, with a sufficient degree of probability, the number of primitive series of animals and of plants respectively, and in arranging correctly the forms belonging to each series in the many ramifications of their relationship. Centuries more must elapse before science will be able to trace back these series to their origin, and to discover the primitive forms of each. And centuries of research will be required before men will find a satisfactory explanation of the causes which control evolution within each series of forms. Shall we therefore be contented to say: 'Before we acknowledge the theory of evolution to have a scientific justification, we had better wait until it has accomplished all these tasks?' To do so would be both unreasonable and foolish. On the contrary, we can only wish that as many serious research-students as possible may apply themselves with all zeal to solving the difficult problems connected with the theory. This solution could not fail to benefit philosophy, whilst it would be far more creditable to the theory of evolution for its supporters to proceed thus, than to act like Haeckel and those who share his opinions, and try to popularise the theory

[1] Cf. also p. 176 and p. 282; also R. von Wettstein, *Berichte der botanischen Gesellschaft*, XVIII, 1900, pp. 184–200; E. Koken, *Paläontologie und Deszendenzlehre*, 1902; Ed. Fischer, 'Die biologischen Arten der parasitischen Pilze und die Entstehung neuer Formen im Pflanzenreich' (*Verhandl. der Schweizer Naturforschergesellschaft*, eighty-sixth annual meeting, Locarno, September 1903); *Über den heutigen Stand der Deszendenzlehre und unsere Stellung zu derselben*, Berne, 1904.

to advance their own ends, and make a wrong use of it as a weapon with which to attack the Christian cosmogony.

6. Systematic and Natural Species

Linnæus, who is to be regarded as the originator of our present conception of systematic species, and who, therefore, has been called the father of the theory of permanence, enunciated the following dictum: *Tot species numeramus, quot diversae formae in principio sunt creatae*—we reckon so many (systematic) species as there were different forms created in the beginning.

How must this dictum be worded to make it agree with the theory of evolution? According to it, the systematic species of the present time do not represent the originally created forms, but are the result of a process of evolution, uniting the species of the present and the past in natural series of forms, the members of which are related to one another, and each of which points back to an original primitive form, whence it is derived. If we designate each of these independent series of forms, not related to other series or families, as a natural species,[1] we can still assent to Linnæus's dictum: *Tot species numeramus, quot diversae formae in principio sunt creatae.* We reckon so many natural species as there were different primitive forms created in the beginning.[2] Each of these natural species

[1] A similar view regarding natural species has already been expressed by Father T. Pesch in his *Philosophia naturalis*, II, p. 334, in order to explain the facts supporting the theory of evolution. He quotes a number of passages from St. Thomas Aquinas and from Suarez in favour of his view. Of course we are here speaking of the *species physicae* of natural philosophy, not of the *species metaphysicae* of logic. Almost inconceivable mistakes as to my definition of natural species have been made by many reviewers of the first edition of this work, some of them being experienced zoologists. Escherich in the Supplement to the *Allgemeine Zeitung* for February 10, and 11, 1905, gave it far too narrow an interpretation, and Haeckel, Forel and others simply followed him and made the same mistake, without examining the matter for themselves. Another mistake was made by Friese (*Wiener Entomologische Zeitung*, 1904, No. 10) and Schroeder (*Zeitschrift für wissenschaftl. Insektenbiologie*, 1905, Part 4), who believe my distinction between systematic and natural species to be identical with that between biological and morphological species; the biological and the morphological species are but two different aspects of the systematic species, whilst the natural species comprises all the members of the same line of ancestry or pedigree, and therefore is much wider from the point of view of natural science. I trust that these remarks will prevent further misunderstandings.

[2] For readers who have studied philosophy, it is perhaps needless to remark again (as I do for the benefit of some of my critics), that the creation of the

has in the course of evolution differentiated itself into more or less systematic species. How many systematic species, genera, and families belong to a natural species, cannot yet be stated with certainty in most cases. Still less are we able to say how many natural species there are, i.e. how many lines of ancestry independent of one another. We must leave the decision to the phylogenetic research of future ages, if indeed it ever succeeds in arriving at one.

The varying degrees of capacity for evolution possessed by the primitive forms of the different natural species depend primarily upon the interior laws of evolution impressed upon their organic constitution; we are probably justified in regarding the chromatin substance of the germ-cells as the material designed to transmit these laws.[1] The interaction of these interior factors in evolution and of the surrounding exterior influences, through which many kinds of adaptation came about, have produced the ramifications from the parent stock of the natural species, and they have been affected also by cross-breeding (amphimixis) and natural selection.

But, it may be asked, what is the practical advantage of distinguishing thus natural and systematic species, if we are still unable to determine which forms actually constitute a natural species, and how many such natural species there are? To this question we may answer: *Firstly*, in many cases we are able at the present day to decide in some degree the group of forms which belong to a natural species, *although we may not yet know with certainty its full extent.*[2] For instance we may reckon, as belonging to one natural species, all the varieties of beetle of the *Paussidae* family, from the Tertiary period to the present time;[3] but as the *Paussidae*, even if they are the outcome, not of a monophyletic, but of a diphyletic evolution (cf. Chapter X, § 9), are related phylogenetically to

first organisms is not to be understood as a *creatio e nihilo*, but as a production of organisms out of matter. On this subject see the sections on Spontaneous Generation (p. 193), and on the Philosophical Limitations of the Theory of Evolution (p. 279).

[1] See Chapter VI, p. 169 and p. 177, &c.

[2] I have italicised these words because they were overlooked by Escherich and other reviewers in the former edition.

[3] Cf. *Stimmen aus Maria-Laach*, LIII, 1897, pp. 400 and 520, &c., 'Die Familie der Paussiden'; also 'Neue Beiträge zur Kenntnis der Paussiden mit biologischen und phylogenetischen Bemerkungen' (*Notes from the Leyden Museum*, XXV, 1904).

the *Carabidae*, and these again to other families of beetles, the real extent of the natural species in question is probably much greater. With still greater certainty may all the varieties of *Staphylinidae* belonging to the group *Lomechusa* be regarded as forming a natural species. We may therefore rightly say: All the *Lomechusini* form *one* natural species and not more than one. But we do not mean to limit the extent of this natural species to the *Lomechusini*, for this group of *Staphylinidae* is connected phylogenetically with other groups of the same family, and the whole family of *Staphylinidae* with other families of beetles, &c.

If we consider the numerous genera and species of ants from the earliest Jurassic period to the present day, we can hardly doubt that they are offshoots of one single natural species, and are not several natural species. The same remark applies to the family of termites, with its great variety of fossil and still existent genera and species.[1] If we trace back the history of the primitive varieties of the Palæozoic age, which even then formed several distinct classes, whence our present orders of insects branched off probably in the Mesozoic age,[2] we may succeed perhaps, in course of time, in proving these varieties of primitive insects to be offshoots of some original stock, which possibly is connected with the earliest marine Arthropoda, so that eventually many hundreds of thousands of systematic species may unite to form one single line, one single natural species.

This is at present all a matter of pure hypothesis; but these examples serve to show plainly that the limits to be assigned to the natural species become more and more uncertain the higher the division of the animal system and the more remote the historical period of animal life under consideration. It will therefore be best for practical purposes to describe as natural species only those groups of forms which investigation has shown with sufficient probability to be uniform genealogical series.

Thus, for instance, we may class as one natural species all the present varieties of horse (*Equidae*) and their fossil ancestors, comprising various systematic genera, although we do not

[1] See p. 276.
[2] Cf. A. Handlirsch, *Die fossilen Insekten*, Leipzig, 1906.

yet know how far the limits of this natural species may be extended into the past of which palæontology takes account.[1] Among Molluscs, the Ammonites may be mentioned as a group of forms very rich in systematic families, genera, and species; they can be traced from the Devonian to the Cretaceous period through a long series of geological strata, as a uniform, close line of forms, that we must reckon as all belonging to one natural species, not to many. I might add many other instances, but those already given will suffice to show that the distinction between systematic and natural species is by no means devoid of actual foundation. It is in fact practically necessary, if we are to have a scientific knowledge of comparative morphology and biology.[2]

Secondly: The distinction is of far greater importance from the point of view of philosophy. It supplies us with a firm philosophical basis, upon which the theories of creation and descent can easily be reconciled with one another. It is obvious that the possession of such a basis is of the utmost importance to those concerned with the defence of Christianity. Our monistic opponents are fond of adopting the device of directing their attacks against the theory of permanence, when they are really aiming them at the theory of creation. They declare the two theories to be identical, and hope, by overthrowing the one, to secure the downfall of the other. But their hopes are doomed to disappointment, if we resolutely maintain the distinction just laid down. *If we believe that only the natural species in their primitive forms were created, but that it is left to natural science to determine the number and extent of these series of natural forms, as well as the character of the primitive forms themselves, then the enemies of the Christian cosmogony will no longer be able to taunt us with having to accept the permanence of the systematic species as an article of faith.*[3] What has it to do with theistic cosmogony whether a hare and a rabbit, a horse and an ass are related or not? The recognition of a personal God. the

[1] Fleischmann's criticism of 'the stock instance of the theory of descent' (*Die Deszendenztheorie*, chapter v) seems only to confirm the above statement, and not to prove much against the relationship of the *Equidae* to one another.

[2] Further information on this subject, derived from my own investigations, will be found in the next chapter.

[3] This italicised passage gives the reason for the bitter attacks made by monists upon the 'natural species.'

Creator of all finite beings, is no more inseparably connected with the theory of permanence in zoology and botany than it was with the geocentric system in astronomy.

If the theory of descent holds its ground, and takes the place of the old theory of permanence, the theory of creation, and with it the Christian cosmogony, remains as firmly established as ever. Indeed the Creator's wisdom and power are revealed in a more brilliant light than ever, as this theory shows the organic world to have assumed its present form, not in consequence of God's constant interference with the natural order, but as a result of the action of those laws which He Himself has imposed upon nature.

We see therefore that, in this department also, true science leads us finally to a fuller recognition of God.[1] It is a mere delusion on the part of modern atheism, in its various forms and shades of opinion, to fancy that the theory of evolution has enabled the world to dispense with a Creator; for, the more manifold and the more independent is the evolution of the organic world according to the laws inherent in it, the greater must be the wisdom and power of the law-giver who created this world. The Darwinian, or rather Haeckelian, theory of chance, which derives all the conformity to law in nature from an original lawless chaos, by means simply of 'the survival of the fittest,' may at the present day be said to be discarded by science. But the monistic view of the universe, which professes to find the first cause of the orderly arrangement of the world in the world itself, and not in a personal Creator substantially distinct from it, is no better than the materialistic theory of chance; for the so-called God of monism, whom it identifies with the world and everything therein, proves to be a true medley of irreconcilable and inexplicable contradictions, when considered in the light of sound reason. We are told that God is the most perfect being, having from all eternity the ground of His existence in Himself; but at the same time He is a God who must develop His own being in and through the world. Such a monistic God would be pitiably incomplete and dependent, for His very existence would depend upon the

[1] On this subject see K. Braun, *Über Kosmogonie vom Standpunkt christlicher Wissenschaft*, 1905, especially chapters 8 and 9. Also J. Reinke, 'Darf die Natur uns als Offenbarung Gottes gelten?' (*Türmer Jahrbuch*, pp. 139-167, especially pp. 162, &c.).

existence of every midge, and fly, and creature in which He develops Himself. To have invented such an idea of God and to seek to make it take the place of the theistic conception of Him, are achievements of modern lack of thought, not of modern science. But, on the contrary, the recognition of a personal God, who, in virtue of the fulness of His own being, created the world out of nothing, is still demanded by sound human understanding, and is therefore a true postulate of science.[1] Although God is present and acts in all His creatures, He is essentially distinct from the world and independent of it, and has shone forth from all eternity with the same unchanging purity and perfection. All the ephemeral deities of modern monism must give way to this only true God of Christianity.

At the present day men are fond of attacking the theistic cosmogony by saying it is an 'untenable dualism' to recognise a God as essentially distinct from the world. Nobody has yet proved this dualism to be untenable, though monism certainly is so. I am not one of those who 'prefer the most pitiable confusion to dualism' (C. Stumpf). There is in reality only one true kind of monism, and that is the unity of the first cause of all finite being—God in His infinity.'[2] People are fond of quoting Charles Darwin as an authority in support of the modern theory of evolution, but he did not feel that blind hatred of the Creator which characterises Haeckelism. Although we know from some of his later statements that he inclined to agnosticism, he never altered the closing words of his chief work, the 'Origin of Species.' Even in the sixth edition, published in 1888, after his death, this beautiful passage occurs: 'There is grandeur in this view of life, with its several powers, having been originally breathed by the Creator into a few forms or into one; and that, while this planet has gone cycling on according to the fixed law

[1] The accounts of the theory of creation given in modern scientific works are most inadequate. See, for instance, Lotsy's *Vorlesungen über Deszendenztheorie*, I, 1906, pp. 5-8. Lotsy there rejects the atheistic and the pantheistic hypotheses regarding the origin of the world, but professes himself unable to accept the theistic view, which he seems to prefer, because 'the idea of self-existence is absolutely unintelligible.' This is true only of those who have never opened a book on Christian theodicy.

[2] Cf. the third edition of my work on *Instinkt und Intelligenz im Tierreich*, 1905, p. 276.

of gravity, from so simple a beginning endless forms most beautiful and most wonderful have been, and are being evolved.'

Very similar is the opinion expressed by Lyell, the great geologist, in writing to Charles Darwin, on March 11, 1863. He maintains that the acceptance of a phylogeny of the organic species by no means enables us to dispense with the idea of creation. 'I think,' he says, 'the old "creation" is almost as much required as ever, but of course it takes a new form, if Lamarck's views, improved by yours, are adopted.'[1]

7. Summary of Results

Before I pass on to a closer comparison between the theories of permanence and descent, it will be well to arrange the results at which we have arrived under different headings. This is the more necessary, as various reviewers of the first edition have given an unfair account of the contents of this chapter.

Our consideration of the theory of evolution has shown that :—

(1) Darwinism and the theory of evolution are two quite different things, which ought not to be confused with one another. Darwinism in the narrower sense of the word is Darwin's theory of selection; in the wider sense it is the generalisation of that theory to a so-called Darwinian cosmogony.

(2) Darwin's theory of selection cannot be the chief factor in any hypothetical race-evolution, because it merely accounts for the extirpation of the unfit, and not for the development of the fit; only a theory of evolution, ascribing due importance to the interior causes of evolution, can possibly succeed in doing the latter. The Darwinian cosmogony must be rejected absolutely.

(3) The doctrine of evolution, as a scientific hypothesis and theory, aims at investigating the successive forms of animals and plants that have existed from the earliest Palæozoic age to the present time, and at discovering their causes.

[1] See Francis Darwin, *Life and Letters of Charles Darwin*, II, London, 1888, p. 193.

It is not an empirical science, but it strives to give a uniform account of the facts observed in biology.

(4) The chief philosophical points to be observed in dealing with the theory of evolution are: (*a*) We must assume the existence of a personal Creator as the first exterior cause of the origin of matter and of life; (*b*) We must believe that a special act of creation on God's part was required for the production of the mind of man; (*c*) Finally, we must acknowledge interior laws of evolution to be the chief causes of an orderly race-evolution.

(5) The following points may be regarded as settled with regard to the scientific aspect of a hypothetical race-evolution: (*a*) There is no scientific evidence at all in support of a monophyletic origin of all living things from one single primitive cell; (*b*) A monophyletic evolution of the animal kingdom on the one hand, and of the vegetable kingdom on the other, appears very improbable, when the results of palæontological research are taken into consideration; but the scientific evidence in favour of a polyphyletic evolution of animals and plants is steadily gaining weight. We may therefore accept the polyphyletic evolution of both animals and plants from the standpoint of biology and palæontology alike; but the number of the various lines of descent, and the extent of each, are still very obscure.

(6) Equally obscure, from the scientific point of view, are the causes of this hypothetical race-evolution. We can only say that probably many interior and exterior factors co-operate in various ways to produce it, and that the interior laws of evolution have always been the chief cause.

(7) If we call each of the hypothetical and distinct lines of evolution in the organic world a 'natural species,' we may say: 'There are as many natural species as there were originally different primitive forms, produced at the creation of the organic world.' We must leave it to future biologists to determine the number and extent of these natural species, and the structure of their primitive forms.

(8) As we have viewed it, the doctrine of evolution as a scientific hypothesis and theory is perfectly compatible with the Christian cosmogony. The ideas of creation and evolution are not antagonistic, but the creation of the primitive forms

is the natural basis of the subsequent phylogeny of the organic world. Both together make up a theory of nature founded on Christianity.

(9) What must we think of the theory of evolution as a theory of the universe from the standpoint of the philosophy of nature? The view adopted by monism is wrong and full of contradictions, for it excludes creation and upholds nothing but evolution. But the view adopted by Christian theism is right and logical, for it accepts God's creative action as the starting point for the evolution of the organic world, and then leaves it to natural science to establish the details of that hypothetical evolution.

(10) We must once more carefully distinguish between the scientific theory of evolution, and its philosophical generalisation into a cosmogony founded on Christianity. The former is still a modest little plant, just raising its head above the ground. The latter is a tree, stretching its branches far and wide, and lifting its top to the clouds, but, as we must never forget, its roots are still embedded to a great extent in philosophical speculations, and not in scientific facts. If we bear this distinction in mind, we may calmly assert:—

The Christian cosmogony, that accords with the theory of evolution, reduces the history of animal and vegetable life upon our planet (though it covers hundreds of thousands of years) to a mere line in the book of the natural evolution of the whole cosmos; but on this book's title-page stands written in indelible characters:

'In the beginning God created the heaven and earth.'

In the following chapter I propose to make use of facts as the groundwork for a comparison between the theories of evolution and permanence—a comparison which, as our present survey of the theory of evolution necessarily suggests, will result in our accepting the former and rejecting the latter.

CHAPTER X

THEORY OF PERMANENCE OR THEORY OF DESCENT

(See Plates III–V)

12. Conclusions and Results.
The theories of permanence and descent compared with regard to their value in supplying explanations (p. 426). The latter alone can suggest natural causes to account for the occurrence of expedient adaptations (p. 427). It reveals the Creator's wisdom and power more strikingly than does the theory of permanence (p. 429).

In the previous chapter I suggested some thoughts on the doctrine of evolution, which made it clear that there is a great difference between Darwinism and the theory of evolution, and threw considerable light on the latter from various points of view. I drew especial attention to the connexion between the Copernican system and evolution on the one hand, and between evolution and the theory of creation on the other. Let us now proceed to examine more closely the facts belonging to the theories of permanence and descent. On account of the enormous extent of the scientific evidence at my disposal, I shall limit myself to a few instances derived from my own branch of research, so that I am not dependent upon any extraneous authority.

1. Arguments for the Fixity of Systematic Species

At first sight the great majority of facts in zoology and botany appear to support the permanence of the systematic species. This theory stands in the same advantageous position as the Ptolemaic system did long ago, for it too could adduce almost all our own observation of nature as testimony in its favour. Even at the present day, it might be a difficult task to convince an ignorant country lad that the sun is stationary and that the earth moves round it, because the evidence of his own eyes is to the contrary, and the scientific proofs are beyond his comprehension. It may well be that the theory of evolution is now faring as the Copernican theory did of old. Apparently most of the phenomena in the organic world are against it, and therefore, unless we study the matter closely and test carefully the scientific circumstantial evidence in its favour, we run the risk of arriving at a decision such as the country lad would form.

Even the adherents of the theory of evolution, when they take facts sufficiently into account, confess more or less frankly that the systematic species forms at the present time

x 2

a morphological **and** biological unit. It is a morphological unit, inasmuch as it is a group of individuals, the members of which agree in the so-called 'essential' characteristics, and are regularly marked off from other groups of individuals. It is a biological unit, inasmuch as this group of individuals constitutes a genetic whole, repeating through an unbroken series of generations the same regular cycle of forms, in the phenomena of embryonic development, metamorphosis and metagenesis; and, further, where sexual intercourse takes place, the members of one species can copulate with one another, but not with those of another species, if their union is to be fertile.[1]

These facts can be denied only by those ardent partisans who, in discussing the doctrine of evolution, care more about maintaining their theory than about giving it an objective foundation. By far the greater number of the systematic species of the present animal and vegetable kingdoms, and most of the fossil species too, represent real morphological and biological units, and this fact is generally recognised. In the case of the fossil forms, it is of course impossible to offer direct evidence in support of the biological unity of the species, but it can be deduced from the morphological unity. The organic world of the present, like that of the past, is not a disorderly chaos of minute variations, such as the Darwinian form of the theory of descent would require, with its quite gradual and imperceptibly minute shades of difference (for their average biological unimportance would prevent the 'Struggle for Existence' from ever arranging them in well-defined groups of forms), but it is an orderly system of species, genera, families, orders, classes, and groups. To attempt any further proof of this is quite superfluous, for every student of systematics knows it as a fact, which we may and must assume to be generally recognised. Abundant information on this subject may be found in any textbook of zoology, botany, or palæontology; and therefore it is the more surprising that many over-zealous advocates of the theory of descent seem still to be completely unaware of it.

[1] This last characteristic is not universally applicable; among plants the majority of the hybrids produced by crossing different species are fertile. Cf. J. Reinke, *Einleitung in die theoretische Biologie*, 1901, pp. 536–537.

Professor L. Plate,[1] for instance, in a review of Fleischmann's book on the theory of descent, published in the 'Biologisches Zentralblatt,' says: 'The experience of systematists teaches us, as plainly as we can possibly desire, that a species cannot be sharply defined, because variability is a fundamental phenomenon in organic life.' The testimony of actual facts is directly opposed to Plate's statement. Without fear of contradiction we may make the following assertion: 'The experience of systematists teaches us as plainly as we can possibly desire, that species are generally sharply defined, because variability in organic forms is mostly confined within the limits of the species.' In his zealous defence of the theory of descent against Fleischmann's attacks, Plate has represented what is actually the exception as the rule, and what is actually the rule as the exception. There are, of course, what are called 'bad species,' connected with one another by varieties, but it is precisely for that reason that we call them *bad*, and contrast them with the 'good species,' which are marked off from one another by constant characteristics, and show no transitional forms.

It would be better not to call these 'bad species' by the name of species at all, but to designate them rather as sub-species or breeds, and to limit the idea of systematic species to the sharply defined 'good species.' This, for instance, is the reason why all the more recent scientific writers, who have dealt with the classification of ants, have followed Forel (1874), and divide them into species, subspecies (or races), and varieties. Only in this way can we succeed in grouping the forms systematically, so as to

[1] 'Ein moderner Gegner der Deszendenztheorie' (*Biolog. Zentralblatt*, XXI, 1901, Nos. 5 and 6). The passage quoted occurs on p. 142. It is hardly necessary to remark that I do not agree with Fleischmann in his absolute rejection of the theory of descent. It is extremely kind of Professor Plate to utter the warning that he gives on p. 172: 'Orthodox philosophy and theology will joyfully seize upon Fleischmann's book, and regard it as a sign that the doctrine of creation is resuming its proper place.' Plate is confusing the theory of permanence with the doctrine of the creation. If the former be abandoned, the latter still remains indispensable, as alone accounting for the origin of the first forms. As I showed at the end of the preceding chapter, the doctrine of the creation is a necessary premiss to every reasonable theory of evolution. Cf. also my answer to Plate in the *Biolog. Zentralblatt*, XXI, 1901, No. 22, pp. 689, &c.

correspond with the natural relationships existing between them.[1]

The species *Camponotus maculatus* F. is particularly rich in forms, and is found all over the world. It now contains about fifty subspecies, and within these again over a hundred varieties.

In the same way the systematic classification of Coleoptera, especially in the genus *Carabus*, has been revised by Ganglbauer and Born. In one of Paul Born's recent works on *Carabus monilis*,[2] he distinguishes twenty-one subspecies within the species, many having previously been regarded as distinct species, and these twenty-one subspecies comprise together over fifty varieties. This gives us some idea of the enormous number of forms belonging to some of the species of the genus; but it proves nothing against the existence of good species among animals and plants, and rather confirms their existence, for otherwise there would be no distinction between species and subspecies. No one, for instance, would take it into his head to question the right of *Myrmica rubra* L. and *rubida* Latr. among ants, of *Carabus monilis* F. and *intricatus* L. among *Carabidae*, of *Dinarda dentata* Grav. and *clavigera* Fauv. among *Staphylinidae* to be regarded as distinct species.

In some genera of animals (e.g. *Dinarda*) there are only a few species and numerous subspecies and varieties; whilst in others (e.g. *Camponotus* and *Carabus*) there are a great many species and a correspondingly large number of subspecies and varieties; and finally in other genera (e.g. *Rhynchites*, a kind of weevil) there are a good many genuine species, no subspecies, and only a few, quite unimportant varieties.[3] It would therefore be injudicious and inaccurate to deny with Plate the existence of sharply defined species. An excellent remark is made by Fr. Dahl,[4] who says: 'Students occupied with departments of science in which sharply defined

[1] Cf. on this subject Aug. Forel, 'Über Polymorphismus und Variation bei den Ameisen' (*Zoolog. Jahrbücher*, Suppl. VII, 1904, pp. 571–186).

[2] '*Carabus monilis* F. und seine Formen' (*Insektenbörse*, XXI, 1904, Nos. 6–10).

[3] Cf. Wasmann, *Der Trichterwickler*, 1884, Appendix on the biology and classification of the species of *Rhynchites* and their relations.

[4] 'Die physiologische Zuchtwahl im weiteren Sinne' (*Biolog. Zentralblatt*, 1906, No. 1, pp. 3–5), p. 14.

species do not occur, are apt to believe that there are no good species in existence; and those, on the other hand, who have to deal exclusively with good species, cannot understand that there may be none in other groups of animals. Every scientist, who aims at forming a just opinion on questions connected with the theory of descent, ought to have experience in both kinds of work.'

Is the fixity of the organic species, that prevails at the present time, to lead us to conclude that species are absolutely invariable, and that therefore no evolution can have taken place in their case? Such a conclusion would be premature, for, granted that an evolution of species took place in previous ages, the results of it might be exactly what we see about us in the Alluvial epoch in which we live. An intelligent day-fly, prevented by the shortness of its life from knowing anything of the alternation of seasons, after seeing the trees in blossom for an hour or two, might equally well conclude that the world around it was in an unchanging state of perpetual spring, and had been originally created in this condition; and yet the fly would certainly be mistaken. Let us beware of coming to a conclusion of this kind! Palæontology teaches plainly enough that, in previous ages also, comparatively long periods of fixity have alternated with shorter periods of transformation of organic forms.[1]

If we are at the present moment living in a period of comparative fixity of organic forms, we may seek in vain for actual changes in the species around us; but that circumstance proves nothing against the theory of descent.

However, even now we can observe facts which serve as evidence, direct or indirect, in favour of an evolution of the organic forms. Let us consider first the direct evidence, although it must needs be very scanty.

[1] Cf. K. von Zittel, *Grundzüge der Paläontologie*, p. 15; also O. Heer, *Urwald der Schweiz*, chapter 18. See also p. 287, note 1.

2. Direct Evidence in Support of the Theory of Evolution

It has recently been shown by Hugo de Vries[1] that at the present time many plants are still in a period of evolution, i.e. they are producing new forms which are as sharply defined, as independent, and as free from variations as real, systematic species. According to de Vries, the evening primrose (*Oenothera Lamarckiana*) is now in a period of mutation. There is no trace of any Darwinian natural selection as causing or influencing this mutation; the new varieties come into being simply in consequence of the interior laws of evolution in the form undergoing change, and not in any way through the force of natural selection. This suggests the idea that even at the present time the process of race-evolution is not complete in the case of all species. With regard to Darwin's theory of natural selection de Vries says ('Die Mutationstheorie,' II, p. 667): 'Natural selection is a sieve; it sifts out, but produces nothing, although it is often wrongly asserted to do so. The theory of selection ought not to take into account the origin of what it eliminates.'

Many eminent zoologists and palæontologists, such as Waagen, Koken, Scott, Steinmann, Abel, &c.,[2] have expressed themselves in favour of the theory of mutation in the animal kingdom. In fact, all the authors who accept an 'abrupt' or 'explosive' or 'iterative' development of forms in the evolution of the race, such as Kölliker (1864), Emery (1893), and Bateson (1894), are approximating to the view

[1] *Die Mutationstheorie: Versuche und Beobachtungen über die Entstehung von Arten im Pflanzenreiche*, I, Leipzig, 1901; II, 1903. Cf. also *Biolog. Zentralblatt*, XXI, 1901, Nos. 9 and 10; XXII, 1902, Nos. 16–19; XXIV, 1904, Nos. 5–7. 'Ältere und neuere Selektionsmethoden' (*ibid.* XXVI, 1906, Nos. 13–15, pp. 385–395). J. Wiesbaur (*Kulturproben aus dem Schulgarten des Stiftungsgymnasiums Duppau*, 1904, p. 42) asserts that within thirty years he has twice observed the spontaneous growth of new plants. For a criticism of the theory of mutation see also J. Reinke, *Einleitung in die theoretische Biologie*, pp. 518, &c. According to J. Reinke the range of mutation is extremely limited.

[2] For the bibliography of the subject see especially E. Koken, *Paläontologie und Deszendenzlehre*, Jena, 1902; also W. B. Scott, 'On variations and mutations' (*American Journal of Science*, XLVIII, 1894, pp. 355–374); M. Standfuss, *Experimentelle zoologische Studien mit Lepidopteren*, Zurich, 1898; J. Gross, 'Über einige Beziehungen von Vererbung und Variation' (*Biolog. Zentralblatt*, 1906, Nos. 13–18). Gross rejects mutation as a factor in forming species among animals. (See pp. 555, 561, &c.)

of the theory of mutation held by Korschinsky, de Vries, and other botanists.

I should like to draw particular attention to the opinion expressed by Zittel ('Grundzüge der Paläontologie,' 1903, pp. 14, 15) that periods of rapid and slow transformation often alternate in the evolution of a race, for this opinion probably is nearest to the truth.

Linnæus stated that new forms could be produced by crossing different species, and this is a very suggestive idea as regards the theory of evolution. As far as the vegetable kingdom is concerned, Kerner von Marilaun has proved in his 'Pflanzenleben' (II, 1898, pp. 565, &c.), that at the present time not only new varieties and subspecies, but new systematic species, can be produced in this manner. Even J. Reinke,[1] who has adopted a very critical attitude towards the evidence in support of the theory of evolution, agrees with Kerner von Marilaun on this point, and refers especially to the genera *Rubus*, *Salix*, and *Hieracium* as instances of groups of forms in which new types are still being developed, that behave like genuine species. There is, in fact, among plants a good deal of direct evidence in favour of the theory of descent, although this evidence may not be of a very important nature.

It is impossible to discuss in detail all the modern views on the subject of evolution of species in the vegetable kingdom. Most of these views coincide with Nägeli's; they draw a sharp distinction between organic characteristics and those due to adaptation, and they refer the former to interior, and the latter to exterior causes. I need allude here only to two works, viz. Ed. Fischer's 'Die biologische Arten der parasitischen Pilze und die Entstehung neuer Formen im Pflanzenreich' ('Biological species of parasitic fungi and the origin of new forms in the vegetable kingdom'),[2] and C. Correns' 'Experimentelle Untersuchungen über die Entstehung der Arten auf botanischem Gebiet' ('Experimental investigations regarding the origin of species among plants').[3] Correns' verdict upon the theory of selection is interesting; he says: 'Natural

[1] *Einleitung in die theoretische Biologie*, 1901, pp. 542, &c.

[2] *Verhandl. der Schweiz. Naturforsch. Gesellsch.* (Proceedings of the Association of Swiss Naturalists), eighty-sixth annual meeting at Locarno, September 1903.

[3] *Archiv für Rassen- und Gesellschaftsbiologie*, I, 1904, Part I, pp. 27–52.

selection does nothing but weed out ; it has laid aside innumerable forms, and so has created gaps, but it has never produced anything new.' This opinion agrees fully with my own (Chapter IX, p. 260). In the animal world it is much more difficult to study the problem of mutation by way of observation and experiment than it is in the vegetable world.[1] This may seem strange at first sight, because most successful results have been obtained by the artificial selection practised in breeding the domestic animals. But these triumphs of selection are completely worthless as affording any evidence of the origin of new species, for all the varieties and breeds of our domestic animals, produced by artificial selection maintained for hundreds or even thousands of years, are deficient in the one quality of *fixity*, which alone could give them any positive value as aiding the solution of our problem. There is not one artificial breed, no matter how well defined or how far divergent from the primitive form, that can preserve its characteristics without the help of man ; left to itself, it invariably reverts in course of time to the original wild type.[2] They supply, therefore, no evidence at all of the origin of new species under natural conditions, because natural species must necessarily be constant, whereas all artificially produced breeds are liable to change. I do not mean to imply that the interesting observations, made by Charles Darwin and his

[1] I may incidentally remark that Schmankewitsch's famous attempts to turn the crab *Artemia salina* into a *Branchipus*, by diminishing the amount of salt in the water, can no longer be regarded as furnishing trustworthy evidence. Cf. Ad. Steuer, 'Der gegenwärtige Stand der Frage über die Variationen von *Artemia salina* Leach' (*Verhandl. der k. k. Zool. Botan. Gesellsch.*, Vienna, 1903, pp. 145, &c.). The result of Steuer's investigations is given on p. 150: 'Just as under natural surroundings no *Artemia* can ever become a *Branchipus*, or *vice versa*, so, most certainly, no one will ever succeed in transforming one creature into the other by artificial means in an aquarium.' On the subject of the alleged capacity for transformation of *Artemia salina* see M. Samter and R. Heymons, 'Die Variationen von *Artemia salina*' (Supplement to the *Verhandl. der Preuss. Akademie der Wissenschaft*, 1902); Cesare Artom, 'Note critiche alle osservazioni del Loeb sull' *Artemia salina*' (*Biolog. Zentralbl.* 1906, No. 7, pp. 204–208). I do not propose to discuss the very interesting experiments on the influence of heat on the colour of butterflies (Dorfmeister, Weismann, Standfuss, Urech, Fischer, von Linden, &c.), and on that of cochineal insects (Chr. Schröder), as the range of variation scarcely exceeds that of 'Saisondimorphismus' under natural circumstances. These experiments, however, prove sufficiently that the direct action of exterior causes is of great importance in the phylogeny of the forms in question.

[2] A very good summary and criticism of facts and statements on this subject is given by Yves Delage in his book, *La structure du Protoplasma et les théories sur l'hérédité*, 1895, pp. 295–298.

followers, on the methods and results of artificial selection are without bearing upon the question of descent; on the contrary, they are of great value in this connexion, but they tend to prove the exact opposite of what the followers of Darwin desire. Instead of showing that new species can be formed on the lines of artificial selection, they have proved that this never occurs. At the present time scientific men are becoming more and more convinced that facts do not justify the comparison, set up by Darwin and his adherents, between artificial selection and the processes whereby new species are formed under natural circumstances. This comparison has found its scientific expression in the theory of selection. If we want to find actual evidence of the evolution of new species in phenomena of our own day, we must begin by setting aside as useless all artificially produced breeds, and we must limit our observation to the processes of natural and independent formation of new varieties. But this is easier said than done! For where can we discover such processes, seeing that we are living in a period when the organic forms are fixed?

3. The Evolution of the Forms of *Dinarda*

As proof that nevertheless such processes are still going on, though they are not of frequent occurrence, and can be regarded as satisfactory evidence only after very minute observation of facts, I may refer to an instance that I discovered in the course of my own research-work.

As a full account of it has already appeared in the *Biologisches Zentralblatt*,[1] I shall only refer shortly to the most important points connected with it.

In the nests of ants living in northern and central Europe are found various kinds of beetles of the genus *Dinarda*, *Staphylinidae* of the sub-family Aleocharinae. In shape these beetles are broad and flat in front and sharply pointed behind, and they belong to the offensive type (Trutztypus) of ant-inquilines,

[1] 'Gibt es tatsächlich Arten, die heute noch in der Stammesentwicklung begriffen sind? Mit allgemeineren Bemerkungen über die Entwicklung der Myrmekophilie und Termitophilie und über das Wesen der Symphilie' (*Biolog. Zentralblatt*, XXI, 1901, Nos. 22 and 23).

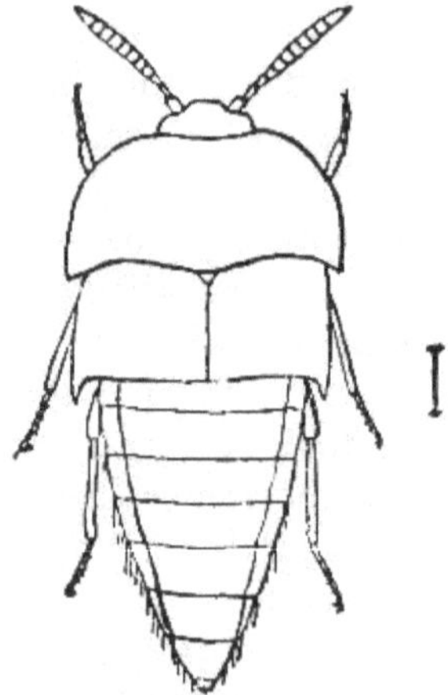

FIG. 29.—*Dinarda Maerkeli* Ksw. (original).

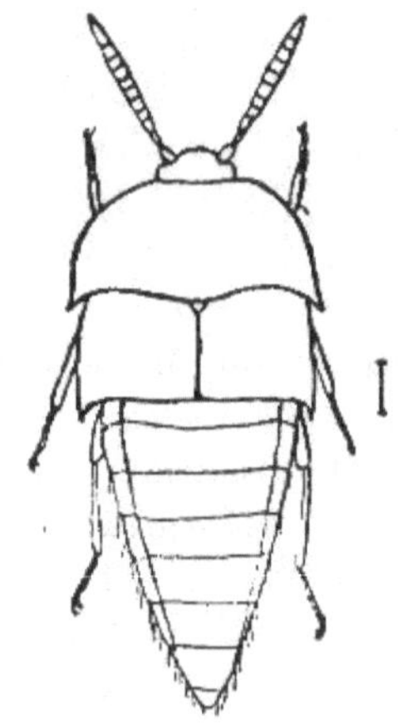

FIG. 30.—*Dinarda dentata* Grav. (original).

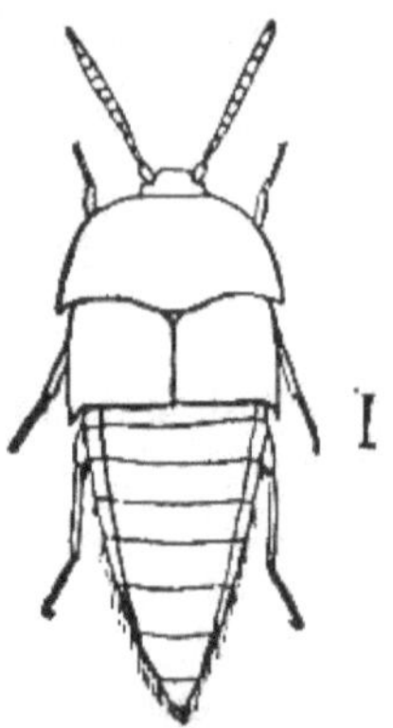

FIG. 31.—*Dinarda Hagensi* Wasm. (original).

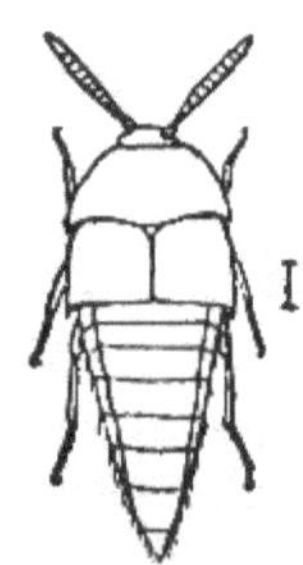

FIG. 32.—*Dinarda pygmaea* Wasm. (original).

The accompanying illustrations represent the four species of Dinarda that occur in central Europe, and show their relative size and shape. In colouring they resemble their hosts, viz. they are red and blackish. In *D. Maerkeli* and *D. dentata* the wing-sheaths and the sides of the prothorax are reddish brown; in *D. Hagensi* they are of a brighter red, and this colour extends further, as far as the base of the antennae and of the abdomen. In the smaller *D. pygmaea* the wing-sheaths are of a dark reddish brown, with a black spot round the scutellum; the sides of the prothorax have a brownish tinge at their edge only. The rest of the body is almost black, with the exception of the legs.

i.e. their structure renders them invulnerable to the attacks of their hosts and enables them to defy them, so that the ants tolerate their presence. There is no spot in the *Dinarda's* body that the ants can reach with their jaws, if they wish to attack them. The whole genus *Dinarda* belongs to this offensive type, but the various species assume various forms adapted to the peculiarities of their hosts, for each species of *Dinarda* has its own especial host. *D. dentata* (fig. 30) lives with the red ants (*Formica sanguinea*), *D. Maerkeli* (fig. 29) with the wood-ants (*F. rufa*), *D. Hagensi* (fig. 31) with *Formica exsecta*, *D. pygmaea* (fig. 32) with *F. rufibarbis*, and especially with a small, dark-coloured subspecies known as *F. fusco-rufibarbis*. A series of observations and experiments, carried on for many years, enabled me to establish the fact that the differences existing between these four species of *Dinarda* might be very simply referred to the following principle:— The larger species of *Dinarda* always lives with the larger species of *Formica* and with such as build large ant-hills; the smaller species of *Dinarda* lives with the smaller species of *Formica*, and with such as occupy simple nests in the earth. *F. rufa* and *exsecta* build ant-hills, and *rufa* is considerably bigger than *exsecta*; therefore the biggest and broadest species of *Dinarda*, *D. Maerkeli*, lives with *F. rufa*; the smaller *D. Hagensi* with *F. exsecta*. The latter *Dinarda* is almost as large as *D. dentata*, which lives with *F. sanguinea*, although this ant is considerably bigger than *F. exsecta*, but *sanguinea* generally constructs simple nests in the earth, which have at best a little heap of vegetable matter at the top, whereas *F. exsecta* builds real ant-hills. *F. fusco-rufibarbis* is the smallest and darkest of all the above-mentioned kinds of ants, and it always makes simple nests in the earth; therefore *D. pygmaea*, that lives with it, is the smallest and darkest of all the *Dinarda* family.

As the *Dinarda* are inquilines of the offensive type, and are tolerated with indifference because of their normal invulnerability, it follows that only smaller *Dinarda* can live among small ants than among large ants, for the larger the *Dinarda* in proportion to its hosts, the more easily can they seize it by its antennae or legs, hold it fast, kill and devour it. I have established this fact by actual experiments. In the same way among ants living in simple nests in the earth only a smaller *Dinarda* can make its way, than among those ants in whose spacious ant-hills there must be many convenient hiding-places for the beetles. But why does the darkest *Dinarda* live with the darkest ants ? For the same reason. Because the *Dinarda* are the largest inquilines of the offensive type, and therefore attract the ants' attention in an especial degree, there must be a certain amount of similarity in colouring between them and their normal hosts, in order that they may more easily escape notice. Now all the above-mentioned species of ants are of two colours, red and black, and so all the four corresponding species of *Dinarda* wear the same livery, and *F. fusco-rufibarbis*, being the ant darkest in colour and most nearly approaching uniformity in tint, is the host of *Dinarda pygmaea*, which is the darkest beetle, and the one most nearly approaching uniformity in tint.

For the facts just stated I can offer no explanation but the following, that our four species of *Dinarda* are four different forms of one and the same generic type, and their differences are due to adaptation to the four kinds of guest-ants. If we assume that within the genus *Dinarda* an evolution has taken place, we must acknowledge that this evolution was determined by the characteristics of the guest-ants, and took place in the way described above. The result of a race-evolution of *Dinarda* could be no other than that which we can observe at the present day.

But has such a race-evolution really occurred ? Yes, for there is important evidence to show that this evolution is not yet ended, but is still going on before our eyes.

The following facts bear out the above statement. In the the first place, there are certain regions in central Europe in which the four forms of *Dinarda* live side by side, after the fashion of genuine systematic species, having their points of difference fixed. Each inhabits the nests of the ants to which it corresponds. Secondly, there are other districts in northern and central Europe, in which only two forms of *Dinarda* (*dentata* and *Maerkeli*) occur, living with their respective ants (*F. sanguinea* and *rufa*), whilst *F. exsecta* and *fusco-rufibarbis* have no *Dinarda* as guests in those regions. Thirdly, there are other regions in central Europe occupying a position between these two extremes, inasmuch as *F. sanguinea* and *rufa* possess their proper kinds of *Dinarda* (*dentata* and *Maerkeli*); whilst *F. exsecta* entertains a transitional form midway between *dentata* and *Hagensi*, and among *F. fusco-rufibarbis* occur forms connecting *dentata* and *pygmaea*. This can be observed best in the case of the *Dinarda* that is the guest of *F. fusco-rufibarbis*. The very small, dark *D. pygmaea*, which is completely adapted to this ant, is connected by a series of transitional forms, having a different geographical distribution, with *D. dentata*, that lives with *F. sanguinea*.

In many parts of central and northern Europe no special kind of *Dinarda* is found living with *F. rufibarbis*, but in other places there is a kind that scarcely differs from *D. dentata*. In other districts again there is the *D. dentata* var. *minor*, which is already distinguished as a variety of *dentata*, and in others the *D. pygmaea* var. *dentatoides*, which closely approximates to the typical *pygmaea*; finally, in other districts the genuine *D. pygmaea* is found, either alone, or as well as the var. *dentatoides*. In order to understand this geographical distribution, we must not lose sight of the fact that, in each district, *Dinarda* occurs among *F. rufibarbis* with greater regularity and frequency the more widely the *Dinarda* form corresponding to the ants in that locality diverges from the *dentata* type, and the more closely it approximates to the *pygmaea* type.

As a science, natural science cannot avoid seeking the fixed pole about which phenomena revolve; it must needs try to discover the laws underlying the multiplicity of phenomena. The law contained in the foregoing account of the distribution of *Dinarda* may be stated as follows:—The specific evolution of the forms of *Dinarda* has reached different stages in different parts of geographical distribution. The adaptation of *D. dentata* to *F. sanguinea* and of *D. Maerkeli* to *F. rufa* is complete all over northern and central Europe, but that of *D. Hagensi* to *F. exsecta* and of *D. pygmaea* to *F. fusco-rufibarbis* is still incomplete; in fact, the last-named adaptation is in progress, being complete in some localities, having advanced half-way in others, and in some places having scarcely begun or even not begun at all. Recent discoveries show that the adaptation of *Dinarda Hagensi* to *Formica exsecta* has advanced further in England and in the Siebengebirge on the Rhine than in other parts of central Europe.

If we wish to determine more exactly the topographical localities corresponding to the different stages of evolution in *Dinarda*, we must distinguish general and particular local influences. As a rule, the four forms of *Dinarda* seem to be most sharply marked off from one another in those districts of central Europe which first became free of ice and water at the close of the last glacial period of the Pleistocene epoch, such as the Rhine valley above the Siebengebirge, southern England, Bohemia, Silesia, &c. The fact that only two species of *Dinarda* appear to occur in the central Alps and in northern Europe agrees with this view. On the other hand special local circumstances may contribute sometimes to a quicker and sharper marking off of the species of *Dinarda* living with *F. rufibarbis*. So, for instance, on the glacis of the old fortress of Luxemburg, on a plateau with steep edges, where there are many nests of *F. rufibarbis*, but none of *F. sanguinea*, I have found *D. pygmaea* var. *dentatoides* in the *rufibarbis* nests, many specimens approximating very closely to the typical *pygmaea*. I observed the same thing on the steep hills of Pulvermühl near Luxemburg, where similar local conditions favour the development of *Dinarda pygmaea*. But on the long ridges of hills between Luxemburg and Treves, I found several *Dinarda* scarcely differing from the typical *dentata*, in nests of *F. rufibarbis* at Ober-Anven ; the evolution of a special *Dinarda* form among *F. rufibarbis* in this district has probably been hindered, because the *Dinarda*, in passing from one ants' nest to another, have had opportunities of crossing with *D. dentata* living in the neighbouring nests of *F. sanguinea*. If the *rufibarbis* nests are circumscribed by the configuration of the locality, the evolution of a particular form of *Dinarda* is doubtless facilitated, although it does not appear to be absolutely necessary that the nests should be isolated ; for at Exaten in Dutch Limburg for many years I used to find in a nest of *F. rufibarbis* var. *fusco-rufibarbis* specimens only of *D. dentata* var. *minor*, with no transitional forms to the typical *D. dentata*, although only about thirty yards away, on the same flat stretch of ground, there were several nests of *F. sanguinea*, inhabited by the typical *D. dentata*.

The objection may be raised that these phenomena are arguments for an evolution within the species only, and not for an evolution of new species from others. In this case what is meant by a 'species'? Is it a *natural* or a *systematic* species?[1]

That our four parti-coloured forms of *Dinarda* belong to one natural species is a matter of course, as soon as they can be proved to be of common origin. But if we ask whether they ought to be reckoned as belonging to one systematic species, the answer is not so simple. In case they are all declared to be only systematic subspecies of *D. dentata* —an opinion that I put forward as long ago as 1896[2]—they are nevertheless subspecies constituting different stages on the way to the formation of genuine species. *D. dentata*, which stands nearest to the hypothetical primitive form, and *D. Maerkeli*, which was the earliest to branch off from it, are already quite as sharply differentiated from one another as are many other systematic species. *D. Hagensi* and *pygmaea* are at a less advanced stage of evolution, and have been differentiated as independent forms only in some of the localities occupied by the ants that are their hosts. It is, however, quite immaterial to the question under discussion, whether we declare the four parti-coloured forms of *Dinarda* occurring among the Fauna of northern and central Europe to be real systematic species, or only races or subspecies at different stages on the way to forming species, for in neither case is it possible to avoid the assumption that we have here a real instance of evolution, the aim of which is the production of forms adapted to a particular way of life, and destined finally to split up into distinct species.

The process of evolution extends even to the generic characteristics of *Dinarda*. In *Dinarda Hagensi* of the Siebengebirge (von Hagens) and southern England (Donisthorpe), the edge of the wing-sheaths is not convex and carinated, as it should be, according to the systematic description of the genus *Dinarda* and of all the genera of *Dinardini*,

[1] For the distinction between these two ideas see pp. 296, &c., in the preceding chapter.

[2] '*Dinarda*-Arten oder Rassen?' (Vienna, *Entomolog. Zeitung*, XV, Parts 4 and 5, pp. 125-142).

Y

but it is simply curved, as it is in the other cognate Aleocharinæ.[1] In other specimens of *Hagensi*, from Linz on the Rhine, the edge of the wing-sheaths is convex and carinated, as it is in *D. dentata*. There are also forms of *Hagensi*, standing midway between the two to which I have referred, with respect to the formation of the edge of their wing-sheaths. This shows plainly that the generic characteristics also of *Dinarda* have only a relative value, and that they are affected by the same laws of natural evolution as those that differentiate species and subspecies within the genus. I shall be able later on to establish this conclusion more firmly by means of a comparison with the *D. nigrita* of southern Europe. How can any one seriously maintain that the phenomena which I have observed in the evolution of *Dinarda* serve only as arguments in support of an evolution within the systematic species?

Some one may, perhaps, grant that within the genus *Dinarda* such a process of evolution is actually still going on, but he may say that he does not see what it has to do with our acceptance of the theory of evolution in general, as possibly this is merely an exceptional case. It is quite true that we have here an exception to the usual fixity of systematic species, and it would be a great mistake to assert that all, or even most, genera of animals are still forming new species in the same way as the *Dinarda*. It would, however, be equally wrong to deny that these phenomena have any weight as evidence in support of the theory of evolution, because exceptions must not be taken as a rule. If it is once granted that the four parti-coloured species of *Dinarda* are really connected by having a common origin, we cannot avoid comparing them with the black *D. nigrita* of southern Europe, which lives with a black Myrmicide ant near the Mediterranean (*Aphaenogaster testaceopilosa*). This species differs so widely from its northern relatives, that Casey has recently decided, with much reason for so doing, that it ought to be regarded as a distinct genus *Chitosa*, and yet it is undoubtedly related to the genuine *Dinarda*, for, when we possess more information as to its mode of life, we shall probably find that the most

[1] Cf. Wasmann, 'Beispiele rezenter Artenbildung bei Ameisengästen und Termitengästen' (written in honour of Rosenthal, *Biolog. Zentralblatt*, Nos. 17 and 18, pp. 565-580). See especially p. 566.

important morphological characteristics distinguishing *D. nigrita* are due to adaptation, just as we have already found them to be in the case of our parti-coloured species of *Dinarda*. That the differences in the former instances are much greater than in the latter can easily be accounted for, inasmuch as *D. nigrita* lives with an ant that is not only generically different from *Formica*, but belongs to another subfamily, whereas our northern *Dinarda* all live with species of one and the same genus *Formica*. Moreover, *D. nigrita* resembles its northern relatives in those systematic characteristics which are independent of the offensive type (Trutztypus), especially in the formation of the parts of the mouth and in the peculiarly shaped tongue. We must therefore assume that it is descended from the same primitive form as our *Dinarda*, and has acquired its present form by a process analogous to that which has produced the northern *Dinarda*, viz. by adaptation to the ants that are its hosts.

It would plainly be inconsistent to admit that the differentiation of our parti-coloured *Dinarda* was the result of a real process of evolution, and to deny that in all probability an identical process of evolution has led to the differentiation of the genera *Dinarda* and *Chitosa*. This comparison certainly proves that in certain cases the principle of evolution may, and even must, be applied to systematic genera of the same family.

A few remarks must be made in order to avoid misunderstandings, to which my account of the evolution of *Dinarda* might possibly give rise.

In all that is essential, the same factors of adaptation, which caused, and are still causing, the parti-coloured *Dinarda* to be differentiated from one another, led to the differentiation of the genera *Dinarda* and *Chitosa* from one common primitive form, but in the latter case the evolution was less slow and gradual than in the former. The great difference existing between the two genera of guest-ants, *Formica* and *Aphaenogaster*, must have brought about a more rapid differentiation of the *Dinardini* that were adapting themselves to them. We shall the more readily accept this statement if we remember that in the Pleistocene epoch, in which this hypothetical process of evolution must have taken place, there was probably a rapid succession of climatic changes, which would facilitate a rapid alteration in the area of distribution of the various kinds of ants.

Let us assume that, in consequence of some climatic change, the southern genus *Aphaenogaster* extended its area of distribution towards the north, encroaching on a locality hitherto occupied by *Formica*, which gradually died out in that neighbourhood, so that the border line of its zone of distribution was drawn further north. A *Dinarda*-like beetle, transferring its quarters from the nests of the *Formica*, that was becoming extinct, to those of the *Aphaenogaster*, that was becoming more common, would be forced to adapt itself to its new hosts, if it were not to be exterminated by them. This circumstance would give a great impetus to the speedy formation of new varieties, or to mutations *per saltum* in a direction favourable to this adaptation ; in fact, the tendency to evolution would receive a fresh impulse. We cannot account for all this, unless we assume the existence of interior laws of evolution,[1] which react beneficially in response to exterior influences ; these laws are indispensable, if we have to recognise the occurrence of advantageous adaptation. We cannot indeed explain how each exterior circumstance acts upon the interior capacity for adaptation in the organism, but we are equally unable to explain how, under the stimulus of light, animal protoplasm is made capable of reacting by forming specks of pigment susceptible to light. The great secret of life is hidden in the capacity for adaptation possessed by living organisms, and we must acknowledge that this secret exists, and not fall into the error of Darwinism, and deny its existence because it is 'mechanically inexplicable.'[2]

If we do not admit this, there is no alternative but to regard the first formation of beneficial modifications as purely accidental ; a theory of chance can never be the foundation of a theory of evolution.

[1] That this assumption is by no means devoid of a material basis has already been shown. See Chapter VI, pp. 177, &c. and Chapter IX, p. 297.

[2] It is a matter for regret that August Weismann, who is otherwise so keen-sighted, in his *Lectures on the Evolution Theory* still brands the assumption of a capacity for adaptation on the part of organisms as 'mystical' or 'extraordinary,' although in discussing what he regards as the smallest units of life (biophors and determinants) he speaks of 'vital affinities,' which is only another name for design inherent in the organism Cf. I, p. 374 and II, p. 36 (Eng. trans.) ; see also p. 176 of this work.

We may, therefore, assume that the process of differentiating the genera *Dinarda* and *Chitosa* from one common primitive form could not have been as gradual as the subsequent process of differentiating the genuine parti-coloured *Dinarda* from one another. The former probably took place *per saltum*, after the fashion of de Vries' mutation theory.

This assumption seems all the more necessary in order to account for the first production of the offensive type (Trutztypus) from the primitive form of the *Dinardini*, for their nearest relatives of the genus *Thiasophila* differ from them so widely that it would have taken hundreds of thousands of years to bridge the gulf between them, if their evolution had been of the gradual sort, such as Darwin imagined. As a matter of fact, however, the primitive form of the *Dinardini* must have come into being in a comparatively short time, at the end of the Tertiary period, or at the beginning of the Pleistocene. This can be proved with a fair amount of certainty from the geographical distribution of *Dinarda*. The genus *Thiasophila* occurs in North America as an inquiline among *Formica*, but the genus *Dinarda* is not found there, although the species of *Formica* are as widely distributed and of as frequent occurrence in North America as they are with us; in fact, they have attained to a more manifold evolution. It follows that the primitive form of *Dinarda* can have been produced only after North America had been completely cut off from Europe and northern Asia by the ocean, which certainly did not take place before the close of the Tertiary period. Otherwise it is inexplicable why the genus *Dinarda* is limited to the northern half of the old world, and does not occur in North America, in spite of the abundance of species of *Formica*, which are mostly identical with our own.

What does this instance of evolution on the part of *Dinarda* really show? That there are cases in which the hypothesis of the theory of evolution assumes a more tangible form and appears more irrefutable, the more closely we examine the details of the facts presented to us. But if we try to trace back the more remote phylogeny of the *Dinardini*, we are involved in obscurity.

The same remark applies to other problems connected with

the theory of descent. As long as they refer to groups of forms within narrow limits, they appear trustworthy, if they are true at all; but when their application is extended to general relationships between higher orders, classes or groups of animals, they are apt to become vague and uncertain, and their charm is often one that attracts only from a distance, as Fleischmann says, in his work on the Theory of Descent ('Die Deszendenztheorie').[1] We may therefore accept the doctrine of evolution without demur,—in so far as it has a scientific basis, and applies to definite groups of forms with a sufficient degree of probability; but, in accepting it, we may decidedly reject, as having no scientific support, those 'Postulates' proposed to us by monism in its name.

And what does this instance of evolution *not* show? That ant-inquilines of other biological types have evolved in the same way and through the same causes as the *Dinardini* belonging to the offensive type (Trutztypus); for precisely because other inquilines do not belong to this type, they are subject to other laws of adaptation, which we shall presently have to consider. No one would be justified in concluding, from what has been said of the *Dinarda* forms, that all species of animals must have been produced in a similar fashion and for the same reasons. If such a conclusion were unjustifiable on no other grounds, it would be quite untenable for the reason that the great majority of the systematic differences between species of the same genus are biologically indifferent, and are neither serviceable nor injurious to their owner; therefore they afford no *points d'appui* for the 'selection of the fittest.' The interior laws of evolution in living organisms, which form the indispensable basis underlying the evolution also of *Dinarda*, have a much greater and more general significance in other departments of the doctrine of evolution than they have here, although it is by no means so devoid of all limitations, as Eimer and other supporters of orthogenesis assume to be the case.

[1] See also *Stimmen aus Maria-Laach*, LXII, pp. 116, &c.: 'Eine Reaktion gegen die Deszendenztheorie.'

4. Indirect Evidence in Support of the Theory of Evolution

Let us now turn to the indirect evidence supporting the theory of descent. In comparison with the direct evidence it is wonderfully abundant and varied, and may be derived from every department of biological research, especially from comparative morphology and comparative morphogeny,[1] from comparative biology, and especially from palæontology, which seeks to establish the relationship between the animals and plants of the present day and the fossils of previous ages. In Chapter IX (pp. 274, &c.) enough has been said to prove the importance of palæontological facts in establishing the occurrence of an evolution of species. As it is not my purpose to write a textbook of the theory of descent, I will only add a few pieces of circumstantial evidence in support of it, taken from my special department of study, viz. from the comparative morphology and biology of inquilines among **ants** and termites.[2]

[1] Particular attention should be paid to the phenomena of parasitic degeneration among animals, for it frequently results in a complete transformation or rather degeneration of the adult animal, so that the place in a natural system, and consequently the connexion of these forms with others derived from the same stock, can be traced only through the larvae, or at a very early stage of development. Instances of this occur among the parasitic Copepods (in the families of Lernaeopoda and Lernaeae), and the parasitic Cirripeds (in the suborder of Rhizocephala). As a rule, degeneration characterises parasitic adaptation, and specific transformation prevails in the symbiotic adaptation of the inquilines of ants and termites to their hosts.

[2] Fuller details may be found in the third and fourth parts of the work: 'Gibt es tatsächlich Arten?' &c. (*Biolog. Zentralblatt*, 1901, Nos. 21 and 22); also in 'Neue Dorylinengäste aus dem neotropischen und äthiopischen Faunengebiet' (*Zoologische Jahrbücher, Abteilung für Systematik*, XIV. 1900, Part 3, pp. 215–289, 275, &c.); 'Termiten, Termitophilen und Myrmekophilen gesammelt auf Ceylon von Dr. W. Horn, mit anderem ostindischen Material bearbeitet' (*Zoologische Jahrbücher, Abteilung für Systematik*, XVII, 1902, Part 1, pp. 99–164, plates 4 and 5); 'Biologische und phylogenetische Bemerkungen über die Dorylinengäste der Alten und der Neuen Welt, mit besonderer Berücksichtigung ihrer Konvergenzerscheinungen' (*Verhandl. der Deutschen Zoolog. Gesellschaft*, 1902, pp. 86–98); 'Neue Bestätigungen der Lomechusa-Pseudogynen-Theorie' (*ibid.* pp. 98–108); 'Zum Mimikrytypus der Dorylinengäste' (*Zoolog. Anzeiger*, 1903, No. 704, pp. 581–590); 'Zur näheren Kenntnis des echten Gastverhältnisses bei den Ameisen- und Termitengästen' (*Biolog. Zentralblatt*, 1903, Nos. 2, 5, 6, 7, 8); 'Ein neuer *Atemeles* aus Luxemburg' (*Deutsche Entomolog. Zeitschrift*, 1904, Part I, pp. 9–11); 'Zur Kenntnis der Gäste der Treiberameisen am oberen Kongo' (*Zoolog. Jahrbücher*, Supplement VII, 1904, pp. 611–682 with plates 31–33); 'Zur Lebensweise von *Atemeles pratensoides*' (*Zeitschr. für wissensch. Insektenbiologie*, II, 1906, Parts 1 and 2);

One thing to be learnt from these phenomena is that it is absolutely necessary to accept the fact of an evolution of the systematic species, and often of the genera and even of the families, within these orders of insects to which most of the inquilines among ants and termites belong. They warn us also to be on our guard against over-hasty generalisations, such as are being made recklessly with regard to the theory of descent. In many cases the occurrence of a real evolution of some particular forms is so strongly borne out by facts, that no thoughtful student of natural science can refuse to accept it, but in other cases there are serious difficulties in the way of accounting for phenomena by means of evolution. It is altogether impossible to apply universally any hard and fast method, like those which some advocates of the theory of descent have adopted and employ as talismans to explain everything.

This is no less true of Weismann's view of the all-importance of natural selection, than it is of Eimer's diametrically opposed theory of orthogenesis. Facts are obstinate things, and refuse to fit in with these theories—what suits one, does not agree with another. The evolution of those inquilines among ants and termites which, like *Dinarda*, belong to the offensive type (Trutztypus) cannot be the result of the same factors as have produced the inquilines of the mimetic type; and these again must owe their peculiarities to a different principle of evolution from the genuine inquilines of the symphilic type.

Nature is intolerant of constraint applied in favour of any particular theory; any one who tries to account for all phenomena in the same way is doomed to failure. Eimer's orthogenesis, according to which interior laws of growth with a definite tendency are the sole causes of evolution, breaks down when applied to inquilines of the offensive and mimetic types, just as Weismann's natural selection theory does when applied to inquilines of the symphilic type.[1]

Beispiele rezenter Artenbildung bei Ameisengästen und Termitengästen (see p. 322, note 1). Works dealing with *Termitoxenia* will be mentioned in § 10 of this chapter.

[1] Cf. the remarks on race-evolution and its causes in Chapter IX, pp. 294, etc. With regard to botany, von Wettstein especially has expressed himself in very similar terms, and has shown 'that it is impossible to refer all the pheno-

The following general considerations are important by way of introduction to a more detailed comparison of the theories of permanence and descent, with reference to the comparative morphology and biology of ant and termite inquilines.

By far the greater number of regular inquilines among ants and termites, that show any marked degree of adaptation to the life of their hosts, belong to the order of beetles. This order is geologically older than either ants or termites, for a number of beetles belong to the Triassic strata, i.e. to the oldest period of the Mesozoic age.

Moreover, this order of insects had attained so high a development by the middle of the Mesozoic age, that in the Black Jurassic are found representatives of almost all our present families and genera of beetles. It was not until the Cænozoic age that ants and termites reached a corresponding height of development. In the Tertiary period they began to form regularly organised states and to play an important part in nature. Before that time, therefore, other insects had no reason for adapting themselves to become inquilines among ants or termites; the conditions that could motive such adaptation were wanting. We must, then, adopt one of two hypotheses:—In the Tertiary period there was a direct creation of a number of new families of beetles, which are exclusively myrmecophile or termitophile, such as the *Paussidae*, *Clavigeridae*, *Gnostidae*, *Ectrephidae*, *Rhysopaussidae*, &c., and of still more numerous myrmecophile or termitophile genera in other families of beetles, among the *Staphylinidae*, *Scarabaeidae*, &c.—and that such a creation took place is from the palæontological point of view most improbable—or else the families and genera of ant and termite inquilines have been evolved from primitive forms,[1] which lived in the Mesozoic age, and only at a later date adopted the myrmecophile or termitophile mode of life.

mena observed in the production of new forms in the vegetable kingdom to the same causes' (*Berichte der deutschen Botan. Gesellschaft*, XVIII, 1900, p. 200). Von Wettstein lays great stress on the distinction between characteristics due to organisation and those due to adaptation, but within the latter group we are forced to distinguish a number of different causes.

[1] These primitive forms belonged to other systematic families and genera of already existing beetles.

The latter hypothesis seems far more probable than the former, not merely for scientific, but for philosophical reasons, as, if we can account for the origin of myrmecophile and termitophile forms by showing them to be natural phenomena according with the theory of evolution, we ought not to have recourse to any hypothesis involving direct new creations.

In order to enable my readers to form some idea of the kind of evidence which a study of ant and termite inquilines affords in support of the theory of descent, I will give a short account of some of these creatures.

5. Hypothetical Phylogeny of the Lomechusa Group

Among the palæarctic and nearctic Fauna, i.e. in the continent of Europe and in northern and central Asia on the one hand, and in North America on the other, is a natural group of closely related genera of Aleocharinae, which I have classed together as the *Lomechusa* group, or *Lomechusini*. They are the most highly developed genuine ant-inquilines of the symphilic type among all the *Staphylinidae* of the northern hemisphere. In Europe and in Asia as far as the tablelands of Tibet they are represented by the genera *Lomechusa* and *Atemeles*. The former lives exclusively with definite species of ants, for instance *Lomechusa strumosa* (fig. 33) is found only in the nests of *Formica sanguinea*, and the ants bring up the *Lomechusa* larvae (fig. 34).

Atemeles, on the contrary, lives with both *Formica* and *Myrmica*; they pass the greater part of their existence as beetles with *Myrmica rubra*, but the larvae are brought up by various species of *Formica*. Throughout North America the *Lomechusini* are represented by the genus *Xenodusa*, and the species found furthest south (*Xenodusa Sharpi* Wasm.) occurs in Mexico. *Xenodusa* lives partly with *Formica*, partly with *Camponotus*, so that it has two sets of hosts, like our *Atemeles*; the larvae are probably brought up by *Formica*.[1]

[1] This supposition has been already confirmed in the case of *Xenodusa cava* Lec. by P. Muckermann's observations in the Prairie du Chien, Wisconsin. This *Xenodusa* causes its larvae to be brought up by a North American subspecies of our red robber-ants (*Formica sanguinea* subsp. *rubicunda* Em.), and, as in Europe, the breeding of these adopted larvae leads to the develop-

The extraordinarily long antennae and legs of *Xenodusa* show a pronounced adaptation on the part of this genus to their mode of life in the *Camponotus* nests. If these extremities were not so long, it would be impossible for the beetles to maintain their friendly intercourse with *Camponotus*, as the ants are much larger than the *Xenodusa*, which are obliged to raise themselves high on their long legs and to stretch up their antennae, whenever they invite one of their huge hosts to feed, and whenever they are fed in their turn.

A very interesting phylogenetic question here arises. With which of the three genera of ants did the primitive form of *Lomechusa* live, with *Formica*, *Myrmica*, or *Camponotus*?

Fig. 33.—*Lomechusa strumosa* F. (5 times the natural size).

Fig. 34.—Larva of *Lomechusa strumosa* (5 times the natural size).

Which of these genera can claim the honour of having trained these genuine inquilines and of having, by breeding, developed their capacity for adaptation and brought it to the highest perfection by amical selection? *Camponotus* is a cosmopolitan genus of ants, and is represented by an immense number of species in the southern hemisphere; in fact, in the south the species are more numerous and more varied than in the north. The genus *Myrmica* belongs chiefly to the palæarctic and nearctic region, but some few species are found in Asia south of the Himalayas, especially in Burma, and one species (*Myrmica aberrans* For.) in Australia. The genus *Formica* is exclusively palæarctic and nearctic. Now the geographical area of distribution of the *Lomechusini*

ment of pseudogynes in the ant colonies. The beetles are found, as a rule, among *Camponotus pennsylvanicus* Deg. and *pictus* For. (Cf. *Neue Bestätigungen der* Lomechusa- *Pseudogynen-Theorie*, p. 106.)

coincides with that of *Formica*, whilst those of *Myrmica* and *Camponotus* are far more extensive. We may, therefore, conclude with great probability that the *Lomechusini* are a product of the symphilic instinct in the genus *Formica*, and that the adaptation of *Atemeles* to *Myrmica* and of *Xenodusa* to *Camponotus* was of later and secondary origin.

This is of course only hypothesis, but it is founded on facts, and is very serviceable as enabling us to understand the morphological and biological peculiarities of *Lomechusini*, as well as their geographical distribution; and without this hypothesis it would be impossible to account for their actual distribution. The remarkable fact that all the species of *Atemeles* still cause *Formica* to bring up their larvae, although at least the smaller of these species (*At. emarginatus* and *paradoxus*) in other respects are better adapted to intercourse with *Myrmica*, suggests the idea that their ancestors continued as beetles to live with *Formica* and not with *Myrmica*. Moreover, a close examination of the morphological peculiarities of the *Lomechusini*, from the biological point of view, would show that fundamentally they are better adapted for intercourse with the genus *Formica*. Therefore the genus *Lomechusa*, which has remained faithful to its original kind of hosts, viz. *Formica*, represents the highest stage of evolution of the symphilic type among the *Lomechusini*.

The theory of permanence is incapable of giving an explanation of any of these phenomena. It can only declare that the various genera and species of the *Lomechusini* were created for their normal hosts. It cannot suggest a reason why the genera *Atemeles* and *Xenodusa* have more than one kind of host, nor can it account for the high development of the tufts of yellow hair and the other characteristics of the *Lomechusini* that are connected with their adaptation to their hosts. Still less can it tell us why the genus *Camponotus* in the southern hemisphere does not enjoy the company of the beautiful *Xenodusa*, whose long antennae and legs are, as it were, created on purpose to fit it for friendly intercourse with *Camponotus*. This is the harder to explain as the larvae of *Camponotus*, like those of *Formica*, spin a cocoon before pupation. The only ants able to render the *Lomechusini* larvae the attention that they require, are those which are in the habit of covering their own larvae with a case made of earth before they enter on the pupal stage. We may therefore safely assert that *Atemeles* are bound to have their larvae brought up by *Formica*, because their other hosts of the genus *Myrmica* have pupae without cocoons, and so cannot help the *Atemeles* larvae in their preparations for pupation. But this is not a valid reason in the case of *Camponotus*. If species of *Xenodusa* and of *Atemeles* occur, nevertheless, within the area of distribution of the genus *Formica*, it can be explained only on the hypothesis that originally all the *Lomechusini* lived exclusively with *Formica*, and afterwards spread to some extent to other genera of ants (*Myrmica* and *Camponotus*), amongst which they now spend the greater part of their imago existence.

Each change of host was accompanied by a further morphological differentiation of the three genera, *Lomechusa*, *Atemeles*, and *Xenodusa*. Those species of *Lomechusini* which remained faithful to one kind of hosts developed into genuine *Lomechusa*, and continued to pass their whole existence in the company of definite species of *Formica* ; whilst those species which accepted the hospitality of two kinds of hosts developed into *Atemeles* and *Xenodusa*, the former being adapted to associate with *Myrmica* and the latter with *Camponotus*, although they returned at times of propagation to the species of *Formica* that could bring up their larvae. This phylogenetic theory gives us the only natural explanation both of the common morphological and biological characteristics of the *Lomechusini*, and also of the differences that we find occurring within this group of beetles.

As an example of evolution of differences between species within the three genera of *Lomechusini*, let us consider more particularly the species *Atemeles*. All the *Atemeles* have, as has been stated, two kinds of hosts ; they pass the chief part of their existence as beetles with *Myrmica rubra*, and in April or May, when they lay their eggs, they migrate to the nests of definite species of *Formica*, with whom they leave their young to be brought up. The newly developed beetles return to the *Myrmica* at midsummer or in the autumn. This migratory life of *Atemeles* is biologically very interesting, and I have therefore kept a record of hundreds of observations made upon it, having studied the creature, partly under normal conditions, and partly as it lived in nests kept for the purpose of research. I cannot do more here than give a brief résumé of the results of my investigations, so far as they bear upon the theory of evolution.

Atemeles lives for one year, and spends the greater part of its life with *Myrmica*, and the smaller part with *Formica*, so that the former may be called the primary, and the latter the secondary host of *Atemeles*. Phylogenetically, however, the relation is reversed, because the adaptation of the *Lomechusini* to *Formica* is of earlier date than the adaptation of one genus of this group, viz. *Atemeles*, to *Myrmica*. It is to this adaptation that the species of *Atemeles* owe their common generic characteristics, which distinguish them from *Lomechusa*. On the other hand, the differences that mark off the individual forms of *Atemeles* as distinct species are due to the differences in the species of *Formica*, amongst which to this day the larvae of *Atemeles* are brought up. In the nests of the various subspecies of *Myrmica rubra*, i.e. among *Myrmica scabrinodis, laevinodis, ruginodis, rugulosa, sulcinodis*, &c., it is not at all uncommon to find several species of *Atemeles* at once; but in the colonies of *Formica* one definite form of *Atemeles* invariably occurs, *Atemeles emarginatus* with *Formica fusca*, *Atemeles paradoxus* with *Formica rufibarbis*, *Atemeles pubicollis* with *Formica rufa*, the *Foreli* variety of *Atemeles pubicollis* with *Formica sanguinea*, and *Atemeles pratensoides* with *Formica pratensis*.

A comparison of *Atemeles pubicollis* with its relatives shows most beautifully that the systematic differences distinguishing the various species of *Atemeles* from one another are really due to adaptation to the particular species of *Formica* with which *Atemeles* lives in summer, and to which it entrusts the bringing up of its larvae.

Atemeles pubicollis resembles its summer host *F. rufa* in size and colouring, and in these respects differs from its smaller and lighter-coloured cousin, *Atemeles paradoxus*, which is the guest in summer time of *F. rufibarbis*, also smaller and lighter in colour. *Atemeles pubicollis* var. *Foreli* was discovered by Forel living among *Formica sanguinea* in the Vosges; it is distinguished from *pubicollis* chiefly by its bright red colour, and this colour distinguishes its host, *Formica sanguinea*, from the darker *Formica rufa*. A comparison between *Atemeles pubicollis* and *pratensoides* is still more instructive. The latter species of *Atemeles* was discovered by me in Luxemburg in 1903, where it occurred in great numbers in an isolated nest of *Formica pratensis*, near the old Roman road which led from Trèves to Arlon through Luxemburg. The ants of this colony are remarkable for being very dark, almost black, in colour, and for being covered with very thick grey hairs; accordingly the newly discovered *Atemeles* differs from *Atemeles pubicollis*, that lives with *Formica rufa*, in being much darker, of an almost uniform blackish brown tint, and by having much thicker hair, especially on the lower side of the abdomen where it curves upwards.[1] I gave this form of *Atemeles* the name *pratensoides* (resembling *pratensis*) because of the remarkable likeness in colour and hair between it and the ants that are its hosts. I was obliged to regard it as a new systematic species, because in its colouring, structure, and hirsute covering it differs from *Atemeles pubicollis* no less specifically than *pubicollis* differs from other species of *Atemeles*. And yet this new species of *Atemeles* is phylogenetically only a highly developed instance of adaptation to *Formica pratensis*, and to a very dark, hairy subspecies of *pratensis*. We have therefore here a very interesting example of the origin of a new species of inquiline, through biological adaptation to a particular ant which is its host, under favourable local conditions. These conditions are the isolated position of the above-mentioned *pratensis* nest; there are no colonies of other species of ants in the neighbourhood, and therefore it is impossible for *Atemeles pratensoides* to breed with other species of *Atemeles* coming from other *Formica* colonies, or to meet them in the neighbouring *Myrmica* nests, where *Atemeles pratensoides* passes the winter and pairs in the early spring.[2]

[1] On this subject cf. Wasmann, 'Zur Lebensweise von *Atemeles pratensoides*' (*Zeitschr. für wissenschaftl. Insektenbiologie*, II, 1906, parts 1 and 2); also *Beispiele rezenter Artenbildung bei Ameisengästen und Termitengästen*, 1906, 46 (568) &c.

[2] I have frequently seen *Atemeles emarginatus* pair with *paradoxus* in my observation nests of *Myrmica*. To this cross-breeding must probably be ascribed the existence of intermediate types of formation of the prothorax standing between the two species. (See 'Beiträge zur Lebensweise der Gattungen *Atemeles* und *Lomechusa*,' in *Tijdschrift voor Entomologie*, XXXI, 1888, 29.)

The formation of a peculiar kind of *Atemeles*, adapted to the very dark and hairy *Formica pratensis*, was favoured by the isolation of the *pratensis* nests in that locality; and, by inheritance and intensification of the characteristics due to adaptation, the special variety became a subspecies, and in course of time a species, which we now recognise as the *Atemeles* resembling *pratensis*, or *pratensoides*.

The accompanying illustration (fig. 35) shows a charming scene, drawn from nature and then reproduced by photography. It represents an *Atemeles pratensoides* being fed by a large

Fig. 35.—*Atemeles pratensoides* Wasm. being fed by *Formica pratensis* Deg. (6 times the natural size).

worker of *Formica pratensis*. In order to reach its hostess's mouth, and to stroke the ant's cheeks with its forefeet as a request for food, and to tickle her head with its antennae, as etiquette among ants requires on such occasions, the guest had climbed on the back of another worker-ant, somewhat smaller in size, belonging to the same nest, which quietly allowed itself to be used as a footstool.

In the account given of *Atemeles pratensoides* we have considered the causes which may have led to the differentiation of the species within the genus *Atemeles*. Let us now turn our attention to some more general considerations which may assist us in giving an explanation of the hypothetical evolution of the whole *Lomechusa* group.

What were the laws which governed the evolution of the *Lomechusini*, and what started the process of evolution? The primitive form was probably one of the Aleocharinae, connected with *Myrmedonia*, a genus that existed in the middle of the Tertiary period and is preserved as fossils in amber from the Baltic. At the present day the *Lomechusa* group of ant-inquilines is sharply divided from the *Myrmedonia*, and no transitional form exists to connect them, but nevertheless there is good reason to suppose that some connecting link between these two genera once existed.

In Schoa (Abyssinia) Antinori discovered a new species of *Staphylinidae*,[1] which answers very fairly to the requirements we should make of a *Myrmedonia* that was in course of approximation to the form of a *Lomechusa*. The antennae are more slender than in *Myrmedonia*, and not thickened like a string of beads. The general shape of the body still resembles *Myrmedonia*, but is decidedly broader and becoming more like *Lomechusa*. The sides of the dark-coloured prothorax are yellowish red, broad and arched as in *Lomechusini*; at the sides of the broad abdomen are small but perceptible tufts of yellow hair. The general colouring is blackish, the antennae and legs being brown. Unfortunately nothing is yet known as to the mode of life of this interesting creature.

Let us now return to the Tertiary period, and to the evolution of our *Lomechusini*. The hypothetical primitive form must in its *Anlage* or tendency to evolution have possessed a capacity for adaptation to a genuine guest-relationship both in organisation and in instinct.

We may suppose that one of the *Staphylinidae*, being a beast of prey and a hostile intruder like most of the *Myrmedonia* to the present day, forced its company upon some species of *Formica* in the Miocene epoch, and, as it possessed this tendency to evolution and adaptation, a genuine guest-relationship gradually grew up, which found its morphological

[1] A coloured representation of the typical example of this species, that is now in the Museo Civico di Storia Naturale in Genoa, was sent me by Dr. R. Gestro, who desired my opinion regarding it. The species is called *Myrmedonia mirabilis* Eppelsheim. I think, however, that it ought to be considered a distinct genus, standing between *Myrmedonia* and *Lomechusa*, and I suggest calling it *Myrmechusa*.

Z

expression chiefly in the greater development of the adipose tissue, in the growth of larger tufts of yellow hair on the sides of the abdomen, in a modification of the prothorax, which became broader and more curved, and in a change in shape of the parts of the mouth and partially also of the antennae. The increased amount of fat in the tissues made it possible for the beetle to emit a volatile substance so attractive to the ants' senses of taste and smell, that they licked it off their guests' bodies. It is in order to enjoy this substance that the ants entertain the beetles as their guests.[1] As it exudes in *Staphylinidae* chiefly between the segments at the sides of the fatty abdomen, it was at these spots that the ancestors of *Lomechusini* were principally licked, and the increased stimulus thus applied was probably the cause of the stronger development of the patches of hair on these parts. When the ant licks these patches, the exudation is emitted, and the hairs facilitate rapid evaporation. As the adipose tissue of the prothorax takes part in the exudation, we can understand why the prothorax has become broader and more curved, as cavities for exudation are thus formed beside the curved edges of the sides. Moreover, the thickening of these edges protects the beetle against the ants' jaws. The change in the shape of the mouth, and especially the increased breadth of the tongue, are connected with the peculiar instinct, possessed by these genuine inquilines, that prompts them to ask food of their hosts by striking them with their antennae and by stroking the sides of the ants' heads with their forefeet, and then to take food from their mouths (fig. 35, p. 336). The bodily modifications due to the growth of a true guest-relationship among the ancestors of the *Lomechusini* must therefore have been accompanied by a corresponding change in their instincts. As the ants took most care of those guests which emitted the fatty substance in greatest abundance, and as they finally brought up the larvae of their friends in the same way as their own young, they were practising a kind of instinctive selection which I have called 'Amical Selection.'[2]

[1] On the subject of the exudatory organs and tissues of the true inquilines amongst ants and termites, see the work mentioned above (p. 327, n. 2), *Zur näheren Kenntnis des echten Gastverhältnisses*, 1903.

[2] Cf. *Biolog. Zentralblatt*, 1901, No. 23, pp. 738, &c. H. Friedmann (*Die Konvergenz der Organismen*, 1904, pp. 187, &c.) has extended the idea of amical

Natural selection, as Darwin understood it, favoured the development of a true guest-relationship on the beetles' part. Those individuals which were capable of resisting the rough treatment that they originally received from the ants,[1] and which could at the same time satisfy the greed of the ants by supplying them with the desired exudation, had undoubtedly a decided advantage in the struggle for existence. But, on the other hand, the same natural selection that promoted the development of a true guest-relationship on the part of the beetles, was opposed to it on the part of the ants, as soon as the latter began to feed the beetles' larvae, for the larvae of the *Lomechusini* are most deadly enemies to the young ants, inasmuch as they consume the lumps of eggs and the young larvae in masses, and finally cause degeneration in the normal instinct of the ant to provide for its own young, so that only deformed pseudogynes are reared. Therefore the colonies of *Formica*, which showed little or no tendency to bring up the beetles' larvae, were certainly better qualified to maintain their existence than those in which the instinctive tendency developed. Hence it follows that natural selection ought never to allow the ants to bring up their worst enemies as true inquilines. Natural selection would inevitably give preference to those female *Formica* in whom that fatal instinct of the worker-ants either did not exist, or existed in a very slight degree. In other words, natural selection would have been bound to oppose amical selection, as soon as the development of a genuine guest-relationship reached a point where it became injurious to the host. As it is, the various species of *Formica* have an inherited instinct, prompting them to entertain as guests definite species of beetles belonging to the group of *Lomechusini*, and to bring up their larvae, in spite of the harm accruing to themselves. Speaking from the point of view of supporters of the evolution theory, we may justly say: Amical selection has triumphed over natural selection, which, in this case, far from being all-powerful, is powerless.

selection so as to include Darwin's sexual selection, and seeks by means of it to explain all the phenomena of direct convergence in the animal kingdom. It seems to me very doubtful whether this is possible.

[1] To this day *Atemeles* and *Lomechusa* are often violently treated by the ants licking them, especially if the guests are old, and their exudatory tissue is exhausted.

Similar conclusions have been reached by the eminent palæontologist Koken, who says:[1] 'The Darwinian principle of selection is not the only one to be taken into consideration, and it appears not to be the most important. In palæontological history we often miss any suggestion of the struggle for existence, and, on the other hand, there is often a tendency to evolution which is not beneficial, and which occasionally is actually injurious to society.'

6. Inquilines among the Wandering Ants

Another proof that the theory of evolution is indispensable to an explanation of the interesting facts of myrmecophily and termitophily is given by a number of *Staphylinidae* belonging to the sub-family of Aleocharinæ, which represent the mimetic type of inquilines among the wandering ants (*Dorylinae*) of the New and Old Worlds (figs. 36, 37). The mimicry on the part of these inquilines is aimed at deceiving the sense of touch possessed by their hosts, who either are blind, or have small and simple eyes, unlike the usual faceted eyes of insects. This mimicry culminates in producing a resemblance between guest and host in the shape of their bodies, and especially in the formation of their antennae; the latter point of resemblance enables the guests to deceive their hosts in an active, and not merely a passive way. This remark is applicable to the companions of the neotropical wandering ants of the genus *Eciton*, as well as to those of the African *Anomma* and its relatives of the genus *Dorylus*, that pursue their prey underground.

If we compare the inquilines of the mimetic type that live among the *Dorylinæ* in both the Old and the New World, we shall find a remarkable similarity existing between the beetles of this biological type that live with the Brazilian and the African wandering ants respectively. This strange similarity is not, however, due to a close systematic relationship between the genera of beetles, and so does not point to there being any direct connexion between them.

Between the genus *Mimeciton* (fig. 36), the highest representative of the mimetic type living in Brazil among *Eciton*

[1] *Palaeontologie und Deszendenzlehre*, 1902, p. 226.

praedator, and the genus *Dorylomimus*, the highest representative of the same biological type living in Africa among *Anomma Wilverthi*, there is an astonishing likeness in *habitus*, i.e. in outward appearance in general; but closer examination shows the likeness to depend only upon pecu-

FIG. 36.—*Mimeciton pulex* Wasm. (S. Paulo, Brazil) (11 times the natural size).

liarities due to adaptation, and not upon the biologically indifferent characteristics, that are totally unlike in the two genera. There can therefore be no question of any close relationship between them. The same result follows from a comparison of the inquilines of the mimetic type living with

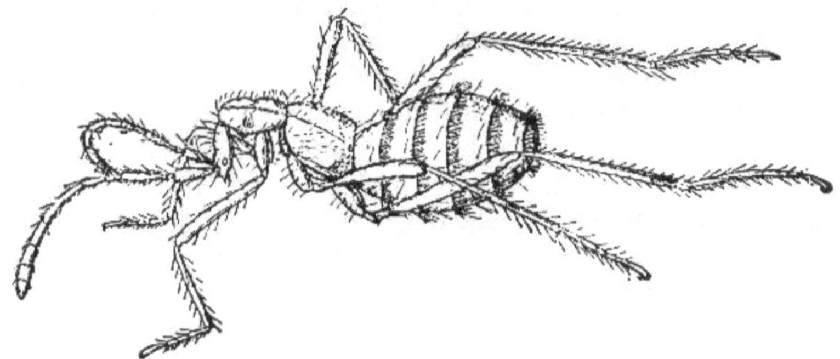

FIG. 37.—*Ecitophya simulans* Wasm. (S. Catarina, Brazil) (7 times the natural size).

various species of one and the same genus of ants, viz. *Eciton*, in tropical and sub-tropical America. In this case again there are striking resemblances in *habitus*, but no close systematic relationship; in fact, these inquilines stand so far apart, that they actually form distinct systematic genera, such as *Mimeciton* (fig. 36), *Ecitophya* (fig. 37), *Ecitonidia*, &c. How can this surprising fact be explained?

The theory of permanence could only make this answer: 'The special genera and species of inquilines were created simultaneously with, and expressly for, the corresponding genera and species of their hosts; the "harmony of the Universe" required this manifold variety on the part of the guests, which have not adapted themselves to their hosts, but were simply created so as to suit them.'

But why is there so great a systematic difference in the representatives of the same biological type, even among the species of the same genus of hosts—a difference which is nevertheless concealed under such a strange likeness of *habitus* that anyone would at once recognise an African *Dorylomimus* as the double of the Brazilian *Mimeciton*? The theory of permanence can give no answer at all to this question—and it is all the more unable to do so because we must undoubtedly refer the systematic species within the same genus of guest-ants, e.g. *Eciton*, to a common stock, from which the present species of *Eciton* were differentiated by a process of natural evolution.

Forms resembling one another so closely as *Eciton Burchelli* (*Foreli*),[1] and *quadriglume, praedator* and *coecum*, cannot possibly be regarded as belonging to species originally distinct; and yet these species have companions, mostly guests of the mimetic type, which generally differ widely from one another, and occasionally even represent distinct systematic genera.

When can these guests have been created? Their existence in their present form would have no meaning until the particular kinds of ants, that are their hosts, had been differentiated into their present species.

We should therefore have to assume, if we accepted the theory of permanence, that the hosts had developed in the course of nature, and that their guests had been subsequently created to match them. How forced and inconsistent such an explanation would be, must be apparent to everyone.

The theory of evolution says on the other hand: 'These inquilines have been produced in course of time from similar, or even from identical primitive forms, amongst which we must

[1] The species formerly known as *Eciton Foreli* Mayr consists of the soldiers and workers of *Labidus Burchelli* Westw. which comprises the males of the same species. For this reason the name *Eciton Foreli* was changed to *Burchelli*.

consider especially the genus *Myrmedonia*, that is geologically very old and widely distributed; their evolution is most closely connected with that of their respective hosts.' The striking resemblance united with a still greater systematic difference, which we can observe in the various genera of inquilines of the mimetic type, is the result of an imperceptibly slow, or rather of a progressive adaptation, occurring among the inquilines of the various genera and species of hosts, but on completely independent lines. The points of resemblance are conditioned by the general laws governing the mimetic type of inquilines among *Dorylinae*; for this type it is essential that the likeness between host and guest in the shape of their bodies should be so great as to deceive the host's sense of touch, and, when the mimetic type reaches its highest point, there is a great resemblance also in the shape of their antennae. The axiom 'when two things are equal to a third, they are equal to one another,' enables us to account for the strange likeness between the highest representatives of the mimetic type of inquilines among *Dorylinae* in different parts of the world. As they all resemble their hosts, they resemble one another. The similar *habitus* possessed by various genera of the mimetic type, that differ systematically (as, for instance, *Mimeciton* and *Dorylomimus*), is to be regarded as a 'phenomenon of convergence,' from the point of view of the evolution theory. The differences, however, are due, partly to the original difference between the primitive forms, partly to differences in bodily formation and way of life on the part of the genera and species acting as hosts, partly to the various ways in which a similarity in the shape of body and antennae can be produced, and partly to the degree of evolution of the mimetic type to which its representatives have attained. Here we have a real explanation of facts, an explanation that is, of course, hypothetical in character, but is nevertheless able to satisfy our requirements. We ought to pay particular attention to the various degrees of evolution of the mimetic type to which the inquilines of the same ants have attained. The guests of *Eciton Burchelli* supply us with good illustrations of these degrees of evolution.

The mimetic type does not stand in sharp contrast to the indifferent type, to which belong inquilines that have retained the original form of their relatives who were not myrmecophiles. There are many instances in which it is doubtful whether we ought to reckon the genus or species of inquilines as still belonging to the indifferent type, or as having passed over to the mimetic type. If a natural process of adaptation has taken place, and the guests have come to resemble their hosts, either by a series of imperceptibly slight variations or by more sudden changes, we can easily understand that we must inevitably meet with the mimetic type at various stages of evolution, and the inquilines remain at each stage until the necessity for adaptation, which varies in the case of various forms, causes a further advance to be made.

If we compare the inquilines of the mimetic type living among *Dorylinae* with those of the offensive type (Trutztypus)

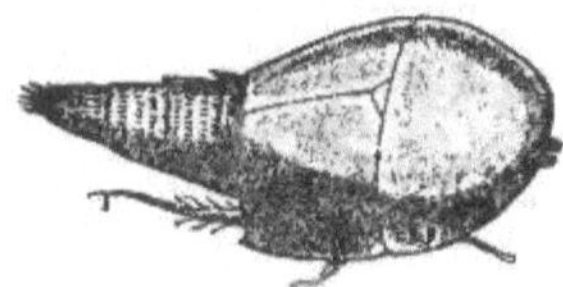

FIG. 38.—*Xenocephalus limulus* Wasm. (Rio de Janeiro) (7 times the natural size).

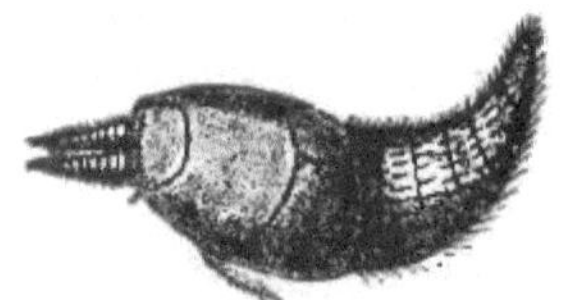

FIG. 39.—*Doryloxenus Lujae* Wasm. (Congo) (22 times the natural size).

(figs. 38 and 39) which belong to the systematic subfamilies *Xenocephalinae* and *Pygosteninae*, a striking difference becomes apparent. The forms of the mimetic type are very numerous and differ systematically, but those of the offensive type are remarkable for their uniformity and for their systematic likeness. The neotropical representatives of the offensive type almost all belong to the genus *Xenocephalus* (fig. 38), and the species of this genus, all being very much alike, live with various species of the genus *Eciton*, whilst the representatives of the same type in the Old World belong to the genera *Pygostenus*, *Doryloxenus* (fig. 39), &c., which also resemble one another very closely, and include groups of very similar species. This peculiar morphological contrast between the manifold forms of the mimetic type, and the uniformity of the offensive type, admits of a very simple and natural explanation according to the principles of the evolution theory.

The inquilines of the offensive type must possess a greater

aggregate of common morphological characteristics, because adaptation aims at producing uniformity; it has favoured the evolution of a definite form of body, not unlike that of a tortoise, the bending round of the head towards the protected lower part of the creature, the shortening and thickening of the antennae, the shortening of the legs and covering them with bristles, &c. The result of this adaptation could not fail to be uniform, as we see in the subfamilies known as *Xenocephalinae* and *Pygosteninae*. But the evolution of guests of the mimetic type was bound to be very various, for their mimicry is designed to deceive the sense of touch in their hosts, and naturally gave rise to forms differing widely in degree of mimicry and in the details of its production. It is true that we cannot do more than offer suggestions as to the course followed in the individual cases by evolution thus directed by adaptation, but the preceding statements are enough to show that in this department the theory of evolution is capable of supplying really satisfactory explanations, whilst the theory of permanence can explain nothing at all.

Let us now compare the *Eciton* inquilines of tropical and subtropical America with the *Atta* inquilines of the same region. *Atta* and *Eciton* both belong to the predominant forms of ant Fauna in the tropics of the New World, and these two genera have stamped their peculiarities on all the other ants; they also play a most important part in the struggle for existence, *Eciton* as prevailing over other insects, and *Atta* over plants, for the *Eciton* are wandering robber-ants, and the *Atta* destroy leaves and grow fungi. The former, as a rule, have no permanent nests, but the latter construct huge nests stretching far under the ground, where they employ the fragments of leaves, that they have carried in, for cultivating a kind of fungus (*Rozites gongylophora*), which they use as food for themselves and their young. As the *Staphylinidae* make their homes preferably in decaying vegetable matter, we should expect the number of exclusively attophile genera to be much greater than that of exclusively ecitophile genera of the same family of beetles. We should be all the more justified in this supposition as the inquilines in the *Atta* nests run much less risk of being eaten by their hosts than do those living with the wandering robber-ants. If, therefore, the guests were originally created expressly for their respective hosts, we should find a great many specially attophile genera of *Staphylinidae* and very few ecitophile.

But what are the facts? They show us a state of affairs that is the direct opposite of this supposition. Of the twenty-one genera of *Staphylinidae* known at the present time which contain species living with *Eciton*, there are twenty genera consisting exclusively of *Eciton* inquilines, and only one genus (*Myrmedonia*) which includes, besides ecitophile species, others living partly with other ants, and partly not with ants at all. On the other hand, there are about twelve genera of *Staphylinidae* containing *Atta* inquilines, and only two of these (*Attonia* and *Smilax*) are exclusively attophile, whilst all the rest include, besides the attophile species, others which live either with other ants, or not with ants at all. These facts speak plainly enough. They show us that the different distribution of the *Atta* and *Eciton* inquilines depends upon the laws of adaptation. Precisely because the wandering ants are rapacious and extraordinarily active robbers, do they have so many peculiar genera of guests, that have adapted themselves to the ants, not merely lest they should be destroyed by them, but also in order to share their booty by allying themselves with the robbers.

And precisely because the *Atta* are peaceful destroyers of leaves and growers of fungi, do they have so few peculiar genera of guests, in spite of the favourable conditions which the *Atta* nests offer to the existence of *Staphylinidae*. The law underlying this apparently paradoxical phenomenon may be expressed as follows in biological language :—Inquilines among *Eciton* are under a much greater necessity for adaptation than those among *Atta*. This greater necessity for adaptation led to increased frequency in its occurrence, and to a higher degree in its attainment, on the part of *Eciton* inquilines as compared with *Atta* inquilines. The theory of evolution can account for this law, but the theory of permanence cannot, for it admits of no modification by adaptation in the systematic species.

It really seems to me that the theory of evolution is not only attractive, as supplying an explanation of facts of this kind, but that it alone is capable of giving a completely satisfactory explanation, although we may not be in a position to describe the processes of evolution as exactly as we were able to do with regard to the differentiation of the species of *Dinarda*.

A few words must be added on the subject of the laws that governed the evolution of *Dorylinae* inquilines of the mimetic type.

The external influence directing the various methods of evolution, which finally culminated in such extreme forms as *Mimeciton, Ecitophya, Dorylomimus, Dorylostethus*, &c., was probably supplied by natural selection, as, among the companions of the wandering ants, those would be most favourably circumstanced which were able to deceive the ants' sense of touch by resembling them in shape and especially in the formation of their antennae. They were not only better protected against attacks on the part of their hosts, but were able to seize a larger share of their booty, consisting chiefly of insects, and incidentally to consume the young of their hosts with impunity. Natural selection alone cannot, however, account for the existence of these methods of evolution, for the material, upon which selection acted, must have been furnished by the already existing tendency possessed by these genera of beetles to adopt certain forms. On the other hand, we must not interpret these tendencies merely in the sense of general laws of growth, as Eimer's orthogenesis does, for the laws of growth governing the original primitive forms of these genera of beetles could not differ much from those governing their nearest systematic relatives belonging to the family of *Staphylinidae*. The general laws of growth of the *Staphylinidae* supply no sufficient explanation of the fact that the inquilines of the mimetic type have differentiated themselves into so many different genera, that are systematically unlike each other and unlike the primitive forms from which they are descended ; we must therefore assume that the capacity for evolution possessed by the earliest forms was influenced and modified by the internal power of adaptation to new biological conditions, so that spontaneous departures from the original form occurred, tending to produce the mimetic type, but this tendency took different directions according to the various genera and species of the creatures amongst which the inquilines lived. The further development of these tendencies to evolution cannot have been the result of a gradual accumulation of innumerable quite trifling variations, as Darwinism maintains, for in that case hundreds of thousands of years would have been required for the production of a single genus such as *Mimeciton*. In the struggle for existence minimum variations are of scarcely perceptible advantage, as they would not enable the guests to deceive the ants' sense of touch. We are therefore forced to believe that the evolution of inquilines of the mimetic type took place by a series of more or less rapid transitions, after the fashion suggested by the mutation theory. Here again Darwin's theory of selection proves to be as unsatisfactory as the directly opposed theory of orthogenesis, put forward by Eimer. I am of opinion that the real solution of this puzzling process of evolution is to be sought in the inward power of adaptation, possessed by the living organism, which power can react beneficially under external stimulus, and can at the same time retain, and perpetuate by transmission, the beneficial modifications once adopted, and even carry them further.

I ought to point out that in *Mimeciton* (fig. 36, p. 341) especially there are certain peculiarities which are explicable neither by natural selection nor by the general laws of growth, such as the change of the faceted eyes into simple ocelli, resembling the simple eyes of its host, *Eciton praedator*, but situated in the hollow at the base of the antennae. This 'excessive mimicry' in the formation of the eyes in *Mimeciton* is the more remarkable, as the beetle often accompanies the ants on their marches even by daylight. It gives the impression that the tendency to evolution of a mimetic type has here exceeded the limits of what is beneficial, as if the process once begun could not be arrested. Brunner von Wattenwyl has given this phenomenon the name of *Hypertely*.

Let us now go back to our comparison between the theories of permanence and descent.

7. Transformation of Wandering Ants' Inquilines into Termite-Inquilines.

(See Plate III, figs. 1, 2)

Some years ago two correspondents of mine in India, Father Heim, missionary in the Ahmednagar district, and Father Assmuth, Professor at St. Francis Xavier's High School in

Bombay, made an interesting discovery. They found in the nest of an Indian species of termite (*Termes obesus* Ramb.) a number of very remarkable inquilines, and amongst them a little beetle of the family of *Staphylinidae*, belonging to the subfamily *Pygosteninae*, and to the genus *Doryloxenus*. This genus represents the most perfect instance of the offensive type of inquiline among the *Dorylinae* of the Old World (cf. fig. 39, p. 344 and Plate III, figs. 1, 2). The tiny creature's spindle-shaped body, that the ants' jaws cannot seize, its short, thick, horn-shaped antennae, and especially its extremely short legs, the tarsi of which are all atrophied and transformed into prehensile organs—all these morphological peculiarities point to a life among wandering ants rather than among termites. Moreover, all the other species of the genus *Doryloxenus*, as far as their mode of life is known, are actually inquilines among the African wandering ants *Dorylus* and *Anomma*. Our new termite-inquiline so much resembles *Doryloxenus Lujae* (see fig. 39, p. 344), from which it differs chiefly in being bigger (2 mm.), that we need only compare the photograph of it (Plate III, fig. 1) with fig. 39, in order to recognise the likeness between them. I have given also on Plate III, fig. 2, an illustration of the forefoot of *Doryloxenus* highly magnified. It is stumpy, not jointed, and covered with long spines and numerous delicate, white, tenent hairs, shaped like funnels, which enable the little beetle to cling to the young of the ants or even to the ants themselves, so that it actually rides when it accompanies the long-legged nomadic ants on their expeditions.[1]

My surprise at discovering a termitophile *Doryloxenus* in India is therefore easily understood. How was it possible that a beetle, whose whole structure proclaims it to be a guest of the wandering ants, and the other members of whose genus actually ride on the ants in Africa, should in India live as a recluse in the clay-dwellings of the termites? When I received the first consignment of Indian termite-inquilines, and found this beetle amongst them, I thought one of my correspondents had made a mistake; I wrote at once to say that he must

[1] Father H. Kohl recently found two distinct species of *Doryloxenus* riding on ants in the Upper Congo, and Luja caught another species on the Zambesi, also riding on an ant that had just crossed a brook. (Cf. *Zur näheren Kenntnis der Gäste der Treiberameisen*, &c., pp. 650, 667.)

have put accidentally an inquiline of the Indian wandering ants into a glass containing termites. But the mistake was on my part. Further parcels sent by my two correspondents showed beyond a doubt that the new *Doryloxenus* was quite a usual, and even a frequent guest among the termites both in the Ahmednagar district and in Bombay. What is the solution of this biological problem?

The only possible solution seems to me to be the following: In India the wandering ants of the subfamily *Dorylinae* at the present time no longer play so important a part biologically as in Africa. It is probable however, that long ago, when in the Tertiary period India and Central Africa were still united and formed a continuous Indico-African continent, the condition of India more closely resembled the present condition of Africa, and in the struggle for existence in the insect world the wandering ants in India were of as great importance as they are now in Africa. The *Staphylinidae*, which had adapted themselves to be inquilines of the offensive type among these ancient *Dorylinae*, and thus had developed into a distinct systematic subfamily (*Pygosteninae*), were doubtless in India also originally the guests of wandering ants exclusively, for no other reason can be given for their characteristics due to adaptation, and especially for those of the genus *Doryloxenus*.

What took place when India was separated from Africa, and the biological importance of the wandering ants there gradually diminished, so that at the present day in India no *Dorylinae* occur that organisé extensive predatory expeditions above ground?[1]

This biological change could not fail to influence the guests of these Indian *Dorylinae*, which share in the expeditions of their hosts and live on their booty. Many of these guests would no doubt find it expedient to seek another refuge. But whither could they go? The wandering ants are fond of attacking and plundering the nests of termites, as the latter with their soft skin can offer but slight resistance to the jaws of the ants, and fall an easy prey to them;[2] and their guests

[1] *Dorylinae* of the genera *Dorylus* and *Aenictus* living underground are still common in India.

[2] This statement is confirmed by E. Luja's observations on the Lower Congo. He found colonies of a *Dorylus* living underground (*D. fulvus-dentifrons*)

accompany the *Dorylinae* on these raids, as they still do in the tropics.

We need only suppose that some individuals of an Indian species of *Doryloxenus* were left behind in a nest of *Termes obesus,* when it was stormed by the ants, and became the ancestors of a new termitophile species of *Doryloxenus.* These little predatory beetles would find plenty of food amongst the young termites; their inherited offensive type was no longer as necessary as before, but it gave them a more than sufficient protection against the jaws of the warriors and workers of their hosts under their new circumstances. Their short legs, with tarsi transformed into prehensile organs, could not be any disadvantage to them in the company of termites, in fact they were useful in the distribution of the species, as the beetles could more easily cling to the winged termites; when these swarmed out of the parent nest to form new colonies This explains why the peculiar formation of tarsi in *Doryloxenus* was retained by the new termitophile species.

This is roughly the hypothetical phylogeny of this interesting Indian *Doryloxenus,* which I regard as a deserter from the company of the wandering ants; that is why I have given it the name *Doryloxenus transfuga.*

Some one may feel inclined to say that this biological metamorphosis, by which an inquiline of the wandering ants is assumed to have become the guest of termites, sounds like a story from the Arabian Nights; it might, perhaps, be compared with some edifying tale from an old Buddhist collection of legends, in which a robber, attacking a peaceful monastery of Bonzes, was converted and remained in the monastery in order to atone for the sins of his previous companions in wrongdoing. Nevertheless, it would be hard to find any other natural explanation, than that suggested above, for the fact that there are in India beetles of the dorylophile genus *Doryloxenus* habitually living as inquilines among termites. The theory of permanence offers no solution for this problem. We have therefore to choose whether we shall regard it as an

at the foot of termite nests (*Acanthotermes spiniger-Lujae*) and occupied in plundering them. Cf. *Zur näheren Kenntnis der Gäste der Treiberameisen,* p. 673. Father H. Kohl has recently made similar observations on the Upper Congo.

incomprehensible natural 'freak,' or acknowledge that in India, within a comparatively short space of time, part of the genus *Doryloxenus* has changed its hosts, and from being an inquiline of wandering ants, it has transferred its quarters to the termites. If such a change can take place, although the modes of life of *Dorylinae* and termites are totally different, or rather diametrically opposed, there is no great difficulty in assuming that the inquilines of ants and termites may have been produced from forms which were originally neither myrmecophile nor termitophile, but have adapted themselves to their hosts by a more or less lengthy process of evolution.

In the case of *Doryloxenus transfuga* the change in its mode of life has not been accompanied by any great morphological modification; as a termite-inquiline the beetle has remained almost the same as it was when a *Dorylinae* inquiline. This is explicable for two reasons—firstly, the change of host did not necessitate any rapid alteration in the characteristics already acquired by adaptation, because the beetle was fairly well suited to its new way of life; and, secondly, its migration from the company of the wandering ants to that of the termites took place after the Tertiary period, i.e. not long ago, from a geological point of view.

Before quitting the subject of *Doryloxenus transfuga*, I must allude to some confirmations of and additions to the hypothesis just laid down.[1]

Other sample nests, subsequently sent from India by Father Heim and Father Assmuth, revealed the surprising fact that not only *one*, but *two* specifically distinct forms of *Doryloxenus* inhabit the nests of *Termes obesus* and its subspecies *T. wallonensis* (*Doryl. transfuga* [cf. fig. 40 and Plate III, fig. 1] and *termitophilus*); in some nests they are very numerous, but they are found chiefly near the young of the termites and in their fungus beds; in this respect they resemble *Termitodiscus*

[1] For the bibliography of the subject see the following works mentioned on p. 327, note 2. *Termiten, Termitophilen und Myrmekophilen aus Ceylon*, p. 158; *Zur näheren Kenntnis der Gäste der Treiberameisen*, pp. 614–616, and 651, 652. (A description of the two termitophile species of *Doryloxenus* and of the new genus *Discoxenus* with its two species may be found in the latter work, pp. 654–656); 'Die phylogenetische Umbildung ostindischer Ameisengäste in Termitengäste' (*Compt. Rend. d.* III *Congr. internat. de Zoologie*, Berne, 1904, pp. 436–448, with plates); *Beispiele rezenter Artenbildung bei Ameisengästen und Termitengästen*, 49 (571) &c.

Heimi and the species of *Discoxenus* to which I shall refer later on.

This fact is a conclusive confirmation of the occurrence of species of *Doryloxenus* in the termite nests of Central India, round Ahmednagar and Bombay, and it completes the account given of their termitophile adaptation.

A close examination of the two kinds of *Doryloxenus* showed that in spite of their having retained the characteristics of their dorylophile adaptation, which they have in common with African species of the same genus living with *Anomma* and

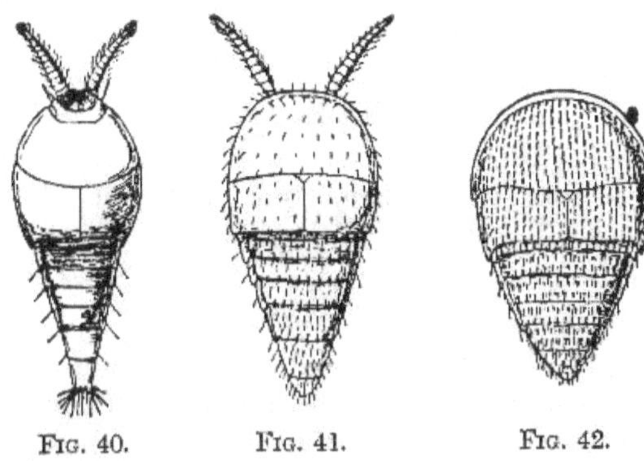

Fig. 40. Fig. 41. Fig. 42.

Fig. 40.—*Doryloxenus transfuga* Wasm. (India) (15 times the natural size).
Fig. 41.—*Discoxenus lepisma* Wasm. (India) (15 times the natural size).
Fig. 42.—*Termitodiscus Heimi* Wasm. (India) (15 times the natural size).

Dorylus, they differ from the latter in several respects, especially in their hairy covering, in the formation of the surface of the body and in the structure of the head. The front part of the head is deeply depressed, as if it were about to turn over to the lower part of the body, as is actually the case in the genera that I am about to mention. Among the inquilines discovered by Father Heim and Father Assmuth in the same termite nests there is also a new genus of *Staphylinidae*, which I described recently, and called *Discoxenus* (fig. 41). In shape it shows a curious cross between the conical body of *Doryloxenus* (fig. 40) and the orbicular form of *Termitodiscus* (fig. 42). This new genus *Discoxenus* contains two distinct species : *Discoxenus lepisma* (fig. 41) and *Assmuthi*. The remarkable feature in this new genus is that it stands (as may be seen from figures 40–42)

2 A

exactly midway between the genera *Doryloxenus* (fig. 40) and *Termitodiscus* (fig. 42), of which the latter represents the most perfect instance of the offensive type occurring among termitophile *Staphylinidae* in India, for the body is round and flat, and affords complete protection to the short extremities of the creature.[1]

In *Discoxenus* (fig. 41) the abdomen is still conical as in *Doryloxenus* (fig. 40), but the front part of the body is already broad and flat, as in *Termitodiscus* (fig. 42). The head is on the lower side of the prothorax as in *Termitodiscus* (fig. 42), but the long spindle-shaped antennae still resemble those of *Doryloxenus* (fig. 40), and project from below the head, whereas in *Termitodiscus* they are very short, broad, and flattened down. In *Discoxenus* the feet have normal tarsi with four joints as in *Termitodiscus*, and are not like those of *Doryloxenus* in having but one joint and being metamorphosed into prehensile organs. *Discoxenus* is therefore, from the point of view of comparative morphology, a transitional form between *Doryloxenus* and *Termitodiscus*.

We have then good reasons for assuming that the Indian termite-inquilines of the genus *Termitodiscus* are descended from ancestors resembling *Discoxenus*, and these again from others resembling *Doryloxenus*. In other words: The evolution of the offensive type of Indian termitophile *Staphylinidae*, which culminated in *Termitodiscus*, probably began among relatives of *Doryloxenus*, which entered termite nests in the course of predatory expeditions made by the wandering ants. The termites, therefore, had to thank these ants for having brought them not only *Doryloxenus*, but also the beautiful genera *Discoxenus* and *Termitodiscus*, as these inquilines were of common origin with *Doryloxenus*.

The process of adaptation, which has resulted in the evolution of the present genus *Termitodiscus* from ancestors that were once guests of the wandering ants, would thus seem to have passed through three different stages; in the first of which there was a likeness to *Doryloxenus*, in the second to *Discoxenus*, and in the third *Termitodiscus* assumed its present form. But

[1] For the description of *Termitodiscus Heimi* see my work: 'Neue Termitophilen und Myrmekophilen aus Indien' (*Deutsche Entomologische Zeitschrift*, 1899, I, 145–180, Plates I, II), p. 147 with Plate I, fig. I.

we must beware of regarding this hypothetical process as consisting of a real series of forms in which our present *Termitodiscus* is the direct descendant of *Discoxenus*, and *Discoxenus* of *Doryloxenus*. We ought rather to regard the process of evolution as composed of three quite distinct processes of adaptation, taking place in different geological periods and absolutely independent of one another.[1]

One proof of this is the fact that *Doryloxenus* has quite rudimentary tarsi, and the other two genera have normal. A form with normal tarsi can never be genetically descended from one with rudimentary, but the reverse must be the case. Therefore the earliest ancestors of *Discoxenus* and *Termitodiscus* must still have had normal tarsi; they cannot have been genuine *Doryloxenus* for this reason, but older relatives of this genus, whilst its tarsi were not yet rudimentary. Further, as we at the present day find the three genera *Doryloxenus*, *Discoxenus*, and *Termitodiscus* together in the same termite nests in India, from the standpoint of the theory of evolution we are forced to assume that relatives of *Doryloxenus* became termite-inquilines in three different epochs. From the last of the three date both the Indian species of termitophile *Doryloxenus*; this transition must, as I have already said, have taken place comparatively recently, perhaps during the Pleistocene epoch, as these species still retain the characteristics due originally to dorylophile adaptation. The genus *Discoxenus*, which differs greatly from *Doryloxenus*, was produced in the second transitional epoch, and this is geologically anterior, and belongs perhaps to the end of the Tertiary period. The first and earliest transition, of which the present genus *Termitodiscus* is the product, is still more remote geologically, and belongs perhaps to the middle of the Tertiary period; for the genus *Termitodiscus*, in spite of having many points of resemblance to *Discoxenus*, displays a much more advanced evolution of the termitophile offensive type. The remote antiquity of this first transition of relatives of *Doryloxenus* to the termitophile mode of life is borne out by the fact that in South Africa

[1] For further information on this subject see the lecture mentioned on p. 352: 'Die phylogenetische Umbildung ostindischer Ameisengäste in Termitengäste.'

there are also two species of *Termitodiscus* (*T. splendidus* and *Braunsi*) living with two different species of termites (*Termes vulgaris* and *transvaalensis*), whilst the genus *Discoxenus* is not yet known to occur in Africa, nor have any termitophile species of *Doryloxenus* been discovered there hitherto. It is possible that further research will fill these gaps in African Fauna. In any case we must assume that the earliest of the three transitions mentioned above, in which the genus *Termitodiscus* was produced, took place before India and Africa were completely separated;[1] otherwise we cannot account for the fact that the genus *Termitodiscus* is common to both continents. If we grant this, we assume that the earliest transition was common to Africa and India, but that the other two transitions of relatives of *Doryloxenus* to the termitophile life occurred only in India.

From the biological standpoint there is no more difficulty in assuming a repeated transition than an isolated instance of transition, and the existence in India of two termitophile species of *Doryloxenus* affords us very weighty grounds for believing this to have occurred.

It is plain that the relationship between the Indian species of *Doryloxenus* found in termite nests, and the allied members of the same genus which accompany the wandering ants, possesses a degree of probability bordering on certainty, and far higher than the relationship between *Discoxenus* and *Doryloxenus*, although this in its turn is more probable than the relationship between *Termitodiscus* and the connexions of *Doryloxenus* through *Discoxenus*. The greater the systematic difference between the forms in question, the weaker are the reasons for assuming that they are of common origin. (See Chapter IX, p. 291.) Nevertheless, we may still regard it as very probable that the Indian and African inquilines of the offensive type, belonging to the class of termitophile *Staphylinidae*, represented by the genera *Termitodiscus* and *Discoxenus*, may be traced back phylogenetically to the intrusion of *Dorylinae* inquilines into termite nests, in the course of predatory expeditions made at various times by the wandering ants.

[1] In the middle of the Tertiary period both ant and termite fauna were already highly developed, and most of our present genera existed, so that there are no palæontological difficulties in the way of this assumption.

I stated this hypothesis at the Sixth International Congress of Zoologists at Berne, in August 1904, since which date it has received very interesting confirmation from a new discovery made in tropical Africa, of which a short account must be given.[1]

In the nests of an African termite which erects peculiar, fungus-shaped structures, *Eutermes* (*Cubitermes*) *fungifaber* Sjöst., at Sankuru on the Lower Belgian Congo, in January 1905, Edward Luja discovered a new termitophile species of the genus *Pygostenus*, which otherwise lives with the African wandering ants, *Dorylus* and *Anomma*, and is closely related to *Doryloxenus*, and belongs to the same subfamily *Pygostenini*.

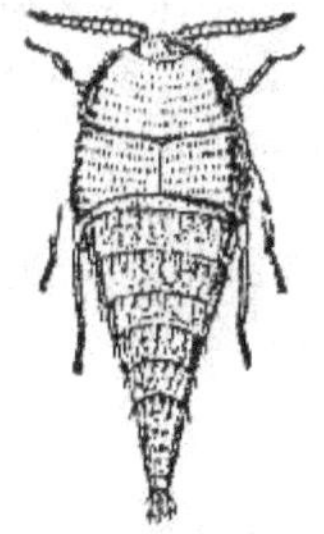

Fig. 43.—*Pygostenus pubescens* Wasm. (Congo) (10 times the natural size).

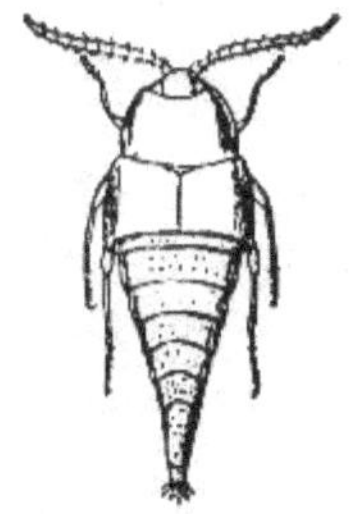

Fig. 44.—*Pygostenus termitophilus* Wasm. (Congo) (12 times the natural size).

I described the new species, giving it the name *Pygostenus termitophilus*.

It is distinguishable from the dorylophile members of the same genus by being more glossy, and by having a less clumsy structure and no hairs on the abdomen; only the tips of it show the usual ring of black bristles. The antennae are longer and the head more arched than in *Anomma* inquilines of the same genus. In order to show these points of difference very clearly, I have given illustrations of *Pygostenus pubescens* (fig. 43), which lives with *Anomma Wilverthi* near the Congo, and of *Pygostenus termitophilus*, side by side, both greatly magnified. The new termitophile *Pygostenus* is marked off from the dorylophile members of the same genus by differences analogous to those which we observed in the Indian termitophile *Doryloxenus*; the modification

[1] For fuller details see *Beispiele rezenter Artenbildung*, 51 (573) &c.

in the form of the head is, however, comparatively slight in comparison with that undergone by the latter genus.[1]

There is, therefore, in tropical Africa at least *one* termitophile species of *Pygostenus*, which may be compared with the Indian *Doryloxenus*, and for whose origin we may account in an analogous way. We must assume that this creature, now an inquiline among termites, was once a guest among wandering ants, for the whole structure of *Pygostenus* is that of the genuine offensive type of *Dorylinae* inquilines, and the other species of the genus—we already know about twenty—are all companions of the wandering ants in Africa. *Pygostenus termitophilus* was not specially created to live with the termites, but it has adapted itself to a termitophile existence; it is like *Doryloxenus transfuga*, a deserter from the company of the wandering ants.

The genus *Pygostenus* represents a decidedly offensive dorylophile type, but one not so highly developed as that of *Doryloxenus*. The body is less like a spindle in shape, and the tarsi are normal and have not become prehensile. The latter point is particularly important. It explains why the *Pygostenus* accompany their hosts on foot, whereas the *Doryloxenus* ride on their backs. Father Hermann Kohl, C.SS.C. has actually observed both these facts on the Congo. The Indian *Doryloxenus*, which have become termite-inquilines, became associated with their new hosts through falling off the ants' backs in the course of a raid upon the termites, and being left behind in the termite nests. The transition to a termitophile existence in the case of the African *Pygostenus* was probably the result of the little beetles' losing sight of the ants during an expedition, and seeking refuge in neighbouring termite nests. Their offensive type would facilitate their securing admission, as the jaws of the termites could not do them so much harm as those of strange ants. When the new guests were naturalised among the termites, a morphological transformation gradually followed, so that in time they became a new termitophile species, viz. *Pygostenus termitophilus*.

[1] There is perhaps a second very small species, *Pygostenus infimus* Fauv. in Gaboon, which is also termitophile, as its shape approximates very closely to that of *Pyg. termitophilus*, but unfortunately we do not yet know precisely where it was discovered.

As the genera *Pygostenus* and *Doryloxenus* are systematically very closely related, and as the former represents a lower stage of evolution of a dorylophile offensive type than the former, it is probable that the above-mentioned connexions of *Doryloxenus*, from which we imagined the termitophile genera *Discoxenus* and *Termitodiscus* to be descended, had more resemblance to *Pygostenus* than to *Doryloxenus*. This is certainly true of the tarsal formation of the earliest deserters; the tarsi must have been normal, as they are still in *Pygostenus*, *Discoxenus* and *Termitodiscus*, and not rudimentary, as they are in *Doryloxenus* at the present time.

Let us now sum up the results of our consideration of the way in which, both in India and in Africa, beetles that once lived among wandering ants have become termite-inquilines.

1. That *Staphylinidae* of the dorylophile offensive type of *Pygosteninae* have passed from the company of wandering ants into that of termites, and in adapting themselves to a termitophile existence have formed new systematic species, has occurred at least twice in the Quaternary period; once among the African species of the genus *Pygostenus*, and once among the Indian species of the genus *Doryloxenus*.

2. The occurrence of these two transformations of ant-inquilines into termite-inquilines we may regard as absolutely proved by facts, for otherwise we can discover no natural explanation of the existence of these isolated termitophile species among the numerous dorylophile species belonging to the same genus. The whole type of the genus is decidedly dorylophile, both in *Pygostenus* and *Doryloxenus*.

3. From these two comparatively recent transformations of ant-inquilines into termite-inquilines we deduce the hypothetical conclusion that two other transformations took place at an earlier date, in the Tertiary period, which resulted in the production of our present termitophile genera, *Discoxenus* and *Termitodiscus*, probably by a similar process, i.e. by the passing over of beetles, that had previously lived with wandering ants, to a termitophile existence. Of these two hypothetical transitions, we must believe that the later—that of *Discoxenus*—took place in India, the earlier—that of *Termitodiscus*—in the Africo-Indian continent.

4. The termitophile species of the genera *Doryloxenus*

and *Pygostenus* may be regarded as direct evidence of a recent formation of species, whilst the termitophile genera *Discoxenus* and *Termitodiscus* supplement this evidence, and enable us to extend it to the explanation of the origin of new genera.

8. The Family of *Clavigeridae*

(See Plate III, figs. 3–6)

Let us now turn to the family of the *Clavigeridae* (Plate III, figs. 3, 5, 6), and see what support they can give to the theory of evolution or to that of permanence.

The little yellow *Claviger testaceus* Preyssl. (Plate III, fig. 3) is the genuine ant-inquiline, whose way of life has been known to us longer than that of any other similar creature among our native Fauna. As long ago as 1818, P. W. J. Müller[1] published his classical observations regarding the relations existing between this beetle and the little yellow field-ant (*Lasius flavus*); but we may remark incidentally that in spite of our long acquaintance with *Claviger testaceus,* we still do not know where and how its larvae live. Its relatives already number over a hundred described species, belonging to every part of the world and divided into about thirty distinct genera. All the members of this family are genuine ant-inquilines, hospitably entertained by the most widely differing varieties of ants. At the end of the book the reader will find a photographic reproduction of our native *Claviger testaceus* (Plate III, fig. 3), and also of two very remarkable *Clavigeridae* from Madagascar, *Paussiger limicornis* (fig. 5) and *Miroclaviger cervicornis* Wasm. (fig. 6). The last is the largest member of the whole family, and is 4 mm. in length ; a giant among its kinsfolk, and distinguished by its antennae shaped like antlers.

The appearance of all the *Clavigeridae* proclaims them to be genuine inquilines (cf. Plate III, figs. 3, 5, 6). All the species are bright reddish yellow or red, and glisten with fat, thus possessing the true symphilic colouring of genuine inquilines ; they have stunted antennae, and the number of joints in them is considerably reduced.

[1] 'Beiträge zur Naturgeschichte der Gattung *Claviger*' (*Germars Magazin der Entomologie,* III, 1818, pp. 69–112).

At the base of the abdomen, the first segment of which is larger than all the others together, they have a more or less extensive hollow or pit for exudation, surrounded or almost concealed by the tufts of yellow hair on the base of the abdomen and the tips of the wing-sheaths (cf. especially Plate III, fig. 6). All these family characteristics of the *Clavigeridae*, which distinguish them from their nearest systematic connexions, the *Pselaphidae*, are due solely to their adaptation to the position of true inquilines. As a representative of the *Pselaphidae* we may take *Pselaphus Heisei*, whose photograph will be found on Plate III, fig. 4. This beetle has very long and highly developed maxillary palpi, but among the *Clavigeridae* they are greatly stunted, the reason for this being that the long palpi are useful to creatures seeking and examining their own food, but they would be useless to the *Clavigeridae*, which are fed by their hosts, and so are relieved from the necessity of procuring food for themselves. The number of joints in the antennae of *Clavigeridae* is much less than in those of the *Pselaphidae*, because the former use their antennae chiefly as a means of communication with the ants, and so it is convenient for the antennae to be short and strong; they are often shaped like a sceptre, a baton or a club (cf. Plate III, fig. 4 with figs. 3, 5, 6), whence the name *Clavigeridae*, *clava*—club. A diminution in the number of joints in the antennae increases the force of the blows that they can give, as they are less pliable when they have fewer joints; and as the ants often seize their tiny guests by the antennae and drag or carry them away, the reduced number of joints in the antennae renders them less liable to be broken off. The tufts of yellow hair and the pit at the base of the abdomen in *Clavigeridae* are unmistakably characteristics due to adaptation (see Plate III, figs. 3, 5, 6),[1] for these hairs assist in the emission of the substance that is so attractive to the ants as to make them lick their guests to obtain it. It is probably some kind

[1] On the photograph of our little yellow *Claviger* (Plate III, fig. 3) the large tufts of yellow hair at the points of the wing-sheaths can hardly be seen. They are quite visible, however, on the photograph of the staghorn beetle from Madagascar (fig. 6); two large tufts of yellow hair screen the semicircular exudatory hollow at the base of the abdomen; two other tufts are situated on each side at the point of the wing-sheath, and a row of small hairs runs round the side edge of the abdomen, and even the feelers have rings of stiff yellow bristles round their lower half.

of ether derived from fat, or some other volatile product of the adipose tissue and peculiar glandular tissue lying immediately beneath the hairs.[1]

In the same way the beetle's glossy yellow colouring is a direct result of its possessing a great abundance of that exudatory tissue, which is anatomically the foundation of its position as a true inquiline. Finally, the remarkable enlargement of the first free segment of the abdomen is connected with the same fact, as, the larger this segment is, the larger can the exudatory hollow belonging to it become. We are therefore fully justified in saying that all the systematic characteristics distinguishing the *Clavigeridae* from the *Pselaphidae* prove on examination to be simply due to their adaptation to the position of genuine inquilines.

Now there are a number of transitional forms connecting the *Clavigeridae* and the *Pselaphidae*, so that in many exotic genera of the latter we can trace a striking approximation to the former family. For this reason Raffray[2] and others regard the *Clavigeridae* as merely a systematic subfamily of the *Pselaphidae*, although the typical *Clavigeridae* are extremely unlike the typical *Pselaphidae*.

Viewed from the standpoint of the theory of evolution this is all quite intelligible. If the *Clavigeridae* originally branched off from the *Pselaphidae*, it was by way of progressive adaptation. The various genera of *Clavigeridae* are so many stages or modes of adaptation on the part of former *Pselaphidae* to the position of inquilines among ants. But the theory of permanence is incapable of assigning any reason for the above-mentioned morphological phenomena. It simply accepts them as facts, and assumes that the various genera and species of *Clavigeridae*, like their normal hosts, were all originally created exactly as we see them to-day. This hypothesis is supposed to exalt the wisdom and power of the Creator, but, in my opinion, they are revealed in a fairer light, if we accept the theory of evolution, and believe that the wonderfully manifold and beneficial morphological and biological peculiarities of the

[1] For a more precise anatomical and histological examination of the exudatory tissues in *Claviger testaceus* see 'Zur näheren Kenntnis des echten Gastverhältnisses' (*Biolog. Zentralblatt*, 1903, No. 5, pp. 201–206).

[2] 'Genera et Catalogue des Pselaphides' (*Annales de la Société Entomologique de France*, 1903–1904).

Clavigeridae are real adaptations to the genuine guest-relationship, brought about by natural causes.

The theory of evolution will not be able to tell us much regarding the precise manner in which the genera and species of *Clavigeridae* have been evolved, until we have a complete knowledge of the mode of existence of all the *Clavigeridae* of the present time, and of their special relations to the ants that are their respective hosts, and until we have, moreover, discovered all the extinct representatives of the same family as fossils. It would be unreasonable to require the theory of descent to account for the origin of genera and species, in the present state of our knowledge. We may remark incidentally that we already know one of the *Pselaphidae* (*Tmesiphoroides cariniger* Motsch.), belonging to the middle of the Tertiary period and found in the Baltic amber in East Prussia,[1] which, by having antennae with a reduced number of joints, appears to be a transitional form standing between the true *Pselaphidae* and the true *Clavigeridae*.

If we are asked to account phylogenetically for the extraordinary antler-shaped antennae of *Miroclaviger cervicornis* (Plate III, fig. 6), that bear no resemblance to the ordinary club-shaped antennae of other *Clavigeridae*, we may reply that this kind of beetle lives with some very large ants in Madagascar (*Camponotus Radamae* var. *mixtellus* For.); the elongation of its antennae is probably due to its living with such long-legged hosts; if it is to reach the ants' heads and ask for food, it needs very long antennae. This does not, however, explain their remarkable shape, for which at present no reason can be suggested, although the same antler-like formation occurs in another of the *Clavigeridae* of Madagascar (*Apoderiger cervinus* Wasm.) as well as in several *Paussidae* in the same island, viz. *Paussus dama* Dohrn, (Plate IV, fig. 6), *elaphus* Dohrn and *cervinus* Kr. Why in Madagascar the ant-inquilines belonging to various families of beetles have antennae tending to resemble antlers, is one of those problems in animal geography for which biology has hitherto found no solution. It is certainly no mere freak of nature, although we cannot account for this

[1] Cf. von Motschulsky, *Études Entomologiques*, V, 1856, p. 26 with plate, fig. 5. Cf. also W. L. Schaufuss, 'Preussens Bernsteinkäfer' (Pselaphiden) (*Tijdschr. voor Entomologie*, XXXIII, 1890, 1–62), pp. 13, &c.

strange phenomenon. However, it does not affect the result to which our previous considerations led us, according to which we regard the *Clavigeridae* as phylogenetically descended from *Pselaphidae*, the differences between them being due to a gradual, or perhaps a somewhat rapid, process of adaptation to the conditions of life of true inquilines.

9. The Hypothetical Phylogeny of the *Paussidae*

(Plate IV)

I have already in a previous article[1] dealt with the family of the *Paussidae* at considerable length. I arrived at the conclusion that it was impossible for this family of beetles to have been developed according to the Darwinian theory, but at the same time I showed that, nevertheless, we must assume a hypothetical evolution of the *Paussidae*, based ultimately upon interior laws of evolution, but directed by exterior circumstances necessitating adaptation, and leading to the production of the various genera and species of *Paussidae* belonging to the Tertiary period, and thence, by a continuation of the same process of evolution, to the production of the present genera and species of the same family. Here again the theory of permanence proves useless, whilst the theory of evolution supplies us with a natural explanation of the origin of those characteristics due to adaptation, which have made the *Paussidae* genuine ant-inquilines.

Let us once more shortly review the phenomena in question. (See Plate IV, at the end of the book.) The *Paussidae* are called ant-beetles because they live in ants' nests; the most important feature characterising them as a family is the great development of their antennae. They are found all over the world, and we are acquainted with thirteen living and three fossil genera (two of the latter being identical with still existing genera) and almost three hundred species.[2]

[1] *Stimmen aus Maria-Laach*, LIII, 1897, pp. 400, &c. and pp. 520, &c.

[2] Cf. R. Gestro, 'Catalogo sistematico dei Paussidi' (*Annali d. Museo Civico d. Genova*, [2] XX, 1901, pp. 811–850). To this catalogue must be added a number of new species from Africa and India, which I described in *Notes from the Leyden Museum*, XXV, 1904, pp. 1–82 with 6 plates ('Neue Beiträge

The *Carabidae* are the nearest natural relatives of the *Paussidae*. Such is the opinion expressed by Burmeister, Raffray, Ganglbauer and Escherich, and it is confirmed by my own anatomical examination of *Paussus cucullatus*, as a series of sections that I made of this beetle showed the ovaries of *Paussus* to resemble those of all the other *Adephaga* in possessing meroistic, polytrophic egg-tubes; in other words, egg-tubes in which chambers containing eggs and nutriment are arranged alternately.

As the *Paussidae* live with ants, their evolution out of the *Carabidae* type cannot have taken place until the family of ants had assumed an important biological position, viz. in the first half of the Tertiary period, for before that time the natural conditions requisite for the evolution of ant-inquilines did not exist.

All the peculiarities which distinguish the *Paussidae* from other beetles, and especially from the *Carabidae*, prove to be due to adaptation to a myrmecophile existence; this accounts for the development of their massive antennae with a diminished number of joints, and also for the formation of various organs of secretion, which enable the beetles to attract the ants and to live as their guests.

As I explained in a previous article,[1] we can distinguish three, or rather four[2] chief groups of *Paussidae*, classifying them according to the number of joints in the antennae of the various genera, and these chief groups represent as many stages in the process of evolving a true guest-relationship between the beetles and ants. That the *Paussidae*, like the *Carabidae*, originally had antennae with eleven joints is rendered very probable by the fact that the genus *Protopaussus*, found in Burma and China, still has such antennae. Next in order come the genera with ten joints, viz. *Homopterus*, *Cerapterus*, *Arthropterus* and *Pleuropterus*. According to Motschulsky's description, the fossil genus *Paussoides*, occurring in amber from the Baltic, had antennae with seven joints;[3] and

zur Kenntnis der Paussiden, mit biologischen, und phylogenetischen Bemerkungen'). The latter work forms a supplement to the account given in this chapter of the phylogeny of the *Paussidae*.

[1] *Stimmen aus Maria-Laach*, LIII, 1897, Part 5, pp. 522, &c.

[2] Four, if we reckon the genus *Protopaussus* as belonging to the genuine *Paussidae*.

[3] Cf. von Motschulsky, *Études Entomologiques*, V, 1856, p. 26 with plate,

the genera *Pentaplatarthrus, Ceratoderus,* and *Merismoderus* have six. The fourth group consists of the genera having antennae with two joints, viz. *Lebioderus, Paussomorphus, Platyrhopalus, Paussus,* and *Hylotorus.* Photographs of some representatives of these groups will be found on Plate IV. Fig. 1 represents *Pleuropterus brevicornis,* a new species from German East Africa, having antennae with ten joints ; fig. 2 represents *Pentaplatarthrus natalensis* from Natal, having six joints ; fig. 3 *Lebioderus Goryi* from Java with two joints ;[1] fig. 4 shows *Paussus howa,* and fig. 6 *dama,* both from Madagascar, and fig. 5 *Paussus spiniceps,* a new species from Sierra Leone in West Africa.

Comparative morphology and biology both show that, as a rule, *Paussidae* with fewer joints and more complicated development of antennae within any one genus approximate more closely to perfection as inquilines, for the development of the exudatory organs increases proportionately in beetles which are true inquilines, and culminates in the genus *Paussus.* In this genus we find an enormous variety of extraordinary formations of the antennae, and also a great development of tufts of yellow hair, of reddish yellow down and bristles, and of exudatory pores and hollows. These latter assist in the secretion of a peculiar substance, which the ants greedily lick off their guests' bodies,[2] and which is the return made by them for the hospitality that they receive.

My anatomical and histological investigations of *Paussus cuculatus*[3] showed the glandular tissue producing this aromatic secretion to be situated chiefly in the hollows of the antennae, under the pores on the brow, under the exudatory hollow of the prothorax, and under the tufts of yellow hair at the extremity of the abdomen. In *Paussus spiniceps* (Plate IV, fig. 5) the organs of exudation are still better developed,

fig. 6. It is possible that there were only five joints, and the illustration almost seems to suggest this, as the first three joints together greatly resemble the first joint in the antennae of *Ceratoderus* or *Paussus,* and the four others form a thick club.

[1] The specimen sent me had been pierced with a needle, hence the dark round spot on the right wing-sheath in the photograph.

[2] On this subject see K. Escherich's observations in his work 'Zur Anatomie und Biologie von *Paussus turcicus*' (*Zoolog. Jahrb. Abt. f. System,* XII, 1898, pp. 27–70, with Plate II).

[3] 'Zur näheren Kenntnis des echten Gastverhältnisses,' &c. (*Biolog. Zentralblatt,* 1903), pp. 232–248.

as the hollow of the antennae is serrate at the edge and provided with yellow hairs, and the hollow of the prothorax is filled with rolls of yellow hairs along the sides; the ring of long reddish yellow hairs at the extremity of the abdomen is so conspicuous in this species that we may be sure *Paussus spiniceps* is a very sweet guest, warmly welcomed by his West African hosts.

Paussus howa (fig. 4) has no tufts of yellow hair, but to compensate for their absence the shell-like hollow in the antennae contains an abundance of sweet substance. In this species the two exudatory pores on the brow and the clefts of the prothorax can be seen very plainly. In *Paussus dama* from Madagascar (fig. 6), not only is the hollow of the prothorax filled with yellow hairs, but the whole body and even the antler-shaped antennae are covered with bristles facilitating exudation, and there are large exudatory furrows on the head. In many other kinds of *Paussus*, especially in *Paussus armatus* and its relatives, a hollow horn crowned with a tuft of yellow hairs projects from the top of the head, and from it the ant drinks its nectar, as once the heroes in Walhalla drank their mead.

The position occupied by the genus *Paussus* among its related genera cannot perhaps be better described, from the standpoint of comparative morphology, than by a comparison that I have already used in this connexion.[1] 'The other genera of this family, which are very numerous, though poor in species, resemble the various halting places in the upward course of the evolution of the *Paussidae*. In the genus *Paussus* an open plateau seems to have been reached, offering abundant scope for the development of the most varied kinds of ant-inquilines. This genus actually contains more species than all the rest together (171 as compared with 118). Finally the genus *Hylotorus*, with its short, almost deformed antennae and legs, may be called a debased type, displaying degeneration connected with excessive parasitism. If we continue our simile, it represents a downward movement from the height of the plateau on the further side of the mountain.'

We now have to face the question: 'Is this evolution of the *Paussidae* real or only imaginary? Was each systematic

[1] *Stimmen aus Maria-Laach*, LIII, 1897, Part 5, p. 524.

species of this family created separately by God, as well in the Tertiary period as in our own day? Or are the genera and species of the *Paussidae* the result of natural evolution of the race, originating at the beginning of the Tertiary period with a form like that of the *Carabidae*, and passing through various stages of adaptation to a myrmecophile existence, until the present multiplicity of forms was attained?' Whether we consider the question from the point of view of philosophy or of natural science, we shall, I think, have to accept the latter theory, as it alone is capable of supplying a natural explanation of the phenomena we have observed.

Of course there is no direct evidence that such an evolution has taken place; we cannot prove that at the present day, from a beetle having ten joints in its antennae, one with six joints may be evolved, nor that one with only two joints may be descended from one with six. But if we are asked whether, in course of the hypothetical phylogeny of the *Paussidae*, a diminution in the number of joints in the antennae may not have been produced in many genera by the joints growing together in pairs or groups—this is quite another matter, and this question must be answered in the affirmative.[1]

Let us consider the shape of the antennae in *Lebioderus Goryi* (Plate IV, fig. 3). Most of my readers would say that there were six joints in this beetle's antennae, but they would be wrong, for the last five joints have grown together so as to form one, although the original divisions between the joints are still marked by deep depressions. We have, therefore, here an unmistakable example of the manner in which a two-jointed form of antennae can be produced from a six-jointed form by the end joints growing together. Of course God could create a *Lebioderus*, having the second joint of its antennae looking exactly as if it were the result of five distinct joints having grown together, but it savours too much of occasionalism for me to be able to adopt this view, and I prefer the phylogenetic explanation, according to which the club at the end of the antennae in *Lebioderus* has really been formed by the coalescence of five joints.

[1] Escherich is mistaken when he asserts (*Zoolog. Zentralblatt*, 1899, No. 1, p. 9) that I ever questioned this possibility. I only maintained that at the present day it is no longer possible actually to observe such a reduction in the number of joints.

In the present state of our knowledge we can hardly expect exact details regarding the hypothetical phylogeny of the *Paussidae*; we can hope to discover them only after both the living and the extinct members of this family have been studied with some approach to completeness. Hitherto only scanty remains of three varieties of fossil *Paussidae* are known to us from Baltic amber.[1] These may be referred to the three genera *Arthropterus*, *Paussoides*, and *Paussus*; thus we already have fossil representatives of three chief groups among our present *Paussidae*, viz. those with ten-jointed antennae, those with six (occasionally seven or five), and those with two. We are therefore justified in concluding that, even in the middle of the Tertiary period, the family of *Paussidae* was well developed, at least in its principal groups.

Fossil *Paussidae* having antennae with eleven joints, analogous to the present genus *Protopaussus*, have not yet been discovered, but this is not surprising, as only two very rare species of the living representatives of this genus are known to exist. We know nothing with certainty regarding the previous history of the Tertiary *Paussidae*, and can only suppose that they are phylogenetically connected with the *Carabidae* of the Lias or earliest Jurassic epoch. Taking into consideration the fossil remains of *Paussidae* from the Miocene epoch, we must further regard it as probable that the unknown hypothetical primitive form of the *Paussidae* divided into the present four chief groups in the first half of the Tertiary period, acting partly under the influence of internal differentiation, and partly under that of adaptation to a myrmecophile existence; these four chief groups being the *Protopaussus* group with eleven joints in the antennae, the *Arthropterus* group with ten,[2] the *Paussoides* group with five or six, and the *Paussus* group with two.

The hypothetical phylogeny of the *Paussidae* took the form probably of a tree with four chief stems, splitting up into many smaller branches and twigs. The *Paussidae* of the

[1] See von Motschulsky, *Études Entomologiques*, V, 1856, p. 26; C. Schaufuss, 'Preussens Bernsteinkäfer' I (*Berl. Entomolog. Zeitschrift*, XXXVI, 1891, pp. 53 and 64) and II (*ibid.* XLI, pp. 51–54). The ants and *Paussidae* of the Baltic amber do not belong to the Miocene, as was formerly believed, but to the older Oligocene epoch. Cf. Handlirsch, *Die fossilen Insekten*, Leipzig, 1906–1908.

[2] I have designated the groups according to the names of the oldest genus in each.

Tertiary period show that the four chief stems have followed each an independent line of evolution, not standing in any close relationship with the branches of other stems. Therefore the present *Paussus* is not directly descended from the present *Lebioderus*, nor is *Lebioderus* descended from the present *Arthropterus* or *Homopterus*, and least of all is *Arthropterus* descended from *Protopaussus*. The four great stems differ greatly in the number of their branches and twigs. On the lowest, the *Protopaussus* stem, we find only one genus with two species; on the *Arthropterus* stem, four genera with about eighty species; and on the highest, the *Paussus* stem, five genera with over two hundred species.

By asserting that the chief stems of the *Paussidae* trunk have continued an *independent* evolution ever since the early part of the Tertiary period, I do not mean to deny the existence of a remote connexion between the chief stems. The genus *Lebioderus*, with its apparently six-jointed, but really two-jointed antennae, is an interesting example of a 'collective type,' marking the transition from genera with six-jointed antennae to those with two; but the actual time of the transition is to be sought not in the Quaternary period at all, but probably before the middle of the Tertiary.

Each of the four chief stems of the *Paussidae* trunk has its own hypothetical phylogeny, and this has been influenced in various ways by different internal tendencies and by different degrees of adaptation to a myrmecophile existence. A few examples will show what I mean.

Like the *Carabidae*, the genus *Protopaussus* has eleven joints in its antennae, and the thickening of the joints is very slight in comparison with the other *Paussidae*. On the other hand, the broad, deep cavity of the prothorax and the yellow tufts at the extremity of the abdomen show unmistakably that this genus occupies a relatively high position as a genuine inquiline. We have to distinguish the characteristics due to organisation from those due to adaptation; the retention of eleven joints in the antennae is a characteristic due to organisation inherited from the *Carabidae*, but the peculiar formation of the prothorax and its means of secretion are due to adaptation, and are characteristics acquired by this genus. The genera with ten-jointed antennae are very different. *Homopterus* and *Arthropterus* have enormously broad antennae, and their massive shape and the diminution in the number of their joints both are due to adaptation to the myrmecophile existence, although they do not mark out these genera as true inquilines, but, like the often very considerable thickening of the legs in these creatures, suggest rather the offensive type, as these peculiarities would serve to protect the beetles from being attacked by the ants.[1]

Unmistakable evidence of adaptation to the position of true inquilines is given first in this group by the genus *Pleuropterus* (Plate IV, fig. 1), in which the prothorax shows a shell-like cavity, and becomes a large uneven exudatory hollow, provided as a rule with yellow hairs, and at the same time traces of yellow exudatory bristles appear more plainly on the antennae. Within the *Paussoides* group, with antennae having five or six joints, we meet with similar signs of independent differentiation in various directions. The genus *Pentaplatarthrus* (Plate IV, fig. 2) has developed in a manner wholly unlike the genera *Merismoderus* and *Ceratoderus*. In it the prothorax appears as an extraordinary labyrinth of exudatory cavities and protuberances, pointing to a high degree of adaptation to the position of inquiline, whilst the long, flat antennae suggest the *Arthropterus* type.

In *Merismoderus* and *Ceratoderus*, on the other hand, the prothorax is only slightly modified, but in the formation of their antennae and in other points these two genera approach the *Paussus* type. Among the genera having antennae with two joints, *Lebioderus* (Plate IV, fig. 3) and *Platyrhopalus* stand fairly close together; to each belong a number of species bearing a certain resemblance to the genus *Paussus*, yet not so great a resemblance as to justify our believing this genus to be directly connected with the other two. *Hylotorus* is likewise very closely related to *Paussus*. Within the genus *Paussus* the evolution of the same generic type proceeds along two lines, and we find a series of species having the prothorax undivided, and others in which a deep cleft divides the prothorax into two parts, between which is situated the large exudatory cavity of the thorax.[2] The latter branch in particular splits up into

[1] Cf. *Stimmen aus Maria-Laach*, LIII, 1897, Part 5, pp. 521, 522.

[2] The species depicted on Plate IV, figs. 4–6, belong to the second group.

2 B 2

a number of smaller branches, representing a considerable number of systematic species, that stand in close relationship to one another. The different species display an immense variety in the shape of the second joint of the antennae and in the development of the yellow hairs and other organs of secretion; I shall refer to these points again later on.

One more remark must be made with reference to the hypothetical phylogeny of the *Paussidae.*

The above account suggests that it is monophyletic in character, originating in one single pre-Tertiary primitive form. But the genus *Protopaussus,* which I have designated the oldest stem of the one trunk, and nearest to the original (although its existence in the middle of the Tertiary period has not been proved), may possibly have had an independent origin, and be descended from a kind of *Carabidae* differing from the ancestors of the other three chief groups of *Paussidae.* Its origin may be of more recent date than that of the other groups which existed even in the Miocene epoch. This supposition would explain why the antennae of *Protopaussus* resemble those of the *Carabidae* more closely than those of the genuine *Paussidae.* If this view is correct, the evolution of the family of *Paussidae* is not monophyletic but diphyletic. The two lines of descent have been quite independent of one another; one originated in a pre-Tertiary form of *Carabidae,* and early in the Tertiary period produced the genera *Arthropterus, Paussoides* and *Paussus*; the other originated later, perhaps in the second half of the Tertiary period, from another kind of *Carabidae,* and produced only the present genus *Protopaussus.*

It is not possible as yet to say with certainty which of these two suppositions is more correct, whether we are to believe the evolution of the present *Paussidae* to have been monophyletic or diphyletic; perhaps future palæontological discoveries will settle the matter.

I have discussed the evolution of the *Paussidae* in detail, because it may be useful in correcting some false impressions, which prevail in many quarters, regarding the relationship existing between genera and species of the same family. It removes also many difficulties raised against phylogenetic hypotheses by those who have not made a special study of the subject.

Respecting the causes of the hypothetical evolution of the *Paussidae*, we have to content ourselves with a few suggestions, as we know very little about them. Undoubtedly the interior capacity for modification, possessed by the primitive form, must be regarded as the first and most indispensable cause of the evolution of the *Paussidae*; otherwise their adaptation to a myrmecophile existence would have been impossible, and still less would there have been any possibility of an adaptation so varied and so complete as to transform the whole bodily structure of these beetles that were once *Carabidae*, to reduce the number of joints in their antennae, whilst rendering them thick and massive, and to equip them with very various organs of exudation and the corresponding tissues.[1]

The evolution of the *Paussidae* was probably neither so slow nor so gradual a process as the Darwinian hypothesis would require it to have been; it is likely that in many cases the changes were effected suddenly, and were such as the theory of mutation assumes to have occurred. This is suggested not only by the fact that many genera of *Paussidae* at the present day are separated from one another by wide intervals, but also by the circumstance that the three chief groups of this family are represented among the fossils of the Tertiary period. That a *per saltum* modification was probably possible in this case—such a modification as a growing together of definite pairs of joints in the antennae—is seen in *Lebioderus Goryi* (Plate IV, fig. 3), which, with regard to the formation of its antennae, stands on the border line between the six-jointed and the two-jointed forms. The absence of transitional links between many genera and species of *Paussidae* can be accounted for more easily, if we believe the progressive modifications to have been effected *per saltum*, than if we assume the process of change to have been extremely gradual. A very gradual change may have taken place within some groups in the case of very closely allied species, e.g. the species of the group *Paussus denticulatus* Westw., but it is hardly possible to see how such

[1] The latter are adipose glandular tissues and are, strictly speaking, metamorphosed hypodermic cells. Cf. 'Zur näheren Kenntnis des echten Gastverhältnisses,' &c. (*Biolog. Zentralblatt*, 1903), pp. 68, 232, &c.

a gradual process could have resulted in the production of the chief genera of *Paussidae*.

Of course, in considering the hypothetical evolution of the *Paussidae*, we must ascribe great importance to the exterior as well as the interior factors of evolution, for all the morphological peculiarities that distinguish the *Paussidae* from their nearest relatives, the *Carabidae*, are due to their adaptation to a myrmecophile existence. The unusual breadth of the antennae and the diminution in the number of their joints are characteristics due to adaptation,[1] as is the wonderful variety in the shape of the antennae in the genus *Paussus*, each tending to make the flagellum firm and convenient for the ants to seize with their jaws, and use as a means of picking up their guests and carrying them about, whilst at the same time in most cases each modification makes the flagellum a more perfect organ of exudation, whence the ants can lick their favourite dainty.[2]

Other characteristics due to adaptation are the different kinds of hair connected with the secretion of this substance; there are tufts of yellow hairs, downy hairs of a reddish yellow colour, bristles, &c., situated on various parts of the bodies of these true inquilines, on the antennae, on the horn, in the cavity of the prothorax and at its edges, at the edges of the wing-sheaths or on their surface, at the extremity of the abdomen, and even on the legs. Other marks of adaptation are the manifold exudatory pores, the horns on the head, and the cavities and furrows on the prothorax. Others again are the peculiar exudatory tissues, which are connected with the external organs of secretion, and, as adipose glandular tissue, approximate partly to the adipose tissue and partly to the common glands of the skin, and furnish the aromatic secretion of which the ants are so fond, that in order to obtain it they keep the beetles in their nests. We may therefore say: 'Adaptation to a myrmecophile existence, and especially adaptation to various degrees in the progressive evolution of true inquilines, is the leading idea governing the evolution of the whole family of *Paussidae*.' But at the same time we must

[1] Cf. on this subject *Stimmen aus Maria-Laach*, LIII, 1897, Part 5, pp. 521,&c.
[2] Cf. *Stimmen aus Maria-Laach*, LIII, 1897, Part 5, pp. 525–528, and 'Zur näheren Kenntnis des echten Gastverhältnisses' (*Biolog. Zentralblatt*, 1903), pp. 242–248.

acknowledge our present inability to explain how this idea has been carried out in the individual cases; how exterior causes, such as the hospitality shown by the ants and natural selection, have co-operated with the interior factors that produce tissues and organs, so as to effect adaptation so varied in form and of so high a degree of completeness.

Let us turn our attention to one of the most interesting phylogenetic problems, viz. to the differentiation in the shape of the antennae within the genus *Paussus*, which contains almost two hundred species (cf. Plate IV, figs. 4–6). To what natural causes can we ascribe the extraordinary variety in the shape of the flagellum in this genus? It is at first sight an unaccountable freak of nature; and it seems as if some skilful artist must have produced almost every conceivable shape, working quite arbitrarily and without any definite purpose; fashioning the flagellum now in the shape of a lens, now like a ball, a club, a sabre, a triangle, a leaf, a rod, a horn, a shell, an antler, adorned with all manner of zig-zags, furrows and points, each being a miniature work of art, given by the Creator to be the plaything of the ants, His favourites in the insect world.

If we study the biology of the *Paussidae*, we shall soon come to the conclusion that these various shapes of the antennae in the genus *Paussus* are by no means useless playthings, but are all different solutions of the phylogenetic problem, 'How can the nose of a beetle (for the antennae are primarily organs of smell, or movable noses) be at once beneficially and pleasantly applied to another biological purpose?'[1]

Or the question may be worded more precisely thus: 'How can the nose of an ant-inquiline be changed into a means of transport, by which the ants can seize their guest with their jaws and carry him away without injuring him? and, further, how can the nose at the same time be made into an organ for exudation, whence the ants derive their delicious nectar?' In other words: the object aimed at in the characteristic metamorphosis of the flagellum of the *Paussus* antennae is fitness to discharge two biological functions, to be at the same time means of transport and of exudation;

[1] Cf. *Stimmen aus Maria-Laach*, XL, 1891, pp. 79, 207, 320, 406, &c., also LIII, 1897, pp. 520, &c.

and the *Paussus* antennae fulfil these two requirements with greater perfection, the higher the stage of genuine guest-relationship attained by their owner.

The species with lens-shaped antennae are those members of this genus which have retained the simplest form, nearest to the original, and they have as a rule their exudatory organs only slightly developed. Antennae shaped like rods, sabres, or antlers belong to species in which the guest-relationship is at a higher stage, and it is highest in those in which the flagellum is hollowed out, so as to form a cup to contain the secretion, especially if, as in *Paussus spiniceps* (Plate IV, fig. 5), this cup is surrounded by notches bearing tufts of long, yellow hair.

It is possible therefore to discover both a biological and a phylogenetic explanation of the idea controlling the morphological variety of form in the *Paussus* antennae. It cannot be denied that natural selection at first sight seems likely to have encouraged this variety, as it might select such antennae as best fulfilled the above-mentioned biological requirements —these antennae being the result of the action of the interior laws of evolution belonging to the various species. Closer examination, however, shows that Darwin's natural selection cannot give a satisfactory account of the actual specific multiplicity of shape in the antennae of *Paussus*.

If natural selection were the controlling factor in the specific evolution of the antennae within the genus *Paussus*, their form would have to originate in accordance with a strict necessity for adaptation, which would eliminate antennae of other shapes as less capable of existence, for this is precisely what is implied by the ' Survival of the Fittest in the Struggle for Existence.' Consequently natural selection would lead to the production of one definite form of *Paussus*, having antennae of one fixed shape, and living with one particular species of guest-ant. The shape of the antennae would be determined by the mechanical necessity for their adaptation to the shape and size of the ant's head, the length and breadth of its upper jaw, and its manner of seizing the beetle, carrying it and licking it. Moreover, the varieties of *Paussus*, living with allied species of the same genus of ants, could differ from one another only as far as was absolutely necessary to adapt them to various species of hosts; for otherwise they would perish in the struggle for existence, as being less capable of life. Let us see how the actual facts stand in relation to the Darwinian hypothesis. They simply do not tally with it at all; about two-thirds of the almost two hundred species of *Paussus* hitherto discovered live exclusively with ants of the genus *Pheidole*, the workers and warriors of which genus resemble one another very closely even in different species; and yet in the *Pheidole* nests occur species of *Paussus* with antennae of all the above-mentioned shapes, except perhaps those with long antler-like antennae, which probably have larger ants as their hosts. Moreover, within the genus *Pheidole* there are a good many species which entertain a considerable number of kinds of *Paussus*, having antennae of very various shapes. For an instance I may refer to *Pheidole megacephala* in South Africa, which has over a dozen species of *Paussus* as inquilines, and of these, according to observations made by Dr. Hans Brauns and G. D. Haviland, nine live with *Pheidole megacephala* var. *punctulata*. Among them are, according to the species in my collection, *Paussus Klugi* and *Curtisi* with rod-like antennae, *Paussus cultratus* and *granulatus* with the flagellum shaped like a knife, and *Paussus cucullatus* and *Elisabethae*, in which it is shaped like a shell.

There are at least five species of *Paussus* with antennae of different shapes living with *Pheidole latinoda* in India, and as many with *Pheidole plagiaria* in Java.

I believe therefore that Darwin's theory of natural selection cannot give a satisfactory explanation of the specific differentiation of the antennae in various species of *Paussus*. On the contrary, the multiplicity of their shapes gives us an impression that the phylogenetic evolution of the antennae in *Paussus* has freed itself to a great extent from the strict laws of natural selection, which would tend to produce uniformity, and not multiplicity of shape.

But how can the great variety of extraordinary shapes have been produced by natural methods within the genus? Primarily as a result of the action of the interior laws of growth, which involved a particularly high degree of variability in the flagellum of the antennae.

The hypothetical previous history of the genus *Paussus*

suggests a reason why the flagellum in this genus tends to develop into so many different shapes. The present flagellum with a single joint was not originally so simple, but has been produced by a number of original joints growing together; the tendency to vary in shape, displayed by the flagellum of *Paussus*, is due to the combined tendencies to vary, possessed by its original components.

Let us refer once more to Plate IV, fig. 3, and look closely at the antennae of *Lebioderus Goryi*. The flagellum has one joint, but there is no difficulty in seeing that it consists of five separate joints grown together, and these apparently separate joints are formed from nine or ten original joints, that have grown together in pairs.[1]

In a great many kinds of *Paussus* with a highly developed rod or shell-shaped flagellum, there is a row of seven or eight transverse furrows on the back of the flagellum and inside the cavity of the antennae, the furrows being separated by teeth or notches at the edge. (See *Paussus howa*, Plate IV, fig. 4.) The antler-shaped flagellum of *Paussus dama* shows a similar peculiarity (fig. 6). The teeth or notches on the flagellum, separated by transverse furrows, are probably traces of original segmentation.

I have, I think, said enough to prove that the flagellum of *Paussus* is rendered capable of development into many different shapes by certain interior causes. In order to fix this tendency to vary, and limit it to the production of definite forms, another factor is required, which must be exterior. As I have already shown, natural selection can act only in a very restricted manner; in fact, it would be more likely to hinder than to promote the development of the great variety of forms that actually exist. How are we therefore to account for the specific differentiation of the antennae in the genus *Paussus*? A comparison that I have used on a previous occasion[2] may serve to elucidate my view of the matter.

[1] I cannot decide whether the ten-jointed antennae of *Arthropterus, Cerapterus*, and *Pleuropterus* (Plate IV, fig. 1), in which the flagellum consists of nine joints, have been formed from eleven-jointed antennae by the reduction of the second joint to a small connecting link between the scape and the flagellum, or by the amalgamation of the two last joints of the flagellum. Reasons can be adduced in support of both these views.

[2] 'Zur Entwicklung der Instinkte,' pp. 182, &c. (*Verhandl. d. k. k. Zoolog. Botan. Gesellschaft*, Vienna, 1897, Part 3, pp. 168–183).

Man is able, by means of conscious selection, to produce a great variety of breeds among the domestic animals; for instance, he has bred pigeons differing in plumage, in the formation of the crop, tail, &c. In exactly the same way, though unconsciously, the ants have bred inquilines of the genus *Paussus* with antennae of very various shapes. If certain shapes found favour with the ants, this was enough to give an impetus to a further evolution in that direction, for guests with antennae of the attractive shape received better treatment from their hosts than others. In this way varieties of *Paussus*, differing in the shape of their antennae, might develop in the nests of one and the same species of ant. The beetle's capacity for existence was not affected by its having a flagellum of one shape rather than another, hence the struggle for existence cannot be made responsible for the selection of any particular shape. I have designated the instinctive selection practised by the ants in breeding their genuine inquilines 'Amical Selection,' as opposed to Darwin's 'Natural Selection.' [1]

We met with this new form of selection in discussing the hypothetical phylogeny of the *Lomechusini* (p. 338), and here again, in the case of the *Paussidae*, we are induced to accept it, as it is based upon very simple and obvious considerations. If, however, there is anyone to whom it does not commend itself, he is perfectly free to devise a better explanation.

10. The *Termitoxeniidae*, a Family of Diptera

(See Plate V).

In the nests of African and Indian termites are found some remarkable Diptera belonging to the family of *Termitoxeniidae*.[2]

[1] *Biolog. Zentralblatt*, 1901, No. 23, p. 739, &c., Escherich's objections, which appeared in the same paper in 1902, p. 658, were answered in it in 1903, p. 308. I need scarcely say that I do not ascribe to the ants any 'aesthetic sense of shape.' The kind of instinctive selection, practised by the ants in their dealings with the *Paussidae*, depends chiefly upon their sense of touch, but also upon taste and smell, and only incidentally upon sight.

[2] Cf. Wasmann, '*Termitoxenia*, ein neues flügelloses physogastres Dipterengenus aus Termitennestern,' I and II (*Zeitschrift für wissenschaftl. Zoologie* LXVII, 1900, Part 4, and LXX, 1901, Part 2); 'Zur näheren Kenntnis der termitophilen Dipterengattung *Termitoxenia*' (*Verhandl. des V internationalen*

They have been mentioned in previous chapters, and photographs of them will be found on Plate V, figs. 1–6. They form the genus *Termitoxenia* Wasm. and its subgenus *Termitomyia* Wasm. These little creatures, only 1–2 mm. in length, are white or pale yellow in colour, and are some of the most remarkable insects in existence. They have neither males nor females like other insects; they do not go through a larval stage nor have they wings like other Diptera. They are protandric hermaphrodites; a stenogastric imago form takes the place of the larval stage, and very remarkable appendages on the thorax represent wings. In one of the two subgenera, i.e. in *Termitoxenia* in the narrower sense, the whole embryonic development seems to take place in the parent, so that the stenogastric imago form is born alive.[1]

The stenogastric imago (Plate V, figs. 1 and 2) is, however, a walking embryo, for its abdomen especially resembles that of a larva, and the fat-body and the muscular system exist in a very rudimentary form; in very young specimens of *Termitoxenia Assmuthi* I have found even the vitelline sac of the embryo. It is only after the stenogastric imago has seen the light that it undergoes an 'imaginal development,' which takes the place of the usual larval development, and thus it gradually reaches the physogastric imago form, representing the full-grown insect (Plate V, figs. 3 and 6). In each individual the male generative glands ripen first, and the ovaries later; hence we have here an instance of protandric hermaphroditism. The development of the ovaries is accompanied by a steady increase in physogastry, until finally the adult insect resembles a whitish sac attached to the forepart of the body as to a small, black stalk. In spite of the unwieldy size of their bodies, their long, powerful legs enable these

Zoologenkongresses zu Berlin, 1901, Jena, 1902, pp. 852–872 with plate); 'Termiten, Termitophilen und Myrmekophilen, gesammelt auf Ceylon von Dr. W. Horn' (*Zoolog. Jahrbücher, Abt. für Systematik*, XVII, 1902, Part 1, pp. 151–153 with Plate V, figs. 4, 4 a–c, and 5); 'Die Thorakalanhänge der *Termitoxeniidae*, ihr Bau, ihre imaginale Entwicklung und phylogenetische Bedeutung' (*Verhandl. der Deutschen Zoolog. Gesellschaft*, 1903, pp. 113–120 with Plates II and III); 'Neue Termitophilen aus dem Sudan' (*Results of the Swedish Zoological Expedition to Egypt and the White Nile, 1901, under the direction of L. A. Jägerskiöld*, No. 13, Upsala, 1904); see also remarks on *Termitoxenia* in the present work, Chapter II, p. 38, and Chapter III, p. 50.

[1] This statement is borne out by a series of sections made of a specimen of *T. Braunsi*, containing an embryo.

creatures to run quickly, as Father Assmuth observed, when he was studying *Termitoxenia Assmuthi*. In the *Termitoxeniidae* the place of the wings in other Diptera is taken by a pair of oar- or hook-shaped appendages on the mesothorax (Plate V, *ap* in figs. 1, 2, 4, 5), which serve a number of important biological functions, but do not enable the creatures to fly. They act as balancing poles, and maintain the fly's equilibrium when it runs; they are means of transport, by which the delicate little guests can be picked up by their hosts without injury; they are also important sense-organs, for the front branch of each thoracic appendage contains a large nerve, and is covered with tactile bristles; they are finally the chief organs of exudation possessed by these genuine inquilines, for the hinder branch of each appendage is a hollow tube, at the upper end of which is a cluster of large membranous pores (Plate V, *pp* in figs. 4 and 5). As is generally the case in physogastric inquilines among termites, the exudation which is eagerly licked off by their hosts, and which secures the inquilines their position as favoured guests, is a constituent of the blood-plasm.[1]

Behind these appendages of the mesothorax, which answer to the front wings of other Diptera, there is on the metathorax a pair of very diminutive balancers, of very primitive structure, which are essentially equivalent to the genuine halteres of the Diptera.

Let us now consider these interesting creatures from the point of view of the evolution theory. What right have we to assign a place in the Diptera order to them, as they have no wings? Moreover, *Termitoxenia* has an incomplete metamorphosis, and *Termitomyia* has none at all, whereas a true larval form always occurs in other Diptera, even in those that give birth to living pupae; but here, in place of the larva, we have a stenogastric imago. The protandric hermaphroditism of these diminutive beings is a characteristic that does not present itself regularly in any other insect. From the standpoint of the theory of permanence we must say: The *Termitoxeniidae* are a class apart, resembling real Diptera in many respects, such as in the shape of their antennae, in the formation of their

[1] Cf. 'Zur näheren Kenntnis des echten Gastverhältnisses' (*Biolog. Zentralblatt*, 1903), pp. 68, 300, 305.

proboscis (which is used to suck the life out of the young termites), and in having halteres instead of hind wings. But these resemblances are insignificant in comparison with the great differences mentioned above, which distinguish them from Diptera. If therefore the *Termitoxeniidae* were created once for all in their present condition, they ought to be classed as an order of insects resembling Diptera, but not belonging to them.

From the standpoint of the evolution theory we should say: These curious creatures were once genuine Diptera, and all their divergencies from the normal type of that order are due to adaptation to a termitophile existence. The peculiar appendages to the mesothorax (Plate V, *ap* in figs. 1, 2, 4, and 5) are the result of metamorphosis of the front wings which their ancestors once possessed; for these appendages were better adapted than wings to the changed conditions of life within the nests of the termites. As the development of the individual was shortened, the larval stage, that the creature's ancestors had passed through, was omitted and replaced by the stenogastric imago form, and in the subgenus *Termitomyia* the process is still more abbreviated, and the stenogastric imago form does not enter the world as an egg but as a living creature.

This abbreviation and simplification of the development of the individuals belonging to this genus is phylogenetically to be referred to the fact that the conditions for nourishing themselves and their young were very favourable in the termite nests. It is a general rule that the number of eggs in an insect stands in inverted ratio to the number of eggs and larvae that develop successfully: the less favourable the external circumstances, the greater the number of eggs laid by an insect to assure the propagation of its species; and the more favourable the conditions, the fewer the eggs. Hence the number of eggs was very small in the case of the *Termitoxeniidae,* and consequently each egg-cell could be supplied with a greater abundance of nourishment (cf. Plate V, fig. 6 *ov*). The result of this was a quickening of the development of the individual, and an abbreviation and simplification of the cycle of reproduction. This explains why in *Termitoxenia* the larval stage fell out and was replaced by the stenogastric imago form, and also why in the subgenus *Termitomyia* the imago form does not proceed from the egg, but appears at once alive. All this is only a consistent continuation of the abbreviation and simplification of the individual development. The hermaphroditism of *Termitoxenia* is a phenomenon that appeared later in the phylogeny of these little Diptera. As they live inside the termite nests, no crossing could occur between the occupants of different nests, when once their wings had suffered metamorphosis, and served other biological purposes than that of flight. When they became able to dispense with the advantages of crossing, the distinction of the sexes gradually ceased, for its chief object is to cause union between individuals differing as widely as possible within the species. Under similar circumstances among other insects parthenogenesis has taken the place of sexual propagation, but *Termitoxenia* developed hermaphroditism, which is to some extent a still more advanced simplification of the method of propagating the species.

Thus we see that the theory of evolution really enables us to understand how the *Termitoxeniidae* have phylogenetically been evolved out of ordinary insects with two wings, and at the same time this theory suggests why we may rightly class them with the Diptera. Certain morphological points of agreement between the *Termitoxeniidae* and the *Muscidae* on the one hand, and the *Phoridae* on the other, lead us to regard the *Termitoxeniidae* as a branch of the Diptera stock, connected originally with the *Muscidae* and *Phoridae*, but having adopted a line of evolution peculiar to itself, in consequence of its thorough adaptation to the termitophile existence.

Many points in this explanation may still appear very doubtful, but it must be granted that it supplies us with a real, scientific means of accounting for the morphological and embryological peculiarities of *Termitoxenia*, which stand in very close connexion with its biology. Unless we assume that these creatures are of common origin with true flies, we are not justified in including them among the Diptera; we should be forced to say with the theory of permanence: 'These creatures are *entia sui generis*, created in their present form to be the inquilines of certain species of termites, which were likewise created exactly as we see them.' In this way an apparently satisfactory account is given of the facts before us, inasmuch as they are referred to the Creator's wisdom and power as their immediate cause. Nevertheless, I prefer the other interpretation, which refers to the Creator's wisdom and power only indirectly, and seeks to discover the natural causes, through which God in His wisdom and power has produced these beneficial adaptations by means of phylogenetic evolution, for this hypothesis is based upon a logical application of the fundamental principle: 'God does not interfere directly in the natural order when He can make use of natural causes, and the natural laws laid down by Him are already in force.'

There is one point in the ontogeny of *Termitoxenia* that we must discuss shortly, as it is of particular importance to the phylogenetic account of these inquilines, viz. the development of the appendages on the mesothorax, that take the place of wings. In *Termitoxenia mirabilis* Wasm. of Natal (cf. Plate V, fig. 2, *ap*), which belongs to the subgenus *Termitomyia*, these appendages are shaped like hooks, and consist of two tubes resembling tracheae, and only partially grown together. This formation, which somewhat resembles the breathing tubes of insect larvae living in water, remains unchanged from the earliest stenogastric to the latest physogastric imago form. The tissues contained in these tubes also are unchanged throughout the whole period of imaginal growth; the front branch is always an organ of touch and contains a nerve, the back branch is connected with the circulation of the blood and with exudation. In the sub-genus *Termitoxenia*, however, in *T. Havilandi* of Natal, *T. Jägerskiöldi* of the White Nile, *T. Heimi* and *Assmuthi* of the East Indies, the original tubes grow more closely together, and in the earliest stenogastric imago (cf. Plate V, fig. 1, *ap*) they might almost be taken for small, stunted wings, but later on they gradually draw together so as to form the oar- or style-shaped horns which are seen in the adult physogastric animal (cf. the photograph, greatly enlarged, on Plate V, figs. 4 and 5). In the stenogastric individuals of the three species at present known of the subgenus *Termitoxenia*, these growths resemble one another very closely, but they differ in the physogastric specimens according to their species. In *T. Heimi* (Plate V, fig. 4), even in their final form they bear more likeness to wings than they do in the other Indian species, *T. Assmuthi* (fig. 5), in which they gradually become like rods, and lose their early resemblance to wings (fig. 1). It is very remarkable that one species found in what used to be the Orange Free State, viz. *Termitoxenia* (*Termitomyia*) *Braunsi* Wasm., is a perfect connecting link between *T. mirabilis* and the other four species, as far as the appendages on the thorax are concerned. Still more striking is a discovery that I made when cutting under the microscope a series of sections of a very young stenogastric specimen of *T. Heimi*. I found the appendages to be at a stage of development at which real wing-veins occur all round the hind branch, but they are suddenly suppressed and are absent in slightly older specimens.

What do we learn from these facts considered in their bearing upon the theory of evolution? They tell us that the subgenus *Termitomyia* (*mirabilis* and *Braunsi*), which is viviparous, departs furthest from the original Diptera type in the formation of the appendages on the thorax, whilst the subgenus *Termitoxenia* (*Havilandi, Heimi, Assmuthi,* and *Jägerskiöldi*), which is oviparous, stands nearer to the genuine Diptera in this respect. This enables us to understand why in the latter subgenus, at a particular point in its ontogeny, there is a genuine but transitory atavism, during which the ancestral wing-veins appear, as if in memory of the past, and then vanish. In other words: The tendency to produce real wings, which in the ancestors of *Termitoxenia* continued without interruption, is still present at the beginning of the ontogeny of our *Termitoxenia*, but is suddenly broken off and diverted to other channels, leading to the formation of appendages on the thorax that are quite unlike wings. In the subgenus *Termitomyia*, especially in *T. mirabilis*, the development of these appendages proceeds uninterruptedly on the new lines, and does not pass through a stage of resemblance to wings. This subgenus dates from an earlier period and is further removed from the Diptera type. This explanation, which the theory of evolution supplies, seems to me the only scientific mode of accounting for the facts, which are an inexplicable 'freak,' when considered with reference to the theory of permanence.

In order to study the anatomy, growth, and mode of life

2 C

of these interesting little termite-inquilines, I have cut 10,000 microscopical sections from sixty specimens of five different species of *Termitoxeniidae*, and from them I have obtained much evidence in support of the theory of descent. Without any exaggeration we may assert that this family of Diptera is perfectly incomprehensible both morphologically and biologically, unless in studying it we take evolution into account. It is almost impossible to dispense with the theory of descent, if we attempt to give a reasonable explanation of the scientific facts.

No detailed argument is needed to show that the hypothetical evolution of the *Termitoxeniidae* is not to be understood in the Darwinian sense. The theory of selection shows us the external reason why the better adapted forms survived, whilst others less capable of existence died out, but it cannot suggest any internal reason for the origin of these beneficial modifications and their regular and progressive development. If the Diptera ancestors of these curious creatures had possessed no interior capacity for adaptation to a new mode of existence, they could never have become *Termitoxeniidae*, and the termites would never have enjoyed the company of these pretty and interesting guests.

Unless we believe in the occurrence of variations with a definite aim among the chromosomes of the germ-plasm, it is simply impossible to explain the complete and thorough changes in the whole organism, mode of propagation and development, that take place in these tiny termitophile Diptera.

11. The History of Slavery amongst Ants

Slavery is an ominous word when used in the history of mankind ; it is a little word, but it conveys the idea of boundless injustice and cruelty, of misery and degradation. But when used with reference to ants the meaning of the word is different, and if we study the subject we gain an insight into these creatures' wonderful instinct, and are filled, not with horror and indignation, but with astonishment and admiration.

In the foregoing sections of this chapter I have reviewed a number of beetles and flies, living as inquilines amongst ants, and have shown that our present systematic species, genera, and sometimes also families of these inquilines must be regarded as the result of phylogenetic adaptation to the myrmecophile or termitophile existence.

Let us now consider an example which ought to throw some light upon the phylogenetic evolution of the instincts.

Previous articles published in *Stimmen aus Maria-Laach* [1] have made my readers familiar with the fact that among ants some are slave-holders, which steal the workers of other species as pupae, carry them to their own nests, and there bring them up to work for them. That the red robber-ant (*Formica sanguinea*) and the red Amazon ant (*Polyergus rufescens*) behave thus, has been known in Europe for the last hundred years, ever since Peter Huber published his classical studies; and later observations have considerably enlarged Huber's discoveries, and have extended them to the American connexions of our robber-ants.[2]

[1] 'Aus dem Leben einer Ameise' (XXXI, 1886, 413–741); 'Die Lebensbeziehungen der Ameise' (XXXVII, 1889).

[2] The chief works on this subject are: Pierre Huber, *Recherches sur les mœurs des fourmis indigènes*, 1810, nouvelle edit., Geneva, 1861. J. Hagens, 'Über Ameisen mit gemischten Kolonien' (*Berl. Entomol. Zeitschr.*, XI, 1867, 101–108). Aug. Forel, *Les fourmis de la Suisse*, Bâle, &c., 1874; 'Études myrmécologiques; Miscellanea myrmécologiques,' I (*Strongylognathus Christophori*), (*Revue Suisse de Zoologie*, XII (1904), 1–52); 'Sklaverei, Symbiose und Schmarotzertum bei Ameisen' (*Mitteilungen der Schweiz. Entomol. Gesellschaft*, XI, 1905, Part 2, 85–89); 'Miscellanea myrmécologiques,' II (*Annales de la Société Entomologique de Belgique*, XLIX, 1905, 191, &c.) (*Wheeleria Santschii*); 'Mœurs des fourmis parasites des genres *Wheeleria* et *Bothriomyrmex*' (*Revue Suisse de Zoologie* XIV, 1906, fasc. 1, 51–69). John Lubbock (Lord Avebury), *Ants, Bees and Wasps*, London, 1904. H. C. McCook, 'The shining slavemaker (*Polyergus lucidus*)' (*Proceed. Acad. Nat. Sci.*, Philadelphia, 1880, 376–384). Gottfr. Adlerz, *Myrmecologiska studier*, II, Stockholm, 1886, and III, Stockholm, 1896. Ch. Janet, *Conférence sur les fourmis*, Paris, 1906 (pp. 27–28 on *Anergates*); *Rapports des animaux myrmécophiles avec les fourmis*, Limoges, 1897 (p. 57 on *Anergates*). M. Ruzsky, 'Neue Ameisen aus Russland' (*Zoologische Jahrbücher Abt. für Systematik*, XVII, 1902, 469–484), (*Myrmoxenus*); 'Die Ameisenfauna der Astrachanischen Kirghisensteppe' (*Horae Societatis Entomologicae Rossicae*, XXXVI, 1903, 1–25, published separately). E. Wasmann, *Die zusammengesetzten Nester und gemischten Kolonien der Ameisen*, Münster, 1891; *Vergleichende Studien über das Seelenleben der Ameisen und der höheren Tiere*, Freiburg i. B., 1900; 'Neues über die zusammengesetzen Nester und gemischten Kolonien der Ameisen' (*Allgemeine Zeitschrift für Entomologie*, 1901, 1902); 'Ursprung und Entwicklung der Sklaverei bei den Ameisen' (*Biolog. Zentralblatt*, XXV. 1905, Parts 4–9, Supplement in Part 19, pp. 644–653); 'Wie gründen die Ameisen neue Kolonien?' (Paper read in the natural science section of the Görresgesellschaft at Bonn, on September 27, 1906, published in the Wissenschaftliche Beilage to the *Germania*, No. 44, November 1). W. M. Wheeler, 'The compound and mixed nests of American ants' (*American Naturalist*, XXXV, 1901, Nos. 414, 415, 417, 418); 'Three new genera of inquiline ants from Utah and Colorado' (*Bullet. American Museum of Nat. History*, XX. 1904, 1–17); 'A new type of social parasitism among ants' (*Bullet. American Museum of Natural History*,

2 c 2

Let us imagine that on a hot July afternoon we are standing beside a little mound in the grass, containing a nest of Amazon ants (*Polyergus rufescens*) with their slaves (*Formica rufibarbis*).[1]

A few minutes ago only reddish grey slaves [2] were running busily about the entrances to the nests, occupied with making earth-works, or were coming home laden with honey after a visit to the aphides, or were dragging dead insects into the nests as their booty, but suddenly the scene has changed. A number of large red Amazon ants have come out on to the surface of the nests. They hurry to and fro, clean their heads and antennae hastily with their fore feet, and the rest of their bodies with their middle and hind feet, and in doing so they make comical leaps, and even turn head over heels. Then they spring at one another, and strike one another on the head with their antennae. Now they are ready for their warlike expedition. Some Amazons take the lead, and are followed by a whole army of several hundreds or thousands, all in rapid march. Like a long red snake the robber band marches in a narrow line, scarcely broader than a hand, straight upon a nest belonging to their slave species (*Formica rufibarbis*), some thirty yards away. Tidings of their approach have already been brought, but too late; a desperate resistance and an attempt to barricade the entrances are of no avail. The Amazons quickly make their way into the nest and seize the pupae, killing only such opponents as continue to offer resistance or refuse to loose their hold upon the pupae that they are trying to save. With one bite the Amazon can drive its sharp, sabre-like jaws through an enemy's head and pierce to the

XX, 1904, 347–375); 'An interpretation of the slave-making instincts in ants' (*Bullet. American Museum of Nat. Hist.* XXI, 1905, 1–16); 'On the founding of colonies by queen ants, with special reference to the parasitic and slave-making species' (*Bullet. American Museum of Nat. Hist.* XXII, 1906, 33–105). K. Escherich, *Die Ameise, Schilderung ihrer Lebensweise*, Brunswick, 1906, 145–155.

[1] *Polyergus rufescens* has as slaves either *Formica fusca* or *F. rufibarbis*, but very seldom both at once. Near Exaten in Dutch Limburg I have always found *F. fusca* as slaves, but near Mariaschein in Bohemia, near Vienna in Austria, and in Luxemburg I have found only *F. rufibarbis*. The above description refers to a day in July, 1892, when I was making some observations in Lainz, near Vienna. In Switzerland Forel found both *fusca* and *rufibarbis* living as slaves with *Polyergus*, but only in one instance in the same colony.

[2] *Formica rufibarbis* is grey, with some red in the middle of its body. It varies, however, very much in colour, for which reason I have described it simply as a reddish grey ant, to distinguish it from the greyish black *Formica fusca*.

brain. In a few minutes the troop of red robbers emerges from the plundered nest; each Amazón is carrying in her mouth an ant-cocoon, containing a pupa. The procession returns to the robbers' nest, though not with such speed and discipline as were displayed when they were marching to the attack. The stolen pupae are adopted by the ants of the same species, who are already slaves, and are brought up in the Amazons' nest. When they develop, the ants, though born in a robbers' nest, follow their own innate instincts as if they were at home; there is no compulsion, no tyranny on the part of their masters. The whole 'slavery' consists in the fact that the service, otherwise performed with a view to the preservation of their own species, now benefits a race of strangers. They not only attend to their young, but clean and feed the Amazons themselves, for in their own home the latter are such helpless creatures that they have forgotten even how to feed themselves! Thus, in the slave-making instinct of the Amazons there is a cheerful as well as a gloomy side, in fact the latter is the inevitable result of the former. Just as the sabre-like jaw of the Amazon ant is an excellent weapon in fighting, but quite useless for domestic work, so their talent for warfare has been highly developed at the cost of losing their normal instinct for self-preservation.

My object in laying this description before my readers is not to amuse them with a highly coloured picture of the raids of the slave-making ants, nor to discuss the psychological value of their instinct,[1] but rather to give a historical account of slavery among ants, an account which includes in broad outlines all the phenomena that have been observed, and traces both the origin of slavery and its development from the simplest beginnings to its fullest perfection, and thence to its lowest parasitical degeneration. The records that supply the materials for this history are not written in any volumes, but on the pages of the living book of nature; they are biological facts, that we must carefully compare and cautiously combine, in order to learn from them the history of the slave-making instinct, which has been developing from the early Tertiary

[1] On this subject see *Die zusammengesetzten Nester*, &c., section 3, chapter i; also 'Die psychischen Fähigkeiten der Ameisen' (*Zoologica*, Part 26, Stuttgart, 1899); *Vergleichende Studien über das Seelenleben der Ameisen*, chapter ii; *Instinkt und Intelligenz im Tierreich*, 1905, chapters viii, ix.

period to the present time. As long ago as the Miocene epoch in the middle of the Tertiary period, there were a great many genera and species of ants, resembling those of the present day in their caste system and social organisation; but forms representing our slave-making ants have so far not been discovered either among the fossils of Radoboj in Croatia, or among those found in amber from the Baltic and Sicily.[1]

We therefore cannot say exactly when the slave-making instinct arose among ants, but as Europe, Asia, and North America all possess, in common, several slave-making genera and species of ants, differing only slightly in the development of their instincts, we are led to the conclusion that at the end of the Tertiary period, when the great continents of the northern hemisphere were finally separated, the slave-making instinct was already present, although it may have developed further after their separation. We have nothing to rely upon but the phenomena of comparative biology, when we attempt to search into the manner in which slavery originated, and to find out through what stages of evolution it has passed. It is obvious therefore that the history of slavery, as sketched here, is of a hypothetical character. It is a biological hypothesis, but one that is based upon a solid foundation of facts, and offers us a very natural explanation of them.

Fifteen years ago, in the last chapter of my work, 'Die zusammengesetzten Nester und gemischten Kolonien der Ameisen,' I discussed the origin and growth of the slave-making instinct, and said that the problem seemed to be insoluble. Observations made in the last few years in Europe, North America, and the north of Africa have, however, revealed a number of facts throwing considerable light upon the matter, and bringing us at least a step nearer to the solution.

(*a*) *Survey of the Biological Facts connected with Slavery*

The biological material that we have to take into account consists of the following nine chief groups of facts.

[1] Cf. G. Mayr, 'Vorläufige Studien über die Radoboj-Formiciden' (*Jahrbuch der k. k. geolog. Reichsanstalt*, XVII, 1867, Part 1); also by the same author, *Die Ameisen des baltischen Bernsteins*, Königsberg, 1868; C. Emery, 'Le Formiche dell' ambra siciliana' (*Memorie d. Reale Accad. d. Scienze*, Bologna, 1891, ser. 5, vol. I.).

1. A very great majority of the 4000 species of ants hitherto described form new colonies thus: After the copulation flight single impregnated females settle down alone, and independently, without the assistance of strangers, bring up their first brood. Among our native ants that found colonies in this way are the greyish black *Formica fusca* and its relative *F. rufibarbis*. All these species, if not interfered with,[1] live in simple colonies, i.e. in such as contain only ants of one species. This method of founding colonies is undoubtedly the oldest and most primitive.

2. There are certain species, particularly in the genus *Formica*, of which the impregnated females after the copulation flight cannot establish the new colonies alone, but need the assistance of workers. Within this group we must distinguish two forms of colonisation; in one, the workers belong to the same species (2*a*), and in the other to a different one (2*b*).

2*a*. Our red-backed wood-ants (*F. rufa*) and the black-backed meadow-ants (*F. pratensis*) have many large and populous nests, because their method of constructing their nests out of dead vegetable matter is very well suited to circumstances of life in cold climates. Old colonies of these ants consequently occupy a large district, which may include several thousand square yards round the nests, and is traversed by ant-tracks in various directions.[2]

If an impregnated female of such a colony alights after the copulation flight on ground within this district, she has no difficulty in finding workers of her own species, who either take her back to her own nest, or proceed to establish with her a fresh branch nest of the colony. This explains why the queens of the species and subspecies belonging to the same group as *F. rufa*, both in the Old and the New World, have lost the instinct and ability to found new colonies independently and alone. But what happens if a queen after her copulation flight meets no workers of her own colony or of

[1] I say: 'if not interfered with,' because *F. fusca* and *rufibarbis* are stolen as pupae by the slave-making ants, and so come to form mixed colonies with them; also because a colony, consisting normally of ants all of the same species, may shelter guests or outcasts of other species.

[2] For an account of giant nests and colonies of *F. rufa* see my *Ursprung und Entwicklung der Sklaverei*, pp. 196, &c.

her own species to assist her? If she is not to perish, she will have to ask a shelter of the workers of some other species. That this actually occurs in the case of our wood-ants (*F. rufa*) I discovered near Luxemburg in the spring of 1906. I found two recently established *rufa* colonies, one of which contained only the queen of *F. rufa* and several workers of *F. fusca;* whilst the other had a *rufa* queen, and over a hundred workers of both species. It seems that *F. rufa* and *F. pratensis* seldom form joint colonies, but *F. exsecta* and the North American *exsectoides* form them more frequently.[1] There are therefore a number of transitions between this group (2*a*) and the following (2*b*).

2*b*. The impregnated females of some comparatively rare species of sporadic occurrence, belonging to the *rufa* group, regularly found their new colonies with the help of the workers of another species of *Formica*, for they make their way into small nests, and force the occupants to accept them as queens. They can do this easily in colonies which have lost their own queen by death. Near Luxemburg I have observed new colonies founded in this way by the red *F. truncicola* with workers of *F. fusca*.[2]

I did not only discover several mixed colonies of *truncicola* and *fusca* at different stages of development, but I was able to prove by actual experiment that a female *truncicola*, wandering about after the copulation flight, was adopted as queen in a *fusca* colony where there was no queen. Wheeler has observed the establishment of new colonies of *F. consocians* in North America, when a female *consocians* has found her way into a nest of *F. incerta* and has been welcomed there.[3]

In the case of other allied species of *Formica*, especially such as, like *consocians*, have remarkably small females (e.g. *F. microgyna, nepticula, impexa,* and *montigena*), Wheeler has shown that in all probability they always establish their colonies with the help of workers of another species. Even among the Myrmicinae of North America there is a species (*Stenamma tenesseense*) which is in the habit of allying itself

[1] I found a colony of *F. exsecta* mixed with *fusca* near Luxemburg, in October 1906.

[2] *Ursprung und Entwicklung der Sklaverei*, pp. 126-131; supplement (Part 19), pp. 650, &c.

[3] A new type of social parasitism.

with the workers of a closely related form (*St. fulvum*), when it establishes new settlements.

The consequence of the adoption of a strange queen by workers of another species is the formation of a so-called 'Adoption Colony.' After the queen's first brood has been reared by the workers, the colony contains workers of two distinct species, and thus becomes temporarily a mixed colony; but after three years the last of the workers who originally adopted the strange queen dies,[1] and the *truncicola* or *consocians* colony again becomes an ordinary colony, containing only one species of ant, and as such it continues to grow, and in the course of ten years it may be inhabited by many thousands of ants.

This method of founding *truncicola* colonies leaves, however, some trace upon the class of workers, the first three generations of whom are reared by *F. fusca*, and the last of these *truncicola* may be still alive in the sixth year after the colony was founded. After the *F. fusca* have all died out, the *truncicola* workers retain a tendency to rear the pupae of *fusca*, although they devour those of other kindred species, or kill the newly developed ants as strangers. This remarkable instinctive preference shown by *F. truncicola* for the *fusca* pupae is due to the fact that the two species once lived together in one colony, as I discovered in 1904 from experiments with a young colony of *F. truncicola* that I kept in a room,[2] and the results were confirmed in 1906 by other experiments with an old colony of the same ants.

Let us now suppose *F. truncicola* to be an ant living chiefly on stolen pupae of other ants. What kind of selection would it make among the pupae stolen under natural circumstances? It would rear the pupae of that species alone by which it had itself been reared, viz. *F. fusca*. We should thus have the identical circumstances which actually prevail in the case of the red robber-ant (*F. sanguinea*). This explains how their extraordinary instinctive choice of the *fusca* pupae may have originated; for *F. sanguinea* also as a rule establishes new colonies with the help of *fusca* workers.

[1] Numerous observations and experiments that I have made, show that the workers of *Formica* live from two to three years.

[2] *Ursprung und Entwicklung der Sklaverei*, pp. 125, 167.

Let us now return to our survey of the biological materials for a history of slavery amongst ants.

3. Among the relatives of *F. truncicola* there are several species whose queens found colonies in the same way as those described as group 2*b*—namely with the help of workers of another species—but the colonies remain permanently 'mixed,' for as soon as the 'primary' assistants die out, they are replaced by 'secondary' workers of the same species as those who participated in the foundation of the colony;

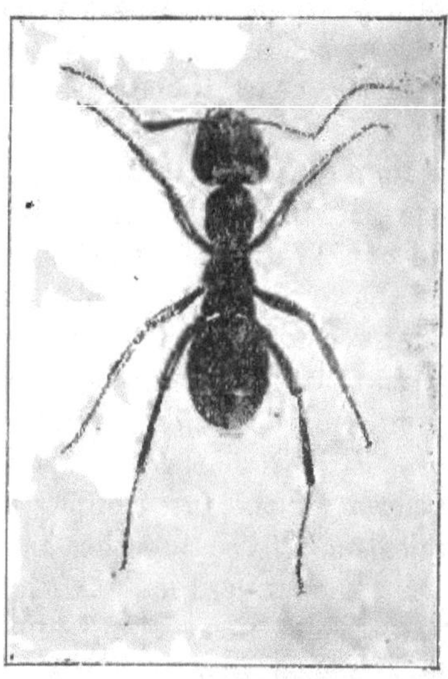

FIG. 45.—Worker of the blood-red robber-ant. (*Formica sanguinea*) (3½ times the natural size).

these secondary workers being obtained by means of a slave-making raid. The red robber-ant, *F. sanguinea*, in Europe and Asia forms mixed colonies of this kind, and so do the North American subspecies *F. rubicunda, subintegra*, &c. It is probable that other North American species, *F. dakotensis* var. *Wasmanni* and *F. Pergandei*, also belong to this class.

My observations, carried on during twenty years, show that the typical *F. sanguinea* (fig. 45) has slaves in almost all its colonies; these slaves mostly are *F. fusca*, seldom *rufibarbis*, and still more rarely both species are mixed.[1]

[1] I have suggested an explanation of the presence of two kinds of slaves in one colony in *Ursprung und Entwicklung der Sklaverei*, p. 209.

Only the most populous colonies—on an average one in forty—contain no slaves, because the assistance of strangers is not required in them. Wheeler thinks that the North American red robber-ant *F. rubicunda* possesses the slave-making instinct in a rather lower degree, and his opinion has recently been confirmed by H. Muckermann, S.J.[1]

Near Prairie du Chien (Wisconsin) Muckermann examined eleven colonies of these ants, and found six containing slaves, mostly *F. subsericea* (which is, like *F. subaenescens*, a variety of our *F. fusca*), and less frequently *F. nitidiventris* or *subaenescens*, whilst the other five nests contained no slaves. The instinctive desire to steal fresh slaves and bring them up seems therefore to cease earlier in this North American robber-ant than in our European *F. sanguinea* ; perhaps it dies out as soon as the colonies have attained a certain size. Forel described a subspecies of robber-ant in Canada, which he believed to have no slaves at all, and named for that reason *aserva* or slaveless. In the United States, however, Wheeler examined eight or nine colonies of these ants and found one containing a few slaves. It is certain that this *sanguinea* subspecies has slaves much less often than its nearest relative *F. rubicunda*. The North American varieties of the blood-red robber-ant still at the present day represent the transitional stages leading to the highly developed slave-making instinct possessed by our European and Asiatic *F. sanguinea*.

The North American *F. dakotensis* occupies a peculiar position midway between the species of the *rufa* group and those of the *sanguinea* group. In biological respects also it greatly resembles the latter. According to the careful observations made by Muckermann and Wolff, S.J., at Prairie du Chien, in thirteen colonies of *F. dakotensis* var. *Wasmanni*, five contained no slaves, the remaining eight had *F. subsericea* as their assistants. All the colonies of *Wasmanni* in which slaves were found were on the left bank of the Mississippi, and all in which there were no slaves were on the right bank. It was not possible to determine which of the former were still adoption-colonies, and which were robber-colonies, that had supplied their need of workers by stealing pupae of another species.

[1] Cf. *Biolog. Zentralblatt*, 1905, No. 19, pp. 651, &c.

In their origin all the slave-keeping colonies mentioned under 3, both those of the red robber-ants in Europe and North America and those of *F. dakotensis* var. *Wasmanni*, are adoption-colonies, arising out of the association of an impregnated female belonging to the ruling species with workers of the auxiliary species.[1]

They differ from the colonies mentioned under *2b* only in becoming subsequently robber-colonies, as the workers of the ruling species procure new assistants of the same species as those which originally helped to form the colony—fresh slaves being obtained by raids, when the first die out, as long as slaves are needed at all to strengthen the settlement. Mixed adoption-colonies are only of a temporary nature, and last but three years, then they give place to more or less permanently mixed colonies of slave-making ants.

In both the temporarily and in the permanently mixed colonies, a new colony is founded by an impregnated female, who makes her way into a nest belonging to another species, and takes up her abode there, whether the workers receive her willingly and promptly, or whether they are forced to accept her against their will and after much hostility. There are a great many different degrees between a peaceful reception and a violent intrusion.[2]

[1] As our *F. sanguinea* has often several nests belonging to one colony, it frequently happens that an impregnated female after the copulation flight forms a new nest with the help of workers belonging to the same colony. In this case there is no new colony, but a new branch of the same colony, as we saw with regard to *F. rufa* and *pratensis* (*2a*). Cf. also *Ursprung und Entwicklung der Sklaverei*, 1905, p. 201.

[2] In 1905 and 1906 Wheeler made experiments with a North American subspecies of the red robber-ant, F. *rubicunda*, and expressed the opinion that the impregnated queen after the copulation flight forces her way into a nest of slave ants, kills or drives out the workers, and takes possession of their pupae, which she brings up as her first assistants. But as Wheeler's experiments were all made with unimpregnated females which he had taken out of *rubicunda* nests, his observations afford no evidence of the manner in which *rubicunda* colonies are formed. According to my experience with European red robber-ants, the young unimpregnated females are very quarrelsome, and occasionally take part in carrying the larvae and pupae in the observation-nests. They have, therefore, characteristics of workers, which are absent in the impregnated females. Moreover, an impregnated female, who may become the foundress of a new colony, has a far better chance of being made welcome in a colony without a queen than an unimpregnated. Experiments with the latter cannot refute the adoption hypothesis. In the case of another North American subspecies of the same robber-ant, *F. subintegra*, Wheeler himself thinks we must believe that it may find admission into weak colonies of the slave species,

There are also many grades between the adoption of a queen belonging to the ruling species in a weak colony of the auxiliary species, and the alliance of a queen of the former species with a queen of the latter, who is engaged in founding a nest and has no adult workers.[1]

The ants belonging to group 3 are distinguished from those of the next group (4), which also live in permanently mixed colonies, by the fact that the masters are still essentially independent of the slaves, and only begin to keep slaves when their colony has attained considerable strength. Among our European *F. sanguinea* this occurs seldom, and only in the most populous colonies, but it is of much earlier and more frequent occurrence among the North American representatives of the species. The latter form a natural link between group 3

F. subsericea. Of course one queen by herself of whatever species—*sanguinea* or *truncicola* or *consocians* or *rufa*—can find admission to the nests of a slave species only under exceptionally favourable circumstances. These conditions seem particularly difficult in the case of *F. sanguinea*, but even here there are undoubtedly many grades between voluntary and compulsory admission, and complete failure to obtain it. During the summer of 1906 I made experiments with fifteen young queens of *sanguinea*, caught directly after their copulation flight, with a view to observing their reception among *F. fusca*, *pratensis*, &c. The result was a confirmation of my adoption theory and a refutation of Wheeler's raid hypothesis. Full details of these experiments will be published elsewhere, as well as of others with *rufa* and *pratensis* queens. In 1904 Emery suggested that new *Polyergus* colonies might be formed if a queen forced her way into a slaves' nest, took possession of their pupae, and reared them as her assistants. As even the *Polyergus* workers have lost their instinct to rear their own young, it seems that this hypothesis is still less probable in their case than in that of *Formica sanguinea.* No such proceeding has been observed under natural conditions on the part of either *Polyergus* or *Formica*, but only on that of *Tomognathus* (by Adlerz). The last-named genus belongs to quite a different subfamily of ants, and their whole slave-making instinct is completely unlike that of the other two genera. (Cf. Group 5.)

[1] Forel's 'Allometrosis' (alliance between queens of different species) supplies at least a possible way of accounting for the origin of mixed colonies of *Formica.* On June 6, 1906, at Ösling, near Hoscheid in Luxemburg, I discovered under one stone a queen of *F. pratensis* (var. *truncicolo-pratensis*) and a queen of *F. rufibarbis* close together, and I took them both away with me. I put the *pratensis* queen in a nest with thirty workers of the *F. fusca*, who for several days pulled her about and ill-treated her. In order to save her life, I took her out, and put her into a nest, where I had the *rufibarbis* queen under observation. She at once approached the latter, began to stroke her with her antennae and to ask for food, as if they were friends. This scene was repeated several times during the day, and on the following day they sat close together, but on the third day the *rufibarbis* queen died of exhaustion, whilst the *pratensis* queen had completely recovered. These observations seem to show that when two queens form an alliance, the one belonging to the auxiliary species may be got rid of in a peaceful manner, after she has reared her first brood of workers so far as to be of service to the other queen in founding her colony.

and the preceding group (2*b*), in which no fresh slaves are procured after the ants that originally helped to found the colony have died out.

4. Closely allied with the genus *Formica*, though differing from it in some important points, is the very interesting genus *Polyergus*, with which we have already made acquaintance (p. 387). The chief difference is in the upper jaw, which

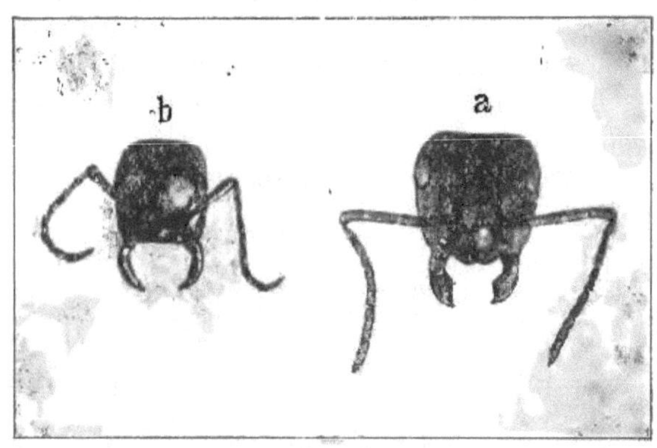

Fig. 46.

a. Head of the blood-red robber-ant (*Formica sanguinea*).
b. Head of the red Amazon-ant (*Polyergus rufescens*).
(6 times the natural size.)

in the Amazon ants (*Polyergus*) is shaped like a sabre, and has no indentations. (Cf. fig. 46*b* with 46*a*.) *Polyergus rufescens* is the European representative of this genus. Fig. 47 shows the ergatoid queen[1] and fig. 48 the worker of the European Amazon. There are four other subspecies in North America, *Polyergus lucidus, breviceps, bicolor*, and *mexicanus*.[2] In their habits they resemble the European Amazon, although closer

[1] The ergatoid queen of *Polyergus* is a real queen in the dress of a worker. She resembles the workers, however, only in the structure of the abdomen and in being wingless. Cf. 'Die ergatogynen Formen bei den Ameisen und ihre Erklärung' (*Biolog. Zentralblatt*, 1895, Nos. 16 and 17, pp. 606, &c.).

[2] Of these subspecies *bicolor* resembles *F. sanguinea* in colouring, and has rather broader jaws than our *P. rufescens*. The genus *Polyergus* is marked off from *Formica* at the present time by a definite morphological distinction, which will be explained in the second part of our examination of the two genera. The slaves of the North American *Polyergus* belong to various species and subspecies of the groups to which *F. fusca* and *pallide-fulva* belong.

investigation may perhaps reveal many instructive degrees in the development of the slave-making instinct in their case, as in the subspecies of *F. sanguinea*. The mode of life of *P. rufescens* is that of which most is known at present, so, in considering the biological characteristics of this group (4), we must limit ourselves to it.

The mixed colonies formed by the Amazons and their

FIG. 47. FIG. 48.

FIG. 47.—Ergatoid queen of the Amazon ant (*Polyergus rufescens*) (3½ times the natural size).

FIG. 48.—Worker of the Amazon ant (*Polyergus rufescens*) (3½ times the natural size).

slaves do not differ essentially from those of the two previous groups in the manner of their formation. They are at first adoption-colonies, resulting from the association of an impregnated female of the ruling species with workers of a definite slave species;[1] and the Amazons subsequently rob the nests of this same species, when they make raids to obtain fresh slaves. Colonies of *Polyergus*, founded with the help of *F. fusca*, afterwards rob *fusca* nests; those founded with the help of *F. rufibarbis* choose *rufibarbis* nests as the normal goal of their slave hunts. In 'Les Fourmis de la Suisse' (pp. 287, &c.),

[1] The colony mentioned by Forel (*Les fourmis de la Suisse*, p. 302, No. 18), in which *Polyergus* and *rufibarbis* were found together, was probably an adoption-colony of this kind. Experiments made by both Forel and myself with isolated *Polyergus* queens have shown that they obtain admission with comparative ease to the nests of the slave ants. (Cf. *Die zusammengesetzten Nester*, 1891, pp. 84–87.)

Forel describes his observations of Amazons with *fusca* and *rufibarbis* slaves, and says that in the same region their military tactics varied somewhat according to the species kept as slaves. Amazons with *rufibarbis* slaves march more quickly and in more regular lines, as they have to attack enemies better able to offer resistance than the *fusca*.

The robber-colonies of the Amazon ants differ, however, from those of the red robber-ants and their relatives in one important respect, viz. that the masters are absolutely dependent upon their slaves. The Amazons can steal slaves, but they are incapable of working and of leading a normal, independent existence. Their helplessness finds expression in their sabre-shaped jaw and in the decay of their domestic instincts, which is carried to such a point that they will starve with food before them, if they have no slaves to put it into their mouths. The absolute dependence of the Amazons upon their slaves is shown also in the biological fact that in their nests the number of slaves is very great in comparison with that of the masters, the slaves being often from five to ten times as numerous. The Amazons steal as many slaves as they *can*, the red robber-ants only as many as they *need* to supply the deficiency in their own numbers. This is the reason why among the Amazons the number of slaves is proportionate to that of the masters in the same nest, whilst in the nests of the red robber-ants the numbers are in inverse ratio ; in the former case—the more masters, the more slaves ; in the latter—the more masters, the fewer slaves.[1] With the genus *Polyergus* we reach the end of the evolution of the slave-making instinct in the subfamily of Formicinae (Camponotinae). It culminates psychologically and morphologically in the Amazons with their sabre-like jaws and their talent for war, but there are in them unmistakable signs of degeneration. We shall be able to trace the different stages leading down to social parasitism, when we compare them with the slave-robbers of the genus *Strongylognathus* among the Myrmicinae, in the sixth and seventh parts of our investigation.

Let us now turn to the subfamily of the Myrmicinae.

5. A quite peculiar and altogether isolated position among slave-making ants is occupied by the northern genus *Tomo-*

[1] *Vergleichende Studien über das Seelenleben*, &c., p. 52.

gnathus, which occurs both in Northern Europe and in North America, and takes its slaves from the closely allied genus *Leptothorax*.[1] The former genus differs from the latter chiefly in having broad mandibles devoid of labium and not indented, whence its name *Tomognathus* (cutting jaw), also it is much larger than any *Leptothorax*. According to observations made by Adlerz in Sweden, the females of *Tomognathus sublaevis*, that bear an extraordinary likeness to the workers, and have no wings, make their way into the nests of *Leptothorax acervorum* or *muscorum*, drive out the inhabitants, and take possession of the young brood, which they rear in the stolen nests. In this way mixed colonies of *Tomognathus* and *Leptothorax* are formed, in which winged specimens of the slave species often occur, and so these colonies differ remarkably from the other robber-colonies of slave-making ants.[2] *Tomognathus* is probably descended phylogenetically from the genus *Leptothorax*, to which its slaves belong. The evolution of its slave-making instinct is therefore certainly not connected genetically with that of *Formica* and *Polyergus*, and perhaps it is not connected with that of the following groups.[3]

6. The Myrmicine genus *Strongylognathus* is a miniature reproduction of the Amazons of the *Polyergus* genus described under group 4. The sabre-shaped jaw accounts for the name sabre or scimitar-ants, which has been given to this genus.

It occurs near the Mediterranean and in western Asia; only one species, *Strongylognathus testaceus*, is found in Central Europe, and the genus is not represented in the north. We are not now concerned with the Central European species, as it does not make slave raids, but we must examine its southern connexions, *Str. Huberi* in the south of Europe and the north of Africa, and *Str. Christophi* in the districts east of the Mediterranean and on the Kirghiz steppes.

[1] *Tomognathus sublaevis* was found not long ago by Viehmeyer in Saxony. It belongs, therefore, to the fauna of Germany.

[2] Near Exaten in Holland, in nests of *F. sanguinea*, I have occasionally, but very rarely, found one or two winged females of *fusca*. (In colonies No. 55 and No. 235, i.e. in two out of 410 colonies.)

[3] It is still doubtful whether the genus *Myrmoxenus*, discovered by Ruzsky on the steppes of the south-east of Russia, is connected with *Tomognathus* or *Strongylognathus*. *Myrmoxenus Gordiagini* always forms mixed colonies with *Leptothorax serviusculus*. We do not yet possess any detailed observations of their way of life.

Like the Amazons, these two species of scimitar-ants are superior to their slaves in size and strength, their colonies also contain a considerable number of both masters and slaves, and they too organise regular slave-hunts, which are, however, in the case of the scimitar-ants directed against the nests of the little turf-ants (*Tetramorium caespitum*). All the species of *Strongylognathus* have some species of *Tetramorium* as slaves in their mixed colonies.

Forel found *Str. Huberi* in the south of the Valais and in Tunis, but he was not able to study their slave-hunts under natural conditions, though he did so in artificial nests. In the summer of 1904 Escherich sent me a colony of these ants from Fully in the Valais, and after I had brought them into contact with some large *Tetramorium* colonies in Luxemburg, they attacked the latter, put them to flight and carried off their pupae. That the same colony under its natural conditions in the Valais had acted in the same way was proved by the fact of its containing, when it reached me, two subspecies of *Tetramorium* slaves, one larger than the other. As in this species the workers of one colony are all of the same size, it follows that the original slaves of the scimitar-ant colony must have belonged to one only of these subspecies, and the other must have been introduced by a raid upon some *Tetramorium* nest of a different subspecies.

Not long ago Rehbinder observed the scimitar-ant of the south-east making a raid under natural conditions, near the monastery on Mount Athos. This ant is the *Str. Christophi* var. *Rehbinderi*, which is remarkable for its size. It is bigger and stronger than Huber's scimitar-ant, and better able to conquer the stubborn *Tetramorium* in a pitched battle. We may therefore assume that, in its development of the slave-making instinct, it stands on a level with *Polyergus*, whereas *Str. Huberi* probably ranks rather lower, owing to its inferior size and strength ; we may, however, regard it as a genuine slave-maker.

7. Let us now turn to the little yellow *Str. testaceus*, the northern relation of the former species, to which it is greatly inferior in size and strength, being no larger than the little turf-ant, with which it lives in mixed colonies.

I have studied its habits both when at liberty and when

living in nests that I had arranged for purposes of observation, in Holland, Bohemia, Luxemburg, and near the Rhine, and I have come to the conclusion that it is no longer capable of making slave-raids. But, if this is true, how can we account for the great number of slaves in its mixed colonies? The number of slaves is relatively much greater than in the *Str. Huberi* nests, although the average number of masters is smaller, often scarcely reaching a hundred, and seldom being more than a few hundreds. In the nests of *Str. testaceus* there are generally from five to ten times as many slaves as masters.

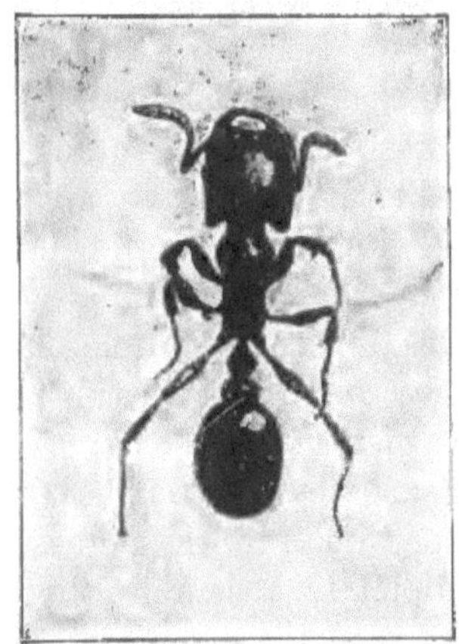

FIG. 49.—Yellow scimitar-ant (*Strongylognathus testaceus*) (12 times the natural size).

How does this little ant obtain its numerous slaves, if it is no longer capable of fighting successfully against the strong colonies of turf-ants, who have a hard chitinous covering and a dangerous sting? What is the solution of this problem?

I was able to suggest a probable solution after making a number of observations near Prague in Bohemia in 1890 and 1891. In two of the mixed *Strongylognathus* and *Tetramorium* colonies there, I found a queen of the latter species;[1] in one of these colonies there were even pupae of the winged males and females of both species. I drew the following inference from these facts.

[1] It is well known that the *Tetramorium* queens are very difficult to find even in independent colonies; it is possible to discover them only when the nest is in an exceptionally favourable situation.

The little yellow scimitar-ant does not obtain its slaves by stealing them, but after the copulation flight an impregnated female seeks the company of a *Tetramorium* queen, who has withdrawn under a stone in order to establish a new colony there. The young workers of *Tetramorium* thus rear the brood of the *Strongylognathus* queen as well as that of their own queen. As the *Strongylognathus* males and females are much smaller than those of *Tetramorium*, the latter show preference to their larvae and neglect their own, which require more food and attention. The little larvae of the workers of both species are reared with equal care. That the number of *Strongylognathus* workers is small in comparison with that of *Tetramorium* workers is best explained by the fact that this caste is no longer necessary to *Str. testaceus* for the preservation of its species, and is therefore gradually approaching extinction. In every colony the winged males and females of *Strongylognathus* are actually in the majority.

The mixed colonies, formed by our northern yellow scimitar-ant with the turf-ant, are therefore the result of an alliance between two queens of the different species. But were they not formerly robber-colonies? Otherwise what is the meaning of the scimitar-shaped mandibles possessed by these little ants, which so completely resemble those of the southern members of the same genus, as well as those of the Amazons and other ants in which the slave-making instinct is highly developed? Did not our little yellow ant once use these formidable mandibles, when it was itself bigger and stronger, to crush the hard head of an enemy in battle, as its relatives still do? What was the use of these peculiar weapons, if the scimitar-ants have always lived in peaceful alliance with the turf-ants? [1]

There can be no doubt that the ancestors of *Str. testaceus* used to steal their slaves, just as their larger kinsfolk still do in the south. The original home of the *Strongylognathus* genus is in Southern Europe, where four species occur—*Huberi, Christophi, Caeciliae,* and *afer*; our little yellow scimitar-ant is an isolated northern offshoot of this group; and the fact of its migration northward gives us a very simple clue to the reason

[1] This difficulty cannot be removed by a reference to the unnotched mandibles of the males of many genera of ants, for these mandibles are often small and weak, and not at all what we mean by scimitar-like.

why it has lost its slave-making instinct. All our slave-keeping ants without exception hunt their slaves only during the hottest hours in the summer months. If a southern slave-hunter migrated in a northerly direction, its slave-making instinct would be felt at longer and longer intervals until it finally died out altogether. This would be more likely to occur, if at the same time the size of the ant's body diminished, so that it gradually lost the power to seize its enemy's head and crush it between its jaws. In the *Strongylognathus* workers the instinct prompting them to steal the turf-ants' pupae to be their slaves gave place to an instinctive desire on the part of the impregnated females to ally themselves with turf-ant queens in order to establish colonies together.

But is this latter instinct something new in the history of *Strongylognathus* colonies? No, it is very old, for, as I have shown in discussing group 2*b*, it was the motive that led primarily to the development of the slave-making instinct. Even at the present time all robber-colonies of slave-keeping ants begin by being adoption-colonies, and owe their origin to impregnated females of the ruling species, who, having made their way to a weak nest of the subject species, obtain admission to it. It is only at a subsequent period that the descendants of these females procure fresh slaves by plundering the nests belonging to the species that originally co-operated in founding the colony.

The mixed colonies formed by the little yellow scimitar-ant and its *Tetramorium* slaves differ from those of its slave-keeping relatives[1] only in the fact that the queen of the slave species remains alive, and thus the masters are supplied with a constant succession of fresh slaves, but they are all born in the same nest, and the slaves are no longer stolen. The allied

[1] The distinction may be expressed in other words as follows: The queens of slave-keeping ants, when about to found new colonies, choose by preference those colonies of the slave species which have lost their own queen, or rather, as a rule, it is only to such colonies that they obtain admission. The *Strongylognathus testaceus* queen, however, finds admission most readily to a young colony of the slave species, which has only just been established, and where there are no full-grown workers with the queen. In both cases the instinct that prompts the queen to seek out a nest belonging to the slave species is due to the same causes, viz. to an impulse to force a way into a nest of strange ants, and to her sense of smell, which draws her to the nest of her normal slave species as being particularly attractive. All the other conditions of her reception depend, not on the instinct of the queen who seeks admission, but on that of the ants that grant it.

colonies of *Str. testaceus* and *Tetramorium caespitum* are characterised by their remaining permanently at a stage, which is only temporary in the case of the robber-colonies. The loss of the slave-making instinct has caused a reversion to an early stage, preceding the development of the slave-making instinct.

8. There are some remarkable ants that have no caste of workers, but live as parasites with other species. All these belong to the systematic subfamily of Myrmicinæ.

As long ago as 1874, in 'Les fourmis de la Suisse,' Forel mentioned some extraordinary males and females of a *Myrmica*,

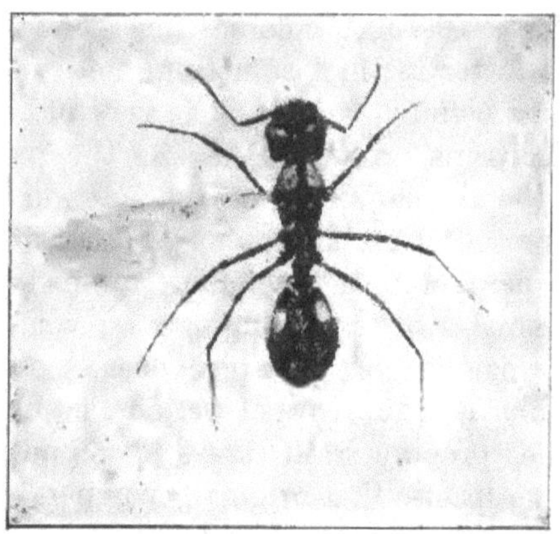

FIG. 50.—*Wheeleria Santschii* (female) (6 times the natural size).

which had been found in the Alps in a nest of *Myrmica lobicornis*, but which were totally unlike the males and females of this species. He proposed calling these peculiar creatures *Myrmica myrmicoxena*, and expressed the opinion that they lived as parasites in the colonies of other species of *Myrmica*.

It is still somewhat doubtful whether *Myrmica myrmicoxena* is really a parasitic species, but Santschi's observations in Tunis have recently revealed the existence of a very interesting parasitic ant in North Africa, described by Forel under the name of *Wheeleria Santschii*, which lives as a parasite in the nests of *Monomorium Salomonis* and its varieties.[1]

[1] Cf. Aug. Forel, *Miscellanea Entomologiques*, II, 1895, and *Mœurs des fourmis parasites*, 1906. See also list of works on p. 387, note 2.

The species consists only of males and females, both winged. The female (fig. 50)[1] loses her wings after impregnation, and then enters a *Monomorium* nest, whence at first she is often driven out; but she persists in returning and finally obtains admission. Thereupon the *Monomorium* workers, preferring to wait upon the new little queen than upon their old large one, kill the latter, and devote themselves to rearing the brood of their parasite. The *Wheeleria* males and females, brought up in a *Monomorium* nest, pair with one another within the nest, and then the impregnated females depart, in order to force their way into other colonies of *Monomorium* and to exact service from their occupants. The mixed colonies of *Wheeleria* and *Monomorium*, as Santschi's observations and experiments have shown very clearly, were originally adoption-colonies.

In North America there are three genera of the subfamily of Myrmicinæ, of which males and females are known, but no workers. According to Pergande and Wheeler's observations, they all live as parasites in the nests of other species belonging to the same subfamily. *Epoecus Pergandei* lives with *Monomorium minutum* var. *minimum*, *Sympheidole elecebra* with *Pheidole ceres*, *Epipheidole inquilina* with *Pheidole pilifera* var. *coloradensis*.

We must assume that the North African genus *Wheeleria*, and the three genera just mentioned of North American parasitic ants, formerly possessed a caste of workers, like all normal ants. History is silent as to the loss of this caste, but it must have been lost in one of two ways. Either the ants were once peaceful guests living with their hosts, as *Formicoxenus nitidulus* still does in Europe, and *Leptothorax Emersoni* and *Symmyrmica Chamberlaini* do in North America, or they used at one time to be robbers, stealing the pupae of their slave species, but subsequently forming permanent colonies in alliance with them, as *Strongylognathus testaceus* does with the turf-ants (group 7). We must leave it to future research to determine which of these two explanations is correct, but in either case the fact that these ants lived with others of a different species would render the preservation

[1] The illustration is from a photograph of a specimen kindly sent me by Santschi.

of their own workers superfluous. The disappearance of this caste has reduced them from the position of guests or masters to that of parasites, living upon their former hosts or slaves.

They have not, however, reached the lowest degree of parasitism, for their winged males and females are normal, although they already show peculiarities which may be regarded as indicating degeneration.[1]

9. Let us now return to European ants.

All over Central and Northern Europe there occurs—though it is very rare—a strangely degenerate little ant, *Anergates atratulus*, which lives in mixed colonies with workers

Fig. 51.—Male of *Anergates atratulus* (12 times the natural size).

of the turf-ant (*Tetramorium caespitum*). This genus is called 'worker-less' (*Anergates*) because it possesses no workers. The winged, black females are fairly normal, but when, as queens, they have lost their wings, their bodies gradually assume the circumference of a small pea, and they pass into a state known as physogastry.

The little yellow males are thoroughly degenerate; not only have they no wings, but in shape they resemble an ant pupa rather than an adult. As Adlerz and I have frequently observed, copulation takes place within the *Tetramorium* nest, and the impregnated females then fly away, in order to discover new *Tetramorium* colonies and obtain admission to them. That very few of the hundreds of females issuing

[1] Cf. Emery, 'Zur Kenntnis des Polymorphismus der Ameisen' (*Biolog. Zentralblatt*, 1906, No. 19, pp. 624–630).

from a colony of *Anergates* and *Tetramorium* succeed in this endeavour, is proved by the rarity of *Anergates*.

Various hypotheses have been brought forward and numerous experiments have been made, to account for the origin of mixed colonies of *Anergates* and *Tetramorium* at the present day, but no one as yet has succeeded in actually observing the establishment of such a colony. It is probable that these mixed colonies are adoption-colonies, like those mentioned in group 8. An impregnated *Anergates* female forces her way into a weak *Tetramorium* nest, where there is no queen, or into a branch nest of a larger colony, and is there adopted as queen. It may be that the turf-ants kill their own large queen after adopting the little parasite; if so, their action is analogous to that of *Wheeleria* as observed by Santschi (see p. 406).

It is difficult to say how such a colony can be maintained permanently unless fresh workers of the turf-ant are produced, for in the mixed colonies of *Anergates* and *Tetramorium* no queen and no worker pupae of the latter species have ever been discovered.[1]

Moreover, it is still uncertain whether such a colony lasts for over three years, and also whether the *Tetramorium* workers may not live longer than that time. As far as we know so far, we have to regard these colonies of *Anergates* and *Tetramorium* as adoption-colonies, that remain permanently at a stage which is only temporary in the case of colonies of slave-keeping ants.

Was *Anergates atratulus* always a parasite, possessing no workers? Were the males always wingless creatures, resembling pupae, and showing unmistakable marks of degeneration? Were these ants originally created in this state of absolute dependence upon their slaves, or are they descended from another genus, capable of an independent existence? It is impossible not to decide in favour of the latter alternative, although the history of *Anergates*, and the process which has led to its parasitical degeneration, are still very obscure.

[1] In 1904 I placed some worker pupae from other *Tetramorium* nests in a colony of *Anergates* and *Tetramorium*, but the slaves of the latter species would not rear them, and either ate them or threw them away.

Let us compare this parasitic ant without workers, with the little yellow scimitar-ant (group 7). Both live as parasites in the nests of the same turf-ant. *Str. testaceus* is not found in the north of Europe, although *A. atratulus* occurs there. The former has not yet lost its worker caste, but the workers are far less numerous than they are among the southern slave-keeping representatives of the genus, and *Anergates* has none at all. Let us imagine that some species of ant, at the same stage of development as *Str. testaceus*, penetrated northwards in some remote age, and the loss of activity and energy, due to the colder climate, led to degeneration in a creature coming from the south. The dependence of the masters upon their slaves would constantly increase until finally the workers of the former species died out, having ceased to be necessary for the preservation of the species. Thus there would exist between our ant and the turf-ant, with whom it lives, a relation similar to that now existing in North Africa between *Wheeleria* and *Monomorium*. The males and females of the parasitic ant would correspond to the normal winged males and females of other ants, as they do in *Wheeleria*, and if this genus sank into still deeper parasitical degeneration, they would finally resemble *Anergates*.

The *Anergates* males are so little able to move, being wingless and like pupae, that they cannot leave the nest, and thus many are saved from the destruction that overtakes most ants on the occasion of their copulation flight. The degeneration of the males becomes in this way a hindrance to the extinction of the species. On the other hand, the physogastry of the female increases her fecundity. Both peculiarities—absence of wings and resemblance to pupae on the part of the males, and physogastry on the part of the females—serve the same end, and indicate a last desperate attempt to preserve the species.

Need this hypothetical account of the evolution of *Anergates* be regarded as purely fantastic? No; for if we once allow that this parasitic ant was not created in its present degenerate condition, we have no choice but to admit that it has reached this condition by a retrograde evolution, produced by a series of either perceptible or imperceptible changes.

(b) *Inferences respecting the Development of the Slave-making Instinct*

We have now completed our survey of the biological facts relating to slavery among ants. What conclusions may we draw from these materials?

They have been clearly indicated in the foregoing pages. If we attempt to give a natural account of the origin of the conditions described in groups 1–9, and still actually existing, we cannot possibly avoid regarding them from the point of view of the evolution theory. It alone is able to give us a clue that will guide us to an understanding of the various phenomena.

Not only are the colonies of slave-keeping ants adoption-colonies in their origin, but they must phylogenetically be descended from similar colonies, mixed for a time only, such as we have considered in group 2*b*. This overthrows once for all Darwin's very ingenious, but unsuccessful attempt to account for the origin of slavery by assuming that the stolen pupae of strange species chanced to be reared by mere accident,[1] and it substitutes a much more probable and intelligible explanation.

The progressive development of the slave-making instinct must have passed through the phylogenetic stages presented to us at the present day by *Formica* and *Polyergus* in groups 3 and 4 respectively. After the culminating point was reached, retrogression must have set in, on the analogy of groups 5–9, and have led to the lowest depth of parasitism, after which nothing remains but the extinction of the species. We know from the evidence of palæontology that in the course of the world's history many thousands of species have perished, though few perhaps have had so easy a death as that which awaits *Anergates*, possibly after some thousands of years.

I may be asked whether we are to regard the history of the slave-making instinct in ants, illustrated by groups 1–9, as a uniform process of evolution, uniting the present

[1] A fuller proof of the futility of this theory may be found in chapter i of my 'Ursprung und Entwicklung der Sklaverei bei den Ameisen' (*Biolog. Zentralblatt*, 1905, Part 4).

representatives of these nine groups into one single genealogical line. This would mean that the present slave-stealing genus *Strongylognathus* (group 6) was descended from *Tomognathus* (group 5), and this again from the present Amazons of the genus *Polyergus* (group 4)! In fact, we ought to regard *Anergates* (group 9) as the descendant of a species of still-existent *Formica* belonging to group 2!

No thoughtful biologist has ever imagined, or ever will imagine, anything of the kind; for *Polyergus* and *Strongylognathus*, *Formica*, and *Anergates* belong to distinct subfamilies of ants, and cannot be closely related to one another. Another suggestion is, that within the same subfamily the present representatives of the successive biological groups may be directly descended from one another. Shall we, for instance, derive the Amazons of the present day from the red robber-ants of the present day, and these again from *F. truncicola* of the present day?

An attempt to do this would display complete misunderstanding of the process of evolution that I have described. I have suggested that our present Amazons once passed through a stage in the history of their race, resembling the present stage occupied by the red robber-ants, as far as the slave-making instinct is concerned. Also, I think that the red robber-ants, which now form with their slaves permanently mixed colonies maintained by slave-hunts, once passed through a stage resembling that at which *F. truncicola* has now arrived, when the mixed colonies were only temporary. We must view in a similar way the connexion between the other successive groups that we have considered as furnishing materials for an account of the growth of slavery amongst ants. This is plainly quite a different theory and is free from the objections mentioned above.

The development and growth of the slave-making instinct, from its simplest beginnings to the parasitical degeneration due to it, may be illustrated by the nine groups that we have considered, but I must again lay stress upon the fact that they do not form one single sequence in evolution, and are not descended directly from one another. The common historical origin of the whole family of ants, and their historical connexion with other families in the order of Hymenoptera is based upon other arguments, supplied by comparative morphology, and has nothing directly to do with our biological question.

Slave-keeping ants of the subfamily Formicinæ can phylogenetically be derived only from other Formicinæ, that formerly led an independent existence; and in the same way slave-keeping or parasitical Myrmicinæ can only be derived from other Myrmicinæ, that once led an independent existence. For this reason, at the close of my account of the fourth group, I drew attention to the fact that the development of slavery in the subfamily of the Formicinæ culminated with the Amazons of the genus *Polyergus*, and at the same time it reached its end, for in this subfamily slavery has not been further developed, and the representatives of the decay of slavery and its degeneration to the lowest social parasitism all belong to another subfamily—the Myrmicinæ. In their case, however, no trace remains among living Fauna of the first half of the process of development, leading up to the culminating point, and we can only supply it hypothetically on the analogy of groups 2–4, which belong to the Formicinæ. We may venture to say that the slave-making instinct possessed by the southern species of of *Strongylognathus* is very like that displayed by *Polyergus*, and probably passed through similar phylogenetic stages, such as we can still observe in the mixed colonies of *F. sanguinea* and in those of *F. truncicola*. In this way we may combine two different parts, so as to form one complete picture, the materials for one part being derived from the subfamily of the Formicinæ, and those for the other part from the subfamily of the Myrmicinæ, and thus they supplement one another, and we have a hypothetical history of the slave-making instinct among ants.

What bearing has this upon what is probably the actual history of slavery among ants? It shows that the instinct appeared in the Formicinæ in a geologically much later age than in the Myrmicinæ; for this is the reason for the absence, in the case of the Formicinæ, of all real evidence bearing upon the second half of the evolution of slavery, and in the case of the Myrmicinæ of all real evidence bearing upon the first part. Among the Formicinæ we still meet with many progressive and preparatory grades in the development of the slave-making instinct, which culminates in *Polyergus ;* among the Myrmicinæ there are almost exclusively descending grades, so that in this subfamily the instinct seems to pass from its culminating point down to complete parasitic degeneration in *Anergates.*

The instinct prompting ants to steal workers of other species to be their slaves developed at least twice in the history of ants, and its appearances occurred in different ages, within two different subfamilies, and quite independently one of the other.

But within these two subfamilies the history of the slave-making instinct is not one single line of evolution, but several lines, beginning among various genera and species, that originally led an independent existence; these lines having very various development and belonging to different periods.

At the conclusion of the fifth section of our biological survey, I pointed out that the robber-ants of the *Strongylognathus* genus (group 6) are probably not closely connected with those of the *Tomognathus* genus (group 7). The slave-making instinct seems to have developed in the ancestors of these two genera independently of one another, and later in the ancestors of the northern genus *Tomognathus* than in those of the southern *Strongylognathus.* Within the latter genus we find a uniform evolution, connecting the slave-making species of the south with the parasitical species of the north. Nevertheless, we must not assume that our present yellow scimitar-ant (*Str. testaceus,* group 7) is the direct descendant of its present relatives in Southern Europe (*Str. Huberi,* group 6), but rather of an extinct species, which formed the starting-point for the subsequent evolution of all our present species of *Strongylognathus*; the southern representatives of this stock became and are robbers, stealing their slaves, whereas the northern branch of the same stock has lost the slave-keeping instinct, and has degenerated into a parasitical condition.

Among the ancestors of *Anergates* (group 9) the slave-making instinct probably developed and perished much sooner than in *Strongylognathus* ; for the parasitic *A. atratulus,* that has no workers, shows the utmost degradation of the slave-making instinct, whilst *Str. testaceus* is still far removed from it. Even if the Tertiary ancestors of both these genera were identical, or very closely connected, we must nevertheless assume that, in the branch of the stock whence our *Anergates* is descended, the slave-making instinct began to develop at an earlier epoch of the Tertiary period than in the branch which gave rise to the present genus *Strongylognathus*.

The American parasitic ants belonging to the genera *Epoecus*, *Sympheidole*, and *Epipheidole* (group 8) represent theoretically a stage preceding that complete parasitic degeneration which we find in Europe in *Anergates* (group 9). But there is apparently no close connexion between the American and the European genera, and in all probability neither is closely connected with the North African genus *Wheeleria*, with which Santschi's recent observations in Tunis have made us fully acquainted.[1] It stands in the same sort of relation to *Monomorium* as *Anergates* to *Tetramorium*, but the *Wheeleria* males are normal and have wings, and are not degenerate creatures such as the pupa-like males of *Anergates*. In this way the gap that has hitherto existed in the fauna of the Old World between *Strongylognathus* and *Anergates* is filled up, but not in the sense that *Wheeleria* is to be regarded as standing phylogenetically midway between these two genera. The striking analogy between *Wheeleria* and *Anergates* is due perhaps only to 'biological convergence,' and the resemblance in their way of life may be a coincidence. Moreover, Santschi has recently discovered in Tunis temporarily mixed colonies belonging to the subfamily of Dolichoderinæ, formed by the intrusion of *Bothriomyrmex* females into *Tapinoma* colonies. This form of symbiosis is very like that which we have considered in group 2*b*, as existing between *Formica truncicola* and *fusca*, and between *consocians* and *incerta*; it is, however, phylogenetically quite independent of the evolution of similar alliances in the other subfamilies of ants.

The subfamily of Formicinæ or Camponotinæ was much later than the subfamily of Myrmicinæ in developing its present genera and species, which are very numerous. This is proved by the fossil representatives of the two subfamilies, which have come down to us in amber from the middle of the Tertiary period. Hence, it is only natural that the Formicinæ should develop the slave-making instinct later than the

[1] See Forel, 'Mœurs des fourmis parasites des genres *Wheeleria* et *Bothriomyrmex*' (*Revue Suisse de Zoologie*, XIV, fasc. 1, 1906, pp. 51–69).

Myrmicinæ, and this is borne out by facts that we considered in our biological survey of the various groups. Among the Myrmicinæ of the present day we find almost exclusively descending grades of slavery, and among the Formicinæ many preparatory and ascending grades of the slave-making instinct, leading up to its culminating point.

Let us once more turn our attention to these forms.

The Amazons of the genus *Polyergus* (group 4) represent the development of this instinct at the highest point which it reaches in the subfamily of the Formicinæ. Phylogenetically they may be traced back to the genus *Formica*, and so they form one real line of evolution with groups 2 and 3. But by this expression I do not mean that the genus *Polyergus* is directly descended from one of the present species of *Formica* belonging to group 3, for instance, from the red robber-ant, for there is a clear morphological and biological distinction between *Polyergus* and the present robber-ants of the genus *Formica* (cf. note 2, p. 398). We must therefore assume the phylogenetic separation of these two genera to have taken place in some remote past, probably in the second half of the Tertiary period, when Europe and Asia were not yet cut off from North America. At that time there was a species of *Formica*, resembling our *F. sanguinea* in its mode of life, having developed the slave-making instinct in a higher degree than other species, and this species became the ancestor of the famous race of Amazons, which exists in several subspecies, all belonging practically to one single species, in both hemispheres.

At a later period, when the division between the eastern and the western continents was going on, but was still not complete, there arose the red robber-ants (*F. sanguinea*), being descended from a race resembling our present *F. truncicola* both morphologically and biologically. *F. sanguinea* has not developed the slave-making instinct so highly as *Polyergus*; and this fact suggests that in the former the instinct made itself felt later; moreover, the subspecies in both continents have developed it in different ways, and those in North America are still behind those in Europe.[1]

Of still later origin than our present *F. sanguinea* is *F.*

[1] Cf. group 3 in the biological survey, p. 394.

truncicola, which did not branch off from *F. rufa* as a distinct subspecies until North America was quite cut off from Europe and Asia, that is to say, probably during the Pleistocene epoch, for it does not occur in North America, although the species and subspecies of the *rufa* group are more numerous and varied there than here. On the other hand, other representatives of the same group are found in America, which, like *F. truncicola*, form temporarily mixed adoption-colonies with the workers of other and smaller species of *Formica*.

The instinct prompting the females of various branches of the large and ancient *rufa* group, to found their new colonies with the help of workers belonging to some other smaller species of the same genus,[1] is the starting-point for the development of the slave-making instinct, among the slave-keeping ants belonging to the genera *Formica* and *Polyergus*. But the adoption-colonies formed at the present day by *F. truncicola*, *consocians*, &c., are only the modern counterparts of similar adoption-colonies, whence at a much earlier date the robber-colonies of our present red robber-ants and Amazons originated phylogenetically.

We see, therefore, that what was probably the real history of slavery amongst ants breaks up into a number of distinct processes of evolution, originating at different times and attaining various degrees of completeness. We may compare the evolution of slavery among ants with a tree, sending out many boughs and branches from its trunk. The oldest bough, shooting off near the root, is the genus *Anergates*; the blossoms once borne by its branches have withered long ago, and the bough itself is dying. The bough with its branches that represents the genus *Polyergus* is in full blossom; it springs from a higher point halfway up the trunk. Above it we see other, younger branches, which are the slave-making species of *Formica* of the present day. They bear buds showing the slave-making instinct to be growing, but as yet these buds are not fully opened. At the top of the tree are some still younger shoots, on which the buds have only just formed; these are the species of *Formica* living temporarily in mixed

[1] I shall return further on to this subject, and give the reason why it is in the species allied to *F. rufa* that the females have lost the instinct to found new colonies independently.

colonies, but stealing no slaves. But will they become real slave-robbers at some future time? We can only offer conjectures on this subject, for the evolution of species and their instincts does not depend solely upon the interior laws of evolution,[1] which supply the *Anlage* or tendency to produce new forms, but also upon the exterior circumstances of life, which condition the realisation of this possible evolution, and co-operate as causes producing it. If the circumstances under which a species lives are persistent and regular, it is probable that there will be no change in the species itself and its instincts; but if the external conditions of life are altered owing to climatic and other changes, it is probable that modifications will ensue in the mode of existence of the species in question, and in the organs and instincts concerned.

Geology teaches us that in both the Tertiary period and the Pleistocene epoch gréat and far-reaching climatic changes have repeatedly occurred in the northern hemisphere. These changes could not fail to affect the ants within this region, and led to modifications in the structure of their nests, in their way of procuring food, and in all the circumstances of their life. We shall therefore probably be right in connecting the repeated origin of and the various degrees in the evolution of the slave-making instinct among ants, with the different climatic changes that took place during the Cænozoic age.

To many persons this hypothesis will perhaps seem very daring, yet there are good foundations for it in fact. In discussing the sixth and seventh groups, I was able to show that our little scimitar-ant, *Strongylognathus testaceus*, had most likely lost the slave-making instinct under the influence of our northern climate, whilst its southern connexions still retain it. It is unimportant whether we are to regard this as a result of the northward migration of an ant, formerly living in the south, or as a consequence of a gradual diminution in the summer heat in a locality already occupied by that species. The decay of the slave-making instinct in a species that once kept slaves can easily be explained by climatic changes; but can the origin of this instinct be accounted for on similar lines

[1] With regard to the nature of these laws, which on its material side depends upon the constitution of the chromatin-substance in the germ-cells, see pp. 176, &c.

in a species that once had no slaves? This is the next question that we have to answer.

We have already seen in speaking of the first and second groups (pp. 391, &c.) that we must regard, as a preliminary to the evolution of that instinct, a habit possessed by certain kinds of ants, of not forming new colonies for themselves, but the impregnated females after the copulation flight are adopted in the nests of ants belonging to another species (group 2*b*). The existence of this habit proves that the queens have lost the instinct prompting them to found new and independent colonies, and, instead of settling down by themselves, they seek out the workers of another species. What can have caused such a lack of independence on their part?

It would be produced most readily in a species that not only is very abundant, but possesses very populous colonies, living in huge nests, so that the surrounding district is dominated by the inhabitants of the colony. In such a district the queens, after their copulation flight, would be sure to meet workers ready to welcome them, and thus they would be relieved of the necessity for founding new settlements alone.

These are exactly the circumstances under which live our northern wood-ant, *F. rufa*, and its nearest connexions of the *rufa* group in Europe, Asia, and North America,[1] and they represent a form of adaptation to life among the forest Flora of an Arctic climate. The genus *Formica* has literally a circumpolar distribution, and the *rufa* group, that builds high heaps, predominates more and more, the further north we go. These huge nests secure to their occupants a high and even temperature, and so protect them against the severity of the climate, and render it possible for the young to be reared even in dense, damp forests. Not only do the decaying vegetable substances, of which the heaps are constructed, produce heat, but the heaps are so shaped as to catch the rays of the sun, and their dry domes are raised well above the damp earth—and all these are marks of adaptation to life in an Arctic forest. In this way we can understand how, in the species belonging to the *rufa* group, the queens may have lost the instinct prompting them to found new colonies, and

[1] On this subject see the details given under group 2*a*, p. 391.

this loss would be an indirect consequence of the adaptation of these ants to life among Arctic forest Flora.

If the instinct was once lost, the descendants of these ants would be devoid of it, even supposing the species phylogenetically descended from these wood-ants to become rarer, in which case the opportunity would more rarely present itself for the queens to meet workers of their own species, and form new colonies by their aid. They would have to seek a home with some other common species of *Formica*, and thus arose the adoption-colonies of *F. truncicola, exsecta*, &c., in Europe, and of other members of the *rufa* group in North America. Therefore the formation of these temporarily mixed adoption-colonies, which represent a preliminary stage leading to the formation of permanently mixed robber-colonies, is connected with the adaptation of their ancestors to life in the Arctic forests. We may even go so far as to pronounce it probable that, in consequence of a gradual change in the climatic conditions which had been most suitable to the genuine wood-ant *F. rufa*, fresh subspecies branched off from the original stock, and took up their abode outside the forests, as is the case with *F. truncicola, consocians*, &c.

But a further question presents itself: 'How can altered climatic conditions cause a slave-making instinct to arise in an ant that at first lived with its assistant ants in only temporarily mixed colonies?' Biological facts give us many indications that will aid us in answering this question. Let us take as an instance our *F. truncicola*, which employs the workers of *F. fusca* in founding new colonies. What prevents it from stealing slaves? There is no direct reason for its doing so. Like *F. rufa*, *F. truncicola* lives chiefly by keeping aphides, and does not catch insects, although occasionally it carries flies and other insects into the nest. It will, however, readily eat the pupae of other kinds of ants if they are given to it. Let us now imagine that in some district, occupied by *F. truncicola*, climatic changes gradually replaced the northern forest Flora by that of steppes covered with heather. As aphides gradually became less abundant on the trees and bushes, the ant would be forced to live on insects more than it had done previously, and as it is a large, strong ant, and its colonies, if long established, become very populous, it would be able to find food easily by stealing the pupae of other smaller ants living in the same region. The commonest of the smaller species of *Formica* is *F. fusca*, and if there are any *fusca* nests in the neighbourhood, *young truncicola* colonies, containing workers that have been brought up by *F. fusca*, would rear at least some of the stolen pupae to be their slaves.[1] I have actually found confirmation of this theory, and as soon as this process occurs, we have a robber-colony.

Our red robber-ant *F. sanguinea* is really an ant living on steppes and moors, and feeding on insects and the stolen pupae of other ants. It belongs as much to the moors of the north as *F. rufa* does to the forests. In *F. sanguinea* we have an instance of a regular slave-breeder, stealing and rearing as slaves the worker-pupae of *F. fusca* or of *F. rufibarbis*. But these are the species of *Formica* by whose help the females of *sanguinea* found their new colonies. Therefore, each individual colony of this robber-ant is for a time a mixed adoption-colony, before it becomes permanently a mixed robber-colony. Are we not justified in believing that the race has developed in a similar way by passing through a *truncicola* stage?

Apparently one link is still missing in the chain of evidence. If one individual *truncicola* colony begins to steal and rear *fusca* pupae, and repeats its raids upon the neighbouring *fusca* nests every year, it does indeed become a new robber-colony, but this does not explain how the slave-making instinct has become hereditary in the whole species, as it is in *F. sanguinea*.

Let us see why this is so.

We must begin by noticing that by no means all the colonies of ancestors of *sanguinea* resembling *truncicola* adopted the practice of stealing slaves suddenly and simultaneously. Some adopted it earlier, and some later, according to their external circumstances. This is suggested by the fact that at the present time the North American subspecies of *F. sanguinea* have developed the slave-making instinct in a lower degree than the European variety of the same species. It would be a mistake therefore to imagine that the ancestors

[1] See group 2*b*, p. 392, and also my 'Ursprung und Entwicklung der Sklaverei bei den Ameisen,' p. 167.

of the red robber-ant suddenly acquired an hereditary instinct prompting them to make slaves.

The transmission of instincts, like that of bodily qualities, is effected by means of the germ-plasm. The impregnated females of *sanguinea* transmit the slave-making instinct, not the workers, which do not normally aid in propagating the species.[1] Let us examine closely the changes that must have taken place in the hereditary plasm of the queens of that *truncicola* type, from which our *sanguinea* is descended.

The females inherited an instinct prompting them to seek *fusca* nests, and to unite with the workers in them for the purpose of founding new colonies. As I showed on p. 417, their ancestors of the *rufa* group (2*a*) had already lost the power of founding new colonies independently, and therefore the *truncicola* queens (group 2*b*), after the copulation flight, have to wander about until they find admission into a nest of the commonest ants of another species of *Formica*, and these happen to be *F. fusca*.

A young *truncicola* queen, forming an adoption-colony with workers of *F. fusca*, needs no 'new instinctive *Anlage* or disposition' in her germ-plasm. Nor do we need to assume the existence of any in order to account for the origin of the instinct prompting the workers to steal the pupae of other ants, for *truncicola*, like other species of the *rufa* group, lives at any rate partially on stolen insects, and when impelled by want of food, it will attack and plunder weak colonies of other species of ants, especially *fusca* colonies, as these ants are remarkable for their cowardice. We need not therefore assume the existence of any 'new instinctive *Anlage*' in the germ-plasm

[1] I say 'normally' in contrast to parthenogenesis, which, however, in *Formica* produces only males. According to observations that I made in Luxemburg, colonies of *F. pratensis* having no queens, but existing under natural conditions, go on for two or three years producing thousands of males, all being hatched from the unfertilised eggs of the workers. If on the copulation flight these males pair with females from other colonies, a transmission of the properties of the workers, that produced the males, to the workers of the next generation is quite possible, through the male germ-plasm. This point has not hitherto received as much attention as it deserves, although it throws considerable light upon the difficult problem of the development of instincts among the workers of the social insects. In some species of *Lasius*, workers appear to be produced directly by parthenogenesis. In giving the above account of the development of the slave-making instinct, I have left parthenogenesis out of consideration, because under normal circumstances the queens, not the workers, lay the eggs from which the males also are produced.

of the *truncicola* queens, in order to account for the origin of the plundering instinct in the workers, nor for their habit of rearing only the *fusca* pupae from among all those that they steal; the workers were themselves reared by *fusca*, and for years formed a mixed adoption-colony with the workers of this species; for this reason the pupae of *fusca* workers impress the individual *truncicola* ants, through their sense of smell, as being familiar companions and not strangers.

Here we have all the preliminaries requisite for gradually producing a definite hereditary instinct for making slaves. The chromosomes of the impregnated females' germ-cells, which are the material bearers of heredity, need only favourable combination in order to secure the transmission of the slave-making instinct. I cannot discover any difficulty as to fixing a combination of elements already existing. The *truncicola* queen already possesses the instinct to unite with *fusca* workers in founding her colony, and she may transmit this instinct to her offspring of the working caste, but in a form adapted to their character as workers. This would strengthen, in the robber-ants reared by *fusca*, an already existing inclination to ally themselves with workers of that species, and, as soon as they are aware of a dearth of workers in their colony, they make expeditions in quest of *fusca* pupae.

Here we see, fully developed, the hereditary slave-making instinct of our red robber-ants.

But, it may be asked, what has Darwin's Natural Selection to do with this evolution of the slave-making instinct? No allusion at all has been made to it. Can we not assign to it at least a subordinate part in the evolution? Yes, we may justly assign to it the part of the executioner, as it wipes off the face of the earth those colonies of ants which have shown themselves incapable of maintaining existence, and thus it averts unfavourable variations in the germ-plasm of the queens. This is, however, the limit of its action, it is not concerned with either the origin or the further evolution of slavery. It is an interesting fact that the theory of natural selection proves to have no more than this to do with the evolution of the slave-making instinct, which Darwin in the 'Origin of Species' considered capable of explanation by means of natural selection (cf. p. 411).

'Nature will not be robbed of her veil of mystery, and what she refuses to reveal, you will not extort from her by using screws and levers.' Certainly screws and levers are of no more avail than unprofitable theoretical speculations. But much may be learnt by careful observations and experiments, and by cautious deductions from them. Perhaps I have succeeded in making such use of the newest materials supplied by biology as to raise, at least in some degree, the veil of mystery that has hitherto enveloped the history of slavery among ants.

We may hope that, as biological research advances, more light will gradually be thrown upon the details of the phylogenetic evolution of the slave-making instinct. The sketch given above is only a modest attempt to solve this very interesting problem. Let us now sum up shortly the results of our examination of this instinct, and consider its bearing upon the theory of evolution.

The development of the slave-making instinct is a matter of hypothesis and not of fact; but the hypothesis proceeds directly from the facts, if we compare them carefully with one another and investigate their genetic connexion. It is a well-grounded hypothesis, as it supplies us with a uniform and satisfactory answer to the question how the actually existent forms of slavery and social parasitism among ants could have been produced by natural causes.

A close examination of the slave-making instinct has shown how quite new instincts may arise in animals from simple foundations, how they may develop to an astonishing point, and how finally they can degenerate and disappear. If we fix our attention only upon the culminating point of this development, e.g. upon the conspicuous degree in which *Polyergus* possesses the slave-making instinct, we are inclined at first to say: 'This instinct must have been implanted in the Amazon ants at their creation, for they cannot exist without slaves; therefore it is impossible for their instinctive desire to steal slaves to have arisen through evolution.'

My answer to this objection is that it is undoubtedly an absolute necessity for the genus *Polyergus* in its present form to possess the slave-stealing instinct, as otherwise it would cease to exist. But if *Polyergus* is phylogenetically descended from the genus *Formica*, which contains other slave-keeping species not so completely dependent upon their slaves, and possessing the slave-making instinct in various degrees, it is possible to give a simple and natural explanation of the origin of the same instinct in *Polyergus,* though this ant possesses it in a far more perfect form. The species have undergone morphological changes as their instincts have developed; and our examination has shown us that the instincts of these ants supply precisely the biological impetus causing modifications in their forms, and producing new species and genera.

The development of the slave-making instinct marked off the red robber-ant (*Formica sanguinea*) as a species distinct from another belonging to the same genus, but not yet possessing this instinct; and as a result of its further development, the genus *Polyergus*, which differs greatly from *Formica* in the formation of its mandibles, branched off from a Tertiary species of *Formica.* The decay of the slave-making instinct in the genus *Strongylognathus* resulted in the production of a new species *Str. testaceus.* The influence of a parasitic existence has led to the formation of a number of new genera, such as *Wheeleria, Epoecus, Anergates,* &c., which differ widely from their nearest systematic relations in the form of their males and females, as well as in having no workers. In short, the history of the evolution of the slave-making instinct has afforded us an opportunity of learning, from clear examples, how new species and genera of animals may come into existence, as their instincts develop.

12. Conclusions and Results

I might bring forward a number of similar instances of evolution occurring among the inquilines of ants and termites, and among ants and termites themselves, but they would all lead to the same conclusion as those already considered. We cannot avoid accepting the hypothesis of a race-evolution, both of species and of their peculiar instincts, but this evolution is not on the lines of Darwin's hypothesis. This result is not new. Twenty years ago I wrote a paper on the evolution of instincts in the primæval world,[1] in which I arrived at the

[1] 'Die Entwicklung der Instinkte in der Urwelt' (*Stimmen aus Maria-Laach,* XXVIII, 1885, p. 481).

same conclusion, although not with as much clearness and certainty as at the present time. No change has taken place in my opinions on this subject, but they have become more definite, after twenty years devoted to my special branch of scientific research.

Let us now once more sum up briefly the results of our criticism and comparison of the theories of permanence and descent, with which we have been occupied in this section.

Of the two contrasted theories, the former, which maintains the fixity of species, is apparently supported by the great majority of facts coming immediately under our observation, because the evolution of many species is complete at the present time, and that of others advances so slowly as to be imperceptible. It is therefore only in exceptional cases that we find species in which we can show evolution to be still going on. As an instance from my own department, I was only able to refer to the evolution of *Dinarda* forms (pp. 315, &c.), which seems to be still incomplete in two of the four species, or rather subspecies, belonging to this group.

We are able somewhat more frequently to discover cases in which the formation of new species has been recent, i.e. has occurred in the last geological period. I discussed in detail some of these cases (pp. 348, &c.) which may be regarded as direct evidence in support of the theory of evolution, and I considered at some length the change in the habits of the beetles belonging to the genera *Doryloxenus* and *Pygostenus*, which were at first inquilines among the wandering-ants, and then found hospitality among the termites. It must be acknowledged, however, that there is comparatively little direct evidence in favour of the evolution of species.

Facts which, on the surface, seem to support the theory of permanence, prove on scientific examination to supply evidence in favour of the theory of evolution, as soon as we bring comparative morphology, biology, and embryology to bear upon them, even if we disregard palæontology.

I referred to a number of instances showing that the systematic peculiarities, distinguishing the species, genera, and families of inquilines among ants and termites from their relatives leading an independent (not myrmecophile or termitophile) existence, are all to be regarded as characteristics due to adaptation to a myrmecophile or termitophile mode of life. These characteristics become intelligible only when we can assign their causes to them, and this necessitates our admitting that an evolution of the systematic species of the same stock can take place.

The theory of permanence can offer a satisfactory account of these characteristics only in as far as it accepts, simply as existing facts, the very various beneficial morphological and biological conditions that present themselves, and does not seek into causes, and demands no explanation beyond this—'the various species of inquilines were originally created in their actual form at the same time as their hosts were created, and expressly for them.' This explanation may satisfy one who is a teleologist and nothing more, but not a scientific student of nature, for his thoughts may, and inevitably must, pass on to the further question: 'Is it not possible to assign to natural causes the origin of these beneficial adaptations?' If he takes his stand on the theory of evolution, he can answer this question in the affirmative, although he need not be under any optimistic delusion regarding the hypothetical character of the various attempts hitherto made at explanation.

In considering the history of slavery amongst ants, we found an instance of the evolution of an instinct, which confirms the above statement. It appeared that only the theory of evolution in a modified form enabled us to arrive at a real comprehension of the origin of these biological conditions.

Let us once more return to our discussion of the doctrine of evolution. In Chapter IX (pp. 272, &c.) I showed that the recognition of an evolution of the systematic species belonging to one stock was closely connected with the Copernican theory of the universe. The geological evolution of our planet is intimately related with a biological evolution, which appears in a succession of various Fauna and Flora, extending from those which are the objects of palæontological research, to those of the present day, and, according to the fundamental principles of the Christian cosmogony, we are perfectly justified in admitting that natural causes may explain this succession. We shall therefore cease to regard the Fauna and Flora of the present time as fixed in number, distinct from and absolutely independent of their predecessors, to account for whose existence it was enough to refer to the Creator's almighty power. On the contrary, we shall consider our present plants and animals as representing the close of a process of natural evolution, and we shall try to penetrate into the secrets of the differentiating methods of nature, which have given rise to this process. As I have shown in the examples already discussed, this attempt is by no means a barren and unprofitable speculation, based on nothing but vague suppositions; on the contrary, the final results are in such astonishing agreement with the hypotheses supplied by the method adopted, that it is hardly possible to avoid the conclusion that we are now on the right road towards solving this difficult problem in nature.

Of two hypotheses in natural science or natural philosophy, put forward as offering an explanation of one and the same series of facts, it behoves us always to choose the one which succeeds in explaining most by natural causes, and on this principle we can hardly hesitate to choose the theory of descent in preference to that of permanence.

I trust that I have now made clear the practical importance of the distinction drawn in Chapter IX (pp. 296, &c.) between systematic species and natural species. I stated then, that if we accepted a modified theory of evolution, we must class together definite series of systematic species, which probably are of common origin, as forming one natural species, and trace them back to one common primitive form. If we wish to account for the origin of these primitive forms, we must have recourse to the old doctrine of creation and say: 'The natural species were originally in their primitive forms produced by God directly out of matter.' The theory of permanence maintains that the present systematic species were originally created in their present form.

I believe therefore that no blow has been struck at the Christian dogma of the creation by all our preceding discussion of the theories of permanence and descent with reference to ants and termites and their inquilines. It is, for instance, a matter of perfect indifference to the Christian cosmogony whether each individual systematic species of the *Clavigeridae* was directly created, or whether we may include in one natural species all the systematic genera and species of the *Clavigeridae* as well as the genera and species of the subfamily of the *Pselaphidae*, so that this one natural species would include a very large family of beetles, consisting of several hundred genera and many thousand systematic species.

In the same way it is indifferent to Christian cosmogony whether we regard the species of the family of *Termitoxeniidae* as directly created, or as forming one natural species with the *Muscidae* and *Phoridae*, two families of Diptera.[1]

The ascertained facts, which I have described, suggest that the latter course is the more correct, and we may follow it without any danger of wrecking our faith as Christians. Indeed, my own conviction is that *God's power and wisdom are shown forth much more clearly by bringing about these extremely various morphological and biological conditions through the natural causes of a race-evolution, than they would be by a direct creation of the various systematic species.*

In the sixth edition of his 'Gottesbeweise,'[2] Father von Hammerstein writes as follows: 'If the Creator did not create each single species of animal in its present form, but caused it to acquire its present appearance and instincts by means of an independent evolution, carried on through a long line of ancestry, His wisdom and power are manifested the more clearly. Therefore if the theory of evolution is proved to be true within definite limits, it by no means sets aside the Creator, but, on the contrary, an all-wise and all-powerful Creator becomes the more necessary and indispensable, as the First Cause of the evolution of the organic species. A simile will bring out the truth of this very clearly. A billiard player wishes to send a hundred balls in particular directions; which will require greater skill—to make a hundred strokes and send each ball separately to its goal, or, by hitting one ball, to send all the ninety-nine others in the directions which he has in view?'

[1] I may here repeat what I said before (see p. 297), and state clearly that I have no intention of defining the whole extent of these natural species, which may be much greater than I have said

[2] Trèves 1903, p. 150

CHAPTER XI

THE THEORY OF DESCENT IN ITS APPLICATION TO MAN[1]

(Plates VI and VII)

PRELIMINARY OBSERVATIONS.

Great importance of this question (*p.* 431).

1. IS THERE ANY JUSTIFICATION FOR TAKING A PURELY ZOOLOGICAL VIEW OF MAN?

No, for it overlooks the chief point—his intellectual and spiritual life. For this reason psychology has the best right to judge of the nature and origin of man (*p.* 433). A purely zoological view of man is one-sided and based on false premises (*p.* 434). Karl E. von Baer on the materialistic explanation of the intellectual life (*p.* 435). Only an act of creation can have produced the human soul (*p.* 436). What are we to understand by the creation of man? St. Augustine on this subject (*p.* 437). Philosophical reflections on the idea of the creation of man (*p.* 439). The Thomistic doctrine of the sequence of various forms of being in the individual development of man. Its application to the theory of descent (*p.* 440). How far is zoology competent to judge of the hypothetical phylogeny of man? (*p.* 442)

2. WHAT ACTUAL EVIDENCE IS THERE OF THE DESCENT OF MAN FROM BEASTS?

(*a*) *A Glance at the Comparative Morphology of Man and Beasts.*

Wiedersheim's 'testimony' to it (*p.* 443). Skeletons of apes and men. Rudimentary organs (*p.* 445).

(*b*) *The Biogenetic Law and its Application to Man.*

Haeckel's anthropogeny and the 22 or 30 phylogenetic stages in the

[1] Objections have been made in several quarters to my adoption of the term 'theory of descent' to designate the theory of the evolution of the organic species from their original stock. 'Descent' implies derivation from some earlier stock, and, according to the theory of evolution, definite series of systematic species are related through being derived from a common stock, and the systematic species of the present day are descended from other extinct species belonging to previous ages, and thus the name 'theory of descent' seems to me very suitable. We need not abandon the word, or the idea which it conveys, because they have been put to a bad use by the Monists. Moreover, the name 'theory of descent' has been generally adopted, at least in scientific circles, to designate the evolution of organisms from an earlier stock. I do not think that anything would be gained by our carefully avoiding this name, and substituting for it 'theory of evolution,' 'transformation theory,' or 'adaptation theory.' If we did so, our opponents might reasonably regard it as a sign of weakness to be afraid of a *word*, after we had accepted the *thing* that it denotes. The particular form of the theory of descent, that I have shown in the preceding chapters to be acceptable from a scientific point of view, is not monophyletic but polyphyletic; nevertheless it does not seem expedient to reject the word 'theory of descent' and replace it by 'polyphylogeny.' Cf. on this subject the remarks in the preface to this edition.

Preliminary Observations

Before we end our examination of the comparative merits of the theories of permanence and descent, I must answer one more question, which has probably occurred to many of my readers. 'If we give up the fixity of the systematic species, and substitute for it an evolution of the species within definite series of forms, each constituting a natural species, must we not apply the same law of evolution to the highest of the systematic species, i.e. to *Homo sapiens*?'[1]

I do not intend to discuss this point in its dogmatic and exegetical aspect, but I may make a few remarks that will throw some light upon it.

The question with which we are now concerned is so important, and has so vast a bearing upon the highest interests

[1] Cf. Chapter IX, p. 296, and X, p. 428.

of mankind, that it cannot be dismissed with mere cut-and-dry phrases. I should describe as a phrase of this kind the statement which the materialists generally make in support of the descent of man from beasts: viz. that zoologically his descent from beasts is self-evident!

Against this statement I may say:

(1) It rests on the tacit assumption that zoology is the only science entitled to judge of the origin of man.

(2) It rests further on the tacit assumption that the descent of man from beasts has already been actually proved by means of zoology.

We cannot, however, tolerate tacit assumptions on a subject of such gravity and having such important consequences. Therefore, we must examine it critically, and find answers for the two following questions: (1) Is zoology really the only science entitled to form an opinion regarding the origin of man? (2) What actual evidence is supplied by zoology in support of the descent of man from beasts?

1. Is there any Justification for taking a purely Zoological View of Man?

If man at the present day were actually nothing more than a higher animal,—if there were no essential difference between man and beast, it would, perhaps, be an obvious answer to give, when asked whether man is descended from beasts: 'He must have come from a Tertiary mammal, as he could not have come into being otherwise.' This answer would not be quite scientific, for it would not be supported by evidence derived from facts, but it would at all events be psychologically near the truth. In fact, this answer, which for the sake of brevity I will call the purely zoological answer, would be given without hesitation by all those who regard the zoological aspect of the question as the only one worth consideration. Unhappily I am forced to admit that not a few of our modern zoologists seem to assume zoology to be our sole source of information regarding the nature and origin of man.[1] For this reason they reject the results of other sciences, if they do

[1] For a criticism of this view see also J. Grasset, *Les limites de la biologie*, Paris, 1902.

not agree with this assumption. But it is based upon a very one-sided opinion, and it would be most desirable if, in this case, we had somewhat more of that freedom from bias of which we hear so much. Although I am myself a zoologist, and esteem zoology and its scientific adherents very highly, I feel inclined to compare a zoologist, who judges man from a purely zoological point of view, with a printer's apprentice, who judges of the nature and origin of one of Mozart's compositions merely as so much printers' ink.

But what other sciences, besides zoology, have any claim to be heard on the subject of the nature and origin of man?

Quite apart from theology, there is above all philosophy, and especially psychology, the branch of philosophy which deals with the spiritual life of man. It teaches us to observe our own spiritual activities, and, by a process of logical deduction, it traces them back to an immaterial and simple principle that we call the rational soul of man. It teaches us to compare our own spiritual life with the manifestations of the animal soul, that is limited to matters of sense, and thus to recognise the great difference between man and beasts. A brute has no power of intellectual abstraction, and therefore it has no free will, and it cannot manifest what it does not possess. It cannot express its perceptions and feelings rationally by means of language; and, having no reason, it is impossible for it to possess any science, religion or morality. Man alone possesses a sensitive and spiritual soul essentially different from the merely sensitive animal soul.[1]

It is very easy simply to deny the existence of this distinction between man and beast, as unhappily superficial thinkers often do at the present time; but such a denial can only be based upon the annihilation of psychology as an independent science, for the purely zoological method is assumed to be the only form of comparative psychology for which any justification exists. Such thinkers concentrate their attention upon the points common to men and beasts, and try to account for all the differences between them by asserting that each point of difference must have been gradually evolved from what was

[1] On this subject see my earlier writings: *Instinkt und Intelligenz im Tierreich*, Freiburg, 1905, and *Vergleichende Studien über das Seelenleben der Ameisen und der höheren Tiere*, Freiburg, 1900; also *Menschen- und Tierseele* Cologne, 1906.

2 F

at first purely animal, as otherwise it could not exist at all. Here we have what I have called the one-sided view of the compositor's apprentice betraying itself again. It is tacitly taken for granted that the zoological view of man is the only possible one—and on this false assumption is based a very common opinion regarding human psychology.

For those who take the purely zoological view, human religion and morality exist only in as far as they have developed naturally from animal origins. Everything beyond this is designated ' mythical,' ' childish,' ' savouring of intellectual slavery,' &c. Of course, the objective element in every religion disappears, and with it all higher motives for human morality. There can be no mention of dogmas, with the exception, of course, of purely zoological dogmas, such as the biogenetic law. Belief in a personal God and Creator seems completely overthrown, and the mere suggestion that the existence of a personal Creator, superior to the universe, may be proved from zoological facts is rejected with indignation, as bringing in a metaphysical element that would destroy the ' purity ' of zoology.

Here again we encounter a lamentable one-sidedness in dealing with the subject.

One who thinks simply as a zoologist is either an agnostic, denying the power of thought to go beyond the limits of what zoology teaches,—and in that case he condemns himself to this intellectual limitation and fetters his own reason ; or he is a monist, venturing beyond these bounds and asserting that the *monos* has in man attained the highest form of animal existence,—and in that case he has ceased to think purely zoologically, and is combining zoology and metaphysics, no less than those do who from zoological facts prove the existence of a personal Creator superior to the universe. The whole difference between them is that the theist arrives at a correct, and the monist at a false, conclusion. Neither the agnostic nor the monist can rightly claim to possess scientific freedom from prejudice.

We may once for all dismiss the purely zoological view of man. I have dwelt upon it at such length only because I wished to show that it is unworthy of a thoughtful human being. It is quite evident what opinion we ought to form of all the specious statements, made in academic lecture rooms and in periodicals dealing with popular science, and professing to adduce zoological evidence of the descent of man from beasts. They supply no real evidence at all, being too purely zoological, and treating man not as what he *is*, but as what he *ought to be*, according to the purely zoological theory, namely an animal, and nothing more. I wish, however, to rise to a higher level, and to consider not only the animal, but also the spiritual side of man. Man's spiritual soul is essentially different from a brute soul, and can therefore never have proceeded from it by any natural evolution.[1] The soul of a child requires the powers of the senses to be developed before its mental powers, but nevertheless it is essentially different from the soul of a brute, for otherwise the child could no more become a reasonable being than a young ape could.

Karl Ernst von Baer, who is undoubtedly one of the greatest and most thoughtful students of nature in modern times, has made use of some similes which describe the materialists' inability to understand what is meant by spiritual.[2]

Some one hears a horn, and perhaps recalls the tune, but naturally does not believe that it is playing itself. Then a mite, sitting in the horn when it began to blow, exclaims: 'Tune! nonsense! I felt it, it was a horrible hurricane that swept me out of the horn.' But a spider on the outside of the horn declares that there has been neither music nor hurricane, merely vibrations, at one moment rapid, at another slow. The mite and the spider are both right from their respective points of view, but neither understands music.

Again, let us imagine that a traveller in Central Africa loses a musical score. A savage looks at it, and takes it for a bundle of leaves; a Hottentot, who has been in contact with Europeans, recognises it as paper; a European colonist sees that it has to do with music; but only a trained musician perceives that it is Mozart's Overture to the Zauberflöte or one of Beethoven's Symphonies.

'It is the same thing,' remarks Baer, 'with perception of what is spiritual. If a man has no tendency to recognise it,

[1] Cf. Chapter IX, pp. 283, &c.

[2] The following expressions used by von Baer were collected by Stölzle, *K. E. von Baer und seine Weltanschauung*, Ratisbon, 1897, pp. 342, 343.

and no appreciation of it, he can leave it alone, only he must not express an opinion upon it, but be contented with his own personal consciousness. The student of nature is to a certain extent justified in stopping short at the point where what is spiritual begins, because his own observations cease to carry him further, and he has nothing that he can measure, or weigh, or perceive, by means of the senses. He has, however, no right to say that nothing exists, because he cannot see it or measure it, nor that only what has a body and can be measured has a real existence, and that what is called spiritual is only a property or attribute of the body, proceeding from it. Whoever should speak thus would be like the Hottentot, seeing lines and dots, but knowing nothing of music, or like the spider counting, if it could, the vibrations of the horn, but not hearing the melody.'

I should like to commend these words of Ernst von Baer to the consideration of all those who, with L. Büchner, Ernst Haeckel, August Forel, and other materialists, declare the spiritual side of the human soul to be a mere matter of the imagination, because it rises above their one-sided view of the processes of nature.

Because the soul of man is spiritual, it differs from the brute soul essentially and not merely in degree, and therefore it can exist only as a result of creation, not of evolution. Even so prominent an upholder of Darwin's theory of evolution as A. R. Wallace has acknowledged that the spiritual side of man cannot have been evolved from animals.[1] As soul and body together constitute one being, man in his completeness occupies a unique position in nature. Therefore, with regard to philosophy, there can be no objection to our postulating an act of creation, in order to account for the origin of man.

Man is man only in virtue of possessing a spiritual soul, and so the creation of the first man took place when his spiritual soul was created and united with his body of clay. That God could make use of matter previously prepared for such a union by natural causes, so as to form a new being when the union with the soul was effected, we may assume to be *possible*. The dogmatic exegetical question as to how the words of Holy

[1] *Darwinism: An Exposition of the Theory of Natural Selection, with some of its applications*, London, 1889, Chapter 15, pp. 474, &c.

Scripture are actually to be interpreted has nothing to do with this subject, and in this biological study we cannot enter upon a more detailed discussion of it.[1]

Our atheistical opponents often taunt us with imagining the God of the Biblical account of the creation as a sort of 'potter in human form,' fashioning for Adam a body of clay, and then breathing the soul into his face. This anthropomorphic view of God was described as *nimium puerilis cogitatio* by St. Augustine,[2] and it is not shared even by those who are convinced

[1] By far the greater number of theologians believe that the substance which God employed, when creating man, to unite with a spiritual soul consisted of inorganic matter. As the creation of man is primarily a dogma of faith, theologians are justified in clinging to the literal interpretation of the text according to constant tradition and the statements of the ordinary teaching authority of the Church (see p. 442, note 1), until satisfactory proof is given that the text ought to be interpreted otherwise. Natural science is not yet in a position to supply such a proof, as will be shown in the second part of this chapter. The teaching authority of the Church has not determined how we are to understand the details of the Biblical account of the creation of man. We may therefore apply to this difficulty the golden rule laid down by St. Augustine, who says: 'Et in rebus obscuris atque a nostris oculis remotissimis, si qua inde scripta etiam divina legerimus, quae possunt salva fide qua imbuimur alias atque alias parere sententias, in nullam earum nos praecipiti affirmatione ita proiiciamus, ut si forte diligentius discussa veritas eam recte labefactaverit, corruamus; non pro sententia divinarum Scripturarum, sed pro nostra ita dimicantes, ut eam velimus Scripturarum esse, quae nostra est; cum potius eam quae Scripturarum est, nostram esse velle debeamus' (*De Genesi ad literam*, l. 1, c. 18; cf. also *ibid.* c. 19 and c. 21; Migne, *Patr. lat.*, xxxiv, 260–262).

[2] I am indebted to Father J. Knabenbauer, S.J., for having drawn my attention to this passage, which occurs in *De Genesi ad literam*, l. 6, c. 11, 12 (Migne, *Patr. lat.*, xxxiv, 347–348). The following quotations also have some bearing upon this subject. In chapter 11 ('Opera creationis die sexto quomodo et iam consummata et adhuc inchoata'): 'Proinde *formavit Deus hominem pulverem terrae*, vel limum terrae, hoc est de pulvere vel limo terrae; *et inspiravit* sive insufflavit *in eius faciem spiritum vitae, et factus est homo in animam vivam.* Non tunc praedestinatus; hoc enim ante saeculum in praescientia creatoris: neque tunc causaliter vel consummate inchoatus, vel inchoate consummatus; hoc enim a saeculo in rationibus primordialibus, cum simul omnia crearentur; sed creatus in tempore suo, visibiliter in corpore, invisibiliter in anima, constans ex anima et corpore.' According to St. Augustine therefore the material of the human body had been created with the other elements at the beginning of creation. But how did this material become a human body? On this subject St. Augustine says in chapter 12 ('Corpus hominis an singulari modo a Deo formatum'): 'Iam ergo videamus, quomodo eum fecerit Deus, primum de terra corpus eius; post etiam de anima videbimus, si quid valebimus. Quod enim manibus corporalibus Deus de limo finxerit hominem, nimium puerilis cogitatio est, ita ut si hoc Scriptura dixisset, magis eum qui scripsit translato verbo usum credere deberemus, quam Deum talibus membrorum lineamentis determinatum qualia videmus in corporibus nostris. . . . Nec illud audiendum est, quod nonnulli putant, ideo praecipuum Dei opus esse hominem, quia cetera dixit et facta sunt, hunc autem ipse fecit: sed ideo potius, quia hunc ad imaginem suam fecit. . . . Non igitur hoc in honorem hominis deputetur, velut cetera Deus dixerit et facta sint, hunc autem ipse fecerit; aut verbo cetera, hunc autem manibus fecerit,

that the Biblical account of the creation is to be understood literally and not figuratively. The Church has not expressed any final opinion as to the nature of the substance used by God in creating the first man, but we may be sure that the Biblical account of the creation was not intended to give us information regarding the origin of man from the point of view of natural science.[1]

Sed hoc excellit in homine, quia Deus ad imaginem suam hominem fecit, propter hoc quod ei dedit mentem intellectualem, qua praestat pecoribus.' A few lines further on St. Augustine repeats himself and says: 'Nec dicendum est hominem ipse fecit, pecora vero iussit, et facta sunt: et hunc enim et illa per verbum suum fecit, per quod facta sunt omnia (Io. i. 5). Sed quia idem verbum et sapientia et virtus eius est, dicitur et manus eius, non visibile membrum, sed efficiendi potentia. Nam haec eadem Scriptura, quae dicit quod Deus hominem de limo terrae finxerit, dicit etiam quod bestias agri de terra finxerit, quando eas cum volatilibus coeli ad Adam adduxit, ut videret quid ea vocaret. Sic enim scriptum est: *et finxit Deus adhuc de terra omnes bestias* (Gen. i. 25). Si ergo et hominem de terra et bestias de terra ipse formavit, quid habet homo excellentius in hac re, nisi quod ipse ad imaginem Dei creatus est? Nec tamen hoc secundum corpus, sed secundum intellectum mentis, de quo post loquemur.' Hence follows the conclusion. 'Primus homo, non aliter quam primordiales causae haberent, formatus fuit.' Cf. *De Genesi ad literam*, l. 6, c. 15; Migne, xxxiv, 349, 350. St. Augustine is, of course, not thinking of an evolution of the human body in the sense of the modern theory of descent, and I need not dwell upon this point. There seems to be two chief ideas in his mind: (1) The difference in God's manner of creating man and beasts lies principally in the fact that to man He gave an intelligent soul. (2) By means of *primordiales causae* the body of man, like that of every other living creature, was based on *rationes seminales*. The holy doctor does not decide how far the *causae primordiales* and *seminales rationes* effected the preparation of its material. He does not discuss the nature of the material to which God united the human soul, but says simply: 'superflue quaeritur, unde hominis corpus Deus fecerit' (*De Genesi contra Manich.* l. 2, c. 7; Migne, *Patr. lat.*, xxxiv, 200). He devotes twenty-seven chapters, however (*De Genesi ad literam*, l. 7; Migne, xxxiv, 355–371), to the subject of the nature and origin of the human soul, and rightly insists upon man's possession of a spiritual soul as being the chief point of difference between man and beast. Every attempt to separate man absolutely from beasts with regard to his body (brain development, upright walk, &c.), or to raise him, as Bumüller does, to a special position as a branch of the animal kingdom, is doomed to failure, because it substitutes the accidental for the essential. All bodily differences between man and beasts are ultimately due to the fact that the human body is united with a rational soul. For this reason man, as *animal rationale*, towers above the whole animal kingdom, whilst in body he represents the highest class of mammal. Cf. my discussion of Bumüller's work *Mensch oder Affe?* in *Natur und Offenbarung*, XLVIII, 1902, pp. 122–126; see also my little work, *Menschen- und Tierseele*, Cologne, 1906.

[1] As I have already shown, the question of the origin of man is of a mixed character, and revelation and natural science are both concerned with its solution. It is most important to keep the various aspects of the question quite distinct, and not to confuse them. On this subject I may quote the following beautiful and weighty passage from Leo XIII's encyclical 'Providentissimus Deus,' November 18, 1893:

'*Nulla quidem theologum inter et physicum vera dissensio intercesserit, dum suis uterque finibus se contineant*, id caventes secundum S. Augustini monitum

From a purely philosophical point of view we cannot contribute much towards the solution of this problem. It is certainly not an indispensable part of the idea of the creation to believe that man as a whole was created directly by God, through an extraordinary interference with the laws of nature; body and soul may have been created by God in different ways, the former indirectly, the latter directly. All that is essential to the idea of the creation of the human body is that the atoms composing it should have been originally created by God, and that the laws governing the formation of the body from those atoms should also have been imposed upon matter by God's almighty power. We may still say of every human being that he is 'God's creature' both in soul and body, although only his soul is directly created, whereas his body is produced from his parents' germ-cells according to the laws of natural growth.

If we apply this consideration to the creation of the first man, we are confronted with two possibilities. We may regard it as seemly to assume that God created the whole man in full perfection, making use, it is true, of already existing atoms to compose the human body, but creating the spiritual soul, the chief part of man. To others, however, it may seem more fitting to believe that in producing the first man, as in

"ne aliquid temere et incognitum pro cognito asserant." Sin tamen dissenserint, quemadmodum se gerat theologus, summatim est regula ab eodem oblata: "Quidquid," inquit, "ipsi de natura rerum veracibus documentis demonstrare potuerint, ostendamus nostris Literis non esse contrarium; quidquid autem de quibuslibet suis voluminibus his nostris Literis, id est catholicae fidei, contrarium protulerint, aut aliqua etiam facultate ostendamus, aut nulla dubitatione credamus esse falsissimum." De cuius aequitate regulae in consideratione sit primum, scriptores sacros, seu verius "Spiritum Dei, qui per ipsos loquebatur, noluisse ista (videl. intimam aspectabilium rerum constitutionem) docere homines, nulli saluti profutura"; quare eos, potius quam explorationem naturae recta prosequantur, res ipsas aliquando describere et tractare aut quodam translationis modo, aut sicut communis sermo per ea ferebat tempora, hodieque de multis fert rebus in quotidiana vita, ipsos inter homines scientissimos. Vulgari autem sermone quum ea primo proprieque efferantur quae cadunt sub sensus, non dissimiliter scriptor sacer (monuitque et Doctor Angelicus) "ea secutus est, quae sensibiliter apparent," seu quae Deus ipse, homines alloquens, ad eorum captum significavit humano more.'

It follows from these words of Leo XIII that natural science is left perfectly free to investigate the origin of man. If science remains within its proper limits, its results can never come into real conflict with revelation. On this subject see Chr. Pesch, *De inspiratione S. Scripturae*, Freiburg i. B., 1906, pp. 409, &c.; Dr. N. Peters, *Bibel und Naturwissenschaft*, Paderborn, 1906, pp. 11, &c, 36, &c., 42, &c.

producing all other creatures, God employed natural causes as far as they were capable of co-operating towards this aim. The quotations from St. Augustine's 'De Genesi ad literam' (p. 437) may, perhaps, be interpreted in this sense although it would not be easy to grasp the full meaning of his words.

Whilst I am dealing with this subject, I may refer also to the opinion of St. Thomas Aquinas [1] regarding the succession of substantial forms of being in the ontogeny of man, and this from the purely philosophical standpoint, to some extent reveals a possibility of accepting a preformation of the first human body by way of evolution. At the first stage of embryonic development the human embryo would possess a merely vegetative soul, at the second stage an animal (vegetative and sensitive) soul, and not until the third stage was reached would a rational or spiritual soul be created and be

[1] Cf. St. Thomas, *Summa theol.* 1, q. 118, a. 2, ad 2; *Contra gentes*, l. 2, c. 89; *De potent.* q. 3, a. 9. As one of my critics has actually interpreted the first of these passages (*Summa theol.* 1, q. 118, a. 2) in a sense opposed to the idea of a succession of forms of being, it may be well to give an outline of the contents of this *quaestio*. The question raised by St. Thomas is: 'utrum anima intellectiva causetur e semine.' He mentions various reasons in favour of this opinion, but decides against it, and states the view of those who assume that there have been several different forms of being in the development of man (ad 2). He then declares himself clearly and definitely in favour of the succession of such forms, but against their simultaneous existence:

'*Et ideo dicendum est, quod anima præexistit in embryone, a principio quidem nutritiva, postmodum autem sensitiva, et tandem intellectiva.* Dicunt ergo quidam, quod supra animam vegetabilem, quae primo inerat, *supervenit* alia anima, quae est sensitiva; *supra illam* iterum alia quae est intellectiva. Et sic sunt in homine *tres animae*, quarum una est in potentia ad aliam; quod supra improbatum est q. 76, 3. Et ideo alii dicunt, quod illa eadem anima, quae primo fuit vegetativa tantum, postmodum per actionem virtutis quae est in semine, perducitur ad hoc, ut ipsa eadem fiat sensitiva, et tandem ad hoc, ut ipsa eadem fiat intellectiva, non quidem per virtutem activam seminis, sed per virtutem superioris agentis, scilicet Dei de foris illustrantis. . . . *Sed hoc stare non potest.*'

After giving his reasons for regarding the latter view as untenable, St. Thomas concludes thus: 'Et ideo dicendum est, quod cum generatio unius semper sit corruptio alterius, necesse est dicere, quod tam in homine quam in animalibus aliis, *quando perfectior forma advenit, fit corruptio prioris*; ita tamen, quod sequens forma habet quidquid habebat prima, et adhuc amplius; et sic per multas generationes et corruptiones *pervenitur ad ultimam formam substantialem* tam in homine quam in aliis animalibus.' According to the opinion here expressed by St. Thomas, there is in the ontogeny of man (and of beasts) a succession of different forms of being, gradually becoming more perfect, the lower form always ceasing *ex ipso* to exist, as soon as the higher succeeds. It was this thought which I took as the foundation for my comparison with the development of the race. Of course St. Thomas had no idea of such a comparison, for it lies quite outside the range of thought of the mediæval theologians. For this reason, in speaking of the creation of the human soul, St. Thomas adopts the view that the body and soul of the first man were created simultaneously (*Summa theol.* 1, q. 90, a. 4).

substituted for the previous forms, which had prepared matter for its union with the rational soul. It is true that at the present time many theologians have abandoned this Thomistic view, and prefer to believe that the rational soul is created at the moment of conception; but as this succession of forms in the development of the individual is by no means incompatible with the subsequent infusion of the rational soul, there would not necessarily be any contradiction involved, if a hypothetical evolution from a parent stock were assumed to have taken place in the case of the human body likewise.

We must therefore admit that it would be possible for anyone to account for the origin of the human body by assuming God to have created a primitive cell, and to say that the earliest ancestors of man were organisms living as simple cells; later on, as the organs were differentiated, and a nervous system was formed, and a sensitive soul came into existence, they developed into animals. The organism gradually increased in perfection, and, as the brain developed, this soul in course of time prepared a human body, suited to be the dwelling of a rational soul and, through possessing highly developed brain-centres, able to satisfy the conditions of spiritual activity and its verbal expression. Assuming this theory to be true, we may still say that man certainly only became man at the moment of the creation of his rational soul; in the previous stages it would, however, be wrong to say that he was simply a plant or simply an animal,—he was already a man in process of development; and thus in the hypothetical development of the race there would be a process analogous to that which we recognise in the ontogeny of the individual, the final form is the true *forma specifica*, which determines once for all the character of the whole cycle of development. According to this theory, the whole development of man occurred within one and the same natural species, viz. 'man,'[1] although scientific systematics may be obliged to classify the ancestors of man as distinct systematic species, genera, &c. I assign nothing more

[1] This manner of accounting for the origin of the human body through the action of the laws of organic development preserves man's dignity at least as well as the assumption that he was directly formed of inorganic matter. Any objection to the theory on this score may be met by a reminder that man's body even now is produced by germinal development from a fertilised ovum.

than a purely speculative importance to these suggestions, for there is an enormous difference between theoretical possibility and actual reality. Hitherto we have dealt only with the philosophical principles underlying the former; in the second part of this chapter we shall have to discuss the latter.

Let us now sum up shortly the results of the first part of our investigation into the origin of man.

Zoology, regarding man only from the point of view of his body, rightly describes him as the highest representative of the class of mammals, and this is true of his embryonic development also, which resembles that of other mammals. He is higher than the other mammals in the material equipment for the life of the soul, inasmuch as his brain is more perfectly organised and more highly developed. Thus far zoology and comparative nervous physiology are competent to judge of man, and philosophy may even admit that it is not impossible for the human body to have come into existence in the way indicated by the theory of evolution. Zoology and its attendant sciences are not, however, competent to judge of the nature and origin of the human spiritual life, because it is quite beyond their scope. Hence it follows that zoology cannot pronounce upon the phylogenetic evolution of man as a whole. It is limited to the somatic aspect of the question, and even here it cannot express a final opinion, because body and soul are united to form one man. The question of the origin of man is therefore of a mixed character;[1] and psychology, which takes into account his higher part, is best qualified to answer it; zoology and its attendant sciences are of subordinate importance, as they can judge only of his lower part. Psychology tells us that the higher part of man cannot be of animal origin, therefore all that is left for zoology and its attendant

[1] After what has been said above, it is scarcely necessary for me to draw attention to the fact that the question is of a mixed character also for another reason:—because not only the natural sciences but theology is concerned with it, since the creation of man touches a dogma of faith. Dogmatic and exegetical theologians are therefore fully justified in using much caution and reserve when they speak of the theory of descent, as they have to take into consideration both the obvious meaning of the story of creation, and decisions such as that of the provincial council at Cologne in 1860 (tit. IV, c. 14). A zoologist, botanist or chemist, who knows nothing of theology, is certainly no more qualified to express an opinion on matters of faith, than a theologian would be, knowing nothing of natural science, to discuss the evolution of *Ammonites* or *Paussidae*.

sciences is to answer the question of inferior importance: 'Must we nevertheless believe that the lower part of man is of animal origin?'

2. What Actual Evidence is there of the Descent of Man from Beasts?

(Plates VI and VII)

In discussing the theory of evolution in Chapter IX. I was careful to point out that the question how far we may regard the theory of evolution to be based upon facts has nothing to do with mere *a priori* possibilities, but means this: 'How far do facts furnish us with actual evidence in support of an evolution of the race?' We are confronted with this question: 'What actual evidence have we at the present time to show that man in respect of his body is descended from animal ancestors?' And the answer is this: 'The evidence is by no means clear and irrefutable, but in many ways it is obscure and contradictory.'

(a) *A Glance at the Comparative Morphology of Man and Beasts*

We are all familiar with the methods of Haeckel, Wiedersheim, and other upholders of Darwinism, who emphasise in an exaggerated and often quite misleading manner the well-known points of resemblance between man and the higher animals with respect to their bodies, and pass over the divergencies.[1]

'The structure of man as testimony to his past,' as described by Wiedersheim in 1887 and even in 1902 (when the third edition of his work appeared), would be a very weighty argument in support of the descent of man from beasts, if it did not contain so many one-sided and distorted statements; such writing unfortunately is characteristic of the Darwinian style of argument, using the name in its worst sense. If we

[1] With regard to the points of difference between men and apes, see J. Bumüller's little work, *Mensch oder Affe*, Ravensburg, 1900. Zoological reasons prevent me from accepting the author's opinion that, with respect to his body, man forms a distinct group in the animal kingdom. Cf. *Natur und Offenbarung*, 1902, pp. 122–126.

believed Wiedersheim, we should regard man of the present day as a mosaic, patched up of pieces resembling parts of animals, and of rudimentary organs, which he is supposed to have inherited from his noble ancestors. There is scarcely an organ in the human body, which Wiedersheim from his standpoint has not tried to use as testimony to the descent of man from beasts. Like Haeckel, he even depicts the prehuman forerunner of man in most minute details. He knows what his hairy covering was like, how the muscles of his skin were constructed, and how large the movable muscles of his ears were ; he knows that the eyes did not look straight forward, but were set sideways in the head, and that as compensation for this disadvantage, there was a third eye in the upper part of the head, which eye we now call the pineal gland. He has measured the length of the prehuman intestine and found it to be considerably longer than ours, because it served to digest nothing but a vegetable diet. He has traced the development of his protégé, and seen how he ceased to be a vegetarian and adopted a mixed diet, and procured a greater number of incisors and projecting canine teeth, thus transforming himself into a beast of prey, whilst his intestine grew correspondingly shorter. Before the hand of this primitive man could wield the stone axe, his teeth were his weapons, and his huge canine teeth projected like tusks. At the same time new formations developed on the larynx of our worthy ancestor, so that his voice acquired power and compass, and became a means of scaring away his enemies.

Wiedersheim describes our, or rather his, forefathers thus feature by feature, and presents us with a picture not in any way scientific, but absolutely imaginary. If we subject all his 'testimonies' collectively to serious criticism, none of them prove genuine. This was shown conclusively by Hamann [1] in his review of Wiedersheim's compilation, and G. Ranke, in his excellent work 'Der Mensch,' has carefully examined the alleged theromorphic forms of man, and has proved that, wherever they are not purely imaginary, they are to be regarded as formations due to arrest in the typical human development, We need not waste time with any further discussion of the

[1] *Entwicklungslehre und Darwinismus*, 1892, pp. 108, &c.

fanciful dreams of Wiedersheim and Haeckel, which have brought the zoological study of man into disrepute.

That there are many morphological resemblances between man and the higher mammals, and especially the higher apes, is an undeniable fact, that cannot be disputed. These resemblances afford a certain amount of zoological evidence showing that probably man is, in respect of his body, connected with the other mammals, but the evidence does not go beyond a probability. The differences between them are so great as not to admit of our coming to any definite conclusion on the phylogenetic question, and they extend to the fundamental structure of the skeleton. In comparing the thigh-bones of man and of the higher apes, O. Walkhoff[1] comes to the following conclusion: 'The radical difference goes so far as that it is possible to determine analytically from any X-ray photograph of a frontal section, and even from any complete piece of bone, whether it belonged to a man or to an ape; in other words, whether its owner walked upright or not.' The reader is requested to refer to Plate VI at the end of the book, and to compare the human skeleton with that of an orang utang (*Simia satyrus*), one of the highest apes. The great differences in the formation of the trunk and extremities are at once apparent, and there is no need to point them out.[2]

Plate VII shows the crania of man and ape respectively, and the difference between them is enormous. In the ape's skull the animal element is unmistakable, the face occupies a very large part of the head, whereas in man it is smaller, as in man the brain, the instrument of his spiritual life, is of greater importance than the jaws. A glance at Plates VI and VII will do more than pages of description to make the reader realise the differences, which cannot be got rid of by mere speculations and monistic postulates.

A conscientious zoologist will proceed with great caution in dealing with the so-called rudimentary organs, which are

[1] *Studien über die Entwicklungsmechanik des Primatenskelettes*, No. 1; 'Das Femur des Menschen und der Anthropomorphen in seiner funktionellen Gestaltung' (*Biolog. Zentralblatt*, 1905, No. 6, pp. 182, &c., esp. p. 184). See also J. Bumüller, *Das menschliche Femur nebst Beiträgen zur Kenntnis der Affenfemora*, Augsburg, 1899, p. 132.

[2] For a detailed account of these differences, see J. Ranke, *Der Mensch*, I, 437-441, and II, 3, &c., 203, &c.

supposed to afford conclusive evidence of man's descent from brutes. Many organs were at one time regarded as useless and rudimentary, because no one had yet discovered what purpose they served. For instance, the thymus and thyroid glands are now no longer reckoned as rudimentary organs, since investigations made by Kocher, Reverdin, Fano, Schiff, Vassale, and others have shown them to be important organs of metabolism, eliminating poisonous matter from the system, and their removal by operations is often followed by serious morbid symptoms.[1] The pineal gland, another organ formerly called rudimentary, and supposed to be a remaining trace of a third eye possessed by our animal ancestors, has now been recognised by Cyon as an organ securing equilibrium, and regulating the circulation of the blood at the base of the brain. It is quite possible that in course of time other 'rudimentary organs' will be found to serve some definite purpose. In the case of some, e.g. the atrophied muscles of the human ear, it is likely that they were better developed at some early period in the history of the human race, and degenerated later. This may be true also of the famous vermiform appendix of the coecum, at least in as far as a pathological formation is concerned, which often gives rise to morbid symptoms.[2]

(b) *The Biogenetic Law and its Application to Man*

But I may be asked—is it true that man in his embryonic development still passes through all those stages in rapid succession, through which his ancestors have once passed in their phylogeny?—for this is what should occur according to the famous biogenetic law, of which Meckel and Charles Darwin had some idea, although it was first enunciated by Fritz Müller, and afterwards elaborated by Ernst Haeckel (1866).

If we could trust Haeckel, we should have to answer this question in the affirmative. The first and second stages, in

[1] See O. Schulz, 'Neuere und neueste Schilddrüsenforschung' (*Biolog. Zentralblatt*, XXXVI, 1906, No. 21, pp. 754–768).

[2] See W. Ellenberger, 'Beiträge zur Frage des Vorkommens, der anatomischen Verhältnisse und der physiologischen Bedeutung des Coecums, des Processus vermiformis und des cytoblastischen Gewebes in der Darmschleimhaut' (*Archiv f. Anatomie u. Physiologie, Physiolog Abtlg.* 1906, pp. 139–186).

which the human ovum is unicellular, would be a repetition of the Moneron and Amœba stages in the phylogeny of man. The third or Morula stage would be a repetition of the Synamœbae. The fourth or blastula stage would be that of the Planaeada. The fifth or gastrula stage would be that of the Gastraeada, for these imaginary creatures consisted simply of a stomach. The sixth stage in the ontogeny of man would repeat that of the primitive or low worms, the seventh that of the soft worms, and the eighth that of the Chordata. This completes the first half of man's pedigree according to Haeckel. The second half begins with the Ascidia.

Next to the Chordata stage comes the ninth, in which the human embryo resembles the Acrania, or skull-less animals, which are represented now by the famous lancelet (*Amphioxus lanceolatus*). The tenth stage is that of the single-nostriled animals or Monorrhina, when we had round, sucking mouths. The eleventh is that of the primæval fish, when our ancestors had fins and gills, and presented the pleasing appearance of sharks. The twelfth stage is that of the mud-fish, the thirteenth that of the gilled Amphibians, and the fourteenth that of the tailed Amphibians. The fifteenth stage in the embryonic development of man is that of the primitive Amniotes; the sixteenth is that of the primitive mammals or Promammalia; the seventeenth is that of the pouched animals or Marsupials; the eighteenth is that of the semi-apes or Prosimiae; the nineteenth is that of the apes with tails; the twentieth is that of the anthropoid apes; the twenty-first is that of the ape-like men or Pithecanthropi; and finally, at the twenty-second stage, we arrive at *Homo sapiens*, and as such the infant enters the world at his birth.

There is no need to compose a satire upon Haeckel's 'Anthropogeny.' It made its appearance in 1874 and has since passed through several editions. It is enough to enumerate the twenty-two phylogenetic stages which the human embryo is supposed to 'recapitulate' before his birth, and this theory at once reveals itself as a fiction devoid of all foundation.

Some quite superficial resemblances between certain stages in the development of the human embryo and the final forms of other creatures, ranging from unicellular Amœbae to vertebrates, have been taken as the basis of a phylogenetic analogy, that has been drawn with more daring than logical accuracy. The gaps in the line of man's ancestry have been filled up with fanciful creatures, existing only in the imagination and described as primitive gastraeada, primitive amniotes, primitive promammals, primitive marsupials, pithecanthropi, &c., and then we are told to regard this pedigree as a scientific proof of the descent of man from beasts, in accordance with the biogenetic law!

Haeckel's phylogenetic stages in human embryonic development, as set forth in his 'Anthropogeny,' have already increased in number from twenty-two to thirty.

They are given in his lecture on our present knowledge of the origin of man, published in 1899 ('Über unsere gegenwärtige Kenntnis vom Ursprung des Menschen,' pp. 36, &c.) and there they bear the highly scientific name 'Progonotaxis of Man.' In Haeckel's latest work, 'Der Kampf um den Entwicklungsgedanken,'[1] which contains his three lectures delivered in Berlin, we find the same Progonotaxis on pp. 96, 97. It is the same sort of hoax—I know no milder expression applicable to it[2]—which Haeckel has been perpetrating for over twenty years, but it appears in an enlarged and by no means improved form. From the imaginary monera—those non-nucleate organisms that have no existence—he leads us along a series of thirty stages, each one decked out with high-sounding, scientific phraseology, until finally we reach the *Homines loquaces*—the speaking—or, more accurately, the chattering men of the present day. It would be a waste of time to dwell at greater length upon this fictitious series, by means of which Haeckel strives to show that he has successfully applied the biogenetic law to man.

Even if the 'law' had good reason for its existence, such an application of it to man would still be, to say the least,

[1] The title of the English translation is *Last Words on Evolution*.

[2] Some critics, e.g. K. Escherich, in the Supplement to the *Allgemeine Zeitung* (see 'A few Words to my Critics' in the preface to this edition), have found fault with me for having 'disparaged and ridiculed those scientific men who established and developed the theory of evolution.' The reference is no doubt to my use of words such as 'mischief,' 'hoax,' &c., in speaking of Haeckel. If Haeckel does not hesitate to make mischief and to perpetrate hoaxes in the name of science, no reasonable man will take it amiss that I feel bound to describe his methods in such language.

purely arbitrary. But we must now consider whether the biogenetic law has really any justification.[1]

Do facts warrant the assertion that the individual development of every creature is invariably an abridged recapitulation of the history of the race? No, they do not; for the exceptions to this rule are far more numerous than the instances of it. The majority of the stages in the evolution of the individual, through which the various species of animals pass at the present day, do not correspond to the hypothetical stages in the history of the race. Haeckel himself had an inkling of this truth, but he very cleverly tried to avoid the difficulty by distinguishing two elements in the ontogeny of the individual, viz. *palingenesis* (πάλιν-γένεσις), which is a recapitulation of the stages corresponding to the evolution of the race, and *cœnogenesis* (καινὴ γένεσις), which is a collective name applied to deviations from it. According to Haeckel, cænogenesis is a falsified or disturbed development, tolerated by nature under the compulsion of adapting the embryonic development of various organisms to altered circumstances. Haeckel was unhappy in his choice of words when he described the evolution as *falsified;* I should prefer to believe the falsification not to be on the part of nature in dealing with her own laws, but on the part of the prejudiced discoverer of these so-called laws.

It is impossible to maintain that the biogenetic law is a general law, giving an account of the ontogeny of the individual in accordance with the hypothetical phylogeny of the race. Haeckel goes so far as to refer to this 'law' the processes of segmentation, by means of which a multicellular organism is produced from a fertilised egg-cell, and he sees in this process a recapitulation of the phylogenetic development of multicellular animals from primitive unicellular forms. There is no justification at all for this theory, for, as Oskar Hertwig remarks in his 'Allgemeine Biologie' (p. 596): 'The whole nature of a unicellular organism makes it impossible for it

[1] For criticisms of it see especially O. Hertwig, *Allgemeine Biologie*, 1906, chapter 28, pp. 592, &c.; K. Fleischmann, *Die Deszendenztheorie*, 1901, chapters 13 and 14; J. Reinke, *Studien zur vergleichenden Entwicklungsgeschichte der Laminariaceen*, Kiel, 1903, No. 13; *Die Laminariaceen und Haeckels biogenetisches Grundgesetz*, pp. 57, &c.; A. Oppel, *Jahresberichte über die Fortschritte der Anatomie und Physiologie*, XX, 1892, p. 683; Karl Vogt, Beard, Hensen, Emery, Driesch, and others have also expressed their disbelief in the truth of the biogenetic law.

to be changed in any other way than by cell-division ; therefore the ontogeny of every living creature must inevitably begin with a process of cleavage.' This process has nothing whatever to do with the hypothetical phylogeny, for if there were no phylogeny at all, a multicellular organism could develop, grow and propagate itself only by way of cell-division. Consequently there must be some degree of resemblance between the processes of individual development in different organisms, as all alike are subject to the general laws of cell-division. The same idea is expressed by O. Hertwig (p. 595), when he says : 'That certain phenomena recur with great regularity and uniformity in the development of different species of animals, is due chiefly to the fact that under all circumstances they supply the necessary conditions under which alone the next higher stage in the ontogeny can be produced.'

These resemblances in the embryonic development of animals of various species have therefore nothing whatever to do with the hypothetical phylogeny.

Oskar Hertwig (p. 593) proposes to make some modifications in, and to add some elucidations to, the biogenetic law as understood by Haeckel. He says : 'We must leave out the words "recapitulation of forms of extinct ancestors," and substitute for them, "repetition of forms regularly occurring in organic development, and advancing from the simple to the more complex." We must emphasise the fact that in the embryo, as well as in the full-grown animal, the general laws governing the development of living organic matter are at work.' By this statement Oskar Hertwig has not 'modified' the biogenetic law, but has simply overthrown it ; for I cannot discover, in his manner of interpreting it, any suggestion of a recapitulation of the hypothetical phylogeny, but a repetition of general conformity to law in the development of living creatures.

In his 'Morphogenetische Studien' (Jena, 1903) Tad. Garbowski uses very similar expressions. He says : 'Most of what is generally ascribed to the action of the so-called biogenetic law is erroneously ascribed to it, for all that is undeveloped and incomplete must be more or less alike.'

As causal factors in the development of every individual, we have to distinguish three things :—

1. The general laws of growth in living matter, which depend upon the processes of cell-maturation and fertilisation, cell-division and cell-growth.

2. The special lines followed by these processes in consequence of descent from definite ancestors, or, in other words, owing to the direct action of heredity.

3. The special lines followed by these processes of growth in consequence of the adaptation of the organism to exterior influences, these being subsequently fixed by heredity.

The biogenetic law owes its origin to the fact that the second of these three factors has been violently torn from its natural connexion with the other two, and has been raised to the rank of an independent and universal 'law.'

The biogenetic law is not a fundamental law, but only under the most favourable circumstances is it even a partial law. The method by which it has attained its position is —when viewed from the standpoint of the theory of evolution— absolutely one-sided, and therefore altogether wrong, and in the twentieth century men of science should not be slow to perceive this.

E. Koken remarks very justly[1] that the biogenetic law originated in a superficial view of facts. 'The biogenetic law informs us that ontogeny in general is a recapitulation of phylogeny. Phylogeny however tells us that it too does not proceed at random, but is directed by the material on which it works, just as ontogeny is influenced by the plasm of the egg-cell.' Thus, just as in the fertilised ovum the tendency to develop is the real *Anlage* or basis of the individual development, so the tendency to develop, possessed by the primitive forms of the race, is the real *Anlage* or basis of the hypothetical development of the race. This is the true parallel between ontogeny and phylogeny.

Let us now turn once more to human embryonic development. We cannot be surprised if it bears a vague general resemblance, in some of its stages, to what may be permanent forms in the case of other animals. We should indeed expect to find such a likeness, for, in conformity to its inner nature, embryonic development, being dependent upon the processes

[1] *Paläontologie und Deszendenzlehre*, 1902, p. 226.

of growth, must make use of them, and must advance from what is simple to what is compound, and from what is general to what is particular.

It must, therefore, begin with a unicellular stage and pass through various multicellular stages, gradually approximating more and more closely to the final form at which the development aims. The development of the embryo as a whole, as well as of its single parts, must at different stages display different degrees of perfection, until at last the goal is attained. All these processes might occur successively in precisely the same way if no hypothetical phylogeny had preceded them. How can we venture to affirm with Haeckel that human ontogeny is quite unmistakably a recapitulation of human phylogeny ?[1] Such a theory is a mere matter of fancy !

There are, it is true, in the ontogeny of various animals certain stages which can be accounted for causally only by reference to the history of the race. This subject has been discussed in the chapter on the theories of descent and evolution, when, in speaking of the termitophile genus of Diptera known as *Termitoxenia*, I alluded to the temporary formation of real wing-veins in the development of the appendages on the thorax, and said that their presence proved the ancestors of our *Termitoxenia* to have been genuine Diptera.[2]

Similar phenomena occur in higher animals, although very rarely. A century ago (1807), the very interesting discovery was made by Geoffroy St. Hilaire, which has been recently confirmed by Kükenthal,[3] that the embryo of a whalebone-whale has teeth, although the adult whale has whalebone plates instead of teeth. Palæontological discoveries show that the earlier fossil whales of the Tertiary period were all toothed whales, retaining teeth throughout

[1] This overhasty assertion was accepted as true by K. Escherich in his criticism of the previous edition of my book (*Beiträge zur Allgemeinen Zeitung*, 1905, No. 55). The remarks that I have made above may serve as an answer to him as well as to Haeckel.

[2] Cf. Chapter X, pp. 384, &c. ; also 'Die Thorakalanhänge der *Termitoxeniidae*, ihr Bau, ihr imaginale Entwicklung und phylogenetische Bedeutung' (*Verhandl. der Deutschen Zoolog. Gesellschaft*, 1903, pp. 113–120 and Plates II and III).

[3] Cf. R. Keller, *Das Leben des Meeres*, Leipzig, 1893, p. 301, in the chapter on aquatic mammalia.

their whole life. We are therefore not merely justified in concluding, but we are almost forced to conclude, that our present whalebone-whales are descended from toothed whales, and that the foetal teeth are a phylogenetic reminiscence which serves no biological purpose, as the whale embryo, like that of all other mammals, has nothing to masticate.

Instances of this kind go far to prove that the theory of descent is at least probably correct, for they admit of only one interpretation. If it were possible to point to similar stages in the ontogeny of man, admitting of only one interpretation, viz. that they are after-effects of his earlier phylogeny, we should have very weighty evidence in favour of the theory that man, in respect of his body, is descended from brute ancestors. But so far no such phenomena have been observed in the case of man.

If we, for instance, examine closely the so-called 'shark-fins' and 'fish-gills' of the human embryo, we shall find them to be formations playing quite another part in the embryonic life, and having therefore a direct reason for their existence in the circumstances under which the embryo develops. We are certainly not bound to infer from their superficial likeness to real fins and real gills that our ancestors were once fishes. In order to satisfy Escherich and other critics, I should like to say a few words on the subject of the branchial clefts and arches, which are regarded as traces of gills. They occur in man and in all vertebrates, but only in fishes do they develop into real, permanent gills. The embryo of man and other mammals has on its neck four so-called branchial clefts and three so-called branchial arches:[1] the first branchial arch is the largest, and eventually forms the oral cavity and the parts belonging to it; the second arch is less developed, and the third is unimportant. Of the so-called branchial clefts separating the arches, only one has any permanence in man, it forms chiefly the external auditory meatus, the others close up again. The three branchial arches partly are transformed into particular organs, partly they become cartilaginous and change into definite parts of the adult body, either permanent or having some considerable duration.

[1] See Ranke, *Der Mensch*, 1, pp. 145, &c.

They form the Meckel's cartilage on the lower jaw, the two delicate auditory ossicles, known as the malleus and incus, as well as the hyoid bone and the styloid process. In fishes, however, the embryonic branchial arches and clefts remain and form the permanent gills. The pharyngeal arches and clefts in the human embryo bear a superficial likeness to the gills of fish, and so they have been called *branchial* arches and clefts, whereas they are really indifferent pharyngeal extroversions in the embryo, supplying the material for other subsequent formations. Can any one seriously regard them as evidence that our forefathers were once fish, and that the embryonic development 'recapitulates' this former fish-stage?

Every thoughtful reader will see that there is a vast difference between fanciful interpretations of phenomena, such as I have mentioned, and genuinely scientific attempts to account for them.

Again, the young of the black Alpine salamander (*Salamandra atra*) are born as land-animals, breathing by means of lungs, but before their birth, whilst still in the Fallopian tubes of the mother, they have large tufted gills and a tail-fin like genuine water animals. In this respect they exactly resemble the larvae of the spotted salamander (*S. maculosa*), which are born at an earlier stage of development, and are at first aquatic, so that they really use their gills and tail-fin, before they become land-animals. The question naturally occurs: 'Why have the larvae of the Alpine salamander gills and tail-fin, when they never, at any period of their life, can use them?' The only obvious answer is: 'Because, like the larvae of all other Urodela, they were originally intended to live in water, and subsequently, in consequence of the period of development being shortened, they were born as complete land animals.'

The difference is obvious between the *real* gills of these salamander larvae and the *imaginary* gills, which the human embryo is said to possess as a reminder of the time when his ancestors were fish.

Again, if we consider the ontogeny of certain parasitic Copepods among the Crustaceans, e.g. in the genus *Lernaea* (see Chapter X, p. 327, note 1), we shall find that at an early stage these creatures resemble other Copepod larvae, but the adult female's body is simply a bag of eggs, and is shaped like a sausage. It cannot be denied that in this case the ontogeny of the individual suggests unmistakably that the parasitic genus *Lernaea* is descended from Copepods once leading an independent existence, and gradually adapted to a parasitic way of life. But I say again emphatically: in the ontogeny of man we know of no such phylogenetically unquestionable phenomena.

The resemblances between the human embryo and that of the other vertebrates are so superficial that His, W. von Bischof, and even Karl Vogt, and many other recent and thorough students of comparative embryology, have protested against Haeckel's regarding these resemblances as phylogenetically significant identities.[1] Nothing but gross want of knowledge can excuse a man at the present day for bringing forward this *argumentum ex ignorantia* in support of this descent of man from beasts.[2]

* * * * *

We might perhaps close our investigation of the zoological evidence for the descent of man from beasts at this point. It may, however, be well to give a short sketch of the two chief theories on this subject, so that the reader may know how the question stands at the present day.

These two theories are antagonistic to one another. The first is practically only an extension of Karl Vogt's Ape-theory. It assumes a direct relationship between man and the anthropoid apes, the so-called primates, and, with Friedenthal, it proclaims man to be simply a genuine ape. The second theory on the contrary denies that man is directly related to the present apes, but admits the existence of a distant, indirect connexion, inasmuch as it traces the descent of both from a hypothetical common stock, which is supposed to have lived in the Older Tertiary or Pre-Tertiary period.

[1] The story of the three illustrations by means of which Haeckel tried to prove this identity in his *History of Creation*, is too well known for it to be necessary to discuss it here. Cf. O. Hamann, *Entwicklungslehre und Darwinismus* (1892), pp. 26, &c. Also E. Dennert, *Die Wahrheit über Ernst Haeckel und seine Welträtsel*, 1904, chapter iii, p. 16, &c.

[2] On this subject see J. Ranke, *Der Mensch*, I, pp. 152–154.

(c) *The Theory that Man is directly related to the Higher Apes*

Let us now examine more closely the first of these two theories. It is held by many modern zoologists, and the following evidence has recently been adduced in support of it. Selenka discovered that the higher apes resemble man during their embryonic development in having a simple discoid placenta, whilst the lower apes have a bidiscoidal placenta. It would, however, be rash to regard this discovery as a proof of direct relationship between man and the higher apes, the value of the new piece of evidence is not greater than that afforded by a number of other well-known morphological and embryological resemblances between man and apes, for in this case also the question arises : ‘ Are these resemblances the result of close relationship, or are they merely converging phenomena, due, not to community of origin, but to adaptation to similar conditions of life or development ? ’

The following consideration shows how much caution is necessary in regarding the formation of the placenta as evidence for the theory of descent. In the Monotremes, which are the lowest mammals, the placenta is absent, and in the Marsupials it occurs only rarely and in a very imperfect form, but the higher mammals are called placentals, as the possession of this organ distinguishes them from the two former subclasses. On the other hand, as Aristotle discovered, and as Johannes Müller found in the nineteenth century, a placenta occurs in the smooth shark (*Mustelus laevis*) and in its relations belonging to the genera *Mustelus* and *Carcharias*, only its vessels are supplied by the yelk-sac, and not, as in mammals, by the allantois. Quite recent research is believed to have revealed the presence of a placenta even in some Arthropods, Kennel has seen it in the American *Peripatus*, and Poljansky in the Indian scorpion.[1] This shows that the existence of a placenta, and still more its peculiar structure, have not, necessarily, anything to do with a direct relationship between the animals in question. Otherwise we should be obliged to regard the Indian scorpion as the ancestor of the placental mammals, the highest of which is man.

[1] *Zoolog. Anzeiger*, 1903, No. 2, pp. 49–58.

No zoologist would venture to draw such a conclusion, but he would prefer to ascribe the occurrence of a placenta in such diverse kinds of animals to independent convergence, as the formations are merely analogous and not homologous.

Not long ago, Dr. Hans Friedenthal[1] thought that he had discovered fresh evidence proving man to be directly related to the primates. As his communications have attracted a good deal of attention in circles interested in popular science, and will probably continue to do so, I propose to examine them critically.

Friedenthal has made a number of experiments, that are neither complete nor conclusive, with a view to investigating the transfusion and reaction of blood. The blood-relationship, that he professes to have discovered between man and the primates, is based upon his observation that human blood destroys the red corpuscles in the blood of the lower apes, but has no such effect upon that of the anthropoid apes. Whether this is a fact or not is still very doubtful, for not many experiments have been made, and the results of those that were made are not altogether uniform. In some cases the serum of the blood of a lower ape (*Macacus sinicus*) destroyed the red blood discs in human blood, and in other cases it did not. We do not yet know whether the serum of human blood never destroys the red blood corpuscles in the blood of the anthropoid apes, and *vice versa*. Friedenthal acted somewhat prematurely in using some probabilities as the foundation of a general law, according to which he proclaimed man to be a blood-relation of the higher apes.

Antiserum and blood-serum have opposite results in experiments on reaction. Antiserum is derived from animals which have been rendered immune from the destructive action of the blood-serum of another species; and it affects only harmonic or similar kinds of blood, and has no effect upon dissimilar. Nuttall[2] has examined the blood of eighteen kinds

[1] 'Über einen experimentellen Nachweis der Blutverwandtschaft' (*Archiv für Anatomie und Physiologie*, Physiolog. Abt., 1900, pp. 494-508) 'Neue Versuche zur Frage nach der Stellung des Menschen im zoologischen System' (*Sitzungsberichte der Kgl. Akademie der Wissensch.* XXXV, Berlin, July 10, 1902, pp. 830-835.

[2] G. H. F. Nuttall, 'The new biological test for blood in relation to zoological classification' (*Proceed. Royal Society*, London, LXIX, 1901-1902, No. 453, pp. 150-153); *Blood Immunity and Relationship*, London, 1904. Cf. also

of apes in its relation to human blood, and has found that they all showed reaction to the antiserum of human blood, but in very different degrees. Anti-ox-serum showed reaction also, not only to the blood of other Bovidae, but also, though in a less marked degree, to the blood of sheep, goats, antelopes, and gnus, although these animals are systematically not closely related to the Bovidae.

Even if it is definitely proved that human blood possesses certain chemico-physiological properties in common with the blood of the anthropoid apes, whilst these properties are wanting to that of the lower apes and other vertebrates, we shall still not be able to infer from this proof that there is a direct blood-relationship between man and the primates in the sense of the theory of descent. Such an inference would be based upon an obvious confusion of two quite different ideas, viz. resemblance in the chemical properties of two kinds of blood, and identity of phylogenetic origin of two kinds of blood. If anyone confuses these two ideas by skilful jugglery, the blood-relationship between man and the chimpanzee may indeed appear to be proved—but only to an uncritical public. The proof will be logically convincing only if it has been previously established, that a similarity in the chemical reaction of two kinds of blood depends solely upon the existence of direct blood-relationship between the animals possessing this blood, and no one can maintain this to have been established. Friedenthal himself declared not long ago that the haemolysis of the serum of any species depended also upon other factors, quite unconnected with genealogical relationship. In the case of the serum of eel's blood the reaction upon the blood of other vertebrates is greatest, with the serum of the blood of amphibia it is weak, with that of reptiles and birds it is strong. From the chemical reaction of two kinds of blood upon one another it is impossible to draw any inference for or against the relationship of the animals in question. According to Friedenthal's own experiments, the blood of a Crustacean (the common crab, *Cancer pagurus*) or that of a lug-worm (*Arenicola piscatorum*) did not destroy the red blood corpuscles of a sea-mew or a rat ;

E. Abderhalden, 'Der Artenbegriff und die Artenkonstanz auf biologisch-chemischer Grundlage' (*Naturwissensch. Rundschau*, XIX, 1904, No. 44, pp. 557-560).

but surely no one would infer that, for this reason, rats must be directly descended from lug-worms, or seamews from crabs! Nor is there any justification for drawing such an inference when we meet with the same phenomenon in connexion with the blood of man and of the orang-utang. We might in fact reverse the whole argument and say: 'Just as the rat cannot be the direct descendant of the crab, nor the sea-mew of the lug-worm, so man cannot be directly descended from an orang-utang, for his blood reacted upon that of an orang-utang no more than the blood of a crab upon that of a rat, or the blood of a lug-worm upon that of a sea-mew.'

Arguments, that need only to be simply reversed in order to prove the exact opposite of what they are intended to show, are obviously very weak. One and the same phenomenon, viz. the chemico-physiological indifference of two kinds of blood towards one another is interpreted in two different ways in Friedenthal's account of his experiments, according as it suits his purpose. On the one hand, mutual indifference of the blood of man and the anthropoid apes is due to the great similarity between them; on the other hand, mutual indifference of the blood of the lower animals and vertebrates is due to the great dissimilarity between them; the same result is referred to two totally opposed causes according to Friedenthal's subjective requirements!

The experiments made in the last few years by Bordet, Wassermann, Schütze, Stern, Friedenthal, Nuttall, Uhlenhut, and others with the serum and antiserum of the blood of a great variety of animals are no doubt of great scientific interest, and in many cases they supply us with valuable clues towards establishing the systematic relationship of various kinds of animals. Men of science will gradually learn to avoid Friedenthal's mistake of overestimating the importance and bearing of the information thus supplied. All that we can learn from such studies with regard to man is that he stands nearer to the higher than to the lower apes and other mammals in the composition of his blood, just as he has long been known to stand nearer to them in respect of the tissues and organs of his body. This line of research will not reveal more. As soon as an attempt is made to ascertain the phylogenetic relationship of animals from the reaction of antitoxins, the defects in this method become apparent, as well as its advantages. They have both been discussed recently by Robert Rössle.[1] These reactions do no more than furnish 'a standard of slight absolute value for estimating the degree of relationship; the reaction justifies this comparison: 'animal A is more closely related to animal B than is animal C'; but it gives us, strictly speaking, no means of judging how close the relationship is.' It would therefore be a serious mistake to conclude with Friedenthal from the reactions of the blood of men and apes that man is descended from the higher apes, or that he is merely a higher ape himself. Rössle considers that there is no reason for assuming that the chemical composition of the fluids in the body is more constant than the formation, for instance, of the skeleton. If he is right, the chemico-physiological resemblance between the blood of man and that of the primates is less important, from the standpoint of evolution, than the resemblances in the structure of their skeletons. Moreover, we have learnt from the experiments in reaction made during the last few years, that many actual contradictions are involved in the theory that the chemico-physiological resemblance of two kinds of blood, which is known as 'blood-relationship,' really involves identity of origin. Rössle remarks on this subject: 'Again, an antiserum shows us two animals as closely connected, whilst they are far apart in the morphological system.'

Finally—and this point is particularly important in our present discussion,—recent investigations have shown the physiological identity of the blood of man and of primates (which Friedenthal maintains) to be at least very doubtful. At the Anthropological Congress at Greifswald in 1904, Uhlenhut spoke of positive reaction, that he had observed, of human antiserum with the blood of lower apes. Friedenthal himself lately mentioned having obtained positive results by mixing human antiserum with the blood of Lemuridae. These statements destroy the force of any evidence based upon such reactions and adduced in support of the direct relationship between man and the anthropoid apes. It seems as if the wish had been the father of the thought in investigating their

[1] 'Die Bedeutung der Immunitätsreaktionen für die Ermittlung der systematischen Verwandtschaft der Tiere' (*Biolog. Zentralblatt*, 1905, Nos. 11 and 12).

alleged blood-relationship, and more unprejudiced research may altogether remove the enthusiasm with which this discovery was greeted. The latest ultra-microscopical examinations have revealed in human blood certain peculiarities, which were hitherto quite unknown. Raehlmann[1] has examined the blood of man and of various animals, and has discovered very considerable differences in the ultra-microscopical structure of the red-blood corpuscles. In human blood, for instance, within the strongly marked diffraction rings at the outside of the blood corpuscles, there are one or two polar bodies which do not occur in the blood of other animals, but are replaced by quite different formations. Finally, Brumpt has succeeded in establishing the fact that sleeping-sickness, which is conveyed by parasites in the blood (trypanosomes), can be produced in all mammals by inoculating them with the blood of a person suffering from the disease, the only exceptions being a few apes and the pig (*La Nature,* April, 28, 1906, Nos. 17 and 18, p. 339). As this inoculation involves a reaction, just as much as the experiments on blood-relationship, we should have to infer from these results that human blood is 'less closely related' to that of apes and pigs than to that of other mammals. In future more prudence ought to be displayed in drawing inferences of this kind!

It is therefore obvious that the newest 'proofs' of the blood-relationship between man and the primates do not justify the conclusion that has been based upon them, and Hans Friedenthal's triumphant statement, made on the ground of the alleged blood-relationship between man and the higher apes—'We are not merely the descendants of apes, but we are ourselves genuine apes'—is seen to be devoid of all justification.

Hitherto absolutely no real proof has been adduced of the ape-theory, i.e. the theory that man is directly related to the higher apes. I may venture to say that in all probability no proof ever will be adduced, for this theory is quite irreconcilable with the second of the above-mentioned theories regarding the descent of man from beasts, and there is far more evidence in support of the latter.

[1] Cf. W. Berg, 'Ultramikroskopie' (*Naturwissensch. Rundschau,* 1906, No. 28, pp. 353, &c.).

(d) *The Theory of the Remote or Indirect Relationship between Man and Apes*

Let us now turn to this second theory, according to which man is not directly descended from the primates, and is in fact not closely related to them. This theory regards man on the one hand, and apes on the other, as the extremities of two lines of evolution, absolutely independent of one another, but meeting in a purely hypothetical common ancestral form, which existed at the beginning of the Tertiary period, or probably even earlier. This opinion is held by Professor Klaatsch [1] of Heidelberg, M. Alsberg,[2] C. H. Stratz,[3] and many other anthropologists.

What are we to think of this theory ?

In itself it is far more acceptable than the ape-theory. It takes into account the phenomenon upon which much stress has been laid by the most eminent anthropologists, Johannes Ranke, Rudolf Virchow, Julius Kollmann, and others, viz. that the bodily structure of man and apes respectively represents two distinct lines of evolution among mammals, diverging widely at their extremities. In some respects, for instance in the development of the hands, the apes have outstripped man, and left him at a comparatively backward stage. Considered from the point of view of the evolution theory, the human hand bears far more resemblance to that of the zoologically lower apes than to that of the highest anthropoid apes, and the human foot is rendered quite unlike the prehensile foot of an ape by the peculiar position of the big toe. I do not, however, propose to discuss the bodily differences between man and ape in this place. They are stated very fully in J. Ranke's 'Der Mensch,' and Bumüller's little work, 'Mensch oder Affe ?' (Man or Ape ?),[4] contains a very clear description of them.

The more perfect development of the brain and the upright

[1] 'Entstehung und Entwicklung des Menschengeschlechts' (*Weltall und Menschheit*, edited by Hans Kraemer, II, 1903, pp. 1–338).

[2] *Die Abstammung des Menschen und die Bedingungen seiner Entwicklung*, Cassel, 1902.

[3] *Naturgeschichte des Menschen*, Stuttgart, 1904; *Zur Abstammung des Menschen*, 1906.

[4] Ravensburg, 1900. Cf. my remarks on p. 438 and p. 445, note 1.

position that it necessitates, which is connected with further corresponding differences in the structure of the extremities—these are the chief points bearing upon our subject, and, when they are considered in their purely zoological aspect, they justify our regarding man, in respect of his body, as forming a special order among mammals. On this point, but only on this, I agree with Moritz Alsberg,[1] who sums up the results of investigations made by Klaatsch and other anthropologists in the following terms: 'That man is directly descended from apes is inconceivable, and it is possible to speak of relationship existing between man and ape only in as far as both are ultimately connected at the root of their common genealogical tree, and this applies to all mammals.'

Are we then to adopt this view of the descent of man from beasts? I am far from doing so, for the following weighty considerations are opposed to it.

Firstly. Klaatsch assumes the existence, in the Tertiary or Pre-Tertiary period, of a hypothetical common ancestor of men and apes; but such an ancestor exists only in his imagination.[2] The properties ascribed to this original form, that he calls the 'general pithecoid type,' are so vague and indefinite, and to some extent so conflicting, that I cannot help regarding this primitive ancestor of man and ape as a *Universale a parte rei*, incapable of any real existence.

At the Anthropological Congress at Lindau in 1899, in speaking of Klaatsch's opinions, Johannes Ranke remarked: 'Whilst a charming picture of the past and possibly of the future is being shown us, and whilst a fanciful design is being carried out in all directions, we are as a rule in quest of facts, not of theories. The facts, however, upon which Herr Klaatsch claims to base his ingenious theory, do not at present exist, and I must protest against his assuming that they have been really furnished by zoology and palæontology any more than by anatomy. . . . All else is still a matter of hypothesis, and if anyone attempts to use it in order to produce a finished picture, the result is a work merely of the imagination.'

Secondly. In considering the origin of man, we must

[1] *Die Abstammung des Menschen und die Bedingungen seiner Entwicklung*, pp. 77–78.
[2] Cf. also *Stimmen aus Maria-Laach*, LVIII, 1900, pp. 471–477.

have recourse to palæontology as well as to comparative morphology. We must inquire what the former science can tell us of the ancestors of man from their fossil remains, and the further back we set the existence of the hypothetical common ancestor of man and apes, the more forms shall we call upon palæontology to show us intermediate between this common ancestor and the modern representatives of the two lines descended from him.

What answer does palæontology make to our question? She does not merely say: 'The missing link between man and ape has not yet been discovered.' Klaatsch's theory does not indeed admit of the existence of a direct link between the two. But palæontology tells us far more than this, and, relying on the results of most recent investigations, she says: 'We have the pedigree of the present apes, a pedigree very rich in species and coming down from the hypothetical ancestral form of the oldest Tertiary period to the present day. Zittel's "Grundzüge der Paläontologie" gives a list of no fewer than thirty genera of fossil Pro-simiae and eighteen genera of fossil apes, the remains of which are buried in the various strata from the Lower Eocene to the close of the Alluvial epoch, but not one connecting link has been found between their hypothetical ancestral form and man of the present time: **the whole hypothetical pedigree of man is not supported by a single fossil genus or a single fossil species.'**

How extraordinary! If man were really descended from a prehistoric ancestor, common to him and to the apes of the present day, there must surely be some fossil trace left of his branch of the genealogical tree, and not only traces of the branch leading to apes! [1]

I should like to commend this scientific truth to the serious consideration of all those who regard the descent of man from

[1] It might, perhaps, be possible to raise the objection that the evolution of the *prosimiae* and of the true apes was a slow and gradual process, and that of the human race rapid and sudden. This might account for the absence of fossil forms standing between the hypothetical primary form and modern man. But this statement cannot be reconciled with the palæontogical fact that man did not appear upon the earth before the Alluvial epoch. If he had been evolved rapidly and without any long transitional stages from an early Tertiary form, we should certainly find traces of Tertiary man as well as of Tertiary apes. Cf. on this subject R. de Sinéty, 'L'Haeckélianisme et les idées du P. Wasmann sur l'évolution' (*Revue des Questions Scientifiques,* January 1906), reprinted separately, p. 18.

beasts as actually proved, or who hope that it will be actually proved in the near future. As a critical student of nature, I am bound to express my fears that the upholders of this theory will find themselves disappointed.

3. Criticism of Recent Palæontological and Prehistoric Evidence for the Descent of Man from Beasts.

(a) The Upright Ape-man (*Pithecanthropus erectus*)

Let us now turn to the consideration of certain points which have recently been brought forward by students of palæontology and early history as evidence of the descent of man from beasts.

We must consider first the famous ape-man, *Pithecanthropus erectus*, of Java. So far the only remains that we have of him are a cranium, a femur or thigh bone, and two molar teeth discovered in 1891 in Pliocene deposits near Trinil by Eugène Dubois, a Dutch military surgeon, who gave an account of them in an address delivered at the Third International Congress of Zoologists in Leyden, in September 1895. He sought to prove that the creature, which he reconstructed from these remains, was neither man nor ape, and could only be a connecting link between them. Virchow, as president of the meeting, uttered a very courteous but crushing criticism upon the speaker's remarks, and showed that it was by no means certain that the remains had all formed part of the same individual, and that it was still less possible to decide whether that individual was a man or an ape, since the femur resembled that of a man, but the cranium seemed to be more like that of an ape. He went on to say that probably it would not be possible to decide finally upon the systematic place of the *Pithecanthropus* until a complete skeleton was discovered. In spite of all the controversy concerning the ape-man in the years following Dubois' discovery, Virchow's criticism still holds good. It is nothing short of an outrage upon truth to represent scanty remains, the origin of which is so uncertain as that of the *Pithecanthropus*, as absolute proof of the descent of man from beasts, in order thus to deceive the general public.

2 H

It cannot be maintained that the *Pithecanthropus erectus* is a real transitional form connecting man with the higher apes; for, as man and ape, from the point of view of comparative morphology, are the extremes of two widely diverging lines of evolution, there can have been no recent link between them, living as late as the Pleistocene or late Tertiary period. Moreover, although the *Pithecanthropus* possesses many peculiarities which seem to place him midway between ape and man, he has also others of a quite different kind, which seem to assign him a place between the lower and the anthropoid apes of the present day.[1]

Professor Schwalbe would certainly do his utmost to assign a high degree of importance to the *Pithecanthropus,* and to place him as near as possible to man, yet he pointed out these latter peculiarities in the course of his examination of the famous calvaria from Java.[2]

For this reason Klaatsch, Schwalbe, Alsberg and other not over-sanguine anthropologists do not agree with Eugène Dubois in regarding his *Pithecanthropus* as the long-sought ape-man, who was described prophetically by Haeckel a quarter of a century earlier. They prefer to regard him as a lateral branch of the pithecoid stock, which, in consequence of so-called 'convergent phenomena,' approximates to man in many respects. Therefore, the *Pithecanthropus* does not belong to the pedigree of modern man, but to that of the modern apes, and so he ceases to be a witness for the descent of man from beasts. I may refer to a few recent opinions on the subject of the *Pithecanthropus,* given by men who cannot be suspected of partiality.

In his 'Lehrbuch der Zoologie' (seventh edition), Richard Hertwig alludes to the remains of the *Pithecanthropus* and says: 'The fragments were regarded by some as belonging to a connecting link between apes and man, *Pithecanthropus erectus* Dubois; by others they were thought to be the remains of genuine apes, and by others again to be those of genuine men. The opinion that is most probably correct is that the fragments belonged to an anthropomorphic ape of

[1] Cf. also Alsberg, *Die Abstammung des Menschen,* pp. 100, &c.

[2] In his *Vorgeschichte des Menschen,* 1904, p. 29, he again says that the *Pithecanthropus* has no place in the genealogical line of man's direct ancestors.

extraordinary size and an enormous cranial capacity, and with a relatively very large brain corresponding to this cranial capacity (*circa* 850 c.cm.). The structure of the femur suggests that the animal probably walked upright.' [1]

Macnamara has recently submitted the skull of a chimpanzee and the much-discussed *Pithecanthropus* cranium to a very careful comparison and examination, in consequence of which he has arrived at a similar conclusion, namely that the *Pithecanthropus* was a true ape of large size.[2] He examined both crania according to Schwalbe's newest methods of taking measurements. In fig. 53 (p. 469) curve IV represents the contour of the Java cranium and curve V that of the chimpanzee cranium. Almost the sole difference between them is in size, and for this reason Macnamara gives it as his opinion that 'the cranium of an averge adult male chimpanzee and the Java cranium are so closely related that I believe them to belong to the same family of animals—i.e. to the true apes.' [3]

(*b*) *The Neandertal Man and his Contemporaries*

The *Pithecanthropus*, however, no longer stands alone, he has found a companion, rather younger than himself, in the Neandertal man, who likewise is supposed to have been neither a man nor an ape, such as now exist, but something between the two. We owe this discovery to Professor Schwalbe of Strassburg.[4] The remains of the skeleton of the Neandertal man were found in a cave near Düsseldorf in August 1856. The cranium was described by Schaafhausen in *Müller's*

[1] Whether this is the case or not might probably be determined by Walkhoff's method of X-ray photography. It has been suggested that the *Pithecanthropus* possessed the power of speech, because in his cast of the interior of the Java calvaria, Dubois found the third inferior gyrus (Broca's convolution) to be double the size that it is in anthropoid apes, though only half what it is in man (Schwalbe, *Vorgeschichte des Menschen*, p. 18). This discovery on a skull that has been decaying for thousands of years is of a nature no less problematical than is its psychological significance.

[2] Kraniologischer Beweis für die Stellung des Menschen in der Natur (*Archiv für Anthropologie*, XXVIII, 1903, pp. 349–360).

[3] If Macnamara nevertheless asserts that the Java cranium bridges the wide interval between the anthropoid apes and the Neandertal man, his assertion is unjustifiable, for the larger cranial capacity is not enough by itself to justify it.

[4] See G. A. Schwalbe, 'Der Neandertalschädel' (*Bonner Jahrbücher*, 1901, No. 106, pp. 1–72, with Plate I); also *Stimmen aus Maria-Laach*, LXI, 1901, pp. 107, 108.

Archiv for 1858 in an article headed 'Zur Kenntnis der ältesten Rassenschädel.' Fig. 52 is a reproduction, reduced in size, of Schaafhausen's photograph, giving a side view of this famous cranium (1888).

Numerous articles have been written on the subject, and in 1901 another thorough examination of the skull was made by Schwalbe, who finally pronounced the Neandertal man to have been a representative of a distinct genus, standing between ape and man.

We must admire Schwalbe's ingenuity in adding a twelfth

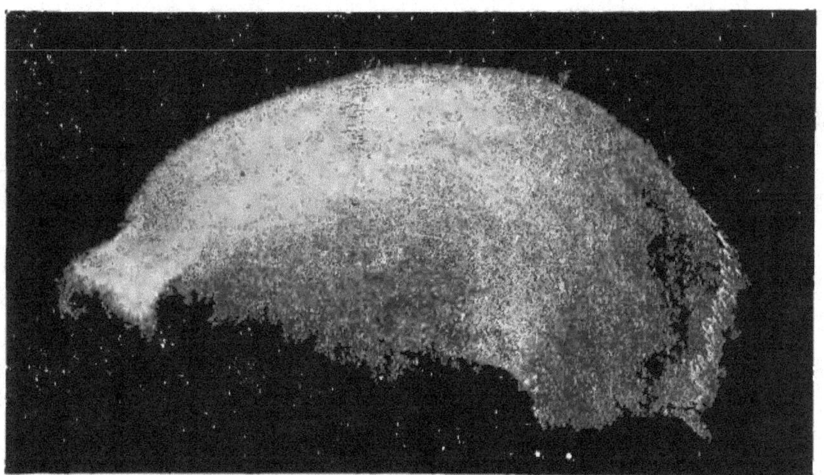

Fig. 52.—Neandertal cranium.

to the already existing eleven opinions regarding the Neandertal man, but he cannot claim any greater authority for his view than the other writers can claim for theirs, which are quite different. It has fallen to the lot of this Neandertal man to be described variously as an idiot, a Mongolian Cossack, an early German, an early Dutchman, an early Frieslander, a connexion of the Australian blacks, a palæolithic man, and a still more primitive ape-man. The remains of his skeleton clearly are of a nature to admit of many interpretations, and each student can make of them whatever he wishes. It would be wrong to assume that a discovery of this kind justifies scientific men in declaring that they have found the long-sought missing link between ape and man.

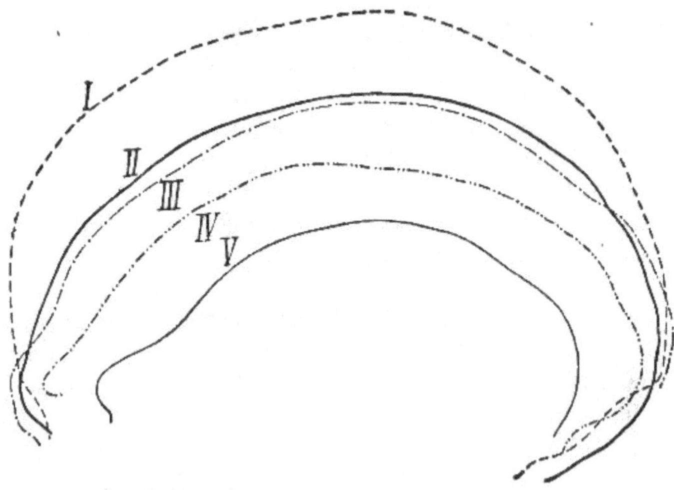

FIG. 53.—Outlines of the sagittal median curves, drawn with Lissauer's diograph :

I. Skull of modern Englishman.
II. Skull of modern Australian black.
III. Neandertal skull.
IV. Pithecanthropus skull.
V. Chimpanzee skull.
(After Macnamara.)

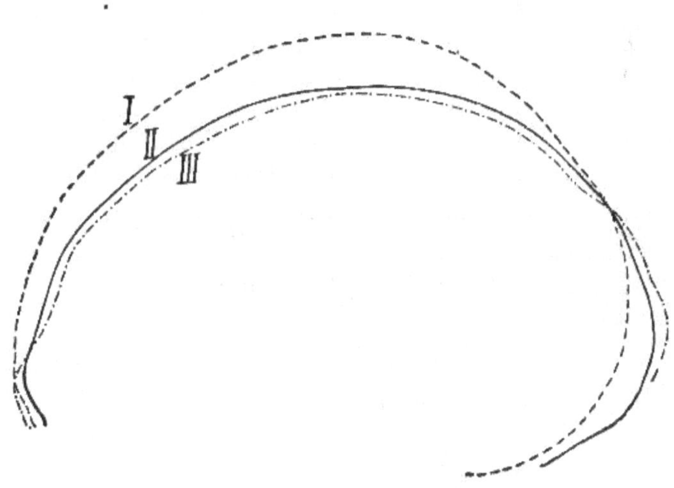

FIG. 54.—Outline of the sagittal median curve :

I. Of the skull of an early brachycephalic Lapp.
II. Of the skull of a dolichocephalic Australian.
III. Of the Neandertal skull.
(After Macnamara.)

The uncertainty regarding the Neandertal remains is increased by the fact that we have no means of judging their geological age; for, as Rauff[1] pointed out recently, no competent judge saw the Neandertal skeleton, in its original position (*in situ*). When Fühlrott, its scientific discoverer, reached the place where it had been found, the workmen in the quarry had already thrown the loam containing the bones out of the cave, and had partially destroyed the wall of rock. For this reason R. Virchow remarked: 'Whether they (the bones) were really in Alluvial loam, as is generally assumed, or not, no one saw. . . . The whole importance of the Neandertal skull consists in the honour, ascribed to it from the very beginning, of having rested in Alluvial loam, which was formed at the time of the early mammals.'[2]

The famous Neandertal man may therefore have lived after the loam was deposited in the cave, and his bones may have become embedded in it later. If this were the case, all speculations as to his importance to the theory of evolution would simply fall to the ground. Virchow said of him:[3] 'We may certainly regard it as decided that the brain-cast bears no resemblance to that of an ape, and even if the cranium is admitted to be a typical race-cranium (which I consider quite unjustifiable), it does not by any means follow that we may deduce from this that it approximates to that of an ape.'

Schaafhausen himself in 1888[4] was content to say: 'In making this discovery we have not found the missing link between man and brute.' Recent investigations on the

[1] 'Über die Altersbestimmung des Neandertalmenschen und die geologischen Grundlagen dafür' (*Verhandl. des Naturhist. Vereins*, Bonn, 1903, pp. 11–90 with one plate). Cf. also on the same subject, H. Schaafhausen, *Der Neandertaler Fund*, Bonn, 1888, pp. 7, &c. Fig. 52 on p. 468 of this book is borrowed from Plate I of Schaafhausen's work. I ought to add that recently a second human skeleton has been found in the Neandertal, but the skull is missing. The fragments are designated *Homo neanderthalensis II*, and are of late Alluvial origin, whereas *Homo neanderthalensis I* is believed to have lived in the early Alluvial epoch, and to have been the real *Homo primigenius*. Cf. Koenen, 'Zur Altersbestimmung der Neandertaler Menschenknochenfunde' (*Sitzungsber. der Niederrheinischen Gesellsch. für Natur- und Heilkunde*, Bonn, June 10, 1901); 'Über Eigenart und Zeitfolge des Knochengerüstes der Urmenschen' (*ibid.* February 9, 1903); 'Die Zeitstellung der beiden Neandertalmenschen' (*ibid.* June 8, 1903).

[2] Quoted from Ranke, *Der Mensch*, II, p. 485.

[3] *Ibid.* II, p. 478. On Virchow's attitude towards the doctrine of descent and especially towards its application to man, see R. Otto, *Naturalistische und religiöse Weltansicht*, Tübingen, 1904, pp. 83–87.

[4] *Der Neandertaler Fund*, p. 49.

subject of the Neandertal man and his Alluvial contemporaries all tend to confirm this statement.

In a paper read on September 23, 1903, at the 75th meeting of German Naturalists and Physicians at Cassel, Dr. Schwalbe discussed the early history of man,[1] and attempted to show that the Neandertal men ought to be considered a distinct *species*, connecting the Miocene apes with man of the present time; he no longer ventured to speak of them as belonging to a distinct *genus*, as he had done in 1901.

Science, however, refuses to accept this new human *species*, which Schwalbe calls *Homo primigenius*, or primitive man, and prefers to see in it merely an ordinary subspecies or breed, such as still occurs in Australia.

N. C. Macnamara, an enthusiastic advocate of Schwalbe's method of examining skulls, has shown still more recently, in the *Archiv für Anthropologie*,[2] that crania, resembling that of *Homo primigenius* in its various characteristics, occur at the present day among the blacks in Australia and Tasmania. In proof of this I may refer the reader to figs. 53 and 54 (p. 469), which are borrowed from Macnamara's work. We see on fig. 53 that the cranium of a modern Australian black (curve II) differs very slightly from that of the Neandertal man (curve III), although both differ greatly from that of a modern Englishman (curve I). In fig. 54 curve I represents the cranium of an old brachycephalic Lapp, curve II that of a dolichocephalic Australian black, and curve III the Neandertal cranium, which is also dolichocephalic. Here again we can easily see that the crania of the Australian black and of the Neandertal man resemble one another far more closely than they resemble the Lapp cranium. Yet no one doubts that Lapps and Australian blacks must both be included in the same systematic species, known as *Homo sapiens*. In comparing the Australian and the Neandertal crania with respect to these curves, Macnamara himself says (p. 358): 'The average cranial capacity of these selected thirty-six skulls (of Australian and Tasmanian

[1] *Die Vorgeschichte des Menschen.* This paper was printed with additions at Brunswick, 1904.

[2] 'Kraniologischer Beweis für die Stellung des Menschen in der Natur' (*Archiv für Anthropologie*, XXVIII, 1903, pp. 349–360).

blacks) is even less than that of the Neandertal group, but in shape some of these two groups of crania are closely related,[1] as is apparent from the drawing of one of these skulls' (fig. 54). [2] We may therefore safely conclude that the Neandertal cranium lies within the limits of variation of the species *Homo sapiens*; *Homo primigenius* represents not a distinct *species* of man, but only an early *race* of man.

In the course of the last few years Professor Gorjanović-Kramberger[3] has very carefully compared *Homo primigenius* of the early Alluvial epoch with *Homo sapiens,* having at his disposal for the purpose the largest collection hitherto available of fossil human remains. He believes *Homo primigenius* (cf. fig. 52, p. 468) to differ from modern man chiefly in the formation of the cranium (see Plate VII, A), with its low, receding forehead and strongly marked supraorbital ridges, in the bent occipital bone and in the large, prognathous lower jaw, devoid of chin. But in all these respects *Homo primigenius* displays numerous transitional forms gradually approximating to modern man.

I may quote Kramberger himself on the subject: [4] 'This short résumé and my previous statements make it perfectly plain that the Alluvial human remains hitherto discovered in the Neandertal, at Spy, La Naulette, Schipka, Ochos, and Krápina, all belong to one and the same species, namely to *Homo primigenius.* What I have said, however, shows further that *Homo primigenius* in almost all his characteristics approximates very closely to *Homo sapiens,* i.e. that there is an unbroken line of development leading from *Homo primigenius,* through the later Alluvial *Homo sapiens fossilis,* to *Homo sapiens* of the

[1] In his table of shapes of crania (p. 357) Macnamara describes as 'closely related' those of which the indices differ by not more than the number 5.

[2] I have quoted this sentence verbatim, because Dr. J. Bumüller, in criticising the previous edition of this work, in the *20 Jahrhundert,* May 28, 1905, asserted that, according to Macnamara, the Australian and Neandertal crania differed enormously, and that I had put a false interpretation upon Macnamara's words quoted above. That Macnamara maintains in general the descent of man from brutes only lends additional importance to his statements on this subject.

[3] 'Der diluviale Mensch von Krâpina und sein Verhältnis zum Menschen von Neandertal und Spy' (*Biolog. Zentralblatt,* 1905, Nos. 23 and 24, pp. 805–812); 'Der paläolithische Mensch und seine Zeitgenossen aus dem Diluvium von Krápina' (*Mitteilungen der anthropolog. Gesellsch.,* Vienna, XXXIV, 1904, Parts 4 and 5).

[4] *Biolog. Zentralblatt,* 1905, p. 810, &c.

present day. This is proved most clearly by the numerous remains found at Krápina, which present many of the characteristic features of modern man, but it is proved also by many peculiarities of *Homo primigenius* that recur occasionally at the present day. Apart from the fact that there are now lower jaws still larger than the largest found at Krápina we may still meet with broad, square dental arches, badly developed chins, and sporadically, among the Australian blacks, even genuine supraorbital ridges (*Tori supraorbitales*); I have moreover in my possession a modern or neolithic lower jaw with a smooth, thick basis, such as we find in the jaws from Spy and Krápina. We occasionally see modern jaws with too many enamel columns near the molars, with no projection at the chin, &c. In fact, even at the present day we can discover a number of features which in the older Alluvial epoch were the general characteristics of mankind, and now occur occasionally by way of atavism, and on the other hand the older Alluvial human remains sometimes present modern characteristics. When all this is taken into account, no doubt can be felt that there has been a continuity in evolution, proceeding from *Homo primigenius* to man of our day.'

Thus far Kramberger. The bearing of his conclusions upon the systematic classification of *Homo primigenius* is far greater than his words imply. If we regard *Homo primigenius* and *Homo sapiens* as two zoological species—and every zoologist would recognise this as a possible way of regarding them—they now cease to be two distinct species, and appear to be merely two races or subspecies of one and the same species, to which, in accordance with the laws of zoological nomenclature, we must give the name *Homo sapiens*. Schwalbe's *Homo primigenius* must therefore be known henceforth as *Homo sapiens primigenius*, to distinguish him from *Homo sapiens fossilis* and *Homo sapiens recens*; he has turned out to be nothing but an earlier race of the one true human species!

If a zoologist discovers a fossil form of wolf having certain constant peculiarities distinguishing it from our modern *Canis lupus*, he describes it as a separate species. Should he, however, subsequently have more abundant material for comparison at his disposal, and find then that none of the distinguishing features are constant, nor limited to one of the two forms under observation; should the characteristics of the fossil wolf recur in some modern wolves, and those of the modern wolf occur occasionally in the fossils, then the zoologist would alter his opinion regarding the systematic value of the two forms, and he would say: 'We have here not two distinct species, but only two races or subspecies of the same species.' Let us adopt the same method and be serious about the 'purely zoological classification of man,' and then we shall acknowledge *Homo primigenius* to be only an older variety of *Homo sapiens.*

Kramberger draws attention (p. 811) to another interesting circumstance in the evolution of Alluvial man. He says that the discovery of the Gally Hill man (in England) seems to him quite extraordinary. The strata in which these remains were found are described as early Alluvial, whilst the remains themselves agree very closely with those of the late Alluvial man found at Brünn. Hence the Gally Hill man cannot be described as *Homo primigenius,* but he must be *Homo sapiens fossilis,* whose remains occur in the Upper Alluvial strata, and who resembles modern man. Kramberger infers from this fact that, ever since the earliest Alluvial epoch, two species of men lived in Europe, one of which, represented by the Gally Hill man, developed sooner and more rapidly, whilst the other remained longer at the *Homo primigenius* stage, and did not become *Homo sapiens* before the later Alluvial epoch. But, as I stated above, the result of Kramberger's investigations really is that *Homo primigenius* was not a different species of man, but only an earlier subspecies of *Homo sapiens.* If therefore the Gally Hill man belongs to the early Alluvial period, we must assume that there were in Europe at that time two contemporaneous subspecies of true human beings.

Kramberger goes on to discuss the relationship between *Homo primigenius* and the *Pithecanthropus* from Java (p. 812). He believes that they belong to the same period, and that as early as the Pliocene epoch the genera *Pithecanthropus* and *Homo* were distinct. This is only hypothesis, and it cannot be proved, as we have no human remains of the Tertiary period; but if it is true, it precludes the possibility that the ape-man of Java might have been an ancestor of man. Kramberger does not recognise the existence of any direct relationship between *Homo primigenius* and the present anthropoid apes; he regards the morphological resemblances between them as nothing more than analogies.

The result of these investigations may be stated in a few words: *Homo primigenius* furnishes us with no evidence in support of the descent of man from beasts.

Schwalbe's *Homo primigenius* therefore began by being the representative of a genus standing between ape and man, then he became an ape-like species of man, and now finally he turns out to be only an early subspecies of *Homo sapiens*. His scientific fate affords fresh confirmation of the notable words used by Schwalbe in the introduction to his work on the early history of Man (' Vorgeschichte des Menschen,' 1904); he says: 'Probably in no department of natural science is the attempt to draw general conclusions from a number of facts more liable to be influenced by the subjective disposition of the student than in the early history of man. On this subject it often happens that upon a few facts theories are based, which are stated with so much conviction as easily to lead those, who have no special knowledge of the subject, to regard them as assured scientific certainties.'

The conflicting character of many of the theories on the history of mankind, which various upholders of the doctrine of descent have propounded, is well illustrated by Kollmann's 'Pygmy theory.'[1] He believes that the tall races are the descendants of pygmies. He does not regard *Homo primigenius* as a distinct species, but only as an offshoot from the tall stock. Kollmann does not think that the *Pithecanthropus* has any connexion with the descent of man, being far too large an ape to have been the ancestor of a human race of dwarf. In his opinion only little Tertiary apes, that walked upright and possessed a high crown to their head, could have been our nearest relatives in the history of our race, but unfortunately there is no evidence whatever to show that these

[1] J. Kollmann, 'Die Pygmäen und ihre systematische Stellung innerhalb des Menschengeschlechtes' (*Verhandl. der Naturforsch. Gesellsch.*, Bâle, XVI, 1902, pp. 85–117); also by the same author, 'Neue Gedanken über das alte Problem von der Abstammung des Menschen' (*Korrespondenzbl. der Deutschen Anthropolog. Gesellsch.* 1905, Nos. 2 and 3). Cf. also R. Weinberg, 'Die Pygmäenfrage und die Deszendenz des Menschen' (*Biolog. Zentralblatt*, 1906, Nos. 9 and 10).

hypothetical links between apes and dwarfs ever had any existence!

We cannot now devote more space to the discussion of Kollmann's 'Anthropogenesis'; [1] its hypothetical character renders it useless for our purpose.

(c) *Conclusions*

The sum total of all these considerations amounts to this: Natural science can tell us nothing with certainty or precision regarding the descent of man from brute ancestors; it is able to offer us only a number of different and contradictory theories, which prove on examination to have in common nothing but the one idea that man must have come into existence 'by natural means,' and for that reason we must insist upon his being the descendant of beasts, although we know absolutely nothing with certainty as to the manner in which this hypothetical process has taken place.

It is no trifling matter to distort truth, as Haeckel and many other supporters of the theory of descent have done in popular lectures and works, when they speak of the descent of man from beasts as 'an historical fact,' thus misleading an uncritical public.[2] Some light is thrown upon this so-called 'fact' by the pedigree of the Primates, sketched by Haeckel in his Berlin Lectures,[3] in 1905. This pedigree is a work of pure imagination, and consists of a mixture of fictitious and of really existing forms, the connexion between them being also fictitious. From an imaginary remote ancestor the *Archiprimas*, Haeckel traces the hypothetical forefathers of our present Lemuridae and apes in an unbroken line, and from a no less imaginary *Archipithecus* he traces the descent of a fictitious primitive gibbon (*Prothylobates atavus*), who was the forefather of a speechless primitive man (*Pithecanthropus*

[1] Cf. Weinberg's article, to which I have already referred, in the *Biolog. Zentralblatt*, 1906, p. 307.

[2] Cf. e.g. Haeckel, *Über unsere gegenwärtige Kenntnis vom Ursprung des Menschen*, Bonn, 1899, p. 30. The English translation bears the title, *The Last Link*, London, 1898, p. 76.

[3] *Der Kampf um den Entwicklungsgedanken*, p. 99. The English translation bears the title, *Last Words on Evolution*, London, 1906. The same pedigree of the Primates, from the *Archiprimas* to *Homo sapiens*, appeared in the previous work already mentioned, in 1899.

alalus) who never existed ;[1] he in his turn was the progenitor of *Homo stupidus*, the stupid man, from whom finally *Homo sapiens* is descended !

If Haeckel hopes that the *Homo sapiens* of the present day will accept his fantastic pedigree, he is mistaken. He might succeed better with *Homo stupidus*, if the race is not yet totally extinct.

At the Fifth International Congress of Zoologists held in Berlin, Professor W. Branco, Director of the Geological and Palæontological Institute of the Berlin University, delivered the closing address on August 16, 1901, and took as his subject 'Fossil Man.' The zoologists among his audience were anxious to learn this competent specialist's opinion of the palæontological evidence for the descent of man from beasts.[2]

Those who had expected to hear strong evidence in support of Darwinism, must have been deeply disappointed, for Branco's lecture was in the main a refutation of Haeckel's controversial opinions expressed in his paper on 'The Last Link: Our present knowledge of the Descent of Man,' read on August 26, 1898, at the Fourth International Congress of Zoologists at Cambridge.

The following were the chief points in Branco's lecture: In the history of our planet man appears as a genuine *Homo novus*. It is possible to trace the ancestry of most of our present mammals among the fossils of the Tertiary period, but man appears suddenly in the Quaternary period, and has no Tertiary ancestors, as far as we know. Human remains

[1] Haeckel does not venture to call him *Pithecanthropus erectus*, because recent research has shown that this fossil ape-man cannot serve as the missing link.

[2] The following statements are based upon the shorthand notes that I made during the lecture. Cf. *Verhandlungen des V. internationalen Zoologen-kongresses*, Berlin, 1902, pp. 237–259. When the reports of the proceedings of the Congress were prepared for the press, however, several of the most important verbal remarks were somewhat modified, or rendered less emphatic. A critic who withheld his name, writing in the *Tiroler Tageblatt* of April 28, 1905, in a feuilleton entitled 'Der fossile Mensch,' stated that in the report given above of Professor Branco's remarks, the Jesuit Father Wasmann had intentionally altered their meaning. The charge thus brought against me is untrue. I wrote to Branco on the subject, and in a letter dated May 10, 1905, he declared that I had reported what he had said accurately on all essential points. Fr. von Wagner, writing in the *Zoologisches Zentralblatt*, 1905, No. 22, p. 699 (see 'A Few Words to my Critics' in the preface to this present edition), calls my comments on Branco's lecture 'frivolous,' but he is, of course, only expressing his own personal feelings.

of the Tertiary period have not yet been discovered, and the traces of human activity, which have been referred to that period, are of a very doubtful nature, but Diluvial remains abound. Man of the Diluvial epoch, however, appears at once as a complete *Homo sapiens.* Most of the earliest human beings possessed a cranium of which any of us might be proud.[1] They had neither excessively long, ape-like arms, nor excessively long, ape-like canine teeth, but were genuine men from head to foot.[2]

Herr Branco regards the Neandertal skull and the Spy skeleton as the sole exceptions known hitherto, and he might have added that these exceptions are of too obscure and problematical a nature to affect the statement that he had just made. Similar exceptions occur often enough among mankind at the present time, as R. Virchow and J. Ranke pointed out long ago. Moreover, I have already shown in the preceding pages that *Homo primigenius,* to whom Branco's remarks about exceptions referred, was merely an early subspecies of man, and not in any sense a brute ancestor of *Homo sapiens.*

In answer to the question: 'Who was the ancestor of man?' Branco gives the following truly scientific reply: **'Palæontology tells us nothing on the subject—it knows no ancestors of man.'** This sentence contains the quintessence of Branco's whole lecture.

We need not be surprised that the lecturer felt bound in conclusion to add some remarks of a speculative character to the scientific dissertation that formed the chief part of his address. In these remarks he said that he was personally convinced for zoological reasons, the weightiest of which was Friedenthal's discovery of the blood-relationship between man and the Primates, that man ought to be regarded simply as the most highly developed animal. Branco was addressing an audience of zoologists, most of whom were probably accustomed to consider man from the purely zoological point of

[1] N.B.—This remark was made before an assembly of eminent zoologists from all parts of the world, whose crania undoubtedly displayed the highest imaginable perfection of development.

[2] On this subject, cf. also J. Ranke, *Der Mensch,* II, pp. 482, 483, where this statement is confirmed in detail. See also H. Obermaier, 'Les restes humaines quaternaires dans l'Europe centrale' (*L'Anthropologie,* XVI, 1905, pp. 385–410; XVII, 1906, pp. 55–80).

view. At any rate, I should like to draw attention to the contrast between the genuinely scientific character of the greater part of Branco's lecture, and the character of its conclusion, in which he dealt with the theory of descent. In the body of his address Branco spoke as a specialist in palæontology, and told us: 'We know of no ancestors of man.' At its end, where he was no longer speaking as a specialist, he weakened this declaration by adding: 'but nevertheless, looking at man from the purely zoological point of view, we must believe him to be descended from apes.'

In the afternoon of August 14, 1901, those who were taking part in the Fifth International Congress of zoologists drove in an almost interminable procession from the Parliament House, where they had their meetings, to visit the Berlin Zoological Gardens, and, as the carriages reached the entrance to the Gardens, the bells of the Kaiser Wilhelm Memorial Church began to toll solemnly in honour of the Empress Frederic, who had just died. Accidentally, therefore, the procession of zoologists was heralded by the sound of a muffled peal, and the sound under these circumstances made a very melancholy impression upon me. It seemed as if the bells were tolling for the death of the Christian cosmogony before the triumphant advance of zoology. Yes, if that purely zoological way of regarding man as nothing more than a highly developed animal is ever generally accepted, there will be no possibility of saving Christianity and the whole modern civilisation that is based upon it. The new cosmogony, upon which the social democrats are even now fixing their longing eyes, will be the unrestrained egoism of higher animals, whose social order stands upon purely brute foundations, and recognises no God, no immortality, and no rewards beyond the grave. When this is the accepted view of life, may God have mercy upon mankind!

But let us hope that zoologists, who think in a truly scientific manner, will see, before it is too late, that the purely zoological way of regarding man takes account only of the lower part of him, and that therefore it is an absolutely mistaken proceeding to apply the theory of descent to him without reserve.

On the occasion of our visit to the Zoological Gardens, to which I referred above, we were met at the entrance by an attendant with two young chimpanzees on his arm, who were to welcome us as comrades. The two little apes grinned at us with cheerful confidence, as if they were fully convinced that we believed in the theory of evolution, and would like to invite us to shake hands in recognition of the bond existing between us. But I thought to myself: 'No, my dear little creatures, thank God, we have not yet come to that!'

I may therefore conclude this examination of the evidence hitherto adduced in support of the descent of man from beasts, by quoting a sentence from J. Reinke:[1] 'The only statement, consistent with her dignity, that science can make, is to say that she knows nothing about the origin of man.'

[1] 'Der gegenwärtige Stand der Abstammungslehre' (*Der Türmer*, V, October, 1902, Part I, p. 13).

CHAPTER XII

CONCLUSION

The rock of the Christian cosmogony amidst the waves of the fluctuating systems evolved by human science (*p*. 481).

The storms at the base of the rock three hundred years ago, and at the present time (*p*. 481).

The rock never can be overthrown by the tempests, because no real contradiction between knowledge and faith can ever exist (*p*. 483).

THE universe may be regarded as a vast ocean having in its midst a mighty rock that has stood there for well-nigh two thousand years. On its summit rises a Gothic cathedral, towering up towards heaven, and within it millions of shipwrecked travellers have found safety. At the foot of the rock surges the sea; the waves sometimes gently lap it, as they play about its base, but at other times they dash wildly against it, and threaten to sweep both the rock and the cathedral away into the deep.

This rock in the sea is the Christian cosmogony upon which the Church of Christ is founded, with her divine revelation and divine teaching, whereby men may be saved. The waves that ebb and flow at the foot of the rock are the ever-changing systems evolved by human knowledge.

Some three hundred years ago a furious storm raged round the rock, for many centuries a peaceful wave had washed its base, seeming to be so calm and friendly as almost to be inseparable from it. Suddenly a mighty tempest arose, and after a conflict of a hundred years a new wave succeeded in driving away its quiet predecessor. The dwellers on the rock trembled at the uproar of the elements; they feared that the rock itself must fall, if the wave that had for so long seemed its inseparable ally were hurled back into the deep, but their fears were groundless. The old wave disappeared, but the rock stood firm, and the new wave, which had at first lashed it in anger, gradually sank to rest, and now rests peacefully at its foot.

The tempest, that I have just described, was the struggle

between the Ptolemaic and the Copernican systems. The former erroneously made our little earth the centre of the universe, with sun, moon, and stars revolving about it. The latter deprived the earth of her central position, assigned to her the moon as her sole satellite, and regarded her as merely one of many planets belonging to one of many suns; reduced her, in fact, to the position of a mere atom in the universe. Many pious minds were overwhelmed with fear lest the rock of Christianity should lose its equilibrium, if the earth really revolved about the sun, but that it does so disturbs no one at the present time. Christianity proved to be far too strong and far too great to be affected by the new theory of the universe. And now this very theory, that once appeared so dangerous, rests peacefully at the base of the ancient rock and even plays about its foundations.

To-day no educated man doubts that the Copernican system is perfectly compatible with Christianity.

Three hundred years passed, and about fifty years ago another tempest arose. The waves of the theory of permanence had long been quietly lapping the rock, and again it seemed as if they were inseparable from it, and many of the inhabitants of the island believed these waves to be indispensable to their very existence, and thought that if they had to give place to other, stronger waves, the downfall of the rock must inevitably follow, and with it the Church built upon its summit must perish likewise. And the new wave came, and like a deluge the doctrine of evolution, originating in England, burst upon the theory of permanence; the conflict between them is still raging, but we can already see what will be its issue; the old wave must pass away and the new wave will remain, until it too has to give place to a stronger.

But the dwellers on the rock need feel no fear; even if the old wave passes away, the rock will stand firm until the dawn of eternity.

On the white crests of the waves that still angrily threaten even the summit of the rock are thousands of tiny bubbles, that seem to fancy themselves about to destroy both rock and Church. They represent modern unbelief, and they imagine that the theory of evolution furnishes them with the best possible weapon against Christianity.

These bubbles, however, deceive themselves. Ere now far more powerful drops have attempted to overthrow the rock, but they have all gone their way and accomplished nothing; and these new bubbles, eager as they are for the battle, will fare likewise. It may well be that ere long the new wave of the evolution theory will lower its proud crest, and sink peacefully to rest at the foot of the ancient rock.

The tide of human knowledge is in no sense a natural enemy of the Christian cosmogony. On the contrary, it is naturally the friend of Christianity, for human knowledge proceeds from the same divine wisdom that created also the rock and the mighty Church upon it.

Between natural knowledge and supernatural revelation no real contradiction is possible, because both have their origin in the same divine Spirit. This fact was defined and clearly stated by the Vatican Council,[1] and the late Pope, Leo XIII, discussed it more in detail in his encyclical 'Aeterni Patris' (August 4, 1879).

If, therefore, the powers of darkness stir up angry tempests which hurl the waves of human knowledge against the rock of the Faith, the waves are not to blame, but rather the powers that make use of them. These storms will never overthrow the rock of Christ: *Non praevalebunt adversus petram!* Whether the waves ebb or flow about its foot, whether the water is calm as a mirror or is lashed mountain-high by hostile forces—the rock of Christianity will stand firm and unshaken to the end of time!

[1] *Constitutio dogmatica de Fide catholica*, c. 4, 'De fide et ratione.'

2 I 2

APPENDIX

INNSBRUCK LECTURES

INTRODUCTION

At the request of the students' association of Innsbruck University I undertook to deliver some lectures there in the middle of October 1909 on the subject of evolution, but I had no idea that they would arouse so much interest in the capital of my native land, as proved to be the case.

According to my usual practice, I spoke extempore, having merely noted down a few headings immediately before each lecture, and I was therefore obliged to write a short summary of the first two lectures for the *Allgemeiner Tiroler Anzeiger* on the morning following their delivery in the hall of the Austria-Haus. The third lecture was given in the Town Hall, before a far larger audience, and on this occasion there were fortunately six shorthand-writers present; I was so completely exhausted by over-exertion that it would have been impossible for me on the day after that lecture to remember what I had said. . . .

As the lectures appeared first in the *Allgemeiner Tiroler Anzeiger*, and as the printers of that paper use rotary presses, no subsequent corrections could be made in the text, and all that I could do was to add a few notes here and there. This explains why the newspaper articles have been reprinted almost unaltered. The first lecture is reproduced in a much abbreviated form, the second somewhat more fully, and the third, having been taken down in shorthand, appears *in extenso*, in fact I have expanded the last section, in which my remarks were much condensed, owing to the lateness of the hour when I concluded my lecture.

Some few repetitions were unavoidable, as, at the beginning of the third lecture, I was obliged to recapitulate what I had said on the preceding evenings for the benefit of many people present, who had been unable to find room in the Austria-Haus. This recapitulation, however, is by no means superfluous, as it contains remarks suggesting new points of view for considering the doctrine of descent as a scientific theory.

* * * * *

My object in publishing these lectures, and thus rendering them accessible to a wider circle of readers, is to supply university students with a short sketch of the scientific doctrine of evolution and its bearing upon monism and Christianity respectively.

The students at Innsbruck in particular are requested to regard this work as a token of my grateful acknowledgment of their efforts to obtain truly scientific information, and I beg them to bear in mind the words with which I concluded my third lecture: The only true monism is that of Christianity; viz. there is but *one* eternal God and *one* eternal truth!

If these lectures serve to confirm and strengthen one among thousands of students in his faith as a Christian, I shall consider myself richly rewarded for all the mental and physical fatigue that they have involved. . . .

ERICH WASMANN, S.J.

LUXEMBURG,
BELLEVUE.

First Lecture, delivered in the Austria-Haus at Innsbruck on Thursday, October 14, 1909

THE THEORY OF EVOLUTION AND THE CHRISTIAN COSMOGONY

THE lecturer began by explaining why he had felt particular pleasure in accepting the invitation to address the students at the university of Innsbruck. Trustworthy information regarding the true value of the theory of evolution and its bearing upon the Christian view of the universe is most necessary in academic circles, as supplying a means of resisting the attacks of monism upon Christianity, since monism employs the doctrine of evolution as ' heavy artillery ' in the strife. The lecturer referred to the discussion aroused in February 1907 by his Berlin lectures on the theory of evolution, and quoted one of his opponents to prove that the freedom to express scientific opinions was jeopardised by the tyranny of ' Monistic beliefs.' ' Free men ought not,' he said, ' to tolerate such tyranny, least of all in the Tyrol.'

The speaker then proceeded to outline the contents of the lectures that he was about to deliver. In the first he proposed to deal with the doctrine of evolution as a theory and hypothesis in natural science, and with the subject-matter of this theory of evolution, the evidence supporting it and its limitations. The various causes of evolution would be discussed in the next lecture.

1. *What is the subject-matter of the doctrine of evolution or descent as a scientific hypothesis and theory?* [1]

Its subject is the investigation of the evolution of plants and animals, from the first appearance of life upon the world to the present time. Man came upon the stage of life as an epigone, and therefore it is only with difficulty that he can decipher the records of life upon our earth, tracing them in fossil remains of creatures long extinct, and comparing them with the organic forms of the present. It is plain that the theory of evolution cannot be an empirical science ; it is only a structure built up of hypotheses for which, both individually and collectively, nothing more than probability can be claimed. To speak of descent from one or other hypothetical ancestor as an ' historical fact,' as Haeckel for instance does in discussing the evolution of man, is wilfully to deceive an uncritical public.

The scientific doctrine of evolution is not concerned with explaining the origin of life from inorganic matter. It assumes the existence of life, and only seeks to ascertain how the living forms of the present have been evolved from those of the past. It has therefore nothing to do with the

[1] For a more complete answer to this question see pp. 267, &c., and *The Problem of Evolution* (Lectures delivered at Berlin), pp. 6, &c.

question of spontaneous generation, nor does it in any way belong to the theory of evolution to decide whether our present forms of animal and vegetable life originated in one single primitive cell, or in a few such cells. It is true that monism maintains a monophyletic evolution of all forms from one common origin to be alone truly scientific, and declares, with great assurance, that it is impossible to accept a polyphyletic evolution from several primitive forms, and that, whoever accepts it, does so under theological influence. But this monistic opinion is not free from presuppositions, and is, on the contrary, thoroughly one-sided and involved in biassed assumptions. Which view we ought to take of the phylogeny of the organic world is not to be decided by the so-called postulates of monism, but solely by a careful examination of facts supplying us with indications. The scientific doctrine of evolution is not a question of dogmas but of facts.

And what do facts tell us regarding the evolution of organic beings? This brings us to the second point:

2. *Actual evidence in support of the theory of evolution.*

This is of two kinds, direct and indirect; the former is naturally very scanty and is derived from relatively slight modifications in species, for the hypothetical evolution of organisms is a process that terminated in some remote past, and only traces of it can be observed by us, who are but newcomers on the earth. There are, however, traces of the formation of new species being actually in progress, or having taken place recently, if we use the word in its geological signification. In illustration, the lecturer referred to instances from his own special department of research, and mentioned particularly the evolution of species within the genus of Dinarda beetles, and the transformation of the guests of East Indian and African wandering ants into termite inquilines, the change in habits having given rise to new species.

Far more abundant is the indirect or circumstantial evidence in support of a race-evolution of animals and plants. It is derived from palæontology, comparative morphology, comparative biology, and comparative ontogeny, or the history of individual development. The lecturer discussed these sources of evidence singly, and illustrated them by a number of instances, taken chiefly from his own branch of biology. In addition to the so-called 'permanent types,' which have remained unaltered for long geological periods, palæontology shows us also certain types that are liable to change, and in the course of time new species, genera, and families have been formed amongst them. Comparative morphology, in conjunction with comparative biology, enables us to recognise the wonderful 'adaptation characteristics,' possessed by the inquilines of ants and termites, as the result of a natural process of evolution, and in the second part of the lecture a number of photographs were shown illustrating this statement. The lecturer showed how comparative biology could account for the growth of the slave-making instinct in ants, and this point too was illustrated by photographs. In speaking of comparative ontogeny, he carefully distinguished the true and the false elements of the so-called 'biogenetic law.' The greatest authorities (Oskar Hertwig, Keibel, &c.) have recently shown that it is impossible to maintain this law to be universally applicable, but nevertheless in many cases the individual ontogeny of an animal furnishes valuable suggestions for the investigation of its phylogeny. This remark is borne out by the appearance of teeth in the embryo of the whalebone whale, and by the development, from formations really resembling wings, of the peculiar appendages on the thorax of the termitophile genus of fly, known as *Termitoxenia*.

3. The lecturer next proceeded to discuss the *limitations of the theory of evolution*. What is proved by all the above-mentioned evidence, direct and indirect? Does it show that the whole animal and vegetable kingdom has developed from one, or even from a few primitive cells, and that the evolution has been monophyletic? No; the advance of phylogenetic research tends to destroy this pleasing fiction, and facts really suggest that the development of both the animal and the vegetable kingdoms has been polyphyletic, i.e. that there have always been many distinct kinds of animals and plants. The names were mentioned of many eminent palæontologists, botanists, and zoologists of the present day who share the lecturer's opinions on this subject.

The idea of the 'natural species' in its bearing upon our acceptance of polyphyletic evolution was the next point discussed.[1]

A natural species consists of the members of one series of forms, connected phylogenetically by descent. This definition of the natural species was given by Neumayr many years ago, and so it is by no means an invention of theologians, as the monists constantly assert. It is true that Neumayr spoke of 'palæontological,' and not of 'natural' species, but he meant exactly the same thing.

At the present day science is not in a position to determine how many such natural species or phylogenetic series we must assume to exist, nor the extent of each series, nor the nature of the primitive forms which gave rise to the natural species. We may, however, confidently expect that more light will be thrown upon these subjects by future research, and this advance in the scientific doctrine of evolution need cause no alarm to theologians nor to any who believe in Christianity. Scientific progress can never contradict our infinitely exalted Christian cosmogony, which is absolutely independent of the fluctuating theories of mankind. The theory of evolution does not clash with the Christian dogma of creation, but completes it in the most beautiful manner. A God who could create a living world capable of evolution is immeasurably greater and higher in His wisdom and power than a God who could only set all living creatures in the world as fixed, unalterable automata. The greatest intellects of the Middle Ages and of antiquity, such as St. Thomas Aquinas and St. Augustine, perceived and expressed this truth, and therefore we may calmly continue to accept the dignified account of the Creation: 'In the beginning God created the heaven and the earth.'

* * * * *

After a short pause a series of about fifty lantern slides was shown. They illustrated the lecturer's particular department of research, and were all original photographs of ants or of inquilines living among ants and termites.

---

[1] See pp. 296, &c., and *The Problem of Evolution*, p. 15.

Second Lecture, delivered in the Austria-Haus at Innsbruck on Saturday, October 16, 1909

DARWINISM AND THE THEORY OF EVOLUTION

We constantly hear most conflicting opinions expressed on the subject of Darwinism. Some maintain that it is dead and buried, others that it is in vigorous health. Some regard it as the outcome of atheism, others as an acceptable scientific theory. One and the same man, Ernst Haeckel, has spoken in very contradictory terms about Darwinism. At one time he declared it to be the 'heavy artillery' of monism in its intellectual struggle with Christianity, afterwards he actually discovered a 'Darwinian Jesuit,' and boldly asserted that the Jesuit Order and the whole Catholic Church had in 1904 gone over to Darwinism. In order to counteract this dangerous flank attack, which threatened the chief stronghold of monism, Haeckel himself gave some public lectures on the subject of evolution in Berlin in 1905. These circumstances add a peculiar interest to the question: 'What are we to think about Darwinism?' It is a very complicated question, and unless we carefully distinguish the various meanings of the word Darwinism, we shall be unable to answer it satisfactorily. Here, as ever, clear comprehension is the mother of truth.

Let us therefore consider Darwinism: (1) From the point of view of natural science; (2) In the sense in which it is used in popular science, and especially in the signification given it by Haeckel and the monists.

1. Darwinism in Reference to Natural Science

Darwinism in this sense is the particular form of the theory of descent which was originated by Charles Darwin, and called by him the 'Theory of Natural Selection.' It differs from the other forms of the theory of descent in the causes and mode which it assigns to evolution. To-day's lecture on Darwinism is therefore, strictly speaking, a continuation of my remarks the day before yesterday upon the doctrine of evolution as a scientific hypothesis and theory. On that occasion I discussed its nature, the evidence supporting it and its limitations; to-day, I have to deal with the causes of evolution and its external manifestation. In this way we shall arrive at a just estimate of Darwinism, from the point of view of natural science.

That Darwinism is not the only doctrine of evolution, but merely one of several such doctrines, and that the name Darwinism ought properly to be applied only to Charles Darwin's theory of natural selection are facts universally acknowledged by scientific men,

Oskar Hertwig was perfectly right in 1900 when he said emphatically with reference to Huxley: 'If Darwinism were swept away, the theory of evolution would stand as it did.' Even Ernst Haeckel in his Berlin lectures in 1905 admitted at last that Darwinism, strictly speaking, was nothing but Darwin's theory of natural selection, although in the course of the same lectures he proceeded to confuse Darwinism with the theory of evolution in his usual fashion.

Darwinism, therefore, is that particular form of the theory of evolution propounded by Charles Darwin in 1859, and by Alfred Russel Wallace at almost the same time, which assumes, in the first place, that natural selection is, if not the sole, at least the chief cause of evolution; meaning thereby that only the fittest individuals survive in the struggle for existence, and, in the second place, that evolution consists of a gradual accumulation of imperceptibly slight 'fluctuating variations' continued through innumerable generations. According to this theory, if we regard natural selection as the chief factor in evolution, enormous periods of time are necessary for one species of animal to be evolved from another.

The lecturer went on to discuss Darwin's Natural Selection more in detail, showing that it was based upon a comparison with the artificial selection employed by man in breeding his domestic animals, which has been so successful in producing new breeds. But in the case of natural selection there is no intelligent breeder directing the process, it secures merely the survival of the fittest, i.e. of the forms best capable of standing their ground in the struggle for existence. It is therefore purely a negative factor, producing nothing new, and having as material for selection only already existing variations. Darwin did not investigate the origin of the beneficial variations, and tacitly assumed that a living organism was by nature capable of evolution. In his opinion the capacity for variation was indefinite and unlimited. It seemed, therefore, to him a matter of chance whether beneficial variations occurred at all, and only by chance again could they be transmitted to succeeding generations. Viewed in this way, Darwin's theory of selection appears to be ultimately a theory of chance.

Darwin was not, however, so extreme a Darwinist as many of his followers, e.g. as Weismann, who, as chief representative of the so-called 'New Darwinism,' proclaimed the all-powerfulness of natural selection. Darwin himself on occasion admitted the claims of the 'Nature of the organism,' and did not deny its capacity for adaptation and the possibility of the transmission of properties acquired by an individual. He accepted also Cuvier's principle of correlation. Nevertheless, natural selection is, and remains, the chief factor of all race-evolution, according to his theory.

What must we, as students of natural science, think of this form of Darwinism? It is thoroughly unsatisfactory, for it accounts neither for the cause nor for the manner of evolution. Natural selection is not a sufficient cause for evolution, because it leaves the origin of what is beneficial unexplained, and able only to account for the extirpation of what is not beneficial. It is a purely negative factor, and de Vries has very aptly compared it with a sieve, that sifts out the unfit, but does not explain the origin of the fit. It may be compared also with a strict examiner, who rejects the badly prepared students, but the reasons why the well prepared candidates pass the examination are to be sought in their knowledge of the subjects set, which the examiner does not invent, but in which he tests others. Again, natural selection resembles a gardener's boy pulling up weeds. His activity is purely negative, and presupposes the existence of the gardener, who has planted in the earth the plants that are to remain untouched.

Pauly says that natural selection is like von Scheffel's 'Hausknecht aus dem Nubierland,' who turns out of the Black Whale in Ascalon any guest unable to pay his bill, but cannot supply money for payment; all he can do is to keep the place clear of unwelcome intruders.

There are other reasons too against accepting the theory of natural selection. It can offer no explanation of biologically indifferent characteristics of animals and plants, although these are of far more frequent occurrence as distinguishing species than the biologically beneficial properties. . . . By assuming that evolution is a process involving an extremely slow accumulation of very slight changes, the theory of natural selection requires, for the evolution of any one species from another, immeasurable periods of time, which are incompatible with geology. It demands also that in the strata containing fossil remains of extinct organisms we should regularly find series of gradual variations, and not sharply distinguished species. Palæontology, however, shows us the actual existence of a contrary state of affairs. Series of very slight variations are an extremely rare exception, not the rule. To try to account for this fact by referring to the defective condition of palæontological records is a hopeless attempt, in view of the positive progress made by modern study of fossils.

We must, therefore, come to the conclusion that we cannot regard natural selection as the chief factor in evolution, for it is scientifically impossible to do so. Must the theory be rejected altogether?

It is an incontestable fact that Hans Driesch and many other scientific opponents of Darwinism have rejected it. Driesch called Plate's attempt to save it 'a funeral oration,' uttered on the principle *de mortuis nil nisi bonum*. Dennert, too, considers that he has already stood by the deathbed of Darwinism and witnessed its last agony. I do not, however, believe this. By far the majority of botanists agreed long ago to set a very modest and greatly modified value upon the principle of selection, and now modern zoologists are doing the same, but they do not wholly reject it, and in my opinion theirs is the only correct attitude towards it. As a subordinate factor among others of much greater importance, Darwin's natural selection still demands recognition, and will continue to do so.

The lecturer illustrated this remark by an interesting example of the hypothetical evolution of the slave-making instinct in ants. The wonderful instinct, prompting them to steal the worker pupae of other species and bring them up as their assistants, is not due to natural selection, as Darwin assumed, but originated in a much simpler, shorter, and more natural manner. It is the result of the establishment by the females of dependent colonies, in conjunction with an alteration in the previous mode of nourishment among the workers. Climatic changes would cause changes in the vegetation, forest flora would be replaced by that of the steppes, and thus ants might be forced to live exclusively on other insects, and preferably on the pupae of other kinds of ants. Of the stolen pupae only those of one particular species were allowed to live, because the females of the robbers had originally founded their colonies by the aid of ants of this kind; hence the latter became the slaves of the former. Thenceforth natural selection might promote the further development of a slave-keeping instinct in the robber-ants (though it would do so only as an exterior subsidiary factor) until this development reached its culminating point, and then degeneration of the slave-keeping instinct began, and led to the lowest state of social parasitism in which the masters are mere parasites dependent upon their former slaves. Such degeneration of the slave-making instinct must lead finally to the extinction of the original masters, and to the dying out of the species. This process was due to interior causes, and continued, although it ultimately proved most destructive to the species; natural selection was unable to check it, and proved in this case powerless and not all-powerful.

The lecturer went on to discuss the other factors of evolution that must be assumed to co-operate in the evolution of a race. The chief factors in the evolution both of a race and of an individual are the interior organic and psychical laws governing the development of organisms. He established the existence of these laws and answered the objections raised by monists and materialists. The working of these interior laws of development is seen, he said, in the capacity for reaction possessed by the simplest little mass of protoplasm, for upon this beneficial capacity for reaction depend the organic functions of nutrition, movement, growth and propagation. Unless we assume these interior factors of development to exist, all development of organic life is impossible. Wasmann's opponents in Berlin could not disprove this statement at the famous discussion on the evening of February 18, 1907; in fact, the eleventh speaker even expressed himself in favour of admitting the existence of these interior factors. When Plate and other opponents of teleology thought they could get rid of these laws by calling them 'mystical,' they were labouring under a false impression due to their absolute failure to understand the nature of these factors. These interior laws of development ought not to be regarded as working automatically like a clock, but as acting reciprocally with the exterior impelling causes and stimuli of evolution. For this reason we cannot accept, in its extreme form, Eimer's 'Orthogenesis,' a theory maintaining that evolution proceeds in an uninterrupted course from interior causes.

The lecturer then referred to adaptation. The purely passive and mechanical adaptation of Darwinism, consisting merely of the elimination of the unfit, is absolutely unsatisfactory as a cause of evolution. Over and above it we need what is of much greater importance, viz. an active and direct adaptation of the organism to the influences of the world around it. Lamarck and Geoffroy St. Hilaire established the principles of direct adaptation early in last century, and these same principles have found their modern expression in such phrases as 'La fonction crée l'organe,' &c. Allusion was made to the close connexion between Lamarckism[1] and the thoroughly sound Neovitalism of Hans Driesch and Reinke, and also to Neo-Lamarckism, which in Pauly and Francé has assumed the form of so-called Psycho-Lamarckism.

The lecturer showed how far these views were justifiable, inasmuch as they recognised in living organisms interior tendencies to evolution; but he criticised very sharply the outgrowths of Psycho-Lamarckism, especially in Francé's works. Francé is unable to avoid acknowledging the existence of a teleological principle of interior design, which must ultimately lead to the recognition of a thinking and intelligent cause, such as Christian philosophy regards as effecting the creation, at the beginning of the evolution of organic life. He, however, prefers to make an unsuccessful and unscientific attempt to represent each cell in a living organism as a diminutive creator endowed with reason. In this way he has placed plant-life on a level with human life in a most uncritical fashion, but nevertheless he has not succeeded in explaining the existing unity in the development of plants and animals from that aggregate of 'cell-souls.' This Psycho-Lamarckism is worse than the most extreme Darwinism from the scientific point of view.

The lecturer discussed briefly the question of the transmission of acquired properties, and the relations between germ-plasm and somatic plasm. He stated any evolution of instinct in the animal kingdom to be, in his opinion, inconceivable, unless this transmission is possible. The difficulties formerly raised against the possibility of inheriting individually acquired properties had, he said, in the case of ants been happily removed by recent investigations.

He went on to speak of the important bearing of climatic changes upon the evolution of species and of their instincts, illustrating his views by instances from the development of slavery and of social parasitism among ants, which he had described more fully in the *Biologisches Zentralblatt* for 1909.

The other factors of evolution were mentioned, which are noticed in R. Wagner's 'Theory of Migration,' in Romanes and Gulick's 'Physiological Selection,' in Roux's 'Histonal Selection,' and Weismann's 'Germinal Selection,' the last two having been introduced to supplement the theory of personal, or, as Darwin called it, natural selection. The lecturer referred also to 'Amical Selection,' a name which he himself had used twelve years previously to designate the instinctive preference shown by ants and termites for certain breeds of inquilines. That this predilection was a factor in evolution had been proved by actual observations. This form of selection differs altogether from both natural and sexual selection, and of all the forms of selection among animals it most closely resembles the artificial breeding practised by human beings.

This part of the lecture concluded with the remark that, if the theory of

---

[1] See also *Geschichte des Lamarckismus* by Prof. Dr. Adolf Wagner of Innsbruck (Stuttgart, 1909).

evolution were to agree with facts, it must avoid all tendency to take a one-sided view of the causes of evolution. Many factors invariably act together, though their participation may vary in degree according to the differences in the lines of evolution under consideration. As proof of this statement, reference was made to the hypothetical evolution of three biological types of guests entertained by ants, viz. the offensive, the mimetic, and the symphilic types respectively, which were illustrated by photographs in the first lecture.

A general survey of the various forms of race-evolution followed. Darwin assumed evolution to be a very slow and gradual process, working by means of fluctuating variations, whereas Kölliker's heterogony and the theories of Korschinsky and de Vries require the changes to have occurred *per saltum*, and Jäckel's metakinesis involves a rapid alteration of forms in the embryonic stage. Heer, Zittel, and de Vries believe periods of change and periods of rest to have alternated in the history of organic life, but care must be taken to avoid adopting any one of these ideas on evolution exclusively, as, in many cases, several kinds of evolution may be at work, sometimes in different, sometimes in one and the same line of evolution.

2. Darwinism in the Wider and more Popular Sense

The word 'Darwinism' is a genuine Proteus; it possesses at least four different meanings. In the first part of this lecture I have been speaking of Darwinism in the correct, scientific sense, viz. Darwin's theory of natural selection. Great confusion has resulted from what we may confidently call the unscientific use of the word in several other senses. By Darwinism people often mean a theory of the universe based upon an absolutely uncritical generalisation of the principle of natural selection, the struggle for existence, that is practically identical with the old materialistic theory of chance, which nowadays calls itself monism, in order to hide its atheism.

A third use of Darwinism is to designate the unreserved extension to man of the theory of natural selection. This results in degrading man to the level of brutes, and overthrows the social order depending upon the principles of Christianity. It has nothing further to do with the scientific evidence of the descent of man from brutes, which I intend to examine in my next lecture.

There is yet a fourth use of the word Darwinism, as synonymous with the theory of evolution in general. Every one knows that in scientific circles Darwinism and the theory of evolution are no longer confused, but in popular language the terms still continue to be treated as interchangeable, and great harm has been done in this way. It was an excusable mistake fifty years ago, when Darwin first became prominent, and his 'Origin of Species' revived the memory of Lamarck's long forgotten ideas regarding evolution, and directed men's attention to the theory of evolution itself. But at the present day there is no excuse at all for confusing Darwinism with the theory of evolution. If the monists persist in doing so, it is because they hope thus to propagate the Darwinian theory of the universe in a by no means scientific, but in a thoroughly unscientific and dishonest way. An article on the further development of Darwinism ('Die Weiterentwicklung des Darwinismus'), published by Francé among Breitenbach's 'Darwinistiche Schriften,' is an instance of what I mean. All the recent progress made in the scientific theory of evolution, even Neovitalism, which is directly opposed to Darwinism, is here represented by Francé as 'further developments of Darwinism.' Not satisfied, however, with thus misleading his readers, Francé has even ventured to falsify a quotation from my works, in order to transform me from a supporter of the theory of evolution into an advocate of the theory of permanence.

This unmistakable falsification was pointed out to him, but, instead of correcting it, he actually repeated it once more. Such a proceeding is not merely unscientific, but absolutely dishonest. Plate's line of action is not much better, for in one of his more recent publications he classes Reinke and myself among the opponents of the theory of evolution, although he knows perfectly well that such a statement is simply a falsehood. If the monists are forced to have recourse to such means as these in their efforts to 'enlighten' the people, and to gain adherents for their new monistic cosmogony, they are much to be pitied.

In what relation does Darwinism stand to Christian philosophy? Christianity has nothing to fear from scientific Darwinism. More than twenty years have passed since Haeckel triumphantly declared that Darwin's theory of natural selection supplied an explanation of finality in nature, and enabled men to do without a 'wise Creator,' but this declaration has proved to be nothing but bombast, and at the present time no one takes it seriously. The Darwinian cosmogony, however, which is based upon a thoroughly unscientific generalisation of the theory of natural selection, has, under the form of Haeckel's monism, revealed itself as barren materialism and atheism, and I shall have to say more about it in the third lecture.

Men of science in years to come will honour Charles Darwin's memory more highly than Haeckel's, for the latter popularised scientific Darwinism with the express purpose of using it as a weapon against Christianity. In so doing he has diminished rather than increased the scientific reputation of the theory of evolution. Allow me to conclude this lecture with the noble words written by Charles Darwin at the end of his 'Origin of Species': 'There is grandeur in this view of life, with its several powers, having been originally breathed by the Creator into a few forms or into one; and that, whilst this planet has gone cycling on according to the fixed law of gravity, from so simple a beginning endless forms most beautiful and most wonderful have been, and are being, evolved.'

Third Lecture, delivered on October 18, 1909, *in the Town Hall at Innsbruck*

THE DESCENT OF MAN, HAECKEL'S THEORIES, MONISM

LADIES AND GENTLEMEN,—

I must begin by thanking you for the extremely hearty welcome that you have given me in Innsbruck. It is all the more pleasant to me because I am myself a native of South Tyrol, I may even say a neighbour of Andreas Hofer's.

As the time allotted for my lecture is very short, although the subject with which I have to deal could receive adequate treatment only in a course of lectures, I must be as brief as possible. My programme is as follows :—

1. Summary of the two previous lectures, rendered necessary by the presence this evening of an audience two or three times as large as on the first night.
2. Display of the most important photographs shown on the previous evenings, and illustrating from my own department my remarks on direct and indirect evidence for the theory of evolution.
3. Discussion of the question : What evidence does natural science furnish of the descent of man from brutes ?
4. After a short interval I shall show you some more photographs belonging to the morphological and palæontological sides of the argument ; and here again I must limit myself to what is most indispensable.
5. In the fifth part I shall have to speak of Haeckel, and throw some sidelights upon his manner of dealing with anthropological problems, especially with reference to the phylogeny of man.
6. In conclusion I shall examine with what right monism claims to have replaced the Christian cosmogony by a new theory of the universe, based chiefly upon scientific principles of evolution.
7. Lastly, there will be a discussion, in which all are invited to take part whose scientific attainments enable them to form an opinion on these questions. I shall avail myself of the opportunity, given me by the discussion, to elucidate two points that seem to me particularly important. All personal feeling shall be set aside, and I intend to speak simply in the interests of truth.

1. Let us begin by reviewing shortly the results of the two previous lectures. My subject throughout is the theory of evolution and the Christian cosmogony, but I am dealing with the latter only in as far as it is necessary to do so, in order to remove the alleged contradictions between the theory of evolution and Christianity. In the first lecture I spoke of the doctrine of evolution as a scientific hypothesis and theory, considering first its nature, secondly the

evidence supporting it, and thirdly its limitations. In the second lecture I discussed the causes of race-evolution, and the particular forms which evolution is supposed by the advocates of different theories to have assumed. I called this lecture 'Darwinism and the theory of evolution,' simply because Darwinism differs from all other theories in the causes and form that it assigns to evolution ?

What are we to think of the doctrine of evolution as a scientific hypothesis and theory ? What really is the theory of evolution ?

It maintains that organic species may be related to one another in virtue of having a common origin, and so they can be arranged in definite lines of descent. This theory contradicts that of permanence, which regards the organic species as unchanging, and received its present form from Ray, Linnæus, and Cuvier. The scientific foundations of the evolution theory were laid in 1809 by Lamarck, in his 'Philosophie zoologique,' and in 1859 Darwin gave it a new form in his 'Origin of Species,' so that this particular form is called Darwinism after him.

It is not the task of the theory of evolution to account for the origin of life, but only to explain the further development of life, taking existing facts as its *points d'appui*. We have, therefore, nothing to do now with the origin of life, and from this definition of the theory it follows that it is not essential to it to trace back all animals and plants to a single primitive cell, nor to assume a common ancestor for all animals and all plants respectively. Whether we are to assume there to have been one or many lines of descent, or, in other words, whether we are to regard evolution as monophyletic or polyphyletic, is a subordinate question, forming no essential part of the theory of evolution. Such questions cannot be answered by the postulates of monism, because the theory of descent, being a scientific hypothesis and theory, has to do with facts and not with dogmas. This may suffice as a short account of what the theory of evolution really is, and it may also remove certain misunderstandings which have crept in, and obscured the definition of the theory, chiefly in consequence of monistic misrepresentations.

We have next to consider what evidence there is for the theory of evolution. What justifies us in believing that any evolution of organic species has occurred among animals and plants ? Men occupy a difficult position with regard to this question, for we are epigoni, appearing at the close of a long process of evolution, begun, perhaps, thousands or even millions of years ago ; it is impossible to fix its duration. We are obliged to gather fossil traces of bygone evolution from geological strata, and to compare these palæontological data with things existing at the present day, in order to connect kindred species in genealogical series.

From its very nature our evidence is circumstantial rather than direct ; to discover direct proofs of the theory of evolution in facts of the present time, or of the not very remote past, is a very difficult

2 K

task, because the hypothetical evolution of the organic world belongs to the most distant ages, in comparison with which thousands of years, as we reckon them, are but a fraction of a second. It follows, obviously, that the theory of evolution can never become an absolute fact, or a branch of empirical science, the results of which can be tested directly by observation and experiments. It never can be more than a structure built up of hypotheses, *i.e.* of more or less probable assumptions. Indirect evidence in support of it may be derived from various sources.

In the first place we have the testimony of palæontology, or, as Steinmann calls it, historical evidence. We must seek the fossils preserved in various strata, and compare them with the still existing forms of animals and plants, in order to discover the relation in which they stand to one another and to our present species.

In the second place we must take into account the results of comparative morphology, which has made great progress in the last few years. We must compare the various organs and systems of organs in animals with one another, and note their points of similarity and of difference, and try to ascertain how far they suggest community of origin. It is true that we must proceed very cautiously, and avoid confusing the so-called phenomena of convergence with phylogenetic resemblances. The former, in consequence of similarity in the mode of life and in the conditions for adaptation, may produce forms showing marked likeness in animals of very different origin. It is safe to draw conclusions from this source only when the evidence derived from morphology agrees with the testimony of palæontology and of comparative embryology.

Comparative biology, by throwing light on the mode of life of various animals and the development of their instincts, becomes our third source of evidence. I illustrated this in my first lecture by discussing the growth of the slave-making instinct and of social parasitism among ants.

Fourthly, we have the comparative embryology of our present animals and plants. This subject is an important storehouse of information in phylogenetic research, and it has made great progress in recent times, for Oskar Hertwig's works have thrown much light upon the embryology of the higher animals, and Korschelt and Heider's upon that of invertebrates. Caution is necessary, however, in making use of this source of evidence, as appears from the history of the biogenetic fundamental law, laid down by Fritz Müller and Haeckel. According to this law, the ontogeny of the individual is an abbreviated and somewhat modified repetition of the phylogeny of the race; but no such general law exists. Here and there the ontogeny of an individual may give some hint that is of importance in the investigation of its probable descent. Instances of this are the occurrence of teeth in the embryo of the whalebone-whale, and the appearance of genuine wing-veins in the imaginal development of the thoracic appendages in *Termitoxenia.*

We come now to the question of the limits of evolution. Do facts constrain us to believe evolution to be monophyletic or polyphyletic? As I showed in my first lecture, there is no scientific proof of the origin of the whole organic world from one primitive cell, nor of the origin of the animal and vegetable kingdoms respectively from one ancestral cell. On the contrary, facts point to a polyphyletic evolution of both animals and plants, and not only palæontology, but also comparative morphology supports this view, as Boveri has shown. What ideas ought we to have of this polyphyletic evolution ? We cannot as yet even attempt to determine the number of lines of descent in the animal and vegetable kingdoms, nor do we know whence they proceed.

It is possible that in another hundred or thousand years we shall know rather more about the phylogeny of living organisms than we do now. All we can do is to continue our researches. In my first lecture I referred to the idea of the natural or palæontological species, which was originated by Neumayr and elaborated by myself. Whoever bears in mind the above-mentioned limits of evolution, which are imposed upon us by actual facts, will certainly not go astray. He will not invent fanciful pedigrees a yard long, which ultimately find favour only with social democrats under the influence of monism, and not with the advocates of the scientific theory of descent.

* * * * *

A series of photographs followed ; the first showed the transformation of guests among Indian ants into termite-inquilines, thus illustrating the formation of new species within comparatively recent times. This picture afforded direct evidence for the theory of evolution, the others supplied indirect evidence by illustrating the formation of new species, genera, and families of beetles and flies in consequence of adaptation to changed conditions of life in colonies of ants and termites.

Let us now pass on to something more important, *Paulo majora canamus !*

We have seen that among plants and animals there is a good deal of evidence in support of evolution, and this is based chiefly upon palæontology. It is more probable that the evolution was polyphyletic, or in many lines of descent, than that it was monophyletic; in fact, the former is the only really probable hypothesis.

What are we to say regarding the descent of man, that all-important question ? Are we to adopt the standpoint of natural science, and say that man, like every other higher vertebrate, has developed from the animal kingdom ?

I must not touch upon either the theological aspect of the subject, or upon the abstract philosophical possibility of such an evolution. I intend to deal with the matter from a practical point of view only, and to discuss : (1) the spiritual evolution of man from brutes ; (2) the bodily evolution.

2 K 2

Let us see to what results science will lead us. In speaking of purely natural evolution, I think we must reject the theory that man on his spiritual side can have been evolved from brutes, and we need have no hesitation in doing so, as our rejection is justified by modern experimental animal psychology. I am not discussing monistic dogmas, nor the altogether unscientific popular practice of ascribing to animals a spiritual life analogous to that of human beings; I am alluding only to the facts of animal psychology, which, in its recent development, so far from bridging the old chasm that Aristotelian philosophy has always recognised as existing between the spiritual life of man and the sensitive life of brutes, has widened it. I repeat: experimental animal psychology, carried on in a critical spirit. Popular psychologists, such as Büchner, Brehm, Marshall, Bölsche and others, are not, I think, to be reckoned among the scientific representatives of animal psychology, the chief of whom are, in America, Thorndike, Kinnaman, Hobhouse, Watson, &c.; in Geneva, Claparède; in Germany, Wundt and Stumpf; and in England, Lloyd Morgan. These are unanimous in saying that we must not ascribe even to the higher vertebrates any capacity for thought, or any power of abstraction in the sense of ability to form rational concepts. The whole life of an animal soul is limited to sense perception, imagination, and instincts. What is called 'animal intelligence' is nothing more than the ability of an animal to learn by the experience of its senses. It does not depend upon reflexion, but upon a repetition of definite sense impressions, upon their combination in the creature's faculty of sense imagination and their reproduction by sense memory. Consequently an animal is taught by sense experience to change its mode of action for its own advantage; in other words, it is able to learn. This is the conclusion at which modern animal psychology has arrived, with reference to the spiritual difference between man and brute. In my opinion it is confirmed by the psychology of ants, and it does not justify us in abandoning the tenets of that ancient philosophy which taught that the psychical endowment of man and brute differed essentially. It is true that there is much of the animal in man, but there is also something higher, viz. the spiritual element in his being. I must not, however, dwell upon this point now.

Let us turn to the aspect of the question with which natural science can deal, and ask: 'What relation exists between man and brutes with regard to their bodies? Is the descent of man from brute ancestors proved or not?'

With regard to the formation of his body, his organs and systems of organs, and the development of his nervous system, man stands undoubtedly very close to the higher vertebrates. This fact cannot be denied. But has natural science any one definite and well-established theory to offer us on the subject of man's relationship with the higher mammals? On the contrary, there are a number of different hypotheses regarding the morphological descent of man. Kohlbrugge has collected them in a thoroughly scientific article and examined them, and the result is : *Quot capita, tot sensus.* We have nothing to guide us but a set of mutually antagonistic hypotheses. This is the simple truth. When monists declare that the descent of man from brutes is 'zoologically evident,' they have no more claim upon our consideration than Haeckel has, when he calls the descent of man from apes 'an historical fact.'

The theories as to the relationship between man and brutes in respect of their bodies may be divided roughly into two classes. Some assume the existence of a direct relationship between man and the higher apes, it is quite indifferent whether the forms in question are still being or extinct. Others maintain the relationship to be less close between man and apes, and seek the hypothetical primitive form of man among mammals of a lower order. Both classes of theories call for critical examination. The former numbers among its supporters many of the more modern zoologists, the latter finds more favour with anthropologists. The former is more intelligible than the latter, which becomes hopelessly embarrassed on the subject of the common ancestors of men and apes. Klaatsch, who is one of the chief advocates of the second class of theories, at a time when he had less clear opinions than he now possesses, used to represent the common ancestor of man and ape as a 'general pithecoid type,' but he did not know where to place him. This type proved to be much too general, and so Klaatsch has given it up again in the last few years.

Stratz, another advocate of the theory of the distant relationship between man and ape, imagined their common ancestor to be a kind of Batrachian called a 'Molchmaus,' but most zoologists are, like myself, still quite in the dark as to what kind of animal that is. Morphologically, man resembles some of the lower orders of mammals, such as the insect-eaters, more closely than the anthropoid apes, but, nevertheless, the 'Molchmaus' seems to me scarcely suited to be a common ancestor of man and ape; in fact, a direct relationship between them would seem much more probable.

I should like at this point to consider briefly the evidence in support of both theories, but especially of that which regards man as the direct descendant of the higher apes. Comparative morphology supplies certain evidence, and it is undoubtedly true that of all animals the higher apes bear most resemblance to man; there are in fact over a hundred points of resemblance; but, on the other hand, we must not overlook the great morphological differences in the formation of the skeleton, of the cranium, &c., to which attention was drawn long ago by Ranke, Virchow, Kollmann, Bumüller, and other anthropologists, who pointed out that in the development of his extremities the ape has outstript man, and that man fits nowhere in the systematic succession of apes, neither at the beginning nor anywhere else No one as yet has been able to explain clearly the descent of man from an extinct form of ape. Even Schwalbe's hypothesis on this subject has met with much opposition from other specialists. I shall have to refer later to the *Pithecanthropus* as a morphological connecting link. Selenka thought that the great likeness between man and the anthropoid apes in the formation of a placenta constituted a trustworthy proof of direct relationship between them. Recently, however, exactly the same placental formation has been shown to occur in other animals, e.g. in a low kind of lemur (*Tarsius spectrum*) found in Madagascar. It follows that this particular kind of placental formation is due to adaptation to the needs of embryonic existence, and, as a result of convergence, it may occur in creatures that are not related. It is impossible to derive any argument in favour of a direct relationship between man and the higher apes from a likeness in their placental formation.

We come now to the evidence derived from comparative embryology. What is known as the 'biogenetic fundamental law' was enunciated by Fritz Müller and elaborated by Haeckel. According to it, the ontogeny of an individual animal is an abbreviated and partially modified repetition of its phylogeny, or the history of its race. In its application to man this law found its dogmatic expression in Haeckel's 'Progonotaxis hominis,' or genealogy of man. It found its dogmatic expression, but nothing more, for, as a matter of fact, precisely at this point there are so many exceptions to the alleged general and fundamental law, that almost nothing is left of it, the exception itself becomes the rule. I may mention, for instance, the extraordinary development of the cerebral vesicles in the human embryo; it would certainly not be possible to find any stage corresponding to them among our alleged ancestors, for any creature possessing so huge a brain in comparison with its other organs would have been a complete monstrosity. At the present day scientific men in general are gradually becoming convinced that it is impossible to claim for the biogenetic law that it is universally applicable. In support of this statement, I may refer to very eminent authorities, such as Oppel, Keibel, and Oskar Hertwig Even Konrad Günther does not venture to call it a law in his work, 'Vom Urtier zum Menschen,' and he acts wisely, for the biogenetic law, when it is logically applied, leads to consequences that turn the doctrine of man's descent from apes simply upside down. The law asserts that the ontogeny of the individual is an abbreviated repetition of the evolution of the race. Now, in the ontogeny of the higher apes there is a stage in the development of the cranium, when the foetus very closely resembles a human being, but there is not, in the case of the human embryo, a stage when, in its cranial development, it resembles an ape. The logical conclusion from this fact would be: Man is not descended from apes, but, on the contrary, apes are descended from ancestors resembling men. This deduction has actually been drawn by a number of eminent men, as Kohlbrugge pointed out. We used to hear a great deal about the descent of man from fish, the theory being based upon the fact that man in the course of his ontogeny is supposed to pass through a fish-like stage. This theory, too, has been shattered by Oskar Hertwig and other embryologists, who have proved that the so-called branchial clefts and arches in the higher vertebrates ought to be regarded as morphologically indifferent *Anlagen*, whence in the lower classes of vertebrates true gills are developed, whilst in the higher classes they furnish material for quite different organs.

We have next to consider comparative blood-reactions, which were believed to afford absolute proof of man's blood-relationship with the anthropoid apes. A few years ago Friedenthal astonished the world by proclaiming, as his discovery, that we were not only descended from apes, but were ourselves genuine apes. He based this statement upon experiments made by himself, Uhlenhuth, Nuttall, and others, on the reaction of different kinds of blood. Let us see what is the real result of these experiments, and whether they actually prove us to be blood-relations of apes, in the sense of being their cousins. I have no hesitation in saying that they do not prove it. The likeness between the higher apes and man in the composition of their blood is indeed greater than the likeness between the lower apes and man. I am quite ready to grant this, but there are a number of questions belonging to physiological chemistry, which throw fresh light upon the significance of these reactions. Not long ago, at the last meeting of the Görres Society at Ratisbon, Dr. Baden, who is a specialist in physiological chemistry, read a paper on experiments in blood-reaction and their bearing upon the subject of phylogeny. The conclusion at which he arrived was, that it was impossible to regard these experiments as affording any actual proof of phylogenetic blood-relationship between man and the higher apes; we might just as well speak of a urine-relationship between man and the higher vertebrates. All that these experiments have proved is that in the composition of his blood—blood being for zoologists only one of the tissues of the body—man resembles the higher apes in many respects more than other animals. It would be a great mistake to infer from this fact that man is directly related in race to the anthropoid apes. Dr. Baden laid particular stress upon the specific difference in the blood of men and apes, and referred to recent works on this subject by Neisser, Sachs, and others.

I was very glad that Friedenthal himself took part in the discussion that followed my Berlin lectures in 1907, and declared that, in using the word 'blood-relationship,' he had never meant anything more than a blood resemblance in the chemico-physiological sense. It was a mistake on the part of writers on popular science to say that by blood-relationship he understood actual kinship, and he protested energetically against having such an idea imputed to him.

In speaking of blood-reactions, from the standpoint of organic chemistry, we are concerned only with the reactions of albumen, with precipitins, haemolysins, &c. It has been observed that the albumen in the lens of the eye shows the same composition in very different kinds of vertebrates, but we cannot derive any phylogenetic inference from this fact. It would be therefore wrong and premature to infer that man is nothing but a genuine ape from blood-reactions, which are likewise only reactions of albumen.

Finally, we have to speak of palæontology, whence most of the evidence in support of the theory of evolution is derived. What does it tell us with regard to brute ancestors of man? What information does it give on the subject of the long-sought missing link between man and apes?

At the fifth International Congress of Zoologists at Berlin in 1901, Professor Branco, one of our foremost palæontologists, delivered a very outspoken address on the subject of fossil man, and his conclusion was that hitherto palæontology has no knowledge at all of any ancestors of man. This was certainly a very honest statement, made by an eminent scholar. Let us now consider more closely the facts bearing upon the subject. For some time it was believed that the missing link between man and the higher apes had been discovered in the so-called *Pithecanthropus erectus*, the ape-man, whose remains were found in Java in 1891. At the third International Congress of Zoologists, held at Leyden in 1895, Eugène Dubois read a very interesting paper about them; the remains found consisted of a cranium, a femur, and first one and then a second molar tooth. Dubois spoke for a couple of hours, trying to construct from these remains a connecting link between ape and man, that was neither an ape nor a man, but an ape-man, standing between the two. Privy-Councillor Virchow was presiding over the meeting, and listened to all that Dubois said with the impenetrable expression of a diplomatist. I wondered what attitude he would assume towards the question. At the conclusion of the lecture Virchow began by thanking Dubois for his kind invitation to be present, and did not allude to the fact that the discovered remains had been shown him only just before the meeting, although he had telegraphed three times, asking to see them. He spoke highly of the lecturer's acumen, but said that in his own opinion it was impossible to decide whether the fragments had formed part of one individual, and still more impossible to ascertain whether they belonged to a human being or an ape. This point could not, he said, be settled until we possessed a complete skeleton. Virchow then pronounced the cranium to be that of a large ape, but he thought the femur and the teeth were probably human. Such was Virchow's opinion on that occasion. Has it been modified subsequently? Further examination of this famous *Pithecanthropus* has led most scientific men to regard him as a genuine ape belonging to the group of Hylobatidae; others, however, consider him at best to be an ideal, but not a real intermediate form between man and ape. I say 'at best an ideal intermediate form,' inasmuch as certain peculiarities in the formation of his cranium and skeleton cause him to approximate more closely to man than do any of the present anthropoid apes. But, on the other hand, there are other morphological peculiarities which suggest his being more nearly connected with the lower apes. Schwalbe has drawn attention to these points, and for these reasons the scientific opinion, which seems most likely to be correct, is that of the zoologists who regard the *Pithecanthropus* as one of the higher apes, representing the end of one side branch of the line of apes. In the case of *Pithecanthropus* we have a repetition of the old comedy; a supposed link in the ancestry of man is at first welcomed with enthusiasm, but finally has to be discarded.

In a subsequent photograph I shall show you how the zoologists assembled at Leyden in 1895 allowed the *Pithecanthropus* to be presented to them as a 'masher,' to enliven them at their banquet.

Here arises the important question of the age of the *Pithecanthropus.*

At first he was believed to have lived in the Tertiary period. As human remains cannot with certainty be assigned to any epoch before the middle Pleistocene—it is at least doubtful whether the Heidelberg lower jaw is really early Pleistocene—we can easily understand why in 1895 it was still possible to seek an ancestor of the human race in the ape-man. More recent investigations made in Java by Voltz and Elbert have transferred the ape-man into the Pleistocene epoch, and, as Branco stated in 1908, he probably lived about the middle of it, and hence he could not have been an ancestor of man, as he was a contemporary of man at that time.

Homo primigenius has played a much more important part than the *Pithecanthropus*, and soon replaced him in the theories of those advocates of evolution who felt it absolutely necessary to discover an intermediate form. This primitive man is in reality the oldest palæolithic man of whom we know anything, and in him science has found true, positive *points d'appui.*

* * * * *

In 1901 Schwalbe submitted the Neandertal cranium to a fresh examination, in consequence of which he added a twelfth to the already existing eleven theories about it. In the *Bonner Jahrbücher* he put forward the hypothesis that the Neandertal man was not a man at all, but the representative of a distinct genus, that ought to be placed systematically between *Pithecanthropus* and fossil man.

The Neandertal cranium was found in a cave in the Düssel Valley near the Rhine about the middle of last century. At the time Virchow considered it to be a pathological formation. He thought that people with similar crania were still to be met with. His opinion was mistaken in one way, but quite correct in another. Modern research has shown that the Neandertal type does not occur amongst Europeans of the present day, although it may be found amongst Australian blacks. As a fossil or primitive man, the Neandertal type represents that of early palæolithic man, who cannot be relegated to a time further back than the middle of the Pleistocene epoch. Obermaier assigns his first appearance to the last third of that epoch, viz. to the last interglacial or Mousterian. To the same date must be assigned the remains found in the South of France, to which I shall refer later, and which archæologically admit of very precise verification.

This does not agree with Schwalbe's view, expressed in 1901, that the Neandertal man was a representative of a distinct genus standing midway between man and the *Pithecanthropus*. Schwalbe himself changed his mind on the subject in 1904, at the meeting of the Association of German Naturalists and Physicians, when he spoke of *Homo primigenius* as a distinct *species* of man, not as a *genus*. Schwalbe believed him to be distinguished both from modern and from later palæolithic man by a number of constant characteristics, of which the chief are a receding forehead, a lower cranium, prominent ridges above the eyes and absence of chin, or rather of the furrow in the lower jaw, which gives rise to the projecting chin common at the present day.

But this opinion also, that *Homo primigenius* represents a particular species, is now untenable, and has been abandoned by almost all scientific men, even by its author, Professor Schwalbe. The first blow was dealt it by the discoveries made at Krápina in Croatia. In 1905 Gorganovic Kramberger, who found the remains there, showed that it was possible to trace a series of gradual transitions between *Homo primigenius* and modern man, and consequently, according to the principles of zoological classification, primitive man cannot be regarded as a different species, but only as an older race of man, who made his appearance in the middle of the Pleistocene epoch. There is most convincing evidence in support of this latter theory, in fact, if we accept Obermaier's redistribution of the glacial epochs, we must assign the appearance of man to the last third of the Pleistocene. No one at the present day can doubt that *Homo primigenius* represents a distinct, early palæolithic race of men, for this has been conclusively proved by palæontology.

We have remains from the Düssel Valley near the Rhine, from Spy and Nalautte in Belgium, from Ochoz in Moravia, from Krápina in Croatia, and from le Moustier and Chapelle-aux-Saints in the south of France, and much has been learnt from them. An important discovery has been made lately in Germany, for at Mauer, near Heidelberg, Schoetensack found a human jaw-bone, which, in his opinion, either is late Tertiary, or belongs to the close of the Tertiary and the beginning of the Quaternary periods. Obermaier and Wilser, however, rightly questioned the accuracy of this date, and showed that the bones found with the jaw were those of animals which might with equal probability be assigned to the Pleistocene epoch. There is very little difference in size and shape between the jaw-bones found at Mauer and Spy respectively; the latter undoubtedly belongs to the Neandertal type. The massive development of the lower jaw in comparison with the smallness of the teeth is certainly remarkable, but exactly the same features occur in a modern

Eskimo skull, shown me a few days ago by Birkner, in the collection of the Munich Institute for Palæontology. The Mauer jaw belongs morphologically to the Neandertal type,[1] and, as I have just said, it is probably not early, but middle Pleistocene.

Owing to the absence of palæolithic stone implements in the Heidelberg deposits, we are certainly not justified in assigning the jaw-bone to any definite period, such as the Chellean or Mousterian. But for this reason to assert, as Schoetensack did, that the owner of the bone was a Tertiary man, and perhaps even a common ancestor of man and anthropoid apes, is too daring a statement, and by no means well established. We can do nothing but wait and see what future research will reveal.

All that we know for certain on this subject at the present time is that an early race of men lived in Central Europe in the latter part of the Pleistocene epoch, and that they were distinguished from the modern inhabitants of Europe by definite, although slight, anatomical and morphological characteristics, such as the strong development of ridges above the eyes, low forehead, receding chin, &c. But, as Klaatsch has proved convincingly, all these peculiarities still occur in Australian blacks. Therefore primitive man, in respect of his body, only belonged to an earlier race of man, and was not a half-ape.

Let us consider the chief of these characteristics somewhat more closely, in order to see whether they really are points of likeness to apes or not. The receding chin is due to a stronger, but quite normal development of the lower jaw. It was only when the lower jaw began to degenerate that the hollow was formed, which causes the chin to project. I cannot now discuss the little bones of the chin, which are morphologically connected with this projection, but the diminution in size of the lower jaw, and the pretty dimple that we now admire, are, considered in their morphological aspect, marks not of progressive but of retrograde development in the formation of the lower jaw. As men became more civilised and adopted a more refined sort of food, their jaws had less hard work to perform than those of primitive men, and consequently diminished in size. With regard to the prominent ridges above the eyes, —the second great peculiarity of the earliest race of men—Klaatsch explained last year, at the meeting of naturalists in Cologne, that they were connected with the size of the eye-sockets, and therefore with the adaptation of early palæolithic man to the life of a hunter. They are a function of the very marked development of his sense of sight, and there is nothing pithecoid about them.

[1] Kramberger has recently shown that in its solid formation the Heidelberg jawbone very closely resembles that of a modern Eskimo skull, the chief difference between them being that the chin is more pronounced in the latter than in the former. This confirms the conclusion that the Heidelberg jaw belonged to a man of the Neandertal type. See 'Der Unterkiefer der Eskimo, als Träger primitiver Merkmale' (*Sitzungsbericht der Preuss. Akad. der Wissenschaften*, 1909).

Some one apparently dreamed (and his dream has been spread far and wide in French newspapers) that primitive man could not walk upright, but advanced, like apes, in a crouching attitude. Klaatsch has publicly called this idea 'nonsense.'

The extremities of the le Moustier man may, by their remarkable shortness, suggest adaptation to cave life, but they are not pithecoid, for apes have much longer arms than we have.

We must investigate the cranial development in palæolithic man somewhat more closely. Did the earliest man known to us stand, with respect to his cranial capacity, somewhere midway between apes and modern men? Certainly not. The cranial capacity of no anthropoid ape reaches 650 cubic centimetres;[1] in the fossil *Pithecanthropus*, a gigantic ape, it amounts to 800-850 c.c. The Weddas, a race of dwarfs in Ceylon, have the smallest cranial capacity among human beings; in their case it is about 960 c.c. In making this statement we are, of course, comparing the absolute measurement of the head of a giant ape with that of a human dwarf. The Neandertal cranium was said to have a capacity of about 1230 c.c., whilst now men in Central Europe (Bavaria) possess on an average a cranial capacity of about 1503 c.c. The capacity of a female cranium is about 200 c.c. less than that of a male, but this does not prove women to be less intelligent than men. Bismarck's skull was enormous, and had a capacity of 1965 c.c., but Virchow discovered one still larger, with a capacity of 2010 c.c., and this skull belonged to a savage in New Britain, not to a civilised inhabitant of Great Britain. This is the largest skull on record.

Where does primitive man stand in comparison? Boule has recently made a very careful examination of the remarkable human remains found at Chapelle-aux-Saints, and, as the cranium was in very good preservation, he was able to test its capacity according to the newest methods, and what was the result? Did he find that it measured about 1230 c.c., the number formerly assumed to be that of the Neandertal type? This would correspond very closely with the cranial capacity of women at the present day. No, the skulls of these oldest palæolithic men vary in respect of their capacity from 1600 to 1700 c.c.; probably 1626 or 1635 c.c. is a safe average to take.

According to the materialistic school, the capacity of the skull affords a direct indication of the mental capabilities of its owner; and if this be so, we are justified in asking what has become of the half-ape? Among human beings of our own time only a few have a cranial capacity greater than that of this fortunate half-ape, not even our most learned university professors, who are rightly considered the *élite* of the human race in respect of intellect.

There seems to be need of greater moderation and caution in

---

[1] Ranke gives 605 c.c. as the maximum for the male gorilla; Topinard thinks the number may reach 621.

accepting the theory that man is the descendant of brutes. We must consult facts, and proceed quietly without reference to the dogmas of monism. I can give an instance of what I mean by consulting facts, connected with the skull from le Moustier. I had opportunity to examine it closely at a lecture given by Hauser to the Anthropological Association at Frankfurt-am-Main in 1908.

Klaatsch's reconstruction of it was noticeably different from the plaster model that stood beside it;[1] the latter bore a strong resemblance, absent in the original, to an ape, especially about the mouth, and this was due to the fact that, through a blunder in taking the cast, in the plaster model the ends of the lower jaw were at a distance of several centimetres from their sockets. In reality, the same relative proportion between the size of the cranium and that of the lower part of the face exists in the le Moustier skull as in *Homo sapiens recens.* You will see this clearly in the photographs of this skull which are copies of those made originally by Hauser and Klaatsch.

May we say then that these palæontological discoveries have given a scientific account of the origin of man? No, we are still far from it. We know that the geologically oldest human beings hitherto known, belonging to the Stone Age of Central Europe, formed a race known as the Neandertal race, but this by no means represents a connecting link between apes and men. We know further that critical investigations made by Boule, Obermaier, and de Lapparent have completely overthrown the belief, based upon Rutot's once famous Eolithic Theory, that even at the beginning of the Tertiary period there existed beings resembling men, who fashioned rough flint implements. De Lapparent not unfairly calls the eoliths 'silex taillés par eux-mêmes,' because they may have been formed by the mere forces of nature. But we do not know if the Neandertal man was really the earliest man, for we cannot tell whence he came. Did he appear as an autochthon in central Europe? Did he migrate hither from the east? As a migration from the east or south can be proved in the case of almost all subsequent European races, it very probably occurred also in the case of *Homo primigenius*, who bears the proud name of first-born among the human race. The negro-like Grimaldi-type of South European, which appears at the close of the Pleistocene epoch, most likely came from the south. In the parts of southern France where remains of the Neandertal type of early palæolithic man are discovered, viz. in the valleys of the Dordogne and of the Vezère, in somewhat higher strata are found traces of a later palæolithic man of the Cro-Magnon type. He belongs to the close of the Pleistocene epoch, and in his cranial formation he is exactly like Central Europeans of the present time. Was he a

[1] Cf. the accompanying illustration, which is a copy of Hauser's original photograph.

descendant of the primitive man, who inhabited the same regions before him? Or did he migrate hither from the east, from western or central Asia? We do not know; nor do we know whether the Neandertal type of man who differed from the latter type in some rough morphological characteristics, was himself a descendant of another, still older race, that migrated from the east about the middle of the Pleistocene epoch.

We have no certain information as to the outward appearance of the oldest man. We cannot tell whether he was like the earliest palæolithic European, or whether he belonged to a higher race, more like modern men, and only acquired the bodily peculiarities of the Neandertal type by adaptation to the life of a cave-dweller and hunter.

The history of the human race is still silent with regard to these points; but we are sure of one thing, that the oldest palæolithic man of whom we have any knowledge, even if he had not attained to a high degree of civilisation, possessed the capacity for being civilised. He discovered the use of fire, and found out how to make the most important implements which we still employ, such as the knife, the axe, and the scraper. In the flint implements of this period we can trace the simplest ideas underlying the construction of our most indispensable tools. He must indeed have been a clever man!

Picture to yourselves a modern civilised human being, bereft of all the means of existence, and devoid of all knowledge how to make tools; I assure you, the poor fellow would probably starve. And yet our ancestor, who is represented as being something between ape and man, succeeded in making his way through the world! He deserves honour, and ought not to be contemptuously spoken of as a half-ape!

I must unfortunately cut short this part of my lecture . . . and will therefore pass on at once to the photographs that I have to show you. They bear upon the comparative morphology of man and ape, and upon primitive man.

The first two photographs represented skeletons of an orang-utang and of a man respectively (from the Army and Navy Medical Museum in Washington); they illustrated the differences between man and ape in the formation of the extremities, the excessive length of the ape's arm and the peculiarity of its foot.

The next two photographs represented the crania of the orang-utang and of a man respectively. In the ape's skull, the skull-cap is very small in comparison with the enormously developed lower part of the face with its powerful jaws. The brain region is insignificant in comparison with the parts concerned in devouring food. In man the case is reversed. The lower part of the face is very small in comparison with the large skull-cap, which contains the brain.

The fifth photograph showed the *Pithecanthropus* as a 'masher,' as he appeared at the banquet given to the Zoologists assembled at Leyden in 1895.

The sixth photograph represented the Neandertal cranium, according to Schaafhausen's illustration of it.

The seventh showed the cranial curves of a chimpanzee, the *Pithecanthropus*, the Neandertal man, a modern Australian black, and a modern Englishman, according to Macnamara. The crania of the ape and of the *Pithecanthropus* were seen to differ only in size ; those of the Neandertal man and of the Australian black resembled one another so closely as both to be within the limits of variation of *Homo sapiens*.

The eighth and ninth photographs were copies from originals, taken by Hauser and Klaatsch, of the skull of the le Moustier man. The size of the cranium, in comparison with the lower part of the face, is relatively almost the same as in modern men, although both are absolutely larger than is the case in most modern skulls. After the lecturer had pointed out on these photographs the characteristics of the Neandertal type, he described the circumstances under which the le Moustier skeleton was discovered. In its case, as in that of the skeleton at Chapelle-aux-Saints, there were unmistakable tokens of solemn burial in the early palæolithic age. The body was laid on its side, the arms and legs being arranged in a definite position. Under the head was a cushion of earth, upon which, at le Moustier, the impression of the dead man's cheek could still be seen. The lecturer said that he had examined the remains found by Hauser, and convinced himself of the truth of this statement. Round about the corpse were arranged the largest and finest stone implements of the period, as Hauser had carefully pointed out. The le Moustier skeleton was that of a young man, whose parents had buried with their child all the precious things that they possessed. Can they have been 'bestial savages,' or 'fierce ape-men'? In a lecture delivered at Cologne in 1908 at the meeting of German Naturalists and Physicians, Klaatsch remarked that the mode of burial of this *Homo mousteriensis* pointed quite plainly to belief in immortality existing in the mind of palæolithic primitive man perhaps 30,000 years ago.[1]

As far as the time at my disposal permitted, I have laid before you what science teaches us regarding our ancestry. And what does it amount to ? We arrive at exactly the same result as Branco did eight years ago, when he stated, at the International Congress of Zoologists at Berlin, that palæontology at the present time knows no ancestors of man. This statement has been confirmed by recent research into the primitive history of the human race. We are acquainted with an early palæolithic race, called the Neandertal type or *Homo primigenius*, but we are not acquainted with any ancestors of man resembling apes. The most remote ancestor of man hitherto discovered by science was both in body and mind a genuine human being, a true *Homo sapiens*.

If this be true, what scientific justification is there for Haeckel's

[1] In speaking of time we are at present unable to do more than offer speculations. We have to estimate the length of periods by changes in the fauna and flora, which again are a result of modifications in climate. The latter, however, especially the alternation of glacial and interglacial periods, are probably connected with the nutation of the earth's axis. For this reason we must assume that the last interglacial period, to which the Mousterian deposits belong, occurred at least 30,000 years ago (Obermaier).

'Pedigree of the Primates,' in which, even in 1907, *Homo stupidus*, the stupid man, appears as the immediate predecessor of *Homo sapiens?* There is no scientific justification at all for it. For the last forty years, Haeckel has been devising such pedigrees of man, and has been proclaiming to the whole world the descent of man from apes—for his Primates are the half-apes and the true apes —as an historical fact, but this cannot be called pursuit of science, but rather mischievous meddling with it.

On February 18, 1907, at the evening discussion that followed my Berlin lectures, Haeckel's assistant, Dr. Schmidt of Jena, came forward and solemnly defended his master against the charge, that I had brought against him, of having published his 'Pedigree of the Primates' as an historical fact. He maintained that Haeckel had never done so, being far too modest and far too ardent a lover of truth; but in my concluding speech there was no need for me to do more than quote one passage from Haeckel's work, 'The Last Link: Our Present Knowledge of the Descent of Man,' in which no one can deny that 'the phyletic unity of the line of primates from the lemurs (or half-apes) to man' is declared to be an 'historical fact.' With such a passage before him, no one could assert that Haeckel never said anything of the sort. Nevertheless, on the following morning, a few daily papers, not, it is true, of the highest class, accused me of having falsified the quotation. This may be called pursuit of science on the lines of monism and social democracy, but it cannot be described as a justification of Haeckel's pedigrees of man.

But Haeckel may possibly have improved lately? Yes, a little, but not much. In honour of the opening of the new Phyletic Museum at Jena in 1908, Haeckel published a large folio bearing the magnificent title 'Progonotaxis hominis.' In this work he has at last corrected some of the false statements to which he had clung so tenaciously. The unfortunate *Homo stupidus* has now vanished from the pedigree of man, and his place is taken by *Homo primigenius*. It was indeed high time, for the latter was discovered fifty years ago! Haeckel remarks too that many geologists consider the *Pithecanthropus* from Java to belong to the Pleistocene and not to the Tertiary period. He ought to have said simply 'geologists,' but nevertheless these words show an advance upon his previous assertions. The advance is, however, only in the text; when we turn to the pedigree of primates, which is given in the appendix, we find that there he has gone backwards rather than forwards. Beside the name *Pithecanthropus erectus* stands, as before, the word 'Pliocene,' i.e. late Tertiary, and *Homo primigenius* is represented as the descendant of this ape-man, although the latter was really a contemporary of man of the Pleistocene epoch. Such is Haeckel's 'scientific spirit!' Elsewhere, too, this scientific work contains manifest contradictions. In the text all the early *races* of men are changed into so many *species*, but on the pedigree of primates they appear again as races, and not as species. How are such blunders possible in a scientific publication of this sort? The only true explanation was suggested to me in Munich a few days ago by an eminent zoologist, who had been a pupil of Haeckel's. He ascribed them to senile decay! But even this explanation breaks down, when we find, on the most recent pedigree, that Haeckel has set the same mark against the ancestors that he has invented in the pedigree of man, as against the fossil forms of extinct primates. The same little cross stands beside both, as a sign that both are extinct. A scientific man really is going too far when he sets purely imaginary forms on a level with real fossils, in order to deceive his reader as to the true value of this human pedigree; to say the least of it, he is playing tricks and juggling with the truth, or, to use plainer language, he is telling lies!

I come now to the charge, which Brass has recently brought against Haeckel, of having tampered with the illustrations of embryos.[1] This charge has attracted much attention, at which I am surprised, for, in the first place, the alleged falsifications of illustrations are by no means the worst falsifications perpetrated by Haeckel. It is far worse that for more than forty years he has been falsifying men's ideas, and so has robbed the German nation of Christianity, and given it instead a materialistic and atheistic cosmogony. To distort the Christian conception of God and represent Him as a 'gaseous vertebrate' is a far worse fraud on Haeckel's part than tampering with a thousand pictures of embryos. In the second place, the charge, brought by Brass against Haeckel, of having tampered with the illustrations, was by no means new.[2] The same accusation was raised against Haeckel by Rütimeyer, a Swiss zoologist, as early as 1868, and by Anton His of Leipzig, a famous anatomist, in 1874, and was then proved to be irrefutable. It is really an 'historical fact' that Haeckel, for the sake of his argument, i.e. in order to convince his readers thoroughly of their descent from brutes, caused the same plate to be printed three times in his 'History of Creation,' and said that it represented three distinct objects extremely like one another. Haeckel himself subsequently acknowledged that he had done so. It is another 'historical fact' that in his 'Anthropogeny' he altered many illustrations of embryos in an arbitrary manner, and assigned to them other names than those which they had originally borne, and thereby he caused His and other colleagues publicly to declare that Haeckel was not seriously carrying on scientific research. In replying to this charge in 1891, Haeckel defended himself in a classical fashion by calling His, Kölliker, and other eminent German embryologists 'a

[1] This subject is treated more fully here than it was in the lecture, when want of time compelled me to be very brief.

[2] On this subject see my article in *Stimmen aus Maria-Laach* for 1909, Nos. 2-4, 'Alte und neue Forschungen Haeckels über das Menschenproblem,'

company of Scribes and Pharisees,' who ought to be described as 'narrow-minded' rather than as 'exact scientists.'

I come now to the famous declaration of forty-six German zoologists on the subject of the dispute between Haeckel and Brass. In all probability this declaration attracted so much attention chiefly because people assumed it to be an 'amende honorable' to Haeckel. Perhaps a consideration of its origin will lead them to form another opinion. In his reply to Brass, Haeckel boldly asserted that if he were to be accused of falsifying the illustrations of embryos, a similar accusation must be brought against hundreds of highly respected embryologists, anatomists, zoologists, &c., for they had had recourse to falsification as much as he himself, and had in many ways 'schematised' their illustrations. This was certainly too daring a suggestion on Haeckel's part. He knew well enough that other scientific men do not 'schematise' in his fashion, for they say what they have done, if they present us with an imaginary form, or alter an existing form to reproduce it in a schematic fashion. Frank acknowledgments of this kind are missing in Haeckel's falsified illustrations of embryos, and so by means of them he has deceived his readers as to the worth or rather the worthlessness of the evidence that they afford of the descent of man from brutes. It was therefore absolutely necessary for Haeckel's German colleagues to adopt some definite attitude in answer to Haeckel's suggestion that they all were guilty of falsification as much as he was. The famous declaration was their reply to this insinuation.

It is obvious that the successors of those exact German scientists, who denounced Haeckel's proceedings so decidedly thirty years ago, and were in consequence called by him 'narrow-minded,' could not in their declaration express approval of Haeckel's action on the point on which Brass challenged him, but only disapproval. This they did in unmistakable terms, but they were afraid of injuring, not only Haeckel's reputation, but also that of the whole scientific doctrine of evolution. For this reason they ostensibly directed their censure chiefly against the 'Keplerbund.' This was a clumsy device on their part, for the 'Keplerbund' is no more opposed to the scientific doctrine of evolution than I am. Moreover, there was no ground for their fear lest a declaration against Haeckel should damage the reputation of science, for no one during the last forty years has done more than Haeckel to compromise the scientific doctrine of evolution in Germany, since he has boldly misused it in his attack upon Christianity. For some reason or other, however, the forty-six zoologists insisted upon the insertion of the clause against the 'Keplerbund' in their declaration against Haeckel, but I do not think that thereby its significance is diminished, in so far as it refers to Haeckel's proceedings.

I am confirmed in this view by the circumstances under which the declaration was issued. It was signed by a very considerable number of German zoologists, some of whom I know personally as men of calm judgment, highly esteemed in the scientific world. In the *Deutsche Medizinische Wochenschrift* for 1909—I think in the eighth number—an article on the dispute between Haeckel and Brass had appeared, written by Professor Keibel of Freiburg i. B., one of our most respected German authorities on the subject of the comparative embryology of man and the higher animals. In this article Keibel criticized Haeckel's illustrations of embryos very sharply, and completely confirmed the disclosures made by Brass regarding Haeckel's so-called 'falsifications.' It is true that Keibel did not speak of falsifications but of inaccuracies. The word, however, is a matter of choice; personally, I believe inaccuracies originating in an intention to mislead the reader are not *mere* inaccuracies. For instance, when Haeckel alters an illustration of the embryo of an ape with a tail, so as to turn it into the picture of one without a tail, and at the same time changes the name of the creature, it can hardly be done unintentionally.

We are not here concerned with Keibel's further statements against Brass in the article to which I have referred. It is true that Brass's work is not free from inaccuracies, but it certainly is free from any intention to deceive the reader.

The declaration of the forty-six zoologists followed Professor Keibel's absolutely crushing criticism of Haeckel in the *Deutsche Medizinische Wochenschrift*, and, in my opinion, in signing the former they expressed their agreement with the latter. In this way the declaration of the forty-six acquires another significance than that hitherto ascribed to it. I regard it as an exculpation, not of Haeckel, but of German science!

I may call attention to the further fact that among the publications of the 'Keplerbund' there has recently appeared a pamphlet written by Director Teudt, in a very calm and impartial spirit, entitled 'Im Interesse der Wissenschaft' (In the Interests of Science). It contains an account of the dispute between Haeckel and Brass, and of the publications dealing with it. It does not, however, connect the declaration of the forty-six zoologists with Keibel's criticism of Haeckel. As the declaration appeared almost simultaneously in a great number of magazines and newspapers, it is quite possible that this connexion, which is certainly to the advantage of the forty-six, has been too much overlooked.

Before leaving this subject, let me say a few serious words on the claim made by monism of having replaced the Christian cosmogony by a new and better theory of the universe. This new monistic doctrine is being actively propagated at the present time, both in academic circles and among the lower classes. It behoves us to ask what monism really is.

The word 'monism' is a genuine Proteus; for all kinds of various meanings are concealed under it, and it is absolutely necessary for us to arrive at some clear conception of what it is, in order to be

2 L 2

able to combat the mischief that is being done with this catch-word ' monism.'

Literally translated it means ' Doctrine of Unity.' This suggests the pantheistic principle of the One ; but we cannot at once adopt this as our definition. In the course of his speech at the evening discussion in Berlin on February 18, 1907, Professor Plate declared that the monist concerned himself only with natural laws, not with what lay behind them, with regard to which different men held different opinions. This would lead us to suppose monism to be synonymous with agnosticism, which denies that God can be known and rejects all metaphysics. But here we have a confusion of ideas rather than a definition. Agnosticism is not synonymous with monism, at least for any one who has had any philosophical training. The essence of monism possibly is, that some of the people who call themselves monists think of the unknown quantity x underlying the natural laws in one way, and others in another. This, however, would not be monism in the philosophical sense of the word, and would be more suitably designated ' confusionism.' What is the real meaning of monism of which we hear so much nowadays ?

Plate, having probably forgotten the definition that he had previously given, offered another in the book written against me : ' Ultramontane Weltanschauung und moderne Lebenskunde, Orthodoxie und Monismus ' (Ultramontane cosmogony and modern views of life; Orthodoxy and Monism). The defective objectivity of this work reveals itself even in its title. It is a faithful reflexion of the line of action—equally wanting in objectivity and equally unsuccessful—adopted by Plate and some others of my opponents at the evening discussion in Berlin, on February 18, 1907. A very sarcastic and shrewd criticism of their proceedings appeared in the Munich *Hochschulnachrichten* (1908, No. 6), which certainly cannot be suspected of clericalism. The writer remarked that they had not treated their guest with any particular consideration, but nevertheless they had not succeeded in positively refuting his statements ; annoyed at the appearance of a Jesuit, these worthy Berlin gentlemen had dragged into the discussion questions that had nothing to do with it, and this deserved notice as a characteristic feature of the times !—Plate and his companions were ill-advised when they attempted to use a discussion of the scientific theory of evolution as an opportunity for attacking the Catholic Church. In the work to which I am referring, Plate solemnly declares that every student of nature must necessarily be a monist ; if he is not, he must be wanting either in ability to reason or in honest love of truth. But what does Plate mean by ' monism ' in this passage ? Something quite different from what he meant before. In this place monism is an effort to obtain as uniform and simple a theory of the universe as possible in accordance with natural science.

If this is monism, Plate is perfectly right in declaring every student of nature, who does not call himself a monist, to be either a fool or a hypocrite. In *this* sense, as aiming at a very uniform and simple explanation of nature, I too am a monist; Father Secchi was a monist when he wrote his 'L'unità delle forze fisiche,' and even St. Thomas Aquinas, Blessed Albert the Great, and St. Augustine were downright monists, for all earnest thinkers in every age, as soon as they have begun to study nature, have striven to find the most uniform and simple explanation possible for its phenomena.

What can we say of this definition of monism given by Plate, a member of the German 'Monistenbund,' of which Haeckel is president?

We can only say that it is calculated to mislead the general public, just after Haeckel's fashion; monism is first defined in such a way that every thoughtful student of nature must be a monist, and then we are told: 'Wasmann and Reinke and all adherents of Christianity are opposed to monism, therefore they are either fools or hypocrites.' About such an argument as this it is not possible to say anything but that it is absolutely dishonest.

You are right in thinking that behind monism, as represented by Plate and the 'Monistenbund,' there lurks something quite different from a desire for a uniform explanation of natural phenomena. It is a name for a number of dogmatic hypotheses, which have nothing at all to do with a scientific account of natural phenomena.

One of these hypotheses is especially connected with the theory of evolution; Plate, Forel, Escherich, Wagner, and other monists maintain that 'scientifically' only a monistic theory of evolution is admissible, i.e. a theory of descent, according to which the whole evolution of the organic world, or at least of the two organic kingdoms, must form one single line, in which the higher forms have proceeded from the lower, and these again from one or a few primitive cells. These representatives of monism ridicule the idea of a polyphyletic evolution of animals and plants, and try to cast upon it a suspicion of being 'theological,' as several of my monistic opponents have done. They are intolerant of my conception of 'natural species,' which groups together as forming a natural unit all the species, genera, and families belonging to one palæontological line of descent. Therefore they maintain the conception of natural species to be theological, and consistent with neither natural science nor natural philosophy. Apparently these gentlemen are not aware that many years ago Neumayr stated his ideas regarding 'palæontological species,' which exactly coincide with my own regarding 'natural species,' and yet Neumayr was neither a theologian nor a Jesuit. Here we have another instance of the monists' vaunted freedom from prejudice! They begin by asserting that only a monistic, monophyletic evolution can have any scientific justification, and they entirely forget that the question of the limits of race-evolution is one of facts and not of dogmas. In the first of my lectures I discussed this point more fully.

Another dogmatic presupposition on the part of monism, as represented by Haeckel and the German 'Monistenbund,' is contained in the assertion that it is indispensable to the unity and simplicity of any explanation of natural phenomena, that the whole natural world should have been evolved in conformity with one and the same law Behind this assertion lurks the further assumption that this universal law must be purely mechanical. Of course, every monist is at liberty to ascribe to each atom in the universe the possession of a 'soul,' which, however, consists merely of an attracting and repelling force; although, as Dubois-Reymond shrewdly remarks, to do so is an insult to all reasonable philosophic thought.

From the scientific point of view, what are we to think of the claims of monism ?

In the first place, monism is absolute dogmatism, and appeals in vain to its 'scientific character.' It is an absolutely dogmatic assumption to declare that one and the same law must necessarily govern the evolution of the inanimate and of the animate world of plants and of animals. No less dogmatic is the further assumption that this sole law governing evolution must have been, and must still be, purely mechanical.

Theories in natural philosophy must be based upon actual scientific results; it is only thus that they can have any scientific foundation. Theories, as is well known, have to square with facts, not facts with theories, otherwise theories become a Procrustean couch for scientific research If, therefore, we find higher laws governing animate nature than the purely physico-chemical laws that govern inanimate matter, we must not deny the existence of these vital laws, through love of any monistic dogma.

If in the psychical phenomena of animal life we find a higher law than purely mechanical and physiological response to stimulus, we must not deny the existence of the psychical life of animals, through love of any monistic assumption.

And if, finally, in the spiritual life of man we find a higher law than in the sensitive life of animals, which sensitive life, in the case of man, forms only the foundation for his spiritual life with its intelligent thought and free will—we must not deny the existence of the human spirit, through love of any dogmatic postulate of monism. To do so would be absolutely unscientific !

In the second place the monistic assumption that in all nature only one law can prevail, and that this law must fundamentally be purely mechanical, is more than mere dogmatism; it is concealed materialism, decked out with Haeckel's 'atomic souls,' in order to render it more attractive to superficial thinkers.

We have now advanced another step towards understanding what is hidden under the catch-word 'monism.' As I said before, the literal meaning of monism is 'doctrine of unity,' or of the One. This is what we must now examine, the kernel in the shell of monism.

As a doctrine of unity, monism is sharply contrasted with dualism of every kind. It not only insists upon there being one sole law governing the evolution of the world, but also upon there being one sole substance. For this reason the monist regards spirit and matter as essentially one, as merely different manifestations of one and the same thing. For the same reason he believes God and the world to be substantially one, for this is the logical outcome of the monistic dogma of unity.

What must we think of this twofold postulate of dogmatic monism? It converts monism, the apparently harmless doctrine of unity, first into concealed materialism, and secondly into concealed atheism.

First into concealed materialism. The monistic theory of identity,[1] which sees in body and soul nothing but two manifestations of the same thing, boasts of not being called materialism, but nevertheless inwardly it does not differ from materialism, for it regards what is psychical only as an unreal, subjective reflexion of the material cerebral processes (Forel), and denies all causality to psychical phenomena. It believes all causality to belong to the material phenomena that accompany the psychical. But where there is no longer any causality, there ceases also to be any reality, and the psychical becomes a mere shadow of the material. This amounts simply to the old materialism dressed up in a new fashion!

Secondly, we come to the monistic identification of God and the world, that aims at banishing the idea of a personal Creator, which is said to be out of date. In the course of thousands of years, pantheism has presented mankind with its doctrine of the One under many different forms, but none has approached atheism so closely as Haeckel's new monism. There is absolutely nothing in this monistic conception of God. It is an empty nut, of which the shell consists of the phrase 'the true, the good, and the beautiful' —that new monistic 'trinity,' as Haeckel called his new God. No less a man than Caprivi openly declared that what was known as monism was simply atheism, and Caprivi was assuredly not a Jesuit!

The inward emptiness of the new monistic conception of God must be obvious to every thoughtful human being. The God of Haeckel's monism is nothing but a shadow of the world, reflected in the cerebral functions of man, the highest vertebrate; just in the same way as in monistic psychology the spirit of man is a mere shadow, a reflexion of the material working of his brain. There

[1] This theory of identity and the whole psycho-physical parallelism of monism have been sharply criticised by two eminent German psychologists, K. Stumpf and L. Busse. Cf. on this subject my own work: 'Die psychischen Fähigkeiten der Ameisen, mit einem Ausblick auf die vergleichende Tierpsychologie' (*Zoologica*, No. 26), 2nd ed., Stuttgart, 1909.

is nothing underlying this conception, in spite of all the fine phrases of the preachers of the new monistic religion.

You will, perhaps, reply that I am surely mistaken in saying that monism is nothing but concealed atheism. Did not Professor Plate, a member of the new German 'Monistenbund,' solemnly declare at the discussion on February 18, 1907, his own personal conviction to be that, if we assumed natural laws to exist, we must also assume the existence of a lawgiver behind those laws? Such a confession is certainly not atheistic!

Plate actually used these words, and his *anima naturaliter christiana* revealed itself in them. I scarcely believed my ears when I heard them, and I made a note of them at once for use in my closing speech, in which I drew attention to the fact that, to my great joy, Professor Plate, a member of the 'Monistenbund,' had that evening publicly declared himself an adherent of Christianity, for a law-giver behind the laws of nature was precisely the personal Creator of Christianity.

A week later, in the course of a lecture delivered in Berlin by Pastor Steudel, of Bremen, who was then president of the 'Monistenbund,' a public rebuke was administered to Professor Plate for this confession of theism. He submitted to the imperious order of the monistic 'Congregation of the Index,' and withdrew what he had said, by appending a note in the printed version of his address to the effect that by these words he had, of course, only referred to 'a lawgiver in the pantheistic sense.'

No logic, not even monistic logic, can justify such a statement! According to the pantheistic conception of God, the lawgiver is identical with the laws of nature, therefore it is impossible for him to be 'behind them.' There is a flagrant contradiction in this monistic trick of hiding a lawgiver somewhere behind the natural laws, who, after all, turns out not to be behind them! It is pitiable to juggle in this way with words, and it is not creditable to the German people. Either let a man frankly acknowledge himself to be an atheist, or let him declare himself a theist, and an adherent of Christianity!

My last words are addressed to the students.—Gentlemen, if ever you have to encounter the perils of modern monism, remember that it behoves you to fight for freedom against the unscientific spiritual slavery of monism. One of my Berlin opponents, Professor Dahl, showed his courage and his love of truth some months later, when, in an article contributed to the Berlin *Naturwissenschaftliche Wochenschrift* for 1907, No. 40, he wrote these words: 'Where is then this freedom for science? I shall be told that in our country science and its teaching are free. They may be so in theory, but those who have to watch over the maintenance of this principle are but men. Adherents of monism have practically power of nomination to all appointments in the department of zoology. What is more natural than that they should nominate only those who are not opposed to monistic doctrines? I am far from suggesting that there is any *mala fides* in question. The men who have to propose names of suitable candidates honestly believe that none but their own views can further the interests of science. Therefore I ask again: where is freedom for science?'

Gentlemen, here we have a free utterance on the part of a free German! Be yourselves free, whether you are Germans or not. Take as your example the heroic struggle for freedom made by the men of Tyrol in 1809. Just as they would not submit to the tyrannical yoke of the Corsican, and remained loyal to their hereditary rulers, so may you declare: 'We will not submit to the unworthy yoke of intellectual slavery which modern monism is seeking to impose upon us! We will abide by our ancient Christian faith loyally and without wavering!'

Yes, Christianity, the old Christian theory of the universe, that is now so often denied, furnishes us with the only true monism, the only true doctrine of unity. There is *one* infinite and eternal God, whose creative power produced all finite beings and preserves them in existence. There is *one* God and *one* truth! Yes, gentlemen, there is only *one* truth, for from the inexhaustible source of everlasting, uncreated truth flow two streams, that of natural knowledge and that of supernatural revelation. Therefore there can never be a real antagonism between knowledge and faith, because there is only *one* truth which cannot contradict itself. For this reason cling with loyalty and courage to your ancient Christian faith!

Before we proceed to the discussion, I venture to make two remarks.

1. Several years ago Professor Blaas, whom I esteem very highly, lectured here on the origin of man. His views were criticised in the press, and the *Innsbrucker Nachrichten* published an article on the descent of man, which went rather too far, and contained several misleading statements. One of my colleagues requested me to send him materials for a refutation, and I referred him to an address on the subject of fossil man, delivered by Professor Branco at the fifth International Congress of Zoologists at Berlin in 1901. I had quoted the shorthand report of this address in my 'Modern Biology and the Theory of Evolution,' and my colleague mentioned this quotation in one of the Catholic papers published in this town. Thereupon, in another Innsbruck paper, the now unfortunately defunct *Tiroler Tageblatt*, I was accused of having intentionally distorted the meaning of Branco's words. I wrote to him at once to Berlin, and asked him to let me know whether the passage in question had been correctly reproduced by me or not. Professor Branco replied that what I had written down whilst he was speaking agreed completely with what he had been saying, but at the present time he should alter a few words in it. He had,

however, really intended to check the tendency to go to extremes. And now people come and accuse me of forgery! I have no desire to be classed with Haeckel!

I feel bound on this occasion to declare explicitly that Professor Blaas has assured me that he was not concerned, either directly or indirectly, with the charge brought against me in the *Tageblatt.* I wish to make this publicly known, for I am a lover of truth.

2. An article by Dr. Franz von Wagner appeared some years ago in the *Zoologisches Zentralblatt*, in which he discussed my 'Modern Biology and the Theory of Evolution.' He acknowledged the value of the scientific sections in which fresh evidence in support of the theory of descent was adduced from guests among ants and termites, my special department of research; but wherever my line of argument did not please him, he remarked: 'You are under theological influence,' and in this way he easily avoided any attempt to refute me. Professor von Wagner must not be offended if I advise him to adopt another line of argument next time. If by personal union, to employ an expression that must be very familiar here in Austria, a man is first a zoologist, then a philosopher, and only in the third place a theologian, it is surely unfair for that reason to cavil at what he says on natural science and philosophy, and for want of a better argument to keep on repeating that he is a theologian. The first thing to do is to show that theological prejudices have influenced me in stating the results of my scientific or philosophical investigations. This remark completes what I have to say . . . and I have only to offer you all, and especially the Catholic students, my most hearty thanks for your attention.

* * * * *

[1] This writer must not be confused with Dr. Adolf Wagner, professor to Innsbruck, to whose work on Lamarckism I have referred on p. 493.

SUPPLEMENTARY NOTES

On Chapter I, p. 5.

The Rev. John Gulick in his book 'Evolution, Racial and Habitudinal' (Carnegie Institution, Washington, 1905), p. 9, defines bionomics in the following words: 'Bionomics is the science that treats of the origin of organic types, and of the relations in which they stand to each other and to the physical environment.' For this definition he refers to Sir E. Ray Lankester's article on Zoology, in the 'Encyclopædia Britannica,' ninth edition. This definition, however, includes the theory of evolution (biogeny), which does not, in my opinion, belong to biology in the restricted sense.

On Chapter VI, p 110, note 2, and p. 169.

On the subject of the accessory chromosomes see also H. Otte, 'Samenreifung und Samenbildung bei *Locusta viridissima*, I' (*Zoologischer Anzeiger*, XXX, 1906, Nos. 17 and 18, pp. 529–535).

On Chapter VI, pp. 130, &c., and p. 134.

On the subject of the conjugation of unicellular organisms see also E. Korschelt, 'Über eine eigenartige Form der Fortpflanzung bei einem Wurzelfüsser, *Pelomyxa palustris*' (*Naturwissenschaftliche Rundschau*, XXI, 1906, No. 38, pp. 503, 504). This little creature, which resembles an amœba, has a complicated method of propagating itself. Numerous gametes are formed within the mother, and subsequently swarm out, and unite in pairs to produce a new individual. At the formation of the gametes, a reduction-division of the chromosomes takes place. The nuclear spindles of the mitotic figures are the result of a division of centrosomes that are very plainly visible.

On Chapter VI, p. 138.

On the subject of parthenogenesis in plants see also O. Rosenberg, 'Über die Embryobildung in der Gattung *Hieracium*' (*Berichte der deutschen botanischen Gesellschaft*, XXIV, 1906, pp. 157–161).

ON CHAPTER VIII, P. 213.

Closely connected with experiments on the regeneration of missing parts of an animal are experiments in transplantation, in which a piece of another animal is grafted on to supply the place of what has been amputated, and the results of the operation are carefully observed. These experiments are very instructive and throw light on the problem of determination. On this subject two very interesting papers were read on September 20, 1906, at the seventy-eighth meeting of German Naturalists at Stuttgart—'Über embryonale Transplantation,' by H. Spemann; and 'Über Regeneration und Transplantation im Tierreich,' by E. Korschelt. (Cf. *Naturwissenschaftliche Rundschau*, 1906, No. 41, &c.)

ON CHAPTER IX, P. 303.

That the doctrine of evolution as a theory in natural science is perfectly compatible with the Christian cosmogony has been repeatedly pointed out by Protestants also. Cf. Dr. Rudolf Schmid. 'Das naturwissenschaftliche Glaubensbekenntnis eines Theologen,' second edition, Stuttgart, 1906, and E. Dennert, 'Bibel und Naturwissenschaft,' fifth edition, Stuttgart, 1906.

INDEX

2 M 2

THE END

To illustrate p. 121 etc.

Plate I.

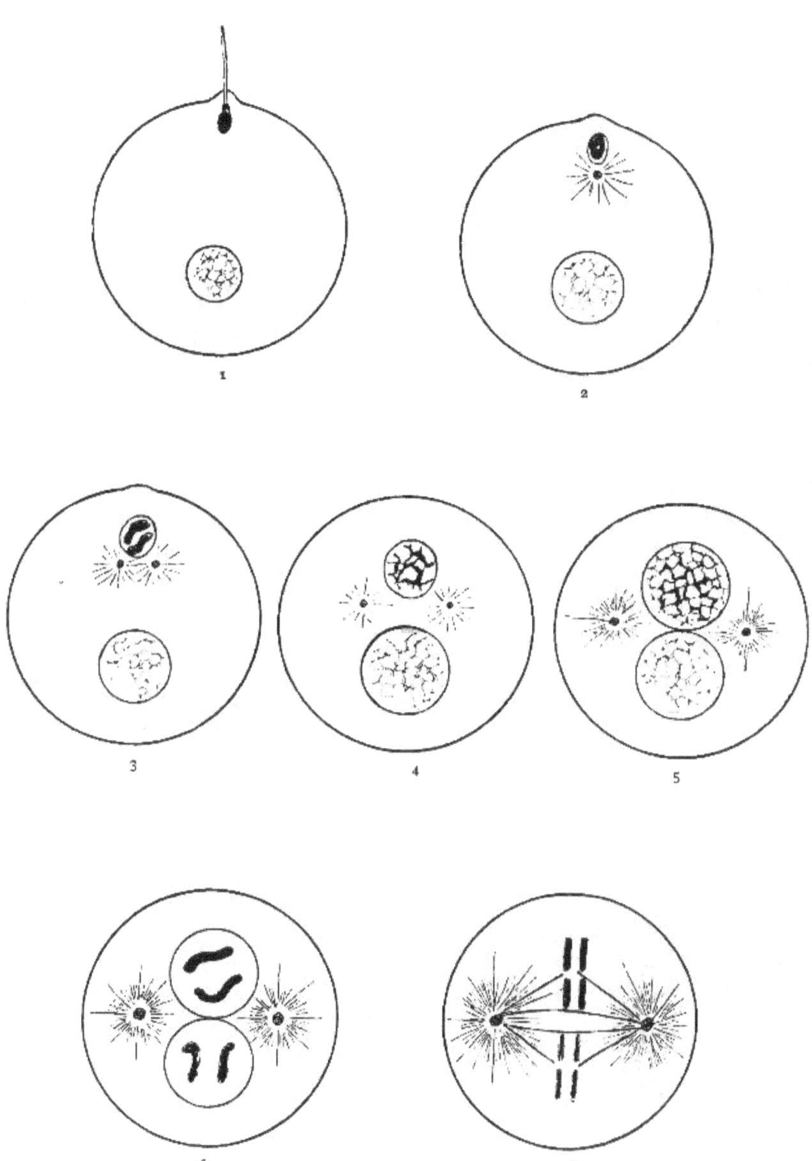

Diagrammatic representation of the process of fertilizing an egg-cell (after Boveri).

See p. 121 etc.

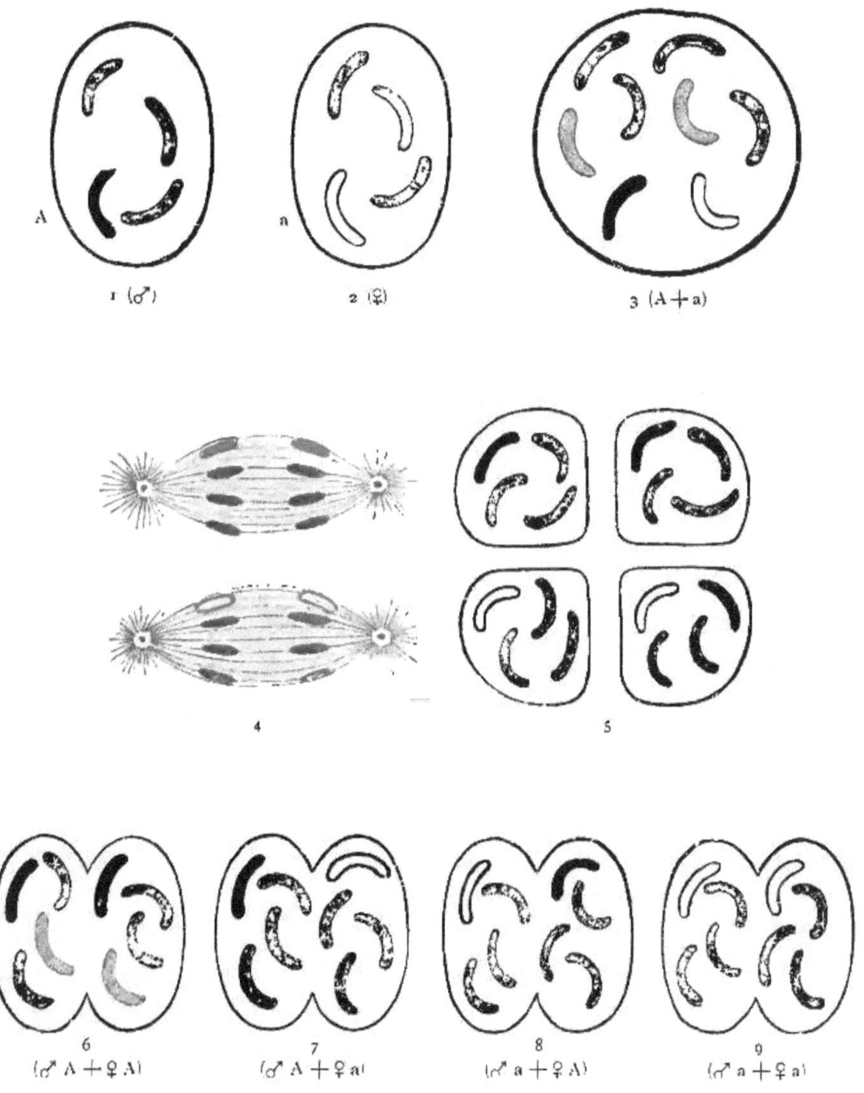

The Chromosome Theory and Mendel's Law of Hybridisation (after Heider).

(The red chromosome A indicates a tendency to produce red blossoms; the red-edged chromosome a indicates a tendency to produce white blossoms; ♂ = male germ-cell; ♀ = female germ-cell.)

Fig. 1 and 2. Nuclei of the parent germ-cells of varieties with red and white blossoms respectively.
Fig. 3. Union of these nuclei in the cells of the first generation of hybrids.
Fig. 4 and 5. Distribution of the chromosomes at the maturation-divisions of the germ-cells of the first generation of hybrids.
Fig. 6–9. Combination of the chromosomes in the cells of the second generation of hybrids.

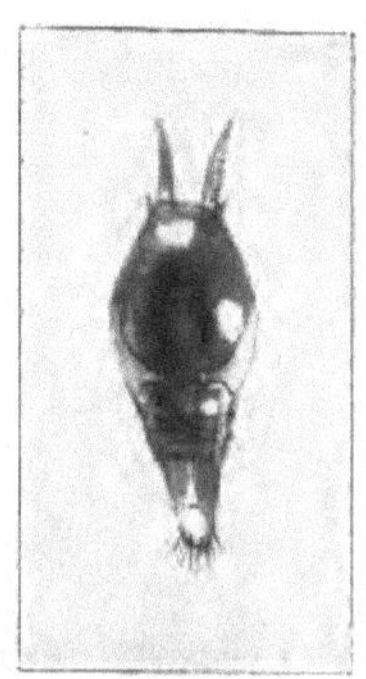

1
Doryloxenus transfuga Wasm. (East Indies.)
12 times the natural size.

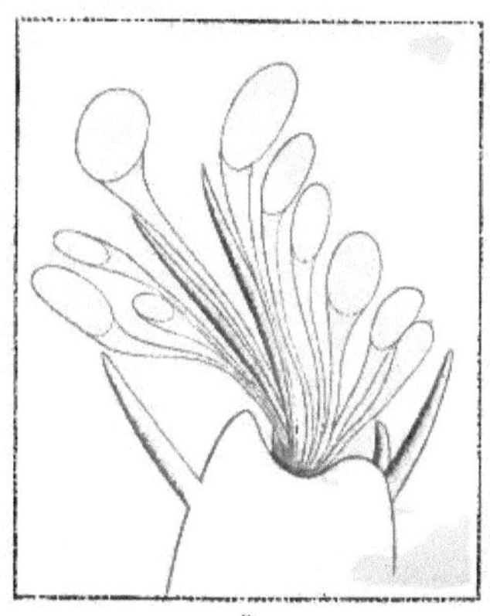

2
Forefoot and tip of tibia of *Doryloxenus*.
500 times the natural size.

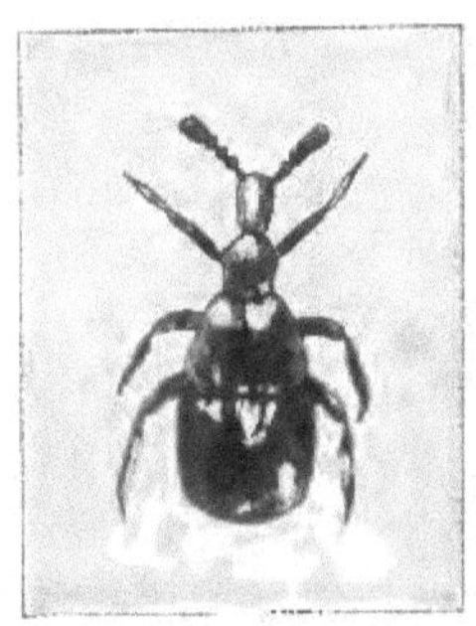

3
Claviger testaceus Preyssl. (Europe.)
12 times the natural size.

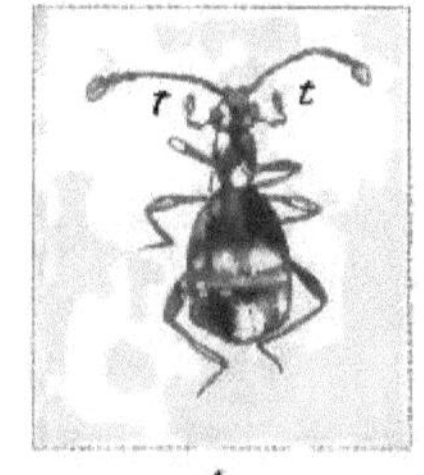

4
Pselaphus Heisei Hbst. (Europe.)
12 times the natural size. t = maxillary palpi.

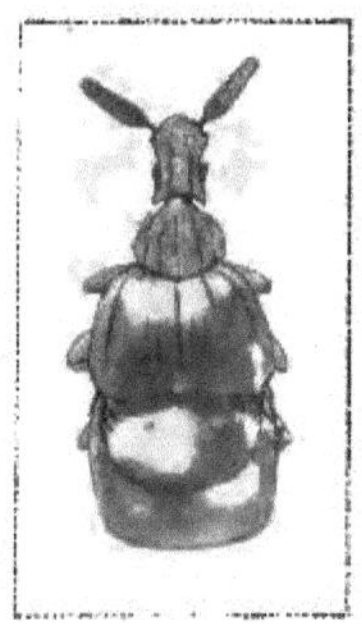

5
Paussiger limicornis Wasm. (Madagascar.)
12 times the natural size.

6
Miroclaviger cervicornis Wasm. (Madagascar.)
12 times the natural size.

1

Pleuropterus brevicornis Wasm. Bagamoyo.
3 times the natural size.

2

Pentaplatarthrus natalensis Westw. Natal
4 times the natural size.

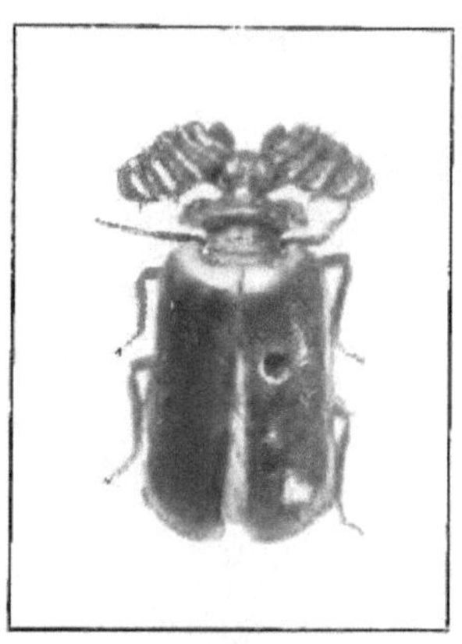

3

Lebioderus Goryi Westw. (Java.
6 times the natural size.

4

Paussus howa Dohrn. (Madagascar.
4 times the natural size.

5

Paussus spiniceps Wasm. (Sierra Leone.
6 times the natural size.

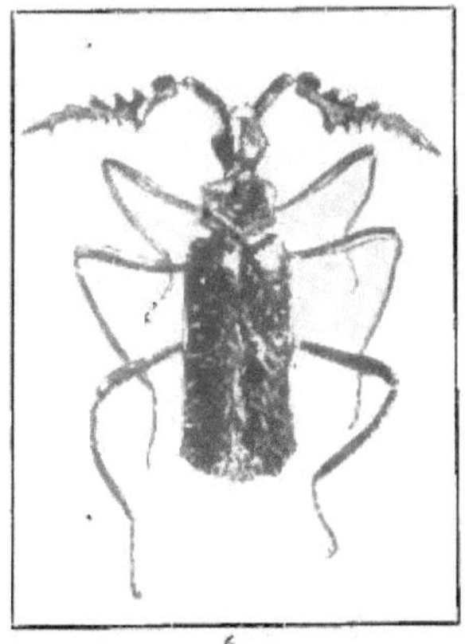

6

Paussus dama Dohrn. (Madagascar.
6 times the natural size.

1
Stenogastric imago of *Termitoxenia Assmuthi* Wasm. (East Indies.)
16 times the natural size.
(ap = appendages on the thorax, taking the place of the front-pair of wings.)

2
Stenogastric imago of *Termitoxenia* (*Termitomyia*) *mirabilis* Wasm (Natal.)
16 times the natural size.
(ap = appendages on the thorax, as in Fig. 1.)

3
Physogastric imago of *Termitoxenia Assmuthi* Wasm. (East Indies.)
16 times the natural size.
(s = point of the abdomen.)

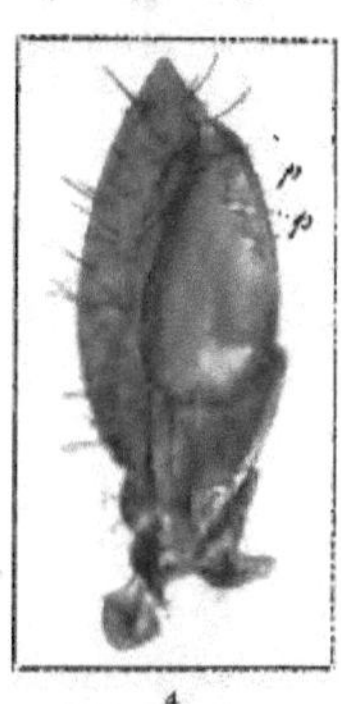

4
Thoracic appendage of a physogastric imago of *Term. Heimi* Wasm. (East Indies.)
115 times the natural size.
(p, p = exsudatory pores on the hinder branch.)

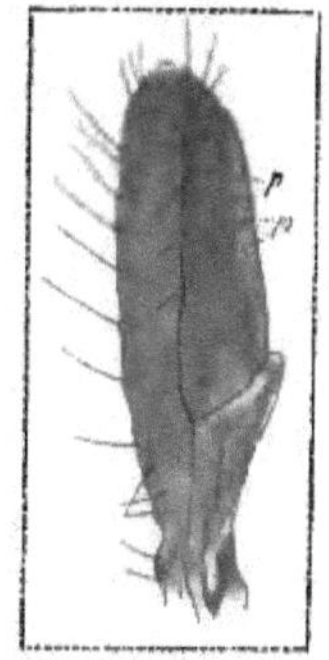

5
Thoracic appendage of a physogastric imago of *Term. Assmuthi* Wasm. (East Indies.)
115 times the natural size.
(p, p = exsudatory pores on the hinder branch.)

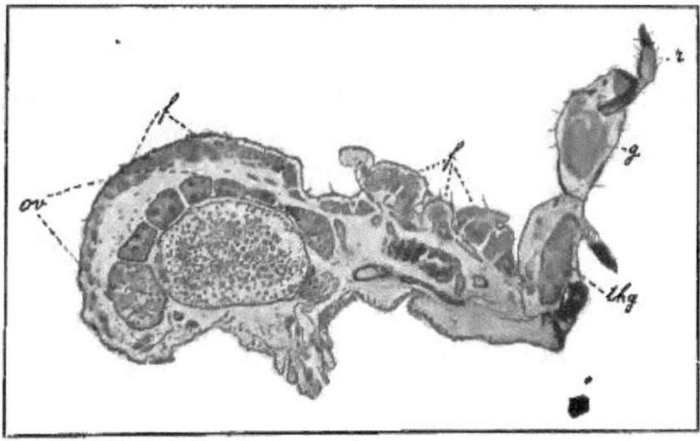

6
Longitudinal section of a physogastric imago of *Termitoxenia Assmuthi* (1/100 mm. in thickness). 32 times the natural size.
(r = proboscis; g = brain; thg = thoracic ganglia united with the abdominal ganglion; f = huge fat cells of the abdomen; ov = ovary, the terminal chamber of which contains a fertilized egg.)
(The antennae, maxillary palpi, thoracic appendages and legs cannot be seen on this section, as they are situated on the sides.)

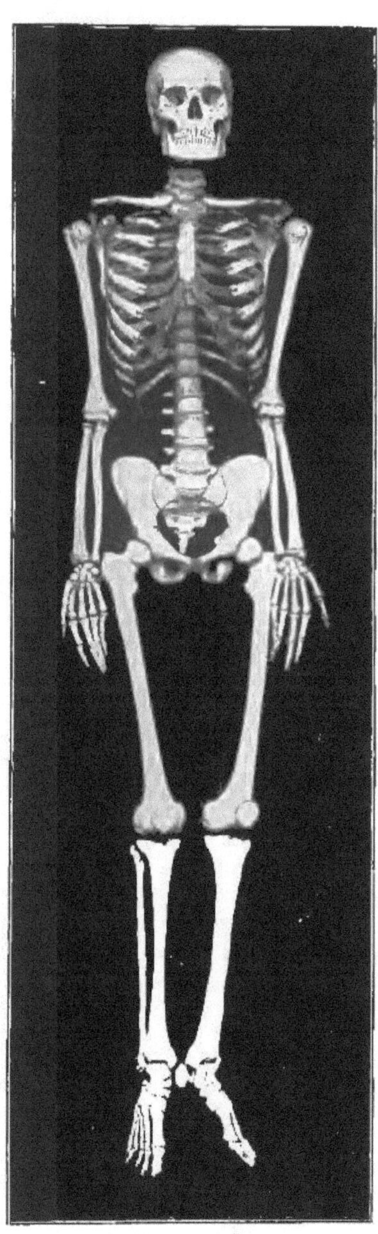

A *Human skeleton*
An adult Frenchman, 30 years of age
1.727 m. in height.
Humerus 28 cm. Femur 47 cm.
Ulna 25 cm. Tibia 37 cm.
Radius 22 cm.

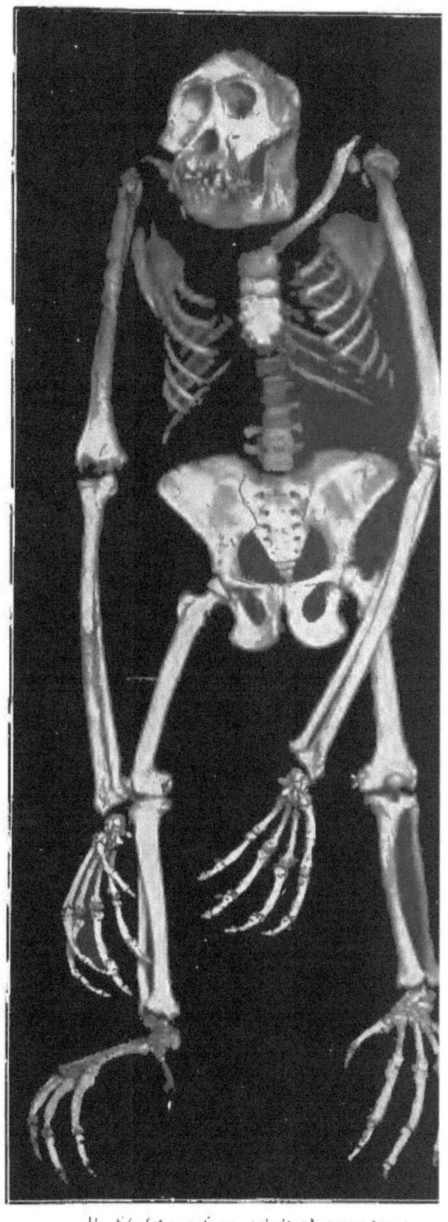

B *Skeleton of an adult Orang-utang*
Simia satyrus L. 1.60 m. in height.
Humerus 36 cm. Femur 31 cm.
Ulna 41 cm. Tibia 25 cm.
Radius 39.8 cm.

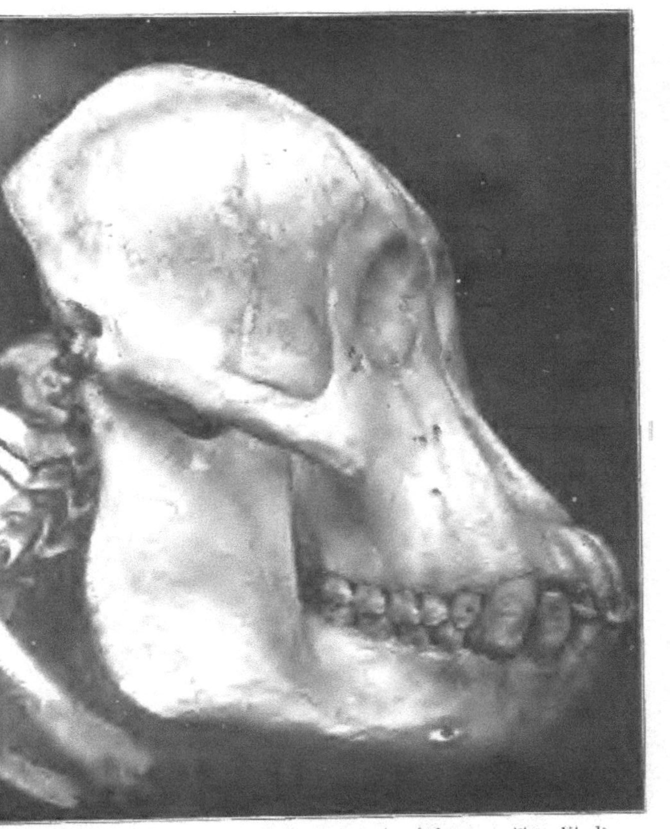

B. *Skull of an Orang-utang* (belonging to the skeleton on Plate VI, B). Circumference of the cranial region 30 cm.

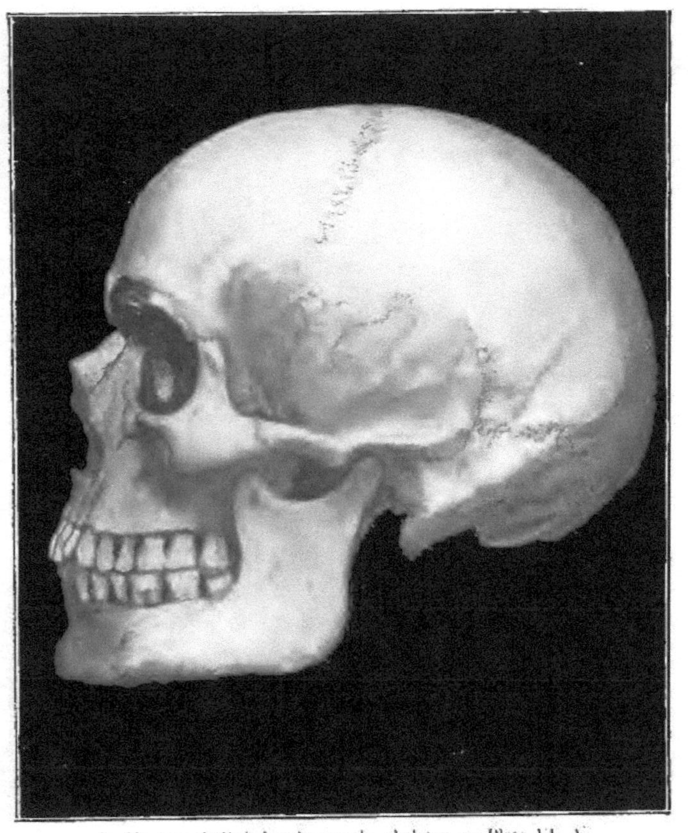

A. *Human skull* (belonging to the skeleton on Plate VI, A). Circumference of the cranial region 51 cm.

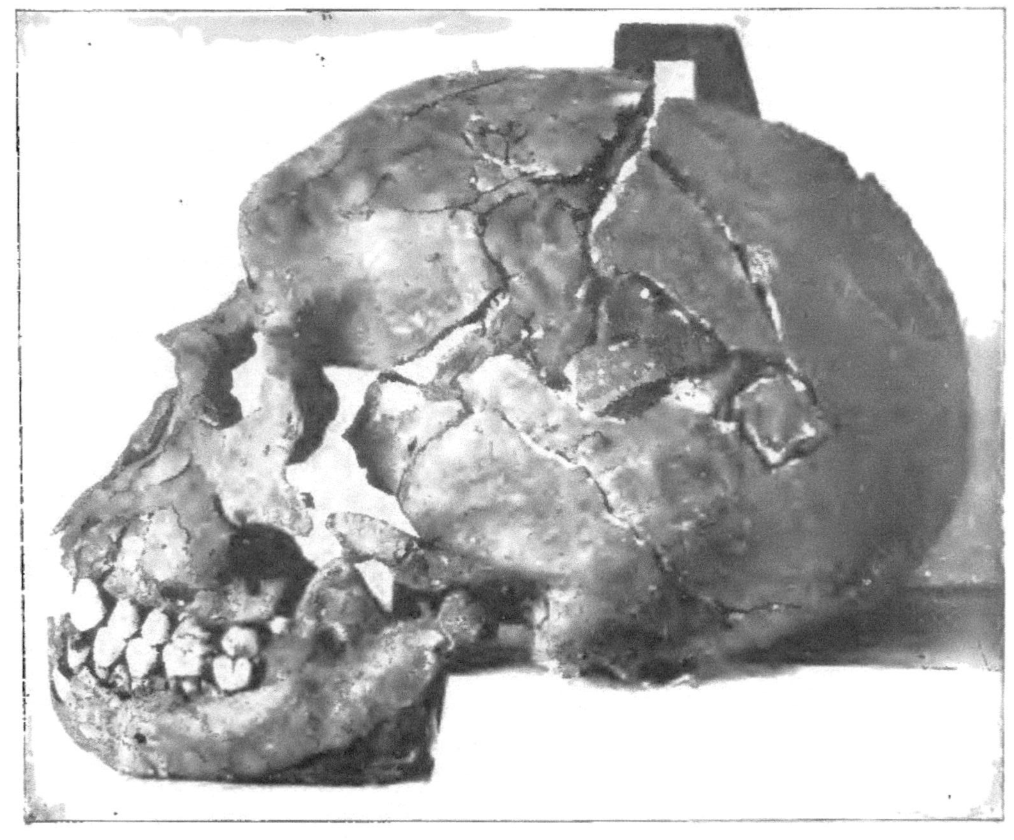

Human skull found at Le Moustier.
(After Hauser and Klaatsch.)

Zeitfracht Medien GmbH
Ferdinand-Jühlke-Straße 7
99095 Erfurt, Deutschland
produktsicherheit@kolibri360.de